Brenden Pidcock

Mining and Rock Construction Technology Desk Reference

Mining and Rock Construction Technology Desk Reference

Rock mechanics, drilling and blasting

Including acronyms, symbols, units and related terms from other disciplines

Compiled by a group of experts from the Fragblast Section at the International Society of Explosives Engineers ISEE, Atlas Copco AB and Sandvik AB

Editor-in-Chief
Agne Rustan, *formerly Luleå University of Technology, Sweden*

Associate Editors
Claude Cunningham, *Consulting Mining Engineer, South Africa*
Prof. William Fourney, *University of Maryland, USA*
Prof. K.R.Y. Simha, *Indian Institute of Science, India*
Dr. Alex T. Spathis, *Orica Mining Services, Australia*

CRC Press is an imprint of the
Taylor & Francis Group, an **informa** business

A BALKEMA BOOK

Cover Illustrations:
Illustration, Front Cover:
Extensive rock reinforcement using rock bolts in combination with wire mesh in the LKAB mine in Kiruna, Sweden. Shown is a Boltec LC rock bolting rig. Photograph: Rob Naylor, 2010. © Atlas Copco. Used with kind permission.

Illustration, Back Cover:
It shows the working stages of the Roofex monitor bolt. © Atlas Copco MAI GmbH. Used with kind permission.

CRC Press/Balkema is an imprint of the Taylor & Francis Group, an informa business

© 2011 Taylor & Francis Group, London, UK

Typeset by Vikatan Publishing Solutions (P) Ltd, Chennai, India
Printed and bound in Great Britain by Antony Rowe (a CPI group Company), Chippenham, Wiltshire

All rights reserved. No part of this publication or the information contained herein may be reproduced, stored in a retrieval system, or transmitted in any form or by any means, electronic, mechanical, by photocopying, recording or otherwise, without prior permission in writing from the publisher. Innovations reported here may not be used without the approval of the authors.

Although all care is taken to ensure integrity and the quality of this publication and the information herein, no responsibility is assumed by the publishers nor the author for any damage to the property or persons as a result of operation or use of this publication and/or the information contained herein.

Published by: CRC Press/Balkema
P.O. Box 447, 2300 AK Leiden, The Netherlands
e-mail: Pub.NL@taylorandfrancis.com
www.crcpress.com – www.taylorandfrancis.co.uk – www.balkema.nl

Library of Congress Cataloging-in-Publication Data

Mining and rock construction technology desk reference : rock mechanics, drilling and blasting, including acronyms, symbols, units, and related terms from other disciplines / compiled by a group of experts from the Fragblast Section at the International Society of Explosives Engineers, ISEE, Atlas Copco AB and Sandvik AB ; editor-in chief, Agne Rustan ; associate editors, Claude Cunningham – [et al.].
    p. cm.
Includes bibliographical references.
ISBN 978-0-415-60043-9 (hard cover : alk. paper) – ISBN 978-0-203-83819-8 (e-book)
 1. Mining engineering – Encyclopedias. 2. Earthwork – Encyclopedias.
 3. Rock mechanics – Encyclopedias. I. Rustan, Agne. II. Cunningham, Claude.
 III. International Society of Explosives Engineers. Fragblast Section.

TN9.M66 2010
622–dc22

2010030666

ISBN: 978-0-415-60043-9 (Hbk)
ISBN: 978-0-203-83819-8 (eBook)

Contents

Editor details — vii
Introduction — ix

Terminology list — **1**

Notes for the user — 3

Terms — 5

List of symbols — **411**

Symbols — 413

Greek symbols — 427

References — 431

Editor details

EDITOR-IN-CHIEF

Agne Rustan

Chairman of the Working Group on Terminology,
Fragblast Section at International Society of Explosives Engineers

Emeritus Associate Professor, Department of Mining
Luleå University of Technology
Lagmansvägen 20
SE-954 32 Gammelstad
Sweden
Agne.Rustan@spray.se

ASSOCIATE EDITORS

Claude Cunningham

Consulting Mining Engineer
Formerly AECI Explosives Limited
Modderfontein House
P.O. Modderfontein
Transvaal 1645
South Africa

William Fourney

Professor, Department of Aerospace Engineering
University of Maryland
College Park
MD 20742
USA

K.R.Y. Simha

Professor, Department of Mechanical Engineering
Indian Institute of Science
Bangalore 560 012
India

Alex T. Spathis

Global Mining Applications
Orica Mining
Services Technical Centre
George Booth Drive
PO Box 196
Kurri Kurri, NSW 2327
Australia

Introduction

This publication extends the scope of the book *Rock Blasting Terms and Symbols published by A.A. Balkema in 1998* and further embraces terminologies from closely related technologies encountered in Mining and Construction Technology. This makes it today a complete Encyclopaedia in Mining and Rock Construction Technology. Effort has also been directed at including common terms not defined in other dictionaries.

Terms from related disciplines commonly used in Mining and Rock Construction literature include *chemistry, detonics, fractography, fracture dynamics, mechanics and strength of materials, micro mechanics, geology, geophysics, image analysis, petrology, physics seismology etc.* Rock types are, however, not included since this would extend the size of the Encyclopaedia too much.

The Encyclopaedia aims to enable newcomers and workers in these fields to communicate and learn effectively by standardising the language of these disciplines. As in *Rock Blasting Terms and Symbols*, it presents not only short definitions of the terms, but also quantifies them and defines their relationship to other parameters. All scientists, engineers, technicians, teachers, practitioners and students working and learning in Mining and Rock Construction Technology should find it indispensable for communicating, reading, writing and teaching through scientific papers, technical reports, etc. These categories will also be better acquainted with symbols, acronyms, abbreviations, shortened forms and units and this knowledge will assist them in publishing in their fields.

The Encyclopaedia will therefore be important for the 'Science and Practical work in Mining and Rock Construction Technology'.

The Chairman, Agne Rustan of the Working Group on Terminology belonging to the Fragblast Section at Society of Explosives Engineers (ISEE) formed an International Group, in collaboration with the ISEE Fragblast Section Chairman *William Fourney*. The group consisted mainly of the following members from the ISEE Fragblast Section; *William Fourney USA*, *Claude Cunningham South Africa* and *Alex Spathis Australia*. To this group joined later on *Prof K.R.Y. Simha* India and *Prof Lusk* USA.

Min Eng Walter Harr, a former specialist at Sandvik AB in Sweden, was engaged to check the additional drilling terms. *Ulf Linder* Atlas Copco AB Sweden, kindly supplied the Terminology Group with an extensive drilling dictionary, from which several terms were taken. To cover the field of Rock Mechanics, *Prof John Hudson* UK was engaged to check most of the new such terms added to the First Edition.

A useful source vocabulary list was the historic 'USBM '*Dictionary of Mining, Mineral and Related Terms*' which contains about 55 000 terms and includes approximately 150 000 definitions (because one term can have several definitions). In current literature, short and often incomplete glossaries or dictionaries relevant to Mining and Rock Construction Technology were found. The most extensive dictionaries in blasting were developed by Atlas Powder, USA (365 terms) and the Institute of Makers of Explosives, USA (293 terms). 109 Rock Mechanics terms not defined in the above mentioned sources were taken form the Five volumes "*Comprehensive Rock Engineering. Principles, Practice & Projects*", edited by John Hudson.

The present Encyclopaedia contains *5127 terms* and, *637 symbols, 507 references, 236 acronyms, 108 formulas, 68 figures, 47 tables, 58 abbreviations and 7 shortened forms.*

The Encyclopaedia of Mining and Rock Construction Technology is arranged for maximum cross-referencing and ease of use. Its features include:

- The use of the standard symbols that will make it much easier to read and understand formulas. Each formula must, however, always be followed by an explanation of the symbols and the units being used when used in papers.
- When a term is shown in the Encyclopaedia, the symbol will be found after the entry of the term together with its recommended unit and with the most common used prefix for the unit. When synonym terms are presented the same information is given.
- When synonyms exist, the entry of the **recommended synonym is given in bold** and the other *not recommended in italics*.
- Short histories of explosives and initiation equipment are given under the term *History of explosives and History of initiation*. Methods to measure fragmentation are systemized and important parameters for fragmentation are highlighted etc.
- Two separate symbol lists are given, one with Arabic symbols and another with Greek symbols.
- By indexing the symbols many different parameters, like different burden distances, can be defined more precisely using distinct symbols.

In the course of the compilation of this Encyclopaedia, four terms were identified that lack international standards for *descriptive* and or *operational* definitions.

1 *Diggability*
2 *Friction sensitivity of explosive materials*
3 *Shape of a particle or fragment*
4 *Storage of explosives*

I would like to express my sincere thanks to all the colleagues mentioned in the preface for volunteering in this extensive work on the extension of the 'Terminology Book in Rock blasting' to an 'Encyclopaedia of Mining and Rock Construction Technology'. Each member of the newly group has applied his special skills and professionalism in the careful examination of the newly added terms and their definitions.

Agne Rustan
Editor-in-Chief

Note: Mining technology is improving fast and new terms are constantly being developed and used. Therefore, it will be necessary to upgrade this Encyclopaedia at regular intervals and *your comments are therefore very much appreciated. Please contact the Chairman of the ISEE Fragblast Section, Working Group on Terminology, Associate Prof. Agne Rustan.* Email: Agne.Rustan@spray.se. Phone +46-920-25 31 29 or, fax +46-920-25 31 29.

Terminology list

Notes for the user

Note 1	SI units are used consistently in this dictionary, where **m** stands for **meter**, **kg** for **kilogram**, **s** for **second**, and **A** for **Ampere**. For some terms and their units a recommendation of the most commonly used prefix is given, e.g. for compressive strength MPa is often recommended instead of Pa.
Note 2	According to the ISO 31-0 Standard, the decimal comma is used instead of the decimal point in numbers. It has been difficulties to get acceptance for this rule especially from English talking countries but Spanish and French talking countries prefer decimal comma. Today it is therefore acceptable to use either decimal comma or decimal point.
Note 3	Weight, is a force and its unit is Newton (N) but it is wrongly used in a large part of the literature on blasting to denote mass (kg). Therefore weight has consequently been changed to mass throughout this terminology list.
Note 4	Details about explosive properties and methods to test explosives are documented in the two books '**Explosives**' by R. Meyer, 1977 and '**Encyclopaedia of Explosives and Related Items**' by Kaye and Herman, 1983.
Note 5	Entries recommended to be used are marked by bold and those not recommended by italics.
Note 6	The **entries are sorted in alphabetic order** even when they consist of two or more words. E.g. feed cylinder, feeder, feed force, feed holder.

abbreviation, a shortened form of a word, e.g. Eng. for Engineering.

ABC (acronym), for '**advanced boom control**'.

Abel heat test, a test method used to assess the chemical stability of an explosive. The parameter which is determined is the time after which a moist potassium iodide-starch paper turns violet or blue when exposed to gases evolved by one gram of the explosive at 82,2°C or 180°F). In commercial nitroglycerin explosives e.g. this coloration should only develop after 10 minutes or more. In a more sensitive variant of the method, zinc iodide-starch paper is employed. Today the Abel test is still used in quality control of commercial nitrocellulose, nitroglycerin and nitroglycol, but it is practically speaking no longer employed in stability testing of propellants. The test was proposed by Abel in 1875. Meyer, 1977. This test can also be used to determine the degree of deterioration of an explosive that may have occurred during the period of storage. McAdam and Westwater, 1958.

Abelite (tradename), an explosive consisting of *ammonium nitrate* (AN) and *trinitrotoluene* (TNT). Bennett, 1962.

Abels equation, a density function $\Phi^{(i)}(l)$ to quantify the statistical distribution of the trace length l of joints (m) on the x_i plane. Hudson, 1993.

abrasion, the wearing and tearing away of particles through friction on the surface of a solid material by any other solid or aqueous material.

abrasion hardness, hardness expressed in quantitative terms or numbers indicating the degree to which a substance resists being worn away by frictional contact with an abrasive material, such as silica or carborundum grits. Also called '*abrasion resistance*' or '*wear resistance*'. Long, 1960. See also '**abrasiveness**'.

abrasion resistance, see '**abrasion hardness**'.

abrasion test, see '**abrasion hardness**'.

abrasion value (AV), see '**tungsten carbide abrasion test**'.

abrasive (noun), a substance used for grinding, honing, lapping, super finishing, polishing, pressure blasting, or barrel finishing. It includes natural materials such as garnet, emery, corundum, and diamond, and electric-furnace products like aluminium oxide, silicon carbide, and boron carbide. ASM Gloss., 1961.

abrasive drilling, a drilling technique where the material is disintegrated by friction of the drill bit against the rock. One example of such drilling method is '*rotary abrasive drilling*' where the drilling is effected by the abrasive action of the drill bit that rotates while being pressed against the rock.

abrasive formation, **abrasive ground** or **abrasive rock**, a rock consisting of small, hard, sharp-corned, angular fragments, or a rock, the cuttings from which, produced by the action of a drill bit, are hard, sharp-cornered, angular grains, which grind away or abrade the metal on bits and drill-stem equipment at a rapid rate. Long, 1960.
abrasive ground (synonym), see '**abrasive formation**'. Long, 1960.
abrasive hardness (A), (**m** or **mm**), resistance of rock against abrasion/wear.
abrasive hardness test, see '**abrasiveness test**'.
abrasive jet drills, see '**abrasive water jet**'.
abrasiveness or **abrasive hardness** (A), (**m** or **mm**), resistance of rock against abrasion/wear.
abrasiveness test or **abrasive hardness test**, abrasion tests serve to measure the resistance of rocks to wear due to friction of surfaces against each other or compression (impact or indentation). These tests include wear when subjected to an abrasive material, wear in contact with metal and wear produced by contact between the rocks. The tests can be grouped into three categories: 1) **abrasive wear impact tests**; 2) *abrasive wear with pressure tests*; and 3) *attrition tests*. See '**Suggested Methods of Determining Hardness and Abrasiveness of Rocks**', ISRM, 1978. One abrasive hardness test employs a rotating abrasive wheel or plate against which the specimen is held. The specimens are abraded for a given number of revolutions, and the mass of material abraded is a measure of the abrasive hardness, see '**tungsten carbide abrasion test**'. There are several other tests suggested in the literature. One index, the *Cherchar abrasivity index* (CAI), is used in full face boring. A steel cone with a cone angle of 90°, tensile strength of 200 MPa and Rockwell hardness 54–56 is pulled over the length of 1 cm of the rock surface. The diameter of the resulting abraded flat on the steel cone, measured in tenths of a mm in a microscope, determines the Cherchar abrasivity index. Quartz and quartzite have CAI values of about 6 (cutter life = CL = 60 km), and limestone about 1 (CL = 2 000 km). CL is approximately inversely proportional to CAI^2. Suana and Peters, 1982.
abrasive rock (synonym), see '**abrasive formation**'. Long, 1960.
abrasive water jet, a mixture of water and abrasive is directed against the rock under high pressure, 70–340 MPa. Two methods for forming water jets with abrasive particles entrained in the jet stream have been developed. One method employs a high pressure, typically 200–340 MPa and a relatively low flow rate water jet system. The other method employs a lower pressure, typically 70 MPa as a maximum but a higher flow rate water jet system. Hudson, 1993.
abrasive wear, see '**abrasion**'.
abrasive wheel, a rotating disk used to sharpen tools and inserts in drill bits.
abrasivity, see '**abrasiveness test**'.
abrasivity index, see '**abrasiveness test**'.
ABS (acronym), (s_{ba}), (**MJ/m³**), absolute bulk strength, see '**bulk strength**'.
absolute bulk strength (ABS), (s_{ba}), (MJ/m³), use '**absolute volume strength**'.
absolute mass strength (AMS), (s_{ma}), (**MJ/kg**), calculated energy per unit mass of explosive. Any such figure must be supported by the name of the detonation code used, the equation of state employed, and the convention for determining energy.
absolute roof, the entire mass of strata overlying a coal seam. Nelson, 1965.
absolute strength value (ASV), (s_{ma}), (**MJ/kg**), is defined as the total strength of an explosive measured in MJ/kg. Lownds, 1986. See '**absolute mass strength**'.
absolute temperature (T), (**K**), temperature quantified by the Kelvin scale. 0°Celsius = 273,15 K. The unit is called Kelvin.

absolute volume strength or **volume strength (AVS)**, (s_{va}), **(MJ/m³)**, calculated energy per unit volume of explosive. Any such figure must be supported by the name of the detonation code used, the equation of state employed, and the convention for determining energy.

absolute weight strength (AWS), (s_{wa}), (MJ/kg), use '**absolute mass strength**'.

abutment, the mass or volume consisting of the walls in an underground excavation supporting the roof of the excavation.

abutment height (H_{ab}), **(m)**, see '**height of abutment**'.

accelerated ageing of a building (due to ground vibrations), appears in a building not founded on solid rock and where there is a settlement with time. If vibration, from for example blasting, is added to this kind of building a faster settlement will occur. Persson, 1993.

acceleration (a), **(m/s²)**, the rate of change of velocity with respect to time.

acceleration due to gravity (g), **(m/s²)**, the acceleration imparted to bodies by the attractive force of the Earth. The acceleration has an international standard value of 9,80665 m/s² but varies with latitude and elevation on Earth. Lapedes, 1978.

accelerometer, an instrument which measures acceleration or gravitational force capable of imparting acceleration. Usually, a piezeo electric or piezeo resistive crystal is used as measuring device.

acceptor charge, a charge of explosive or blasting agent which receives an impulse from an exploding donor charge.

accessories and tools, see '**blasting accessories**'.

access tunnel or **adit**, a near-horizontal passage from the surface by which an underground mine is entered. The adit is therefore only open to the surface at one end.

ACL (acronym), see '**asymmetric charge location**'.

acoustic coupling, see '**impedance ratio**'.

acoustic dispersion, dependence of speed of sound in any material solid, liquid or gas on the frequency of the sound.

acoustic emission (AE), also called 'micro seismic activity', '*rock noise*', '*rock talk*', '*elastic shock*' and '*stress wave emission*'. AE is caused by small failure of grain structures in the rock mass due to increased rock stress. It creates sound waves when the stored stress energy is relieved. The frequency of the waves ranges from less than 1 Hz to more than 10 kHz. The frequency decreases with increasing magnitude of the energy release. This energy can be measured by geophones or accelerometers. When more than four measurement points are used the location of the event can be calculated. Acoustic emission has been measured in several field applications, like tunnel and mine stability, slope stability, underground storage of radioactive waste and finally in the gas and petroleum industry. TNC 1979 and Brady, 1985.

acoustic emission transducer, a transducer measuring the acoustic emission in rock samples. It is attached to the side of a drill core of the rock sample with Canada balsam and may for example be a lead-titanate-zirconate (PTZ) compressional mode disk with natural frequencies from 100 to 300 kHz. Yoshikawa, 1981.

acoustic impedance (Z), **(kg/m² s, Ns/m³** or **Pas/m)**, the product of the P-wave velocity and the density of a material. It characterizes a material as to its energy transfer properties. The acoustic impedance is an important quantity in characterizing rock blastability (critical burden) and rock fragmentation (uniformity index). A classification of acoustic impedance values found in the laboratory and in the field are given in Table 1.

Table 1 Classification of acoustic impedances for different kinds of rock and rocklike materials.

Class of acoustic impedance	Acoustic impedance (×10⁶ kg/m²s)	Examples of materials	Acoustic impedance (×10⁶ kg/m²s)		Expected critical burden
			Lab.	Field	
Very low	0–5	• Light-weight concrete Sweden, 1983	1,12	–	**Very high**
		• Kemmerer coal USA, Hearst et al., 1976	3,11	2,84	
		• Barea sandstone USA, Fourney et al., 1976	4,26	–	
Low	5–10	• Carbonaceous talc Russia, 1994	–	6,88	**High**
		• Alum slate (perpendicular to the bedding planes) Norway, 1990	4,39–11,1	–	
		• Salem limestone USA, Fourney et al., 1976	9,45	–	
Medium	10–15	• Luossavaara magnetite Sweden, 1983	11,9	–	**Medium**
		• Magnetite concrete Sweden, 1983	12,0	–	
		• Charcoal granite USA, Fourney et al., 1976	12,1	–	
		• Öjeby granite Sweden, 1983	13,0	–	
		• Henry quartzite Sweden, 1983	14,8	–	
High	>15	• Dolerite, India, 1973	16,2	–	**Low**
		• Storugns limestone, Sweden, 1983 and 1988	16,4	13,1	
		• Tamtas limestone, Turkey, 1993	16,6	5,65–8,61	
		• Uralite diabase, India, 1973	18,2	–	
		• Steirischer Erzberg dolomite, Austria, 1993	18,6	–	
		• Steirischer Erzberg siderite, Austria, 1993	18,6	–	
		• Tremolite schist, India, 1973	19,1	–	
		• Kallax gabbro, Sweden, 1983	20,4	–	
		• Divrigi hematite, Turkey, 1991	21,6	–	

Note: The P-wave velocities measured in the field are generally lower than those measured for small laboratory samples. The acoustic impedance values measured perpendicular to the bedding planes in Alum slate, Norway vary to a relatively large extent depending mainly on whether the sample is dry or wet. The acoustic impedance values were taken for Austria from Moser (1993), for India from Ramana and Venkatanarayany (1973), for Norway from Lislerud (1990), for Russia from Petrosyan (1994), for Sweden from Rustan et al. (1983), Nie (1988), for Turkey from Bilgin (1991) and Bilgin and Paþamehmetoðlu (1993), and finally for USA from Fourney et al. (1976) and Hearst et al. (1976).

The *impedance theory* describes the energy transmission from one media to the adjacent, and can be used approximately when studying how much energy can be transmitted from an explosive in a blasthole to the surrounding rock. The basic theory assumes a plane elastic wave which impacts at normal incidence to an interface between two dissimilar materials.

$$p_t = Z_R = \frac{Z_i}{Z_t} = \frac{2p_i}{(1+Z_R)} \quad \text{Acoustic impedance theory} \quad (A.1)$$

where the transmitted pressure p_t is given in (Pa), p_i is the incident pressure in (Pa), Z_R is the impedance ratio Z_i/Z_t where Z_i and Z_t are the impedance's for the material of the incidence wave, and the material of the transmitted wave respectively. The *maximum energy transmission* from one media to an adjacent is achieved for normal incidence, e.i. when the acoustic impedance ratio is equal to 1, see '**impedance ratio**'.

acoustic impedance of an explosive (Z_e), (kg/m² s), is defined by the charged density of the explosive in the blasthole multiplied by its velocity of detonation measured at the charged explosive density and confinement.

acoustic impedance of the material of the incidence wave (Z_i), (kg/m² s), the product of the P-wave velocity (c_P) and the density (ρ) of the material of the incidence wave.

$$Z_i = cP\rho \quad (A.2)$$

It characterizes a material as to its energy transfer properties. An acoustic or seismic wave will be reflected and transmitted at a discontinuity which is defined by a change in material properties. The materials at the discontinuity can be either a gas e.g. air, a liquid e.g. water or a solid e.g. rock.

acoustic impedance of the material of the transmitted wave (Z_t), (kg/m² s), the product of the P-wave velocity (c_P) and the density (ρ) of the material of the transmitted wave.

$$Z_t = cP\rho \quad (A.3)$$

It characterizes a material as to its energy transfer properties. An acoustic or seismic wave will be reflected and transmitted at a discontinuity which is defined by a change in material properties. The materials at the discontinuity can be either a gas e.g. air, a liquid e.g. water or a solid e.g. rock.

acoustic log, a continuous record made in a borehole showing the velocity of sound waves over short distances in adjacent rock; velocity is related to porosity and nature of the liquid occupying pores. A.G.I. Supp., 1960. Type of rock and rock mass stress are other parameters important for the velocity of sound waves.

acoustic trace, the curve on the ground vibration record that records the sound level.

acoustic velocity log, see '**acoustic log**'.

acoustic warning, a distinct audible warning used to indicate the progress of blasting operations. AS 2187.1, 1996.

acoustic wave, a mechanical (longitudinal) wave in gases, liquids and solids. The audible part of mechanical waves has a frequency range between 20 and 20 000 Hz. In blasting, the mechanical waves transmitted in air are called air blast waves. The waves may be described in terms of change of pressure, particle displacement, particle velocity, particle acceleration, or density. Acoustic waves can be used to measure the physical properties of rocks, and the composition of gases. Investigations may be made both in situ, and in the laboratory, see also '**seismic waves**'.

acronym, a word composed of the initial letters of the name of something, especially an organization, or of the words in a phrase. Examples of acronyms are NATO, RADAR and TEFL.

active anchor system, rock reinforcement by pretensioned rock bolts.

active subsidence, see '**subsidence**'.

active support system, rock reinforcement by pretensioned rock bolts or cables.

acultural vibration, vibration that is strange and unfamiliar to the observer. Konya and Walter, 1990.

adamantine drill, see '**shot drill**'.

adapter, see '**adapter sleeve**'.

adapter sleeve, the mechanical device connecting the drill machine to the drill rods for the purpose of transmitting axial and rotational forces to the drill rods or tubes. Sandvik, 1983.

adapter threaded, fitting used to connect dissimilar threads. Threaded adaptors called 'subs' are commonly used to connect dissimilar threads in drill tubes. Atlas Copco, 2006.

ADC test, ADC is an acronym for '*Ardeer double cartridge test*' which is a test method to quantify the '**sensitivity to propagation**' of an explosive. See '**sensitivity to propagation**'.

adhesion, holding surfaces together with an adhesive. CCD, 1961.

adiabatic, a thermodynamic state characterized by no heat exchange between the working material, and its surroundings (or among different elements of the working material). In both ideal and non-ideal detonations the expansion of the detonation products beyond the Chapman-Joguet curve (CJ curve) closely approximates an adiabatic process or adiabatic conditions. A curve which represents the adiabatic process in a pressure-volume diagram is called an adiabat, see also '**entropy and isentropic**'.

adiabatic bulk modulus (K_a), **(MPa)**, the bulk modulus determined in dynamic tests when there is no heat exchange with the environment. Hudson, 1993. See also '**bulk modulus**'.

adiabatic compression, a compression process during which no heat is added to or subtracted from a gas volume, and the internal energy of the gas is increased by an amount equivalent to the external work done on the gas. The increase in temperature of the gas during adiabatic compression tends to increase the pressure at the expense of a decrease in the volume alone, and, therefore, during adiabatic compression the pressure rises faster than the volume decreases.

adiabatic exponent or **specific heat ratio** (γ), **(dimensionless)**, the ratio of specific heats at constant pressure and constant volume (C_p/C_v) of the detonation gases, which varies from about 3,0 at the detonation state to 1,3 when the gases are fully expanded. AECI, 1993. The adiabatic exponent (γ) is assumed to be a function of temperature T (°C). Hommert et al., 1987.

$$\gamma = \left(\frac{0,667}{10^4} T\right) + 1,35 \qquad (A.4)$$

adiabatic pressure or **explosion pressure** (p_a), **(MPa)**, the pressure in an adiabatic process. If the process is caused by explosives in a blasthole, the pressure is called borehole pressure (p_b), see '**borehole pressure**'.

adit or **access tunnel**, a near-horizontal passage from the surface by which an underground mine is entered. The adit is therefore only open to the surface at one end.

adobe charge, a mud-covered or unconfined explosive charge fired in contact with a rock surface without the use of a borehole (Synonymous with bulldoze and mud capping). Atlas Powder, 1987.

ADR (abbreviation) for '**European Agreement Concerning the International Carriage of Dangerous Goods by Road.**' It is a European collection of rules for transportation of dangerous goods on **roads**. Wikipedia, 2008.

advance (A), (**m/day** or **m/month**), the linear distance in meters driven during a certain time in tunnelling, drifting, raising or shaft sinking. Fraenkel, 1954.

advanced boom control system (ABC), offers three modes: Basic, Regular and Total. These corresponds to manual, semi-automatic and fully automatic operation of drill rigs. Atlas Copco, 2007.

advanced undercut (block caving), an undercut strategy where the undercut mining face is advanced slightly ahead of a partially developed extraction level below the undercut level.

advance factor (R_{af}), (**dimensionless**), linear advance in meter per meter of drilled depth in drifting, tunnelling or raising.

advance per round or **pull** (A_r), (**% of drilled depth** or **m/round**), see '**pull**'.

advance rate (A), (**m/day** or **m/month**), see '**advance**'.

advancing, mining from the shaft area or other access sites out towards the boundary of the orebody.

adz, a hand cutting tool with the blade set at right angle to the handle. Used for rough dressing of timber. Crispin, 1964. Also spelled '**adze**'. Webster 3rd, 1961.

adze, see '**adz**'.

AE (acronym), see '**acoustic emission**'.

afterblast or **inrush**, an inrush of air during an explosion of methane and oxygen where carbon dioxide and steam are formed. When the steam condenses to water a partial vacuum is created, which causes an inrush of air and this is called an afterblast. Cooper, 1963, p. 195. Most common in coal mines.

afterbreak, in connection with mine subsidence, a movement from the sides, the material sliding inward, and following the main break, assumed at right angle to the plane of the seam. The amount of this movement depends on several factors, such as the dip, depth of seam, and nature of overlying materials. Lewis, 1964, p. 618. Most common seen in coal mining.

aftercooler, *component of an air compressor* used to remove heat liberated from the compression of air. Atlas Copco, 2006.

aggregate, uncrushed or crushed gravel, crushed stone or rock, sand or artificially produced inorganic materials, which form the major part of concrete. Taylor, 1965.

AGI (acronym), for '**American Geological Institute**'.

agitator, set of paddles rotated in a tank to mix two or more substances. Agitators are used for mixing cement. Atlas Copco, 2006.

AMS (acronym), see '**absolute mass strength**'.

air blast, the airborne shock wave or acoustic transient generated by an explosion. In underground workings an explosion is accompanied by a strong rush of air, e.g. when explosives are used, and it is also caused by the ejection of air from large underground openings owing to the sudden fall of large masses of rock, the collapse of pillars, slippage along a fault, or a strong flow of air pushed outward from the source of an explosion. There are four main types of air blast overpressures defined as follows:

APP or air pressure pulse, air blast overpressure produced from direct rock displacement at the face or mounding of the blast collar; *RPP or rock pressure pulse,* air blast overpressure produced from vibrating ground; *GRP or gas release pulse,* air blast overpressure produced from gas escaping from the detonating explosive through rock fractures; and *SRP or stemming release pulse,* air blast overpressure produced from gas escaping from the blown-out stemming. The most damaging cases of air blast are caused by *unconfined* surface charges. The resulting air blast overpressure may be estimated from the following formula;

$$p = 185 \left(\frac{\sqrt[3]{Q}}{R} \right)^{1,2} \tag{A.5}$$

For confined borehole charges, air blast overpressure may be estimated from;

$$p = 3,3 \left(\frac{\sqrt[3]{Q}}{R} \right)^{1,2} \tag{A.6}$$

where p is pressure in (kPa), Q explosive charge in (kg) and R is distance from the charge in (m). Atlas Powder, 1987. *Underground* air blast means the rapid flow of air through an underground opening following compression of the air in a confined space.

air blast focusing, the concentration of sound energy in a small region at ground level due to refraction of the sound waves back to the earth from the atmosphere. This occurs under certain meteorological conditions, for example during inversions.

airblasting, a method of blasting in which compressed air at very high pressure is piped to a steel shell in a shot hole and discharged. B.S. 3618, 1964, sec. 6.

air concussion, see 'air blast'.

airdeck blasting, a controlled contour blasting method where airspace is replacing the column charge. The method was initially developed for controlled contour blasting of large diameter holes on the surface. A plug at the bottom of stemming is used to prevent the stemming from falling into the column volume. Chiappetta and Mammele, 1987. Airspace can also be used as a substitute for solid stemming material between multiple charges in a blasthole.

airdecking (gapped borehole charges), a blasting method where cylindrical charges are spaced by air in blastholes. With this method fragmentation can be improved and explosive consumption reduced and output per man shift increased. Akaev, 1971.

air decks, see '**airdecking**'.

Airdox (tradename), for cartridges of compressed air of very high pressure (70 MPa) charged into 'Airdox' cylinders, which are inserted into the drill holes. Special valves fitted with the 'Airdox' cylinders permit sudden release of the compressed air at the back (bottom) of the borehole. Airdox are employed to break soft rocks (e.g. coal), and do not ignite a gassy or dusty atmosphere.

air gap, the clear vertical distance between the top of a pile of caved ore and the *in situ* cave back.

air gap test, a gap test with air as medium between charges, see '**gap test**'.

air-hardening steel, alloy steels in which a certain degree of hardness has been induced merely by air cooling under controlled conditions. Camm, 1940.

air hose, flexible tube made of rubber or plastic and used for conveying compressed air to a desired point. Atlas Copco, 2006.

air intake, opening, attachment or duct designed to allow the inflow of air. Atlas Copco, 2006.

air leg, pneumatic cylinder attached to the drilling machine with a joint. The pusher (air leg) is standing on the floor and angled forwards in the direction where drilling is going to be performed. The purpose of the pusher is to create the feed force of the drilling machine. For vertical drilling special designed air legs can be used without any joint.

airline lubricator, see '**line oiler**'.

air loader, a 'charging machine' which uses compressed air to transport the explosive into the blasthole. See '**charging machine**'.

air lock, a compartment in which air pressure can be equalized to the compressed air inside a shield-driven tunnel as well as the outside natural air pressure to permit passage of men and material. Bickel, 1982. See also '**shield-driven tunnel**'.

air motor, motor driven by compressed air working against a set of pistons or a rotor to cause the rotation of the out-put shaft. Atlas Copco, 2006.

air outlet, opening, fitting or orifice designed as a connecting point to an air source. Atlas Copco, 2006.

air pressure pulse (APP), (p_{oa}), **(Pa or dB)**, air blast overpressure produced from direct rock displacement at the face or mounding of the blast collar, see also '**air blast**'. Atlas Powder, 1987.

air receiver, specially constructed tank, designed to contain compressed air. Atlas Copco, 2006.

air scrubber, see '**scrubber**'.

air shaft, a shaft used for ventilation of mines, downcast when transferring fresh air from the surface to the underground workings and upcast when discharging exhausted air to the surface.

air shock measurement, the measurement of air shocks in blast holes up to 2 000 MPa can be done with e.g. 'ytterbium stress gauges'. See also '**piezoresistance dynamic pressure transducer**'.

air spacer, the purpose with a spacer could be 1) to divide the charge in a blasthole into separate delays or 2) to introduce a better blasting effect by *airdeck blasting*. One example of a spacer is designed of *two plywood disks* with a diameter little less than the blasthole diameter and hold together with 50 × 50 mm wooden spacer. A hole is made through the two disks for the *detonating cord*, *NONEL* or *electric wires* see Fig.1. Day, 1982.

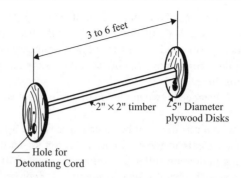

Figure 1 'Air spacer' used to divide the charge into separate delay charges or the introduction of airdeck blasting. After Day, 1982.

air system, collection of valves, controls and air driven components connected by tubes and hoses to perform required functions. The flushing air system on surface drilling rigs. Atlas Copco, 2006.

air turbo lamp, device consisting of an electric lamp and a air driven electric generator. The air turbo lamp is normally used where there is a ready source of compressed air and no electric power available. Atlas Copco, 2006.

air turbo system, air driven turbine. Atlas Copco, 2006.

air valve, valve used to control the pressure or flow of compressed air. Atlas Copco, 2006.

air winch, a motor to create a linear force for the purpose of moving material consisting of a wire winded on a cylindrical drum, which is rotated by compressed air.

ALANFO (acronym), a mixture of **aluminium powder** (AL), **ammonium nitrate** (AN), and **fuel oil** (FO).

alcohol metering device, device used to inject metered amounts of alcohol into an air system to prevent the formation of ice. Atlas Copco, 2006.

alignment deviation or **alignment error** (D_a), (**degrees**, **m/m** or **%**), angular, relative or percentage deviation from the intended angle of drilling in collaring.

Alimak (tradename), Swedish company making mechanical rack driven hoists with the purpose to drive raises or to be used for emergency in mines, see '**raise driving methods**'.

all-around drill bit, general used drill bit, suitable for drilling in a variety of rock formations. Atlas Copco, 2006.

alluvial, materials such as loose gravel, sand, and mud deposited by streams.

alluvial mining or **placer mining**, the exploitation of alluvial deposits by dredging, hydraulick, or drift mining. The alluvial could be deposited in stream beds beach dunes or on the ocean floor.

alteration, change in the mineralogical composition of a rock, typically brought about by the action of hydrothermal solutions. Sometimes classed as a phase of metamorphism but usually distinguished from it because it is milder and more localized; also applied to secondary (supergene) changes in rocks or minerals. A.G.I., 1957.

altered rock, a rock that has undergone changes in its chemical and mineralogical composition since it was originally deposited. Weed, 1922.

alternate pillar and stope, see '**square-set stoping**'.

aluminized ANFO (ALANFO), a mixture of ANFO and aluminium powder (AL). The adding of aluminium powder increases the strength of the ANFO explosive.

aluminized explosive, an explosive to which aluminium has been added in the form of powder, grains, or flakes. This increases the mass strength of the explosive, because of the large amount of heat of oxidation produced when aluminium is oxidized.

ambient noise level (p_{am}), **(Pa or dB)**, existing background noise level. FEEM.

American Society for Testing and Materials (ASTM), a non-profit organization that provides a management system in which producers, users, ultimate consumers, and representatives of government an academia develop technical information in the published form of agreed upon documents called 'voluntary consensus standards'. Current membership is over 29 000 organizations and individuals, world-wide, with a total unit participation of well over 80 000 in 137 technical committees. In addition to the 6 600 standards contained in the 48 volume Annual Book of ASTM Standards, ASTM also provide numerous other technical publications and related material which have evolved from committee activities. ASTM, 1982.

ASTM has developed test procedures for the following tests;
1. Laboratory determination of pulse velocities and ultrasonic elastic constant.
2. Direct tensile strength.
3. Rock bolt anchor pull test.
4. Rock bolt long-term load retention test.
5. Dimensional and shape tolerances of rock core specimens.
6. Modulus of deformation using a radial jacking test.
7. Elastic modulus of intact rock in uniaxial compression.
8. Specific heat.
9. Transmissivity and storativity of low permeability rocks using the constant head injection test.
10. Triaxial compressive strength.
11. Thermal diffusivity.
12. Creep in uniaxial compression.
13. Creep in triaxial compression.
14. Modulus of deformation using flexible plate loading.
15. In situ deformability and strength in uniaxial compression.
16. In situ shear strength of discontinuities.
17. Permeability measured by flowing air.
18. Thermal expansion using a dilatometer.
19. In situ stress by USBM borehole deformation gauge.
20. Rock mass monitoring using inclinometers.
21. Splitting tensile strength of intact rock core.
22. Transmissivity and storativity of low permeability rocks using the pressure pulse technique.
23. Unconfined compressive strength.

American system drill, see '**churn drill**'.

American table of distances (ATD), tables used to determine the appropriate barricaded or unbarricaded minimum distances between commercial explosive manufacturing and connected explosive storage sites and lightly or heavily travelled public highways, passenger railways, and inhabited buildings. It is published by the Institute of Makers of Explosives (IME) as pamphlet No. 2. USBM, 1983.

AMIRA (acronym), for '*Australian Mineral Industries Research Association*'. A large research organisation in Australia including research in mining and mineral processing.

ammonia gelatine, see '**ammonium nitrate gelignite**'.
ammonium nitrate (AN), a chemical compound, NH_4NO_3. Its properties are: molecular weight 80,04, colourless, orthorhombic (−16,3 to + 32,3°C), density 1 725 kg/m³ (at +25°C), melting point +169,6°C, soluble in water as well as in ethyl alcohol. AN is used in crystalline or pelletized form in explosives, and as a fertiliser. Bennett, 1962. **Warning**! AN transported on trucks can detonate if the truck caught fire after about 30–60 minutes. Several cases reported. Mainiero, 2009.
ammonium nitrate and fuel oil (ANFO), a non-water-resistant explosive ideally composed of 94,0 to 94,3% ammonium nitrate (AN), and 5,7–6,0% fuel oil (FO). The quality and density of the AN prills, the quality of the fuel, and the control of the overall mix is important for the explosive performance. In the USA, ANFO is called a blasting agent because it is not cap sensitive.
ammonium nitrate gelignite, explosive similar to straight gelatine except that the main constituent is ammonium nitrate instead of sodium nitrate. In the USA the term ammonia gelatines is used.
amorphous, is defined as '*without form*'. Applies to some rocks and minerals having *no definite crystalline structure*. Fay,1920. The internal arrangement of the atoms or molecules is irregular and which in consequence has no characteristic external form. Anderson, 1964.
amplification factor (due to ground vibrations) for a structure, R_A, **(dimensionless)**, the ratio between the *peak structure vibration* (building, pipe line etc) to the *peak ground vibration* caused e.g. by blasting or rock bursts and measured at the foundation of the structure.
amplitude (A), (**m** or **mm**), the maximum positive or negative value of one period in a cyclic changing quantity.
amplitude of particle acceleration (A_a), (**m/s²** or **mm/s²**), see '**amplitude and acceleration**'.
amplitude of particle displacement (A_d), (**m** or **mm**), see '**amplitude and displacement**'.
amplitude of particle velocity (A_v), (**m/s** or **mm/s**), see '**amplitude and velocity**'.
AMS (acronym), (s_{ma}), (**J/kg** or **MJ/kg**), see '**absolute mass strength**' of an explosive. The term is recommended to replace *absolute weight strength, AWS*.
AN (acronym), for '**ammonium nitrate**'.
analytic model, a modelling of a process or processes based on the natural laws. Mathematics is normally used for the modelling.
anchor (verb), to fasten down or hold in place. Long, 1960.
anchorages, rock bolting, see '**anchoring**'
anchor bolt, metal rod or pipe pushed into drillholes which are normally drilled perpendicular to the rock surface being reinforced. It may be glued by cement or epoxy in the hole or fastened at the bottom by some mechanical device. Anchor bolts normally have a length of a couple of meters.
anchoring, the action and technique to reinforce the rock mass by rock bolts made by steel or (glass fibre) or steel tubes. The anchoring is often combined with surface support by shotcrete.
anchoring equipment, machines designed for the purpose to reinforce the rock by rock bolts. The anchoring machine can today drill the hole, insert grouting if it is used, and finally insert the rock bolt.
anchoring tube, special reinforcement used to support open cuts in clay and other soft formations. Sandvik, 2007. See '**micro piling**'.

anchor prop or **steel prop**, a steel or timber prop fixed firmly between roof and floor at the end of a longwall face and from which a coal cutter is hauled by rope when cutting. Nelson, 1965.

ANFO (acronym), for '**ammonium nitrate and fuel oil**'.

ANFO charger or **(loader)**, a machine consisting of a pressure vessel for explosive storage and a semi-conducting charging house connected to the pressure vessel to be used to charge (load) ANFO (ammonium nitrate and fuel oil) into blastholes, usually upholes underground.

ANFOPS, a controlled contour blasting explosive consisting of **ammonium nitrate** (AN) mixed with **fuel oil** (FO) and **polystyrene prills** (PS). Jimeno et al., 1995. The last component is used in varying proportions dependent on the strength of explosive needed. In Norway the mixture is called ISANOL.

angle-cut, the opening (cut) of a new round in drifting using drill holes that converges so that a wedge is blasted out. This leaves an open or relieved cavity or free face for the following holes to blast against. The latter holes are provided with a later delay.

angled drilling, the use of angled drillholes to the vertical in production blasting in open pits or quarries. In drifting, drillholes directed with an angle to the surface.

angle drilling, see '**angled drilling**'.

angle of break, angel of failure (α_{cav}), (°), see '**angle of caving**' or '**subsidence angle**'.

angle of breakage or *breakout angle* (α_b), (°), the angle between the two side cracks formed by the breakout prism when a hole is blasted parallel to a free face (bench blasting) or perpendicular to the free surface (crater blasting), see Fig. 2. The angle depends on the rock type, and it varies between 120–160° in bench blasting. The angle decreases a little with the size of the burden. Rustan et al., 1983. In half scale blasts, in contrast to model and full scale blast, it has been reported an increase in angle of breakage from 150° to 180° with an increase of burden from 0,35 to 2,3 m. Persson, 1993.

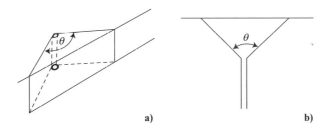

Figure 2 Angle of breakage in a) bench blasting and b) crater blasting.

angle of caving, *angle of break, angel of failure* (α_{cav}), (°), is not a constant angle in underground mining because it is normally an approximation of a curved line bent upwards and defined by a point between the loosening drawbody shape (or loosening ellipsoid shape for fine material in bunkers and silos) and the drawbody, see Fig. 3. When special slide surface occur on the hanging wall it could be almost a straight line. The angle is also changing with depth below surface. The angle of caving is normally approximated by a line between two points defined as follows; the first point is the outmost surface crack registered on a hihger level or on the surface of the mine and the second point is the

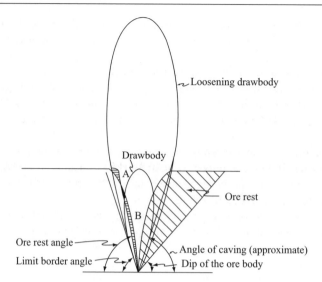

Figure 3 Definition of the terms **drawbody shape** and **loosening drawbody shape** and the following angles; **angle of caving** (α_{cav}) = *angle of break* = *angle of failure* = *angle of subsidence* (α_{sub}) = *angle of draw* and finally the two other angles, **angle of limit border** (α_{lb}), **angle of ore rest** (α_{orer}). After Rustan 2000.

outmost point on the draw level in the same direction as the other point. Rustan, 2008. It has been observed, that the angle of caving of the hanging wall at the Grängesberg mine in Sweden, using block- and sublevel caving, was successively decreasing with increase in mining depth; from α_{cav} = 80° at 138 m depth below surface to α_{cav} = 60° at 300 m below surface. Hoek, 1974.

angle of deposition (δ_{dump}), (°), see '**angle of dumping**'.

angle of dip (α), (°), the angle at which strata or mineral deposits are inclined to the horizontal plane measured in a vertical plane. Synonym for 'dip'. Nelson, 1965.

angle of draw, see '**angle of subsidence**' (α_{sub}) or '**limit angle**' (α_{lim}), (°). (Brady, 1985).

angle of dumping (δ_{dump}), (°), = *angle of deposition* = *angle of pouring* = *angle of rest* = *natural slope angle* is the mean slope with respect to the horizontal at which a heap of any loose or fragmented solid material will stand without sliding or come to rest when the material has been deposited. Rustan, 2008. Typically angles measured at LKAB in Kirunavaara are 35°–40°. The angle of dumping is roughly equal to the angle of internal friction if the cohesion C = 0. Hansagi, 1965.

angle of external friction (δ_{fekin} and δ_{ferest}), (°), the angle between the abscissa and the tangent of the curve representing the relationship of shearing resistance to normal stress acting between soil or rock on a surface of another material, e.g. steel in a chute. Two angles exists, one when the material is sliding against the wall called **angle of kinematic external friction** (δ_{fekin}) and another when the material is at rest at the wall called angle of external friction at rest (δ_{ferest}).

angle of failure (α_{cav}), (°), see '**angel of caving**'.

angle of friction (φ), (°), the maximum angle at which a body will be in rest when the body is placed on an inclined plane, see Fig. 4.

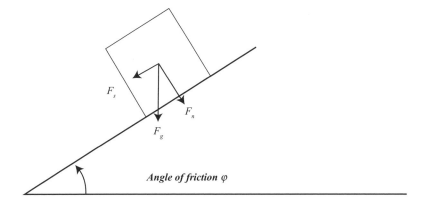

Figure 4 Angle of friction (φ) is defined as the maximum angle without slide of the body on the inclined plane.

angle of internal friction or **shear angle** (ϕ), (°), the angle between the abscissa and the tangent of the curve representing the relationship of shearing resistance to normal stress acting within a soil. ASCE, 1958. See also '**shear angle**'.

angle of limit border (α_{lb}), (°), see '**limit border angle**'.

angle of loading (δ_{load}), (°), the maximum slope with respect to the horizontal at which a heap of any loose or fragmented solid material will stand when it is loaded from the bottom of the heap. This angle is slightly larger than the angle of dumping (angle of repose in dumping).

angle of ore rest (α_{orer}), (°), an approximation of the out flow concave cone by an angle to the horizontal starting at the outmost affected draw at the outlet to the level of interest so a vertical cut with the concave outflow cone is approximated by this line. This approximation of the concave cone is necessary if hand calculations of the ore rest is done. For computer calculations the real concave cone should be used, because it does not exist any precise angle, see '**angle of draw**'.

angle of pouring (δ_{dump}), (°), see '**angle of dumping**'.

angle of repose (in dumping or loading), (δ_{dump} or δ_{load}), (°), see '**angle of dumping**' and '**angel of loading**'.

"*angle of repose (in dumping, deposition or pouring),* see '**angle of dumping**' (δ_{dump}), (°), '. Also called, angle of deposition, angle of pouring, angle of rest or natural slope angle the maximum slope with respect to the horizontal at which a heap of any loose or fragmented solid material will stand without sliding or come to rest when the material has been deposited". The angle is normally between 30° to 60°.

angle of repose (in loading or draining), see **angle of slide** (δ_{load}), (°). The maximum slope with respect to the horizontal at which a heap of any loose or fragmented solid material will stand when it is loaded from the bottom of the heap. This angle is for blasted rock or ore essentially larger than the angle of repose (in dumping). Normal values could be 60° to 90°.

angle of repose (in loading or draining), see '**angle of slide**' (δ_{load}), (°). The maximum slope with respect to the horizontal at which a heap of any loose or fragmented solid material will stand when it is loaded from the bottom of the heap. This angle is for blasted rock or ore essentially larger than the angle of repose (in dumping). Normal values could be 60° to 90°.

angle of rest (δ_{dump}), (°), *see* '**angle of dumping**'.

angle of shear (visual) in biaxial compression (β), (°), the angle between the planes of maximum shear which is bisected by the axis of greatest compression, see Fig. 5. Rice, 1960. Not to be mixed with *angle of internal friction* or *shear angle* (ϕ).

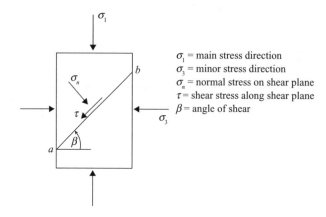

Figure 5 Definition of the angle of shear (β) caused by the failure plane a–b in biaxial compression. Brady, 1985.

angle of slide (δ_{load}), (°), the slope, measured in degrees of deviation from the horizontal, on which loose or fragmented solid materials will start to slide; it is a slightly greater angle than the angel of rest. Bureau of mines Staff. See '**angle of loading**'.

angle of subsidence (α_{sub}), angle of subsidence is the angle made with the vertical by a line drawn from the base of the seam to the nearest point of zero surface subsidence. For UK coalfields, α_{sub} is 35° to the vertical. Its values vary with the mechanical properties of the rocks, being higher for stronger rocks and lower for weaker rocks and soils. The value of α_{sub} will also vary with the resolution of the instruments used to measure subsidence and with the cut-off value taken as being equivalent to 'zero' subsidence. Brady, 1985.

angle of wall friction (δ_{fekin} or δ_{ferest}), (°), see '**angle of external friction**'.

angle of yield (α_{cav}), (°), see '**angle of caving**'.

angle reading instrument, an instrument to determine the angle of, e.g. the drill rod (vertical and/or horizontal) when starting a drill hole, collaring, or in the hole.

angular deviation in drilling, see '**bending deviation**'.

angular frequency (ω), (**rad/s** or °/s), is defined by $\omega = 2\pi f$ where f is the frequency in Hz.

anisotropy, directional dependence of a physical quantity characterizing a material. Almost all rocks are anisotropic. Anisotropy can arise in rock due to the reaction of tectonic stresses, sedimentation, metamorphism or due to the fracturing during blasting and mining activities. Anisotropy can be quantified by measurement of seismic wave velocities in different directions.

ANN (acronym), see '**artificial neural network**'.

ANOL charging equipment, loading equipment for ANFO. In Swedish ANOL is an abbreviation for "<u>a</u>mmonium<u>n</u>itrat och <u>ol</u>ja" and corresponds therefore to the English ANFO.

A stainless steel chamber is filled with ANFO and the steel chamber is pressurized by compressed air. The explosive is transported by the air pressure through a hose from the chamber to the blasthole. The technique can be used for blasthole diameters ranging from 25 to 150 mm. In upholes with diameters >75 mm special techniques must be used to get the explosive fasten in the hole.

AN prills, small spheres or pellets (less than 5 mm in diameter) of **ammonium nitrate (AN)** as opposed to flakes. Konya and Walter, 1990.

anti-freeze container, container or reservoir used for storing an engine cooling fluid that resists freezing at low temperatures. Atlas Copco, 2006.

anti-jamming system, when drilling in fissured or cavitied rock mass, there is a considerable risk that drill bit, steel or rods get stuck in the hole. This risk is substantially reduced by the built-in anti-jamming system that increases the pressure in the rotation circuit for the drill steel. When this pressure exceeds a set value the rotation and feed are reversed.

AN-TNT slurries, an explosive consisting of a mixture of ammonium nitrate solution and trinitrotoluene. (Lewis, 1964.)

apparatus case, box-like structure used to house and protect an apparatus.

apparent density or *compact density* (ρ_a), **(kg/m³)**, the mass of an object or material divided by its apparent volume less the volume of its open pores. ACSG, 1961.

apparent porosity (n_o), **(dimensionless)**, see '**open porosity**'.

aperture, an opening with different shape. When the term is used for characterization of joints it denotes the width of the joint.

APP (acronym), (p_{oa}), **(Pa or dB)**, see '**air pressure pulse**'.

apparent crater, the cavity caused by crater blasting as it exists before the broken material is removed.

appropriate authority, see '**regulatory authority**'.

approved, the issue of a statement in writing by the regulatory authority, setting out details of any act, matter or thing which is approved unless otherwise stated. AS 2187.1, 1996.

approximate potential strain energy release rate (G_m), **(J/m² or MJ/m²)**, energy needed to create new crack surfaces. ISRM, 1988.

apron, the front gate of a scraper body. Nichols, 1962.

aquarium technique, a high-speed photographic technique for examining the development of a detonation and propagation of shock waves in a transparent water tank. Not to be confused with '*aquarium test*' which is different to '*bubble test*' also called '*underwater test*'.

aquarium test, a laboratory test method to measure the pressure of an underwater explosion. Lead and copper membranes are employed, and the estimate of the parameter is based on membrane deformation as a function of the performance of the explosive and of the distance from the explosion site. The measuring apparatus, consisting of piston and anvil, resembles the Kast brisance meter, see also '**brisance**'. Meyer, 1977.

arch height (H_a), **(m)**, see '**height of arch**'.

arching, has two meanings 1) a process which occurs when the electrical current to a blasting cap is too high, and the resistance wire in the fuse head explodes instead of being heated up. This causes a light arch which may destroy the function of the blasting cap. 2) the formation of an, at least temporarily, concave surface of in situ or blasted or caved rock by the transmission of lateral inter-block or inter-particle lateral forces.

arc shooting, a method of *refraction seismic prospecting* in which the variation of travel time with azimuth from a shot point is used to infer geological structure. The term also refers to a refraction spread placed on a circle or a circular arc with the centre at the shot point. A.G.I., 1957.

Ardeer double cartridge (ABC) test, see '**sensitiveness**'.

Ardeer tank fume test, a test method for blast fumes similar to Crawshaw-Jones fume test, see '**Crawshaw-Jones fume test**'. See also '**fume classification test**'.

area (*A*), (m²), the physical quantity, the measure of the size of a two-dimensional surface, or a region on such surface. Lapedes, 1978.

arm, anything projecting from the main body of a piece of equipment. A boom arm, lift arm etc. Atlas Copco, 2006.

armor plate test, see '**projectile impact test**'.

armour stone, large blocks used in sea-defence structures such as breakwaters. Favourable conditions for armour stone production are large burdens, small spacings, low specific charge (0,2 kg/m³), and simultaneous firing of all holes in a row. Wang et al., 1991.

Arrhenius equation, an equation describing the mass rate of reaction through self heating of the explosive due to the decomposition of the explosive or other chemical substances.

$$P_r = (1-\xi)ke^{-W_a/RT} \tag{A.7}$$

where P_r is the mass reaction rate in kg/s, ξ is the fraction of the original mass that is consumed, k is a material constant, W_a is the activation energy in J (Joule), R is the gas constant and T is the absolute temperature in degree Kelvin.

artificial barricade, an artificial mound or revetted wall or soil embankment of a minimum thickness of 0,91 m, in USA. IME, 1981.

artificial neural network (ANN), determination of normally one output parameter in a biological or technological process dependent on several other parameters by inserting some known relations between the input and output parameters and thus building a relation on existing data by learning. It is a computational system, either hardware or software, which mimics the computational abilities of a biological or technical system by using a large number of simple, interconnected artificial neurons. Artificial neurons are simple emulation of biological neurons; they take in information from sensor (s) or other artificial neurons, perform very simple operation on this data and pass result on to other artificial neurons. Artificial neuron network technique has also been used in blasting technology.

ASA (abbreviation), for a mixture of lead **a**zide, lead **s**typhnate and **a**luminium. It is termed as priming charge and is used in blasting caps.

A-scale, the scale of a sound level measurement instrument in which an in-built filter discriminates against low frequencies. It approximates the frequencies which can be heard by the human ear. Konya and Walter, 1990.

ascending cut and fill, a mining method where two levels are first connected, the lower and upper one by a raise, from the bottom of which mining is begun. The work proceeds upwards, filling the mined-out room, but in the fill, chutes are built through which the broken ore is discharged by gravity. In inclined ore bodies the chutes are also inclined and made normally by timber or steel. The lower level drive is protected either by timbering or vaulting, or by a fairly strong pillar of vein fillings.

aspect ratio (R_a), (dimensionless), the ratio between the minimum and maximum width of an opening e.g. excavation or a pore in the rock mass. High aspect ratio (HAR) are defined as $R_a > 0,1$ and low aspect ratio (LAR) with $R_a = 0,1$–$0,0001$. Miguel, 1983.

assembly set, a machine or portion of a machine that is delivered broken down into its component parts and assembled on location. Often an assembly set of parts will be supplied to incorporate design improvement on drill rigs in the field. Atlas Copco, 2006.
ASTM (acronym), see '**American Society for Testing Materials**'.
ASV (acronym), see '**absolute strength value**'.
asymmetric charge location (ACL), eccentric location of a cylindrical and elongated charge in a blasthole.
asymmetric tunnel or **drift**, the roof of a tunnel or drift is made higher at one side to make it possible to load rock directly on to a truck at the face in the tunnel or drift.
ATD (acronym), see '**American Table of Distances**'.
atomizer, see '**line oiler**'.
attachment arm, anything projecting from the main body and used to support or position another object. The swing arm that supports the operating controls on certain types of drill rigs. Atlas Copco, 2006.
attachment axle, rod like piece used to connect something to the main body of the machine in such a way that it is free to rotate. Atlas Copco, 2006.
attachment panel, flat plate-like structure made from thin metal or other hard material used as a means of attachment, usually perforations. The attachment panel is used to attach another object to the main machinery. Atlas Copco, 2006.
attachment plate, flat plate-like structure made from thicker (more than 10 mm) metal or other hard material used as a means of attachment, usually perforations. The attachment plate is used to attach another object to the main machinery. Atlas Copco, 2006.
attenuation, decrease in amplitude of a wave as a function of distance of propagation from its source. There are three causes of seismic wave attenuation: 1) *Geometrical damping* or spreading (divergence) dependent on the decrease of pulse magnitude due to larger and larger space over which the pulse extends and 2) *Material damping* or internal damping decrease of pulse magnitude from viscous damping and 3) *Scattering* losses due to reflection and refraction.
auger, a tool, a hollow cylinder, that cuts and removes clay in rod percussion drilling, Stein, 1977.
auger drill, drill rig specially designed to drill soft materials using rods fitted with a continuous spiral plate that mechanically removes the cuttings from the drill hole without the need to introduce a flushing media. Auger drills can only be used to drill in soft unconsolidated material such as earth or clay. Atlas Copco, 2006.
auger drilling, the process of rotary drilling with a screw shaped drill in loose formations (soil, clay etc) for the purpose of investigating soil properties and reinforcement of the soil. Drilling machines are manufactured for hole diameters from 64 to 600 mm in diameter. Hole depths vary from 10–30 m with light equipment and with heavy equipment, depths up to 175 m can be reached in soft formations. Atlas Copco, 1982.
auger feed loading, the use of the auger principle (a screw) to transport pulverized explosives e.g. ANFO from an explosive truck to the blasthole when loading blastholes on surface. Persson, 1993.
auger mining, a mining method often used by strip-mine operators when the overburden gets too thick to be removed economically. Large-diameter, spaced holes are drilled up to 60 m into the coal bed by an auger. The auger drill is like a bit used for boring holes in wood, this consists of a cutting head with screw like extensions. As the auger turns, the head

breaks the coal and the screw carries it back into the open and dumps it on an elevating conveyor; this, in turn, carries the coal to an overhead bin or loads it directly into a truck. Auger mining is relatively inexpensive, and it is reported to recover 60–65% of the coal in the part of the bed. USBM, 1990.

Australian Institute of Mining and Metallurgy (Aust. I.M.M.), has a similar organizational structure as the Institute of Mining and Metallurgy in London, see '**Institute of Mining and Metallurgy**'.

Australian Mineral Industries Research Association (AMIRA), is a large research organisation in Australia including research in mining and mineral processing.

auto-feed, term used to describe the feed function on a diamond drill that advances the drill head at a set rate. Atlas Copco, 2006.

auto-loader, device containing a number of diamond drill rods that are automatically fed into the rod lifter. Atlas Copco, 2006.

automated rod lifter, device used to automatically add and remove rods on a diamond drilling rig. Atlas Copco, 2006.

automatic bit changer, mechanical equipment on drill rigs that can change bits automatically. Atlas Copco, 2007.

automatic drain valve, valve that automatically opens to expel unwanted fluid. Some types of air compressors are equipped with automatic drain valves that get rid of unwanted water. Atlas Copco, 2006.

automatic feed alignment, see '**automatic feed positioning in drifting or on surface**'. Atlas Copco, 2006 b.

automatic feed positioning in drifting, this function is being used on drill rigs to avoid mistakes setting out feed angle, and cancels out operator error. The operator simply has to press a button in his cab, and hold it until the feed is set to the correct angle. Automatic feed positioning reduces set-up time and ensures parallel holes, resulting in better blasting and smoother bench floors. Atlas Copco, 2006 b.

automatic feed positioning on surface, an inbuilt system on drill rigs on surface to align the drilling beam in the wished geographic angle to the north and to the azimuth at the touch of a button. Atlas Copco, 2006 b.

automatic rod adding, a mechanical system adding or reducing rods continuously on a drill rig when drilling a hole. It enables the operator to drill a hole automatically to a given depth, allowing him to leave the cab to carry other duties such as maintenance checks or grinding bits, while keeping the drill rig in sight. Atlas Copco, 2006 b.

automatic rod adding system AutoRAS (tradename), enables the driller to drill a hole automatically to a given depth, allowing him to leave the cab to carry out other duties, such as maintenance checks or grinding bits, while keeping the drill rig in sight. Atlas Copco, 2006 b.

automatic rod lifter, see '**automated rod lifter**'.

automatic tunnel profiling systems, systems for measurement of the tunnel or drift profile. Most commonly are laser beams being used for measurement of distance. Atlas Copco, 2007.

autonomous tramming, use of load haul dumpers for automatic loading, transport and dumping of ore or waste. Atlas Copco, 2007. Autonomous stands for automatic steering.

AV (acronym), for '**abrasion value**', see '**tungsten carbide abrasion test**'.

average overbreak distance (l_{oa}), **(m)**, overbreak volume outside the projected area of the drift or tunnel divided by the projected area of the two walls and roof. Overbreak can be correlated to RQD and specific charge. Hudson, 1993.

AVOCA mining, a synonym used for the rill mining method, see '**rill mining**'. Avoca is the name of the mine on Ireland where rill mining has been used. Rustan, 1990.

AVS (acronym), (s_{va}), **(J/m³ or MJ/m³)**, for '**absolute volume strength**' of an explosive which replaces '*absolute bulk strength*'. See '**absolute volume strength**'.

AWG (acronym), for '**American wire gauge**' see '**vibration wire sensor**'.

AWS (acronym), (s_{wa}), **(J/m³ or MJ/m³)** for '*absolute weight strength*' that should be replaced by '**absolute mass strength, AMS**. See '**absolute mass strength**'.

axial coupling ratio (R_{ca}), **(dimensionless)**, the ratio between the length or volume of the blasthole being charged to the total available length or volume for charge in a blasthole.

axial decoupling ratio (R_{dca}), **(dimensionless)**, the ratio between the total length or volume of a blasthole to the length or volume of the section of the blasthole being charged.

axial detonation, see '**axial initiation**'.

axial initiation or *detonation*, the initiation is made more or less instantaneous along the cylindrical charge by a high strength detonating cord. Advantages with axial initiation is *shorter rise times of the pulses* in rock and *larger peak strain* than the strain pulses produced by *end initiated (detonated) charges*. Duvall, 1965.

axial notched blastholes, see '**notched blastholes**'.

axial priming, system for priming blasting agents in which the core of priming material for example a detonating cord is placed lengthwise along the charge.

axial strain (ε_a), **(dimensionless)**, in *plane wave propagation*, the axial strain ε_a can be approximated by

$$\varepsilon_a \sim 10^{-3} \, v_{maxP}/c_P \tag{A.8}$$

where v_{maxP} is the *maximum particle velocity* of the P-wave in mm/s and c_P is P-wave velocity (longitudinal wave velocity) in m/s.

B

babbited socket, a part of the drill tool in line drilling to fasten each strand of the wire line into the drill hammer by using molten white metal. Stein, 1997.

babcock socket, is used for fishing with a cable tool rig, to prevent the tools rotating on the wire line. It must be set directly on the cable. The wire line is set in the socket using white metal or lead. Stein, 1997.

back, the ceiling of any underground excavation.

back analysis, a technique which can provide controlling parameters of a system by analysing its output behaviour. In back analysis of rock engineering problems, force conditions such as external loads and/or rock pressures, and mechanical properties of rock, such as modulus of elasticity, Poisson's ratio, cohesion, internal friction angle, etc, are identified from displacement, strain and pressure measured during and/or after construction. Hudson, 1993.

backbreak or **overbreak**, rock volume broken beyond the plane defined by the last row of blastholes.

backbreak depth (l_o), (m), the depth broken beyond the plane defined by the last row of blastholes.

back damage, reduction of the rock mass strength at the back of an excavation by blasting.

back face, the end of the drill bit. Sandvik, 2007.

backfill, material filled into open stopes underground to stabilize the rock and to prevent the subsidence of the ground surface. The backfill material may consist of waste rock, tailings, natural sand etc.

backfilling, the operation to fill the mined open stopes underground with 1) Waste rock excavated from the mine, 2) Waste rock or sand brought from surface through raises and 3) concentrator tailings or sand brought from surface in pipelines. The purpose could be to avoid subsidence on surface or to support the hanging wall in open stopes when they have been fully excavated.

backhammer, device built into a hydraulic rock drill to enable the reversal of the impact force. The backhammer is used to extract a stuck drill string when drilling in bad rock. Atlas Copco, 2006.

back hammering, the process of reversing the impact of a rock drill to free a stuck drill string. Atlas Copco, 2006.

back head, rear portion or segment of an assembly. The rear segment of a rock drill. Atlas Copco, 2006.

backhoe, the most versatile machine (rig) used for trenching. The basic action involves extending its bucket forward with its teeth armed lip pointing downward and then pulling it back toward the source of power. Carson, 1961.

backhoe loader, a boom and digging scoop arrangements mounted on the back of a tractor. Atlas Copco, 2006.
back hole, a roof hole in drifting or stoping. Singh and Wondrad, 1989. Suggested nomenclature for drill holes in development headings is given in Fig. 6.

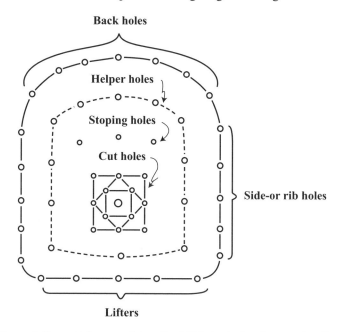

Figure 6 Suggested nomenclature for drill holes in development headings.

back-off shooting, the firing of small explosive charges for releasing stuck drilling tools in a borehole. The shock of detonation causes the joint to expand and unscrew slightly. All rods above the joint can then be removed from the hole. Nelson, 1965.
back stoping, drilling and blasting the roof of a stope or room. Hanko, 1967.
back-up detonator, an extra detonator used for the safe detonation of a charge in a round. The use of back-up detonators increases the probability of detonation if the first detonator fails.
back up equipment, in full face boring the equipment behind the rotational head of the full-face borer.
back wall, final wall in surface blasting.
bad ground or **bad rock**, rock formations in which mine or excavation openings cannot be safely maintained unless heavily timbered or supported in some manner. Long, 1960.
bailer test, a method to quantify the water production in a well by bailing the water from the well. The cable tool driller can make a 17 minute bailer test to check the available yield from the well at any time. Stein, 1997.
balance rope, a steel wire rope, generally of the same weight per meter as the main winding rope, which is attached to the bottom of the cage, and extends down to form a loop in the shaft bottom or sump. Its function is to balance out the difference in weight of the upgoing or downgoing main ropes during the wind. Nelson, 1965.
balance shot, in coal mining, a shot (blast) for which the drill hole is parallel to the face of the coal that is to be broken by it. Fay, 1920. In metal mining the term used is '**slashing**'.

ball cock, self acting cock activated by a hollow ball attached by a long arm that floats on the fluid into reservoir. Atlas Copco, 2006.

ballistic buttons, cemented carbide button inserts used in drill bits, where the protruding portion has a parabolic shape instead of the standard spherical shape.

ballistic disc, a cylindrical charge with a half spherical liner at one end turned with the convex side towards the charge and where the cylindrical and half spherical liners consist of spun sheet metal. The charge may consist of the high energy explosive composition RDX/TNT/wax. It is a shaped charge used for clearing hang ups in underground metal mines. The charge is supported on a telescopic leg stand, that allows adjustments for aiming at the target boulder. Sen, 1995.

ballistic mortar test, see '**ballistic pendulum**'.

ballistic pendulum (mortar) test, a testing device for explosive strength which measures the impulse of an explosive. The laboratory instrument consists of a heavy pendulum mortar in which a standard explosive charge mass is fired, and the angle of recoil is measured. Historically it was used to calibrate explosives strength against that of blasting gelatine. The 10 g explosive sample is, however, insufficient for many modern explosives to develop their characteristic energy release at a certain blasthole diameter.

ball valve, simple hand operated valve, the body of which is a perforated spherical ball that can be turned 90° to shut the valve.

BAM, see '**BAM steel tube test**', '**BAM 50/60 steel tube test**' and '**fall hammer test**'.

BAM fall hammer test, see '**fall hammer test**'.

BAM steel tube test, a shock test of explosive under confinement in a steel cylinder and by initiation from a detonator and booster (RDX/wax 95/5). The sensitivity of a solid or liquid can be measured. This test is needed for the United Nation classification of explosives. Persson et al., 1994.

BAM 50/60 steel tube test, a test procedure to determine the risk of detonation of a material during transport. Persson, 1993.

band, applied to a stratum or lamina conspicuous because it differs in colour from adjacent layers. A group of layers displaying colour differences is described as being banded. A.G.I., 1957 and 1960.

banded (*adjective*), the property of those rocks having thin and nearly parallel bands of different textures, colours, or minerals.

banjo fitting, fitting made from two parts of which one is a hollow threaded tube having a head on one end and a thread on the other, and a second tube having a thread at one end and a ring shape attachment at the other with a perforation at right angles to the tube. The banjo fitting is used to fasten a tube or a hose to a manifold block where there is very little space available, the first parts fit through the perforation. Atlas Copco, 2006.

bank height (H_b), **(m)**, see '**bench height**'.

banking level or **pit bank**, the level at which charged cages or skips come to rest and are discharged after being wound up the shaft. B.S. 3552, 1962.

bar, a drilling, scaling or tamping rod.

bar down, to bend down loose stones from the back and walls.

bar drill, a small diamond- or other type rock drill mounted on a bar and used in an underground workplace. Also called bar and used in an underground *work bar rig*. Long, 1960.

bare, to cut coal by hand. Mason, 1951.

bar grizzly, a series of spaced bars, rails, pipes, or other members used for rough sizing of bulk material passed across it to allow smaller pieces to drop through the spaces. ASA MH4.1, 1958.

bar mining, the mining of river bars, usually between low and high waters, although the stream is sometimes deflected and the bar worked below water level. Fay, 1920.

barney, a small car, or truck attached to a rope and used to push cars up a slope or an inclined plane. Also called *bullfrog*; *donkey, ground hog, lorry, ram, mule and truck*. Fay, 1920.

barren gangue, see '**waste (rock)**'.

barren rock, see '**waste (rock)**'.

barricade (noun), grown- over earth embankments erected in order to protect buildings which may be endangered by an explosion. The top height of the barricade must be at least one meter above the building being protected. The required safety distances between explosive manufacture buildings or storage houses can be halved if the houses are barricaded. Meyer, 1977.

barricaded (adjective), the effective screening of a building containing explosives from a magazine or other buildings, railway, or highway by a natural or an artificial barrier. A straight line from the top of any side-wall of the building containing explosives to the eaves line of any magazine or other building, or to a point 3,6 m above the centre of a railway or highway shall pass through such a barrier. IME, 1981.

barrier branching, when the outgoing P-wave from a blasthole is reflected at the front surface, the reflected PP-wave will interact with the outgoing radial cracks tips from the borehole causing these cracks to turn and run in circumferential direction and they will branch numerous time. The net result is intense fragmentation. Fourney, 1993.

bar rig, a small diamond or other rock drill designed to be mounted and used on a bar. Also called bar drill. Long, 1920.

barring, the end and side timber bars used for supporting a rectangular shaft. The bars are notched into one another to form a rectangular set of timber. Nelson, 1965.

barring-down or **scaling**, removing, with a bar, loose rock from the sides and roof of a mine working.

barrings, a general term for the setting of bars of timber for supporting underground roadways or shafts. Nelson, 1965.

bar velocity (c_P or c_S), **(m/s)**, velocity (P- and S-wave) measured in a bar (rock core) with an aspect ratio (length to diameter) approaching 10:1. If specimens with aspect ratios less than 1:1 are used then correction factors must be used in subsequent calculations. Hudson, 1993.

base charge, the detonating component (usually PETN) in a blasting cap, initiated by the priming charge or the initiation element. It also sometimes refers to the charge loaded into the bottom of a blasthole, usually called **bottom charge**.

basement rock, a name commonly applied to metamorphic or igneous rocks underlying the sedimentary sequence. A.G.I., 1957.

basin, a depressed area with no surface outlet. The term is widely applied, e.g. to a lake basin, to a ground water basin, to a shallow depression on the sea floor, to a circular depression on the moons surface, or to a tidal basin. Bates and Jackson, 1987.

BBSD (acronym), for 'blasted block size distribution' or fragment size distribution after blasting.

BC (acronym), see '**block caving**'.

BDI (acronym), (D_j), **(dimensionless)**, see '**blast damage index**'.

beam, straight heavy piece of steel or other hard material serving as a support for the rocks drilling machine. Some examples on drill rigs are aluminium feed beams and steel support beams used in a rig frame. Atlas Copco, 2006.

beanhole connector, a connector usually made of aluminium, crimped at the end of the safety fuse, and connected to a plastic igniter cord for easy initiation of multiple shot holes. Sen, 1992.

bearing capacity of rock bolts or **shotcrete**, the largest load a rock bolt or layer of shotcrete can withstand without breakage.

bearing sleeve, tubular structure used to cover and protect a bearing. Atlas Copco, 2006.

Becker-Kistiakowski-Wilson (BKW) equation of state, describes the detonation states. It is a correction factor to the ideal gas law $pV = nRT$.

$$pV = nRT\left(1 + \frac{\sum_i n_i k_i}{V(T+\theta)^\alpha} e^\beta\right) \tag{B.1}$$

p is the pressure of gas in Pa, V is the volume of gas in m^3, n_i is the number of moles in the ith product, α, β and θ are the empirical constants and k_i represents the co-volume, R is the gas constant in Pa · m^3/ mole· °K and T the absolute temperature in °K.

bedded deposit, a mineral deposit of tabular form that generally lies horizontally and is commonly parallel to the stratification of the enclosing rocks. A coal seam is a typical bedded deposit; others may contain industrial minerals, and some are metalliferous. Gregory, 1983.

bedding cleavage, a cleavage that is parallel to the bedding. Billings, 1954.

bedding fault, a fault that is parallel to the bedding. A.G.I., 1957.

bedding plane, in sedimentary or stratified rocks, the division planes that separate the individual layers, beds, or strata. A.G.I., 1957 and 1960.

bedding spacing, the width of the layers in a bedded formation. The bedding spacing can quantified, see Table 2. ISRM, 1981.

Table 2 Bedding spacing. After ISRM, 1981.

Description	Spacing (m)
Very thickly bedded	>2
Thickly bedded	0,6–2,0
Medium bedded	0,2–0,6
Thinly bedded	0,06–0,2
Very thinly bedded	0,02–0,06
Laminated	<0,02

bedrock, the solid rock of the earth's crust, generally covered by overburden e.g. gravel, soil, or water.

beds or **bedding**, layers of sedimentary rock, usually separated by a surface or discontinuity. The rock can often be readily separated along these planes.

Beethoven exploder, a blasting machine for multishot firing of series-connected detonators in tunnelling and quarrying. Nelson, 1965.

belled shaft (in pier drilling), is similar to a straight shaft, but once founding level is reached, a "belling" or "under reaming tool" is used to expand the diameter of the hole or base. This increases the allowable load carried in proportion to the enlarged base area without increasing the overall shaft size. Stein, 1977.

bellow, a device that can be expanded or contracted, usually constructed from a flexible material to form an enclosed space. Atlas Copco, 2006. An instrument with an air chamber and flexible sides, used for directing a current or air. In a foundry, small hand bellows are used for blowing parting sand away from the faces of patterns, etc. Crispin, 1964.

bell wire, see '**connecting wire**'. AS 2187.1, 1996.

belt conveyor, a moving endless belt that rides on rollers and on which coal or other materials can be carried for various distances. Kentucky, 1952.

belt cut sampling, the technique to take samples from the conveyor belt and use it for analysis of mineral content, size distribution etc. Onederra, 2004.

BEM (acronym), see '**boundary element method**'.

bench, *bluff or ledge*, a horizontal surface in or at the top of a vertical or near-vertical rock face from which holes are drilled vertically or inclined down into the material to be blasted. The height of a bench varies usually from a few meters up to 30 m, and ~15 m is a common bench height in surface mining. Benches are normally used in surface mining but sometimes also in large underground openings.

bench blasting, an excavation method where benches are blasted in steps. When the method is used on the surface, one or several rows of blastholes are drilled parallel to the vertical or inclined free face. The bottom of the holes are more confined than the column. The confinement is dependent on the inclination of the blastholes, see '**confinement**'. In large area tunnelling or open stoping, the top part of the tunnel or stope is made as a drift (pilot heading), and the bottom part is taken out by benching with either vertical, inclined or horizontal holes.

bench drilling, drilling of the holes along a bench. Atlas Copco, 2006.

bench drilling rig, drill rig specifically designed to drill blast holes on benches in quarries, mines or construction sites. Atlas Copco, 2006.

bencher feed, feed on a drill rig used for drilling blast holes in quarry benches. Atlas Copco, 2006.

bench height (H_b), **(m)**, see '**height of bench**'.

benching, a process of excavation where a highwall is worked in steps or lifts.

bending, process of deformation perpendicular to the axis of an elongated structural member when a moment is applied normal to its long axis. ISRM, 1975.

bending (angular or **in the hole) deviation** or **error in drilling** (D_b), **(mm/m** or **%)**, deviation of the effective axis of the hole drilled from the desired orientation. Bending deviation is mainly caused by dynamic and frictional instabilities of the drilling tool, by the structure of rock, the difference in hardness of different rock types, the gravity forces on the drill bit and the drill steels, the drilling method, and the stiffness of the drill steel.

bend strength (σ_f), **(MPa)**, see '**flexural strength**'.

bell jar, see '**jar collar**'.

bellows, device that can be expanded or contracted, usually constructed from a flexible material to form an enclosed space. Atlas Copco, 2006.

belt cut sampling, the technique to take samples from the conveyor belt and use it for analysis of mineral content, size distribution etc. Onederra, 2004.

bench drilling, drilling of the holes along a bench. Atlas Copco, 2006.
bench drilling rig, drill rig specifically designed to drill blast holes on benches in quarries, mines or construction sites. Atlas Copco, 2006.
bencher feed, feed on a drill rig used for drilling blast holes in quarry benches. Atlas Copco, 2006.
Bergmark-Roos formula, see '**draw body**'.
BERI (acronym), see '**burst energy release index**'.
berm, a horizontal shelf or ledge cut into an embankment or sloping wall, or an open pit or quarry to break the continuity of an otherwise long slope, for the purpose of strengthening and increasing the stability of the slope, or catching or arresting slope slough material. A berm may also be used as a haulage road or serve as a bench above which material is excavated from a bank or bench face. The term is sometimes used as a synonym for bench. USBM, 1968.
Bernoulli equation, the conservation of energy equation applied on a detonation can also be used to describe the energy conservation of a flowing *metal jet created by a shaped charge*.

$$w_i + pv + \frac{1}{2}v_1^2 = const \qquad (B.2)$$

where w_i is the *specific internal energy* in J/kg, p is the *pressure* in the metal jet in Pa, v is the *specific volume* of the metal jet in m³/kg, v_1 is the *inflow velocity* in m/s of the cone material.
BI, (I_{BI}), **(dimensionless)**, see '**blasting index**'.
biaxial compression, state of stress caused by the application of compressive and normal stresses in two perpendicular directions.
biaxial testing, see '**Hoek and Franklin biaxial pressure chamber**'.
Bichel gauge fume test, a test method for blast fumes similar to Crawshaw-Jones fume test, see '**Crawshaw-Jones fume test**'.
Bieniawski's geomechanic classification, a classification system for the rock mass. It uses five parameters and was developed mainly from civil engineering excavations in sedimentary rocks in South Africa. Bieniawski, 1973 and 1976.
  1 Strength of intact rock material (Uniaxial compressive strength of cores or point load index for very low strength rocks)
  2 Rock quality designation, RQD
  3 Spacing of joints
  4 Condition of joints
  5 Groundwater conditions
bifurcation (cracking), the process when a propagating crack is divided into two or more crack tips that continue to propagate.
bifurcation theory, is stating that the equations describing the equilibrium continuation do not provide a unique solution. Hudson, 1993.
big hole or **large hole blasting**, a production method using large diameter holes for drilling and blasting defined as holes with a diameter >100 mm underground and >200 mm on surface. Synonym to 'large hole blasting'.
bighole or **largehole stoping**, an underground mining method using blastholes with a diameter >100 mm.

binary explosive, an explosive based on two ingredients, such as nitro methane and ammonium nitrate, which are shipped and stored separately and mixed at the jobsite to form a high explosive. USBM, 1983.
binary image, a digitized picture in image analysis.
bing hole (historical term), a hole or chute through which ore is thrown. Fay, 1920.
BIPM (acronym), see '**Bureau International des Poids et Mesures**', in English 'The International Bureau for Weight and Measure'.
BIRD (acronym), see '**blast induced rock damage**'.
birdcaged twin steel strand, a rock reinforcement method where two cables and its strands are open and twisted together and fastened together at certain distances along the drill hole where the cables are inserted and grouted, see Fig. 7. Stillborg, 1994.

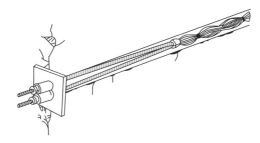

Figure 7 Birdcaged twin steel strand. After Stillborg, 1994.

BIRDI (acronym), see '**blast-induced rock damage index**'.
bit, see '**drill bit**'.
bit blank, a steel bit in which diamonds or other cutting media may be inset by hand peening or attached by mechanical process such as casting, sintering or brazing. Also called *bit shank, blank, blank bit* and *shank*. Long, 1920.
bit disc, a bit with two or more rolling discs which do the cutting. Used in rotary drilling through certain formations.
bit drag or **drag bit**, a bit with serrated teeth used in rotary drilling. Hess, 1968.
bit kerf, flushing water notches in a diamond drill bit. Atlas Copco, 2006.
bit load (F_{bi}), **(N)**, the force applied to a bit in drilling operation expressed normally in tons. Also called '*bit weight*' or '*drilling weight*'.
bit pressure (p_{bit}), **(MPa)**, the pressure applied to a drill bit in drilling operation. Also called '*drilling pressure*' or '*drill pressure*'.
bit shank, see '**bit blank**'.
bit wear index (BWI), a measure of resistance against wear determined by the abrasive hardness, see '**abrasiveness**'. The value is used to predict the lifetime of drill steel equipment. It is a function of the *drilling rate index* (DRI) and the *abrasion value* (AV), see '**tungsten carbide abrasion test**'. Selmer-Olsen and Blindheim, 1970.
bit weight (m_{bit}), **(kg)**, the load applied to a drill bit in drilling operation. Use 'bit mass'. Also called *drilling weight*.
BKW (acronym), an equation of state, see '**Becker-Kistiakowski-Wilson equation of state**'.
black powder, a 'low explosive' consisting of a mixture of sodium or potassium nitrate, carbon, and sulphur. Black powder was first used in underground mines in 1627. Today it is seldom used as an explosive, because of its low energy, poor fume quality, and

extreme sensitivity to sparks. Black powder is used in safety fuse, and as an explosive for quarrying dimensional stones.

black powder fuse, see '**safety fuse**'.

blade bit or drag bit, a drill bit with tungsten carbide inserts. Stein, 1997.

Blake's crusher, see '**jaw crusher**'. LKAB, 1964.

blank bit, see '**bit blank**'.

blanket blasting (shooting), see '**buffer blasting**'.

blank off, to line a specific portion of a borehole with casing or pipe for the purpose of supporting the sidewalls or to prevent ingress of unwanted liquids or gas. Also called *case, casing, case off* or *seal off.* Long, 1920.

blast or **shot** (noun), the process of inducing shock waves and quasi static gas pressure by explosives for breaking rock, welding different metals together, demolishing buildings or initiating seismic waves for geophysical purposes. See also '**blasting**'.

blastability, ability of rock or any material to fragment when being blasted. The most important physical and mechanical rock properties for blastability include; 1) *Acoustic impedance* 2) *Rock structure and friction properties of the discontinuities* 3) *Water content in the rock mass* . Different test methods can be used to determine the blastability: 1) *Determination of the rock constant c*, which is a specific value of the resistance of the rock to the explosive force, determined by single hole blasting (SHB), see '**rock constant**'. Fraenkel, 1954. 2) *Small-scale crater blasting* (cones) is used to determine the optimal charge depth in full-scale crater blasting but are not useful for the design of bench blasting. 3) *Multiple hole blasts on the half scale.* Bergmann et al., 1973. 4) *Critical burden tests.* Rustan et al., 1983 or *Continuous critical burden test.* Whimmer, 2008. 5) *Single hole blasting tests at different burdens* in the laboratory or on the full scale. Rustan et al., 1983 and Rustan and Nie, 1987.

The definition of blastability depends on what we define as the most important result parameter in a blast such as fragmentation, backbreak, throw, muck pile shape, diggability less fines etc. *Blastability is a system property* because it is dependent on both the explosive- and rock properties, the effectiveness of energy transfer between the explosive and the rock mass, the *confinement of the blastholes* and the *stresses in the rock mass*. Normally blastability is coupled with fragmentation. The most important, easily gained *engineering* rock properties, influencing blastability regarding fragmentation known today are the in situ rock density and the in situ P-wave velocity and its product, the *acoustic impedance*, see Müller, 1990 and Rustan, 1983. The acoustic impedance of the rock mass is important for the energy transfer from explosive to the rock mass. The rock mass density is important regarding the heave energy needed to throw the blasted material. Rock mass density also effects the strength of the rock. The in situ P-wave velocity is also a rough indicator of the strength of the rock mass (also affecting the heave of the rock mass). The in situ P-wave velocity will be influenced by the following properties of the intact rock; compressive strength, porosity, joint frequency and type of filling material and finally the water content in the rock mass. The joint orientation in regard to the blast face is also of very large importance for the blastabilty. The P-wave velocity measured in the laboratory on cores is dependent on the following elastic parameters of the rock mass; *Young's modulus*, *density of rock* and *Poisson's ratio*. Historical parameters to classify blastability of rock have been 1) static tensile strength, 2) ratio of compressive strength/tensile strength and 3) toughness.

blast area, the area near a blast within the influence of flyrock or concussion.

blast casting, a blasting method used in surface coal mining where the overburden is casted (thrown) by blasting a larger distance. This makes it possible to mine the underlying coal with less overburden haulage.

blast damage, damage caused by blasting, e.g. excessive ground vibrations and/or air pressure and/or flyrock distances.

blast damage and gas penetration into the remaining rock, blast gases can penetrate through existing joints or discontinuities or earlier created blast cracks crossing the blast hole to other measurement holes behind the blast or to other blastholes. In water filled holes, the pressure can be measured with pressure transducers. Initially there is a negative pressure, explained to be due to the swelling of rock, and this negative pressure is further on in time changed to a positive pressure. LeJuge, 1993.

blast damage index BDI, (D_i), (dimensionless), is defined by the ratio between the induced stress from blasting and the dynamic tensile strength for the rock mass. It is a dimensionless indicator analogous to the reciprocal of safety factor and may vary between 0–3. It is used as an index to quantify the damage from blasting underground on the surrounding rock mass. The blast damage index D_i is defined according to Yu and Vongpaisal, 1996 as follows:

$$D_i = \frac{v_R \rho_r c_P}{k \sigma_{td}} 10^{-9} \tag{B.3}$$

where is v_R is the vector sum of peak particle velocities in three orthogonal directions (m/s), ρ_r is the density of rock mass in (kg/m³) and c_P the P-wave velocity in (m/s), k site quality constant (0–1), and σ_{td} the dynamic tensile strength of the rock mass in (Pa). The correlation between D_i and type of damage see Table 3.

Table 3 Blast damage index D_i and type of damage. After Yu and Vongpaisal (1996).

Blast damage index D_i	Type of damage
≤0,125	**No damage to underground excavations**. Maximum allowable value for key permanent workings, e.g. crusher chambers, shafts, permanent shops, ore bins, pump houses etc.
0,25	**No noticeable damage**. Maximum tolerable value for long term workings, e.g. shaft accesses, rescue stations, lunch rooms etc.
0,5	**Minor and descrete scabbing effects**. Maximum tolerable value for intermediate term workings, e.g. main drifts, main haulage ways, etc.
0,75	**Moderate and discontinuous scabbing damage**. Maximum tolerable value for temporary workings, e.g. cross-cuts, drill drifts, stope accesses, etc.
1,0	**Major and continuous scabbing failure, requiring intensified rehabilitation work**
1,5	**Severe damage to an entire opening, causing rehabilitation work difficult or impossible**
≥2,0	**Major caving, normally resulting in abandoned accesses**

blast damage and gas penetration into the remaining rock, blast gases can penetrate through existing joints or discontinuities or earlier created blast cracks crossing the blast hole to measurement holes behind the blast. In these water filled holes, the pressure can be measured with pressure transducers. Initially there is often a negative pressure, explained by the swelling of rock. This negative pressure turns later on into a positive pressure. LeJuge, 1993.

blast damage and microcracks, the blast influence on microcracks at small scale blasting (0,25–2,5 kg of explosive) has been studied in the field. An exponential relationship was observed between the microcracks (No. of cracks per cm^2) and the charge weight in kg. Olson, 1973. However, full scale tunnel test in Sweden showed no correlation between microcrack frequency in quartz grains (easy to detect the cracks) and distance from the tunnel blast in a fine to medium grained granite. Kornfält, 1991.

blast damage measurement methods, are used to quantify the disturbance to the surrounding rock mass.

1 Overbreak expressed in mean distance of overbreak (l_{oa}), (**m**) or percentage of overbreak compared to the theoretical area of the drift (m or dimensionless).
   - Telescopic rod method is used to measure the distances mechanically. The accuracy of the method is limited to the number of points measured in a section.
   - Light slot method is used to compare the blasted contour with the planned contour. The overbreak area can be calculated by the computer or be evaluated with a planimeter. The accuracy of this method is better than 2%.
   - Laser beam for measurement of distances. Depending on how many readings are taken the accuracy could be made as high as in the slight slot method.
2 Increase of micro fractures.
   - Diamond drillings in the rock mass that might be disturbed is done before and after blasting. The increase of the frequency of microcracks is an indication of disturbance.
3 Increase and new opening of macro fractures.
   - A borehole binocular, camera or video camera can be used to study the increase of number of cracks in special observation boreholes drilled into the rock around the excavated area.
4 Vibrations measurements (a common used method).
   - Vibrations are measured with accelerometers or geophones and is an indicator of the stress level that the rock mass has been subjected to.
5 Movement of individual rock blocks in the rock mass.
   - For this purpose extensometers can be used.
6 Measurement of increase in borehole pressure in boreholes drilled near the excavation.
   - Normally the pressure will be reduced during blasting and its believed to depend on the swelling of the rock mass at blasting creating more space for the gas in the borehole to stream into.

The mentioned methods together with some other methods are summarized below.

1 Vibration monitoring of vertical maximum particle velocity. This is a common used (indirect) method.
2 Half cast factor measurements.
3 Borehole examination, with binocular, camera, TV-camera or video camera for detection of induced macro cracks.

4 Ultrasonic cross hole technique for P- and S-waves in rocks with few joints. Militzer, 1978.
5 Drill cores are examined by acoustic pulse velocity before and after blasting. USBM, 1974.
6 Diamond drilling before and after blasting.
7 Extensometers.
8 Permeability changes.
9 Water head measurement (gas pressure in neighboring holes).
10 Air head measurement.
11 Electrical, resistivity, changes. Hearst, 1976.
12 Pneumatic packer to make impressions from the cracks.
13 Detonators in water filled up holes can be used as a seismic source. Tunstall, 1997. Analysis of changes in wave velocities, Fourier analysis to determine the attenuation of different frequencies etc.
14 Seismic refraction.
15 Surface seismic.
16 Measurement of induced microcracking.
17 Gas penetration into remaining rock mass measured in boreholes behind the blast breakage area.
18 Heave of the surface of the remaining rock mass in a bench blast measured by vertical oriented borehole extensometers.

blast damage quantification methods, see '**blast damage measurement methods**'.

blast damage related to joint frequency, it has been found a correlation between the total joint frequency in the rock mass and the half cast factor measured after a controlled contour blast. The higher joint frequency the lower is the half cast factor. Lewandowski, 1996.

blast damage zone radius (r_{bdz}), (m), the radius of the zone around a blasthole where new cracks and dilation of rock mass is caused during blasting. The blast-induced damage zone is the result of a complex interaction of many parameters like rock quality (e.g. the Barton Q-factor), (linear) charge concentration (kg/m), confinement (e.g. the burden and spacing), delay time between contour holes and decoupling. The extent of the damage zone can be quantified, see '**blast damage measurement methods**'.

The blast induced damage zone at *surface blasting* can be divided into three categories according to Mojtabai and Beattie, 1996. The *heavy damage zone* where the rock is totally shattered and fragmented. Each rock block is completely separated or loosened and an excavator can dig through most of the material with little or no difficulty. Large craters sometimes form on top of the muckpile or near the bench toe. The *medium damage zone* (called only damage zone by Mojtabai and Beattie), is characterized by large, wide-open radial cracks as well as fractures that form parallel to the bench face. Some other fractures have openings as wide as 125 mm. Ground heave and an increase in volume are also apparent. The *minor damage zone is characterized by* small, closed fractures that normally run parallel to the bench face. Very little ground heave may be present. This type of damage may not be considered serious, but it is expected to be more severe in the lower portion of the bench where the explosive column is placed. In Table 4 the vibration levels for four different kind of rock types are shown. See also '**blast damage index BDI** 'for underground damage'.

Table 4 Vibration levels (mm/s) for different rock types and damage classes. After Mojtabai and Beattie, 1996.

Rock type	Uniaxial compressive strength (MPa)	RQD (%)	Minor damage zone (mm/s)	Medium damage zone (mm/s)	Heavy damage zone (mm/s)
Soft schist	14–30	20	130–155	155–355	>355
Hard schist	49	50	230–350	350–600	>600
Schultze granite	30–55	40	310–470	470–1700	>1700
Granite porphyry	30–85	40	440–775	775–1240	>1240

blast design, design of a blast operation which includes the layout of the blasthole pattern, the selection of explosives, decking, delay times, initiation pattern, stemming, and necessary safety measures.

blast disturbance, see '**blast damage**'.

blasted burden, *true or effective burden* (B_b), **(m)**, shortest perpendicular distance from the centre or centre line of the charge to the free surface created by earlier charges fired in the round. The exact shape of the free surface created by earlier firing charges cannot always be determined. The size of blasted burden, after the blastholes have been drilled, depends on the firing sequence.

blaster, see '**blasting machine**'.

blaster (USA), **shotfirer (UK)**, a qualified person in charge of a blast. Also, a person (blaster-in-charge) who has passed a test, approved by some authority, which certifies his or her qualifications to supervise blasting activities. USBM, 1983.

blasters log, a pocket sized book to record the statistical information of each shot.

blasters' galvanometer, see '**blasting galvanometer**'.

blasthole, a cylindrical opening drilled into rock or other materials for the placement of explosives.

blasthole charger, mechanical equipment used to charge a borehole. For example, when using prilled ANFO, a portable unit consisting of a explosive tank is air-activated, and the explosive is discharged into an antistatic charging hose. The blasthole charger permits rapid loading of prilled explosives into blastholes drilled in any direction. For upholes, a special technique is necessary for hole diameters >75 mm. The equipment should be grounded to guard against build up of static electricity and possible accidental initiation.

blasthole dewatering, the process of dewatering boreholes before charging with the help of compressed air operated pumps or electrical submergible impellent pumps.

blasthole diameter (d), **(m)**, the diameter of the borehole into which the explosive is placed.

blasthole drill, any rotary, percussive, fusion-piercing, churn, or other type of drilling machine used to produce holes in which an explosive charge is placed. Also called shothole drill. Long, 1920.

blast hole drill rig, general term used to describe drill rigs designed specifically for surface drilling of blast holes. While the term blast hole drill rig is usually used to describe large rotary rigs, it can also be used to describe crawler mounted rigs. Atlas Copco, 2006.

blasthole inclination (α_{bi}), **(°)**, the angle between the blasthole and a reference plane, normally the horizontal plane. In bench blasting an angled hole is to be preferred because it is easier to break the toe and the fragmentation will be better, less overbreak, reduced drilling and blasting costs due to increased burden and a stable bench edge. In most drill rigs, a facility to drill inclined holes is provided but for reasons of "availability" and "utilisation" vertical drilling of holes is commonly preferred.

blasthole length (l_{bh}), **(m)**, the length of the blasthole as measured along the axis from the collar to the bottom of the hole.

blasthole liner, often a thin plastic envelope to protect non-water-resistant explosives from water in the blasthole. The explosive is filled into the liner in charging. When ANFO is charged pneumatically, plastic borehole liners should not be used because of the risk of initiation of the explosive by static electricity.

blasthole plug or **sealer**, see '**borehole plug**'.

blasthole pressure, see '**detonation pressure**' and '**borehole pressure**'.

blasthole spacing (*S*), **(m)**, see '**spacing**'.

blasthole tubing, stiff plastic tubes used in blastholes to reduce the linear charge concentration or to stabilize the walls of the blasthole.

blast induced damage zone, see '**blast damage zone**'.

Blastine (tradename), an explosive consisting of trinitrotoluene (TNT), ammonium per chlorate, sodium nitrate, and paraffin wax. Bennett 2nd edition, 1962.

blasting barrel, a piece of iron pipe, usually about 12 mm in diameter, used to provide a smooth passageway though the stemming for the miners squib. It is recovered after each blast and reused until destroyed. Fay, 1920.

blasting cartridge, see '**cartridge**'.

blasting compounds, explosive substances used in mining and quarrying. Hess, 1968.

blasting environmental impact, the main impact from blasting are: 1) Ground vibrations, 2) Flyrock, 3) Dust, 4) Pollution of mine water by ammonia and nitrates from explosives due to spillage, 5) Dust explosions, 6) Poisonous blast gases underground, 7) Incomplete detonation and misfires and finally 8) Pollution of air by NO_x gases especially from ANFO-explosives.

blasting in cold regions, blasting in cold climate needs a change of blasting methods. e.g. when blasting in permafrost, the blast design rules has to be modified due to the larger strength of rock and soil under frozen condition. Mellor, 1975.

Blasting index (fragmentation oriented) (BI), (I_{BI}), **(dimensionless)**, Julius Kruttschnitt Mineral Centre in Australia has proposed a *blastability index BI*. Lilly, 1986. BI is determined as follows:

Parameter	Rating
1 **Rock mass description (RMD)**	
• Powdery/Friable(Can be crushed by hand)	10
• Blocky (Breaks into large blocks)	20
• Totally massive	50
(Massive = homogeneous structure without stratification, flow-banding, foliation, schistosity, etc.)	
2 **Joint plane spacing (JPS)**	
• Close	10
• Intermediate	20
• Wide	50

3	**Joint plane orientation (JPO)**	
	• Horizontal	10
	• Dip out of face	20
	• Dip into face	40
4	**Specific gravity influence (SGI)**	
	• Waste rock	25
	• High density ore "magnetite"	50
5	**Hardness (H)**	1–10

According to the definitions above the blasting index BI varies from a minimum of 56 to a maximum of 200. BI is linear related to the specific charge (powder factor) and this means that the variation of specific charge is 1:4 dependent on the rock mass condition. BI is calculated from the parameters above by the following empirical formula

$$I_{BI} = 0{,}5(f_{RMD} + f_{JPS} + f_{JPO} + f_{SGI} + f_{H}) \tag{B.4}$$

Another assessment system for blastabilty is based on the interaction matrix for rock mass structure developed by John Hudson, 1989. This was modified by Latham, 1999 for definition of blastability.

blast-induced rock damage (BIRD), can be grouped into 4 different zones
> **overbreak zone**, zone of rock breakage or significant damage to the rock mass beyond the designed perimeter of the excavation following the blast and the subsequent face dressing.
> **crack widening zone**, the zone where the aperture of existing cracks in the rock mass increases due to expanding gases or the dilatation of rock mass.
> **incipient crack growth zone**, the zone where the fresh cracks are generated in the rock mass due to blast-induced strain. However, the rock does not get dislodged immediately even after dressing. Repeated blast cycles may trigger the crack growth and eventually result into overbreak.
> **intact zone**, the zone where the rock mass strength is not reduced significantly and hence the effect of blasting is insignificant in terms or rock damage.

blast-induced rock damage index (BIRDI), (D_i), (dimensionless), an empirical blast-induced damage index defined by the following formula. Paventi, 1996.

$$D_i = abcd\,(e + f), \tag{B.5}$$

where
> a = considers the reduction in intact rock strength due to micro-fracturing
> b = evaluates the extent of the exposed excavation surface area remaining in place using the post scaling half cast factor
> c = determines the drift condition by assessing the drumminess of the back with a scaling bar
> d = accounts for the amount of scaling arising from damage
> e and f = considers the direction of structure with respect to drift direction to account for the anisotropy potentially caused by structural features at meso- and macro scale.
> See also 'blast damage index'.

blasting, a chemical-physical-mechanical process including the initiation (firing) of explosives for the purpose of breaking materials such as coal, ore, mineral stone or else materials, moving the material, splitting off rock blocks for building purposes, demolishing building and constructions or generating seismic waves. The explosive is often located in drill holes in rock, but sometimes, e.g. in boulder blasting, it may be located on the surface of the rock mass. Compared with placing the charge in a borehole, surface charges require a specific charge up to 10 times in excess, but typically 3 times for correct placing of an appropriate lay-on charge.

blasting accessories, non-explosive devices and materials used in blasting, such as, cap crimpers, tamping bags, blasting machines, blasting galvanometers, and cartridge punches. In South Africa, this term strictly refers to initiating devices for explosives, e.g. capped fuses, igniter cords, detonating cords, electric blasting caps, etc.

blasting agent, any explosive which is composed of non-explosive ingredients and cannot be initiated in unconfined conditions by a standard No. 8 blasting cap. These explosives consist predominantly of ammonium nitrate.

blasting cap or **(elemented cap)**, a small metallic (aluminium) tube closed at one which may contain a small amount of primary explosive ~1 g. For delay caps a pyrotechnical substance is used to create the proper delay time. Blasting caps are initiated by a safety fuse, detonating cord, shock tube of type e.g. NONEL, detonating gas in plastic tubes or electric current. The initiation of the explosive can be achieved directly by the methods mentioned or via a booster. The unit, detonator cap and booster together, is called a primer where a detonator is the elemented cap assembled with the shock tube, the electric wires or the safety fuse. A booster is used when the charge cannot be initiated by a detonator only. 'Seismic' blasting caps needs a high degree of uniformity of firing and therefore they have a reaction time often <1 ms.

blasting cap No. 6, a blasting cap with a base charge of 0,22 g. Sen, 1992.

blasting cap No. 8, a blasting cap containing 0,40 to 0,45 g of PETN base charge at a specific gravity of 1 400 kg/m^3, and primed with standard mass of primer, depending on the manufacturer. USBM, 1983.

blasting cap No. 8 star, a blasting cap with a base charge of 0,88 g. This is the common blasting cap used in most mining. Sen, 1992.

blasting cap test, a method for testing blasting caps. Among the many tests are four major ones 1) Water resistance test, 2) Drop test, 3) Snatch test, and 4) Vibration test.

blasting circuit, the electrical circuit used to fire one or more electric blasting caps. USBM, 1983.

blasting crew, a group of persons who assist the blaster or shotfirer in loading, tying in, and firing a blast.

blasting diary, see '**blasting log**'.

blasting efficiency (η), the ratio of useful energy output to energy input in blasting. In rock blasting the explosive energy is transformed into wave propagation and quasi-static gas pressure energy in the blasthole. The wave energy is used for fracturing (cracking), frictional losses in the rock, and throw, and is finally lost to the surrounding air as noise. The quasi-static pressure is used to bend the rock prism between the two side cracks, to extend the radial cracks starting close to the blasthole wall, and to eject the rock mass.

blasting equipment, tools, instruments and machinery used for charging and blasting operations, e.g. primer prickers, depth gauges, vibration monitors, storage boxes, resistance gauges, ANFO loaders, etc.

blasting fuse, see 'safety fuse'.
blasting galvanometer, an electric instrument to measure the resistance of an electric blasting circuit. It is powered by, e.g. a silver chloride cell, and has an inbuilt current limiting device. Only devices which have been approved by the authorities are permitted to be used for this purpose.
blasting gelatine, a gelatinous mixture of 92% nitroglycerin and 8% nitro-cellulose (nitrocotton). It is a strong (high) explosive, has a high water resistance, and is still one of the most powerful explosives. Used as a standard for explosive strength. Gregory, 1980.
blasting in frozen gravel, techniques to blast in frozen gravel. The technique for drifting is however equal to that of drifting in rock. Both V-cuts and burn cuts have been tested. Dick, 1970.
blasting in hot rock, special explosives and initiation equipment are needed to make the blasting safe. The rock may be cooled by solid CO_2 (dry ice), or a combination of CO_2 and explosive cooling by cooling the blast hole with water before charging the explosive. Special explosives are manufactured for temperatures up to +200°C. Roschlau, 1994.
blasting log, a written record of information about a specific blast as may be required by law or regulation. Atlas Powder, 1987.
blasting machine, an electrical, electromechanical, electronic or non electric device which provides electrical energy to a blasting circuit for the purpose of energizing electric blasting caps. A CD-type blasting machine uses capacitors for the electric discharge. The non electric blasting machine produces a spark for the detonation of the Nonel tube.
blasting machine-generator, a blasting machine generating energy by twisting the handle to a generator.
blasting machine tester, a device used to examine if the blasting machine delivers enough energy to initiate all blasting detonators. The tester consists of a resistance and glow tube. When energy is enough, the tube will glow. Pradhan, 2001.
blasting mat, a covering, usually made of scrap tires, logs, ropes or wire cables, placed over a blast to reduce flyrock.
blasting needle, a pointed tool for piercing the cartridge of an explosive, to permit introducing a blasting fuse.
blasting oil, see '**nitroglycerin**'.
blasting plan, see '**blast design**'.
blasting powder, a powder containing less nitrate, and in its place more charcoal than black powder. Its composition is 65 to 75% potassium nitrate, 10 to 15% sulphur, and 15 to 20% charcoal. In the United States, sodium nitrate is used largely in place of potassium salt. Fay, 1920.
blasting scheme, see '**blast design**'.
blasting site, see '**blast site**'.
blasting stick, a simple form of fuse. Fay, 1920.
blasting strength value (BSV), (MJ/kg), is defined as the *absolute strength value* (ASV) of an explosive minus *brisance energy* and *lost energy*. Lownds, 1986.
blasting supplies, includes explosives, blasting caps, fuses, and ancillary equipment such as charging poles, blasting machines, testers, etc.

blasting switch, a switch used to connect a power source to a blasting circuit. The switch can be located in a surface stand, e.g. if the danger of gas outbursts is given in the underground.

blasting timer, an instrument that utilises a power line as a source of electrical current and which closes the circuits of successive blasting caps with a delay time interval. Streefkerk, 1952.

blasting unit, a portable device including a battery or a hand–operated generator designed to supply electric energy for firing explosive charges in mines, quarries, and tunnels. Also called *blaster, exploder or shot-firing unit*. ASA, 1956.

blasting vibration, the part of the energy from a blast that manifests itself in rock mass vibrations which are radiated from the immediate blast area through the rock mass and overlaying soil. See also '**seismic waves**'.

blast layout, the drilling pattern and initiation delay layout.

blast monitoring, the measurement *before blasting* of rock mass-, explosive-, charging- and initiation properties, *during blasting* the velocity of detonation, throw, stemming ejection and ground vibration and finally *after blasting* the fragmentation, muckpile shape, backbreak, rock damage, angle of breakage, secondary breakage, diggability etc.

blastometer, see '**Nobel blastometer**'. Nelson, 1965.

blast overbreak measurement, see '**perimeter measurement**'.

blast parameter control, see '**blast monitoring**'.

blast pattern, the geometry of the drill holes laid out on the top of a bench or the real pattern after drilling at the bottom of a bench. Also the geometry of the drill holes at the face of a drift or tunnel including the burden and the spacing distance and their relationship to each other. ISRM, 1992.

blast performance analysis or **monitoring**, see '**blast monitoring**'.

blast plan, see '**blast design**'.

blast pressure, see '**borehole pressure**'.

blast pull, see '**pull**'.

blast shelter, a cabin strong enough to protect the personnel from throw of stones in surface blasting. The cabin might be protected by timber mats.

blast site, the location where the explosive material is handled during charging, including the perimeter of blastholes and 15 m (50 ft) in all directions from charged holes or holes to be loaded. In underground mines, 4,5 m (15 ft) of solid rib or pillar can be substituted for the 15 m distance. These rules are used in the USA. Atlas Powder, 1987.

blast vibration attenuation, the general formula for calculation of the blast vibration attenuation is,

$$v_{max} = k \frac{Q^\alpha}{R^\beta} \qquad (B.6)$$

where v_{max} is the maximum particle velocity in mm/s, k is a site constant dependent on distance from the blast, Q is the maximum charge detonated in one delay in kg and R is the distance from the blast in m, and finally α and β are site dependent constants. The values of the constants are dependent on blast site and the blast parameters.

blast vibration monitoring, the technique to measure the blast vibrations in the ground and/or air pressure around a blast. The technique for blast vibration monitoring is described in 'Suggested methods for blast vibration monitoring', see ISRM, 1992.

blind boring, drilling large holes upwards without any guide hole and with the purpose to be used as raise for opening, transport of ore, personnel or mining equipment, escape way in emergency. The minimum diameter will be about 0,5 m. Full-face boring machines can be used for this purpose. MassMin, 2008.

blind drift, a horizontal passage, in a mine, not yet connected with the other workings. Fay, 1920.

blind flange, disc shaped structure, having no central opening, made of a metal or other hard material with perforations around the periphery. A blind flange is fixed by bolts to a mating piece attached to a pipe to prevent flow from the pipe. Atlas Copco, 2006.

blind rivet, rivet made from a soft metal, suitable for installing in a hole drilled only partially through a plate or structure. Often used to install identification and data plates to engines and pumps etc. Atlas Copco, 2006.

blind shaft, an underground shaft, connected to the main shaft by a drift or transfer station. It is therefore a shaft not ending at the surface.

blind shaft boring, a winze boring method that is used where access to the lower level is limited, or impossible. A pilot hole can be drilled and later on reamed to its full diameter or the hole is fully driven in one step. At down hole reaming weights are attached to the reamer mandrel. Stabilizers are located above and below the weight stack to ensure verticality of the hole. Cuttings are removed using a vacuum or reverse circulation system (RC). Atlas Copco, 2007.

blister, in quarrying, an unconfined charge of explosive used to bring down dangerous ground that cannot be made safe by scaling (barring) and that is too inaccessible to bore. South Australia, 1961.

blistering, see '**mud capping**'.

blister shooting, see '**mud capping**'.

block, in geology defined as a piece of rock having a size larger 20 cm. Magnusson, 1963.

blockage, see '**hang-up**'.

Block Cave Fragmentation (BCF), (program), an expert system program developed by Dr. G. S. Esterhuizen for predicting the degree of fragmentation likely to be produced in a block cave. MassMin, 2008.

block caving (BC), a mining method where the ore is broken by *gravity forces*. The block planned to be mined is normally made unstable by undercutting the block at the foot wall and at the bottom and the ore will start to block and the blocks move downwards to a craters draw points or loading ditches. When ore strength conditions are favourable, block caving is a very cheap mining method because of no cost for drilling and blasting of ore. A disadvantage is the long time needed for development of a block. Block caving is a relative cheap method used to mine low grade sulphur mineral ore bodies. It is a mass mining method.

block deformability, change of shape of a block due to stresses acting on the block, see also '**discrete element modelling**'.

block hole, a small hole drilled into a rock or boulder into which an anchor bolt or small charge of explosive may be placed. Long, 1960.

block holing, the breakage of boulders by loading (charging) and firing small explosive charges. Atlas Powder, 1987. See also '**popping**'.

block off, to fill and seal undesirable openings, fissures, or caving zones in a borehole by cementation or by lining the borehole with pipe or casing. Also called *blank off*, *case off and seal off*. Long, 1960.

block volume distribution, a histogram showing the percentage of blocks smaller than a certain volume.

block retreat caving, see '**block caving**'.

block size, rock block dimension or volume resulting from the mutual orientation of intersecting joint sets, and the spacing of the individual joint sets. The block size can be quantified according to standard procedure. Classification of block size, see Table 5. ISRM, 1981. J_v is called the volumetric joint count.

Table 5 Classification of block size. After ISRM, 1981.

Description	J_v (Joints/m³)
Very large blocks	<1
Large blocks	1–3
Medium-sized blocks	1–10
Small blocks	10–30
Very small blocks	>30

block size index (I_b), (m), is estimated by selecting by eye several typical block sizes and taking the average dimensions. ISRM, 1978.

block size in situ, the joints and planes of weakness in a rock mass define the size of the blocks in situ. Several methods have been proposed in the attempt to determine the block size in situ. These methods are the parametrical, the simulation and the computational methods. Wang and Latham, 1990.

blocky hole, see '**block hole**'.

blown-out shot, a shot that dissipates the explosive force by blowing out the stemming instead of breaking down the coal or rock. It may be caused by insufficient stemming, overcharging with explosive or a burden that is too much for the charge to dislodge. Nelson, 1965.

blowout shot, an improperly placed or overcharged shot of black blasting powder in coal (where used) which frequently results in a mine explosion. von Bernewitz, 1931.

blowpipe, a metal pipe which is connected to an air or water hose, or both, to clean out blast holes before charging or to remove stemming in the event of a misfired charge. The blowpipe should be made of copper or any material which will not cause sparks when it is in contact with rock or other materials, because a spark may initiate the explosive.

blunt tool, tool having a worn, thick, dull edge, not sharp. Atlas Copco, 2006.

Blåsjö cut, a fan cut used for the opening of new swell space in tunnelling and drifting. A single V-shape of holes are drilled where all the holes are horizontal and parallel and meet the horizontal holes from the other side at an angle that may be as low as 30°. Developed in Sweden at Blåsjö hydro power station, see Fig. 8. Langefors, 1963, p. 194.

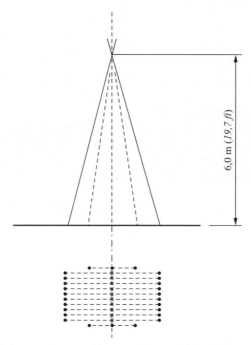

Figure 8 Blåsjö cut, a fan cut with great advance, 83%. After Langefors, 1963.

Bo (abbreviation), for '**boulder**' defining fragments with sizes from 200 to 630 mm.
board, see '**bord**'.
board-and-pillar, see '**pillar and breast**'.
board-and-wall, see '**pillar and breast**'.
board run, the amount of undercutting that can be done at one setting of a coal mining machine, usually 1,5 m, without moving forward the board upon which the machine works. Fay, 1920.
boasting, the rough dressing of stone with a *boasting chisel*. Standard, 1964.
boasting chisel, a flat chisel with an edge 51 mm wide, used in dressing stone. Standard, 1964.
body of motion, see '**drop of motion**'.
body waves (seismic waves), are transverse (shear) and longitudinal (mainly compression but also a small tensile part) waves both also called seismic waves. They propagate in the interior of an elastic solid or in fluids with shear resistance. In water only longitudinal waves can exist because the shear strength of water is zero. The body waves are not related to a boundary surface. A perfectly sharp distinction between body waves and surface waves is difficult to make unless the waves are plane or spherical.
booster, a cap sensitive explosive with high detonation velocity (c_d) used in small quantities to initiate and improve the performance of another explosive, the latter forming the major portion of the charge. A booster does not contain an initiating device.
boot-leg, butt, socket *or gun,* that portion or remainder of a shot hole in a face after the explosive has been fired incompletely after the blast and that still may contain explosives and is thus considered hazardous. A situation in which the blast fails to cause total failure of the rock because of insufficient explosives for the amount of burden, or owing to incomplete detonation of the explosives.
bogey, see '**bogie**'. USBM, 1990.

bogie, under carriage with two, four or six wheels that can swivel to negotiate curves. Atlas Copco, 2006.
bogy, see '**bogie**'. USBM, 1990.
bolt, cylindrical pin, made from metal or other hard material, having a head at one end and a screw thread at the other extremity. The engineering term "cap screw" is also used to describe this device. Atlas Copco, 2006. See also '**rock bolt**' or '**roof bolt**'.
bolt hole drilling (underground), process of drilling holes in underground openings to insert rock bolts. Atlas Copco, 2006.
bolting, has two meanings 1) the process of drilling holes and setting rock bolts and 2) separation of particles of different sizes by means of vibrating sieves. Bennett 2nd ed., 1962.
bolting drill string, arrangement of bit, rod and adapter used to drill holes for the installation of rock bolts. Atlas Copco, 2006.
bolting rig, a combined drilling and bolting machine where all operations can be performed from the drillers cabin. They can handle installation of most types of rock bolts commonly used today such as cement and resin grouted rebars or the swellable tube anchor called Swellex. Up to 80 cartridges, with the purpose to fix the bolt to the rock mass, can be injected before the magazine needs refilling. Also because *meshing* is often carried out in combination with bolting, an optional *screen handling arm* can be fitted parallel to the bolt installation arm, to pick up and install the bulky mesh screens. Up to 5 different pre-programmed cement–water ratios can be remotely controlled. Depending on machine, bolt lengths can vary between 1,5 to 3,5 m or 1,5 to 6,0 m and can be inserted in roof heights up to 8 m. Atlas Copco, 2007.
bolt magazine, storage magazine on a bolting rig used to hold a number of rock bolts. Atlas Copco, 2006.
boltometer, for cement and resin grouted steel bars the quality of grouting as well as the bolt length can be checked by an electronic non-destructive testing device based on the principle of reflected stress waves. Stillborg, 1994.
bolt tester, see '**boltometer**'.
Bond's third theory, is used in crushing to state that the total energy useful in breakage of a stated weight of homogeneous broken material is invariably proportional to the square root of the diameter of the product particles. Pryor, 1963. The reduction from one particle size to another can therefore be described by the following equation.

$$W = 10 w_{Bond} \left(\frac{1}{\sqrt{k_{80out}}} - \frac{1}{\sqrt{k_{80in}}} \right) \quad (B.7)$$

where W is the total energy needed in joule (J) to reduce a size distribution where k_{80in} is the particle size for 80 weight-% passing in microns for the feed coming in and k_{80out} is the particle size for 80 weight-% passing for the products coming out and finally w_{Bond} the *Bond Work index* which is defined as the energy required to crush a solid of infinite size to a product where 80% passes 100 microns.
boom, strong spar or beam projecting from a structure or vehicle, used to support, lift and positioning an object. Drill rig booms have several links or articulation points to allow movement in many directions and are used to support and position the rock drill, feed and other associated components and structures. Atlas Copco, 2006.
boom beam, main ridged portion of a boom. Atlas Copco, 2006.
boom bracket, fabricated structure, attached to the rig frame in such a way as form an attachment for the boom. Atlas Copco, 2006.

boom area coverage (A_b), **(m²)**, the area that can be drilled by one boom, normally at drifting.
boom fork, integrated U-shaped attachment device at the extreme ends of the boom.
boom head, structure at the front end of the boom having attachment points for hydraulic cylinders and the rock drill feed. Atlas Copco, 2006.
boom link, device forming an articulation point in the boom. Atlas Copco, 2006.
boom support, attachment structure for the boom. Atlas Copco, 2006.
boom system, drill boom and all it's components and controls. Atlas Copco, 2006.
booster, something that increases the power or effectiveness of a system or device. Atlas Copco, 2006.
bord or **board**, a passage or breast, driven up the slope of the coal from the gangway, and hence across the grain of the coal. A bord 4 m wide and larger is called a wide bord and one less than 4 m in width is called a narrow bord. Term used in Newcastle. Fay, 1920.
bord and pillar, *pillar and stall, post and stall, bord and wall* and finally *stoop and room*. A *room and pillar mining method* used in coal mines. First bords (roadways) are driven, leaving supporting pillars of coal between. Next, cross drives connect the bords leaving supporting coal as rectangular pillars. Finally, the pillars are mined (extracted, won, robbed) and the roof allowed to cave. The bordroom is the space from which bord coal has been removed. Pryor, 1963.
bord-and wall, see '**bord and pillar**'.
bore, to cut a circular hole by the rotary motion of a cutting tool. Long, 1920. Making a hole, tunnel or well etc. in soil or rock by drilling or digging. Atlas Copco, 2006.
bore, see '**set inside diameter**'.
borability, see '**drillability**'.
borehole, cylindrical hole drilled by a mechanical device or jet-piercing equipment into the rock.
borehole binocular, periscope or strata scope, instrument for singular observation of the borehole wall and in which the illumination is created by a bulb.
borehole camera, an instrument including a camera which can be inserted into a drill hole to take shots of geology, joints and damage to the hole.
borehole casing, a steel pipe lining used in a borehole, particularly when passing though loose, running ground. Flush jointed casing, that is smooth inside and outside may be either screwed or welded. The Swedish diamond core drill casing is flush-jointed, whereas that of the United States is usually coupled. A coupling adds 12 mm on the 114 mm casing, and 25 mm on 508 mm diameter casing. Nelson, 1965.
borehole deformation gauge, a device for measuring the change in diameter of a hole across multiple directions and used to infer the state of stress in the rock.
borehole deviation, see '**drilling accuracy**'.
borehole deviation measurement, the measurement of the deviation of a borehole. Different techniques may be used such as accelerometers (change of inertia), gyroscopes, photography of concentric rings (Maxibor also called ABEM Reflex fotobor), using a light in the hole, or observing a reflex pushed into the hole, etc. For measurement on the surface a vertical borehole deviation device 'Borepak' is commonly used. It consists of a set of interlocking 2 m long and 63 mm in diameter rods containing microchips and inclinometers. A sighting device (binocular or theodolite) at the top of the borehole is

used to mark the position of the collar and the drilling direction. An electronic notebook is used to collect the required data for input into a personal computer to obtain a plot of the hole.

borehole diameter (*d*), (**m** or **mm**), the mean width of the cross-section of the borehole. The borehole diameter varies depending on drilling conditions and equipment. A nominal diameter must be assumed for the purpose of calculating mass per meter of explosive in the borehole. Normally the diameter is larger than the drill bit size, and it decreases with length in upholes, but because of flushing it increases with depth in downholes. In scientific work it is important to determine the actual mean hole diameter. The size of the blasting operation controls the size (small, intermediate or large) of the diameter of the boreholes. A suggestion for classification of borehole diameters is given in Table 6.

Table 6 Definition of borehole size according to the diameter.

Borehole size	Underground mines (mm)	Surface mines (mm)
Small boreholes	<50	<50
Intermediate boreholes	50–100	50–200
Large boreholes	>100	>200

borehole dilatometer (borehole jack), instrument used to measure the deformability of rock. The test method involves expanding the loading platens of the dilatometer against opposite walls of a borehole, and measuring the change in hole diameter resulting from successive increments of loading and unloading. Hudson, 1993.

borehole equilibrium pressure (p_{be}), (*MPa*), see '**equilibrium borehole pressure**' at the term 'borehole pressure'.

borehole impression packer, consists of two plates covered with plastic films capable of deforming to record an impression of the walls of the hole. The impression tool force the two plates of wax or plastic film against the wall of the hole. Most impression tools are designed for collection of geological data or to measure damage to casing and well screens. Stein, 1977. The technique is also used to quantify damage from explosive on the blasthole wall.

borehole investigation methods, there are a large amount of methods available for the inspection of a drill hole like impression blocks, inspection by periscope, camera, TV, video and geophysical logging methods, crosshole seismic etc.

borehole jack, see '**borehole dilatometer**'.

borehole length (l_d), (**m**), the length of the borehole as measured along the axis from the collar to the bottom of the hole.

borehole liner or '**blasthole liner**', a thin plastic sleeve put into the borehole before charging to keep water in the rock from penetrating into water sensitive explosives like ANFO. When ANFO is charged, pneumatically, plastic borehole liners should not be used, because static electricity may cause an initiation.

borehole logging, the determination of the *physical*, *electrical* and *radioactive* properties of the rocks traversed in a borehole. B.S. 3618, 1963.

borehole mining, the extraction of minerals in the liquid or gaseous state from the earth's crust by means of boreholes and suction pumps. Boreholes are used for mining petroleum, gas and for the extraction of liquid solution of salt, sulphur, etc, Nelson, 1965.

borehole plug, material used to block a borehole with the aim of fixing the explosive within a certain length of the borehole. Borehole plugs are made of wood or plastic materials, inflated tyres or gasbags, and sand filled bags. The poly-deck gasbag is a self-inflating bag available in a range of sizes from 75 mm to 400 mm in diameter. The inflation is caused by reaction of diluted hydrochloric acid and bicarbonate producing CO_2 gas. Borehole plugs may be used in open stoping or in vertical retreat mining. They can also be used to separate deck charges or to create air decks.

borehole pressure cells, instruments to measure rock stresses by inserting the instrument into boreholes. See '**rock stress measurement**'.

borehole pressure or **explosion pressure**, (p_b), (**MPa**), the theoretical pressure exerted on the borehole wall by the explosive after the detonation front has passed, and before any expansion of the borehole wall has taken place. The borehole pressure at the maximum expansion of the blasthole diameter but before the breakdown is called the *equilibrium borehole pressure* (p_{be}), (**MPa**). The energy expanded up to this point is defined as the *shock energy* and the typical operating pressure during this phase ranges from the equilibrium borehole pressure of about 1 000 MPa and depending on rock and explosive properties, down to 30 MPa. In the older literature the borehole pressure is calculated before expansion of the diameter. This pressure is calculated for the explosive gas density equal to the density of the original explosive, and in classical theory (ideal explosives) it is quoted as

$$p_b = \frac{\rho_e c_d^2}{8} 10^{-6} \quad \text{(ideal fully coupled explosives)} \tag{B.8}$$

where p_b is the borehole pressure in MPa, ρ_e is the density of the explosive in kg/m³, and c_d the detonation velocity of the explosive in m/s. This formula cannot be used for decoupled charges and non-ideal explosives.

The *borehole pressure before expansion of the borehole* and using radial or axial decoupled charges can with reasonable reliability be calculated according to the following formula:

$$p_b = \frac{\rho_e c_d^2}{8} \left[\sqrt{R_a} \frac{d_e}{d_b} \right]^k 10^{-6} \quad \text{(non-ideal and decoupled explosives)} \tag{B.9}$$

p_b = borehole pressure (MPa)
ρ_e = density of the explosive (kg/m³)
c_d = velocity of detonation (m/s)
R_a = axial decoupling, percentage of explosive column charged (%)
d_e = diameter of the explosive after charging into the blasthole (m)
d_b = diameter of the blasthole (m)
k = a value which has to be determined experimentally on site. The value of k is ~2,6 according to Atlas Powder, 1987. For ideal explosives (military explosives) and decoupled charges, e.g. linear shaped charges used for controlled contour blasting, the value of k is in the range of 4 to 6. Bjarnholt, 1981.

The relation between borehole pressure and volume of the gases follows from the equation of state of the explosive. If the gas pressure is above 100 MPa the equation of state is expressed by $pV^\gamma = const$, where p is the pressure (MPa), V is the gas volume (m^3), γ the adiabatic exponent. (In thermodynamics κ is used as symbol instead of γ).

borehole radar, instrument based on the use of electromagnetic waves to investigate the rock mass properties in tunnelling, see '**radar tomography**'.

borehole rod extensometer, see '**extensometer**'.

borehole slotting (*measurement of rock in situ stress*), a diamond saw is used to make a 1,0 mm vide and 25 mm deep axial slot in the borehole wall at 0°, 120° and 240°. Before, during and after slotting the tangential stress at the borehole wall is measured by a specially developed recoverable contact strain sensor of high sensitivity. The sensor and its application device are part of the borehole slotter. This technique was developed at James Cook University of North Queensland. Hudson, 1993.

borehole spacing (*S*), (m), see '**spacing**'.

borehole TV, a TV-camera which is inserted into the borehole with the aim of observing the borehole wall.

borer, a circular drill bit mounted on a drill steel with smaller diameter than the drill bit and used to drill circular holes.

bore rod, see '**drill rod**'.

boring, the action when a full-face boring machine is used for rotational crushing of the face with hard metal cutters for the purpose of making tunnels (normally only in civil construction engineering) or large diameter holes in rock for both mining and civil construction works (>0,5 m).

boring bar, a rod, made in various lengths, usually with a single chisel cutting edge, for hand drilling in rock. The blows are given with a sledge hammer. Nelson, 1965.

boring head, **drill head** or **swivel head**, the assembly which applies the drilling pressure and rotation to the drill rods. B.S. 3618, 1963.

boring log, the record of the events, like rock types, grain size, water inflow etc, of the formations penetrated in drilling a borehole. Also called 'drill log' or 'drilling log'. Long, 1920.

boring rate (v_d), (m/h), the depth of penetration achieved per unit of time with a given type of rock, full-face boring diameter, and air or water pressure, etc.

boring stroke (A_{max}), (m), the maximum advance for a full-face boring machine in the axial direction after pressurizing the radial hydraulic jacks to create a rigid position for the borer. This makes it possible for the axial hydraulic cylinders to create a thrust at the head of the borer.

BorPak (tradename), a relatively new mini full-face boring machine for blind boring (*boxhole boring*) which climbs up the raise as it bores. It comprises a *guided full-face boring machine*, *power unit*, *launch tube* and *transport assembly*, *conveyor* and *operator console*. Cuttings pass through the centre of the machine, down the raise and launch tube, and onto a conveyor. The BorePak has the potential to bore holes from 1,2 to 2,0 m in diameter at angles as low as 30° from horizontal. It eliminates the need of a drill string and provides the steering flexibility of a raise climber. Atlas Copco, 2007.

bort, diamond material unsuitable for gems because of its shape, size, or colour and because of flaws or inclusions. It also occurs in finely crystalline aggregates and is usually crushed into finer material. Also spelled *boar, boor, boartz, borts and bowr*. USBM I.C. 8200., 1964.

bort bit, synonym for 'diamond bit'. Long, 1920.
Borts (synonym), see '**bort**'. Long, 1920.
bort-set bit (synonym), see '**diamond bit**'. Long, 1920.
bortz (synonym), see '**bort**'. Long, 1920.
bortz bit (synonym), see '**diamond bit**'. Long, 1920
bortz-set bit (synonym), see '**diamond bit**'. Long, 1920
bottom, the floor in any underground operation. Fraenkel, 1954. Lowest part of a structure. Atlas Copco, 2006.
bottom bench, underground bench, blasted after the excavation of the top heading. Nitro Nobel, 1993.
bottom charge (m_b), **(kg)**, the mass of charge used to break the rock mass at the bottom of a blasthole. Normally the bottom charge is given a larger volume strength (*bulk strength*) (the charge concentration should be ~2,5 times the charge concentration in the column), because of the larger confinement at the bottom of the hole.
bottom cutter, a coal cutter for making floor cuts. Also called 'dinter'. Nelson, 1965.
bottom-discharge bit, a bit designed for drilling in soft formations and for use on a double-tube core barrel, the inner tube of which fits snugly into a recess cut into the inside wall of the bit directly above the inside reaming stones. The bit is provided with a number of holes drilled longitudinally through the wall of the bit through which the circulation liquid flows and is ejected at the cutting face of the bit. Also called, *face ejection bit or face-discharge bit*. Long, 1920.
bottom-dump car, a mine car that opens its bottom to discharge. This needs a special designed dumping station. The technique was first used at LKAB in Kiruna.
bottom-dump scraper, a carrying scraper that dumps or ejects its load over the cutting edge. Nichols, 1956.
bottom-dump truck, a trailer or semi trailer that dumps bulk material by opening doors in the floor of the body. Also called 'dump wagon'. Nichols, 1956.
bottomed, a completed borehole, or the point at which drilling operations in a borehole are discontinued. Long, 1920.
bottom heading – overhand stoping, method of blasting used for adits, tunnels, and drifts. The lower part, or bottom heading, is either driven in steps or to the full length, whereafter the upper part is broken out by stoping or slabbing. Fraenkel, 1954.
bottom initiation, initiation of a charge at the bottom of the blasthole.
bottom load, see '**bottom charge**'.
boulder (very coarse soil), oversize rock fragment from blasting to be disintegrated by following blasting (secondary blasting), mechanical or other methods into small sized pieces suitable for further handling. Very course soil is subdivided into *Large boulder* (LBo) >630 mm, *Boulder* (Bo) >200 to 630 mm and *Cobble* (Co) >63 to 200 mm. Boulders can normally not be handled by the loading equipment, and/or by the crusher in use. See also '**particle size**'. SS-EN ISO 14688–1:2002, 2003.
boulder blasting or **secondary blasting**, the breaking down of rocks at quarries or mines too big to be handled by the mining equipment (loaders, chutes and crushers) by a small explosive charge. The charges can be inserted in small drillholes or put on the surface and covered by mud (*mud capping*). See '**secondary blasting**'.
boulder buster, a heavy, pyramidical- or conical-point steel tool, which may be attached to the bottom end of a string of drill rods and used to break, by impact a boulder encountered in a borehole. Long, 1920.

boulder cracker, a heavy iron rod to be dropped upon a rock encountered by the drill in the overburden in well boring. Standard, 1964.

boulder frequency (f_{bo}), **(No./1 000 t** or **No./1 000 m³)**, number of boulders per 1 000 t (metric ton) or 1 000 m³ of rock broken.

bounce, see '**rock burst**'.

Boundary Element Method (BEM), numerical technique based on segmenting the surface of the body area only. Advantageous for the calculation of large movements of rock mass during blasting. The boundary element method provides distributions of stresses and displacements over the boundary of the domain (area) under consideration.

box cut, the initial cut driven into a rock mass, where initially no open sides existed, see Fig. 9. This results in a highwall on both sides of the cut. Swelling of the rock will occur both vertically and horizontally. Kennedy, 1990. In Australia the term box cut is also used for a pattern of holes adopted for blasting to a new level below an existing floor (e.g. in shaft sinking or to create a deeper level in surface mining).

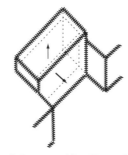

Figure 9 Box cut. After Kennedy, 1990.

box-cut method, a method of opencast mining of coal where the dip of the seam is relatively steep. A box-like excavation is made to the dip, or at an angle to it, and the coal seam worked to the right and left. Nelson, 1965.

boxhole, underground mining opening (raise) excavated from the bottom and up by blind boring and that can be used for transport of ore, waste rock, man and material and for ventilation of air. See '**boxhole boring**'.

boxhole boring or **drilling**, is used to excavate raises where there is limited access, or no access at all, to the upper level like the openings in sublevel caving. There are three methods in use. 1) A mini full-face borer machine is used to drill the full area in one step, see '**BorPak (tradename)**'. 2) Drilling of a pilot hole and ream it up to the wished diameter. One reamer is shown in Fig. 10.

Figure 10 Reamer with two cone rollers used in mechanized raise driving.

The machine is set up at the lower level producing the feed and rotation forces, and the pilot hole or the full diameter raise is bored upward. Stabilizers are periodically added to the drill string to reduce oscillation and bending stresses of the drill string. Cuttings gravitate down the hole and are deflected away from the raise boring machine (RBM) at the lower level. Atlas Copco, 2007. 3) The full diameter is drilled in one step. The boxhole drilling machine was originally developed for drilling raises 90 m long and 1,5 m in diameter in the gold mines in South Africa in 1973.

box thread, the thread on the inside surface of a coupling or tubular connector. Also '**female thread**'.

box-thread bit, a bit having threads on the inside of the upper end or shank of the bit by means of which the bit may be coupled to a *reaming shell*, *core barrel*, or *drill rod*. Long, 1960.

brace, a reinforcement element in tunnelling or mining made of steel or timber with the purpose to take compressive forces. TNC, 1979.

bracet, shelf like brace. Atlas Copco, 2006.

bracing, a method to join two elements together such as steel and cemented carbide. Sandvik, 2007.

braided earthing strap, braided electrically conductive strap like device made of metal and used to earthen and thereby protect a machine or part of machine from unwanted electrical current.

branching of a crack, division of a propagating crack into two or more cracks. Adequate crack propagation energy is a necessary condition for the branching of cracks to occur.

branch line, the length of detonating cord or signal tube running between (and connected to) the trunkline and the primer. AS 2187.1, 1996.

brash, a mass of loose or broken fragments of rock resulting from weathering or disintegration on the spot. Fay, 1920.

Brazilian test, an indirect method for the determination of the static tensile strength of rock, concrete, ceramic, or other material by applying a load vertically at the highest point of a test cylinder laying and which is supported on a horizontal plane. The method was first used in Brazil. Dodd, 1964.

break (strength), see '**failure**'.

break (geol.), a class term including faults, ruptures, fractures, joints, fissure and other discontinuities in rock formations.

break (mech.), mechanical device that decelerates and stops the motion of a wheeled vehicle. There are two main types of breaks; drum- and disc brakes. Brakes may also be used to arrest the motion of other rotating devices such as a winch. Atlas Copco, 2006.

breakage zone or **failure zone**, the zone where the rock mass has zero tensile strength. The rock mass can still take compression loads at an angle to the breakage zone but not parallel to the zone. A breakage zone has still some shear strength.

breakage-, **failure-** or **rupture wave**, when the rock mass breaks, fails or ruptures, seismic waves are created, and these waves can be measured by accelerometers.

break angle (α_{cav}), (°), see '**angle of failure**'.

break cylinder, extendable hydraulic or air cylinder used to exert a force to engage the parts of a brake. Atlas Copco, 2006.

break disc, disc shaped structure usually attached to a moving device such as a wheel or drum, that is frictionally engaged by the caliper brake pads to decelerate or arrest motion. Atlas Copco, 2006.

breaker, a demolition tool having a non- rotating percussive mechanism, generally used with a chisel or moil point to demolish concrete or missionary. Can be pneumatic or hydraulic powered and hand held or rig mounted depending on the size and weight of the machine. Atlas Copco, 2006.

breakout angle or **angle of breakage** (α_b), (°), the angle in *bench blasting* in a plane perpendicular to the blasthole and defining the angle between the two side cracks of the triangular prism broken. In *crater blasting* it will be the top cone angle of the broken material. See '**angle of breakage**'.

breakout gun, a hydraulic or compressed air actuated device attached to breakout tongs used to couple or uncouple drill rods, drill pipe, casing or drive pipe. Also called '*makeup gun*'. Long, 1920.

breakout table, device used on a drill rig to grip and turn threaded drill tubes to uncouple them. Common when using down-the-hole hammers. Atlas Copco, 2006.

breakout tongs, a heavy wrench, usually mechanically actuated, used to couple or uncouple drill rods, drill pipe, casing, or drive pipe. Also called 'makeup tongs'. Long, 1920.

breakout wedge, the prism broken when blasting in a bench. Hudson, 1993.

break shoe, object used to exert a frictional force on a moving part such as a wheel or drum, to decelerate and stop the motion. Atlas Copco, 2006.

break through, the process when two drifts meet each other or a drift is meeting the surface or an underground room during a blast or by full face boring. TNC, 1979.

breast, the face of working. In coal mines, a chamber driven in the seam from the gangway, for the extraction of coal. Fay, 1920.

breast and pillar, a **room and pillar mining method** of working anthracite coal by bords 10 m in width, with narrow pillar, 5 m width, between them, holed through at certain intervals. The breasts are worked from the dip to the rise. See '**bord-and pillar**'. Fay, 1920.

breast board, timber planks to support the face of tunnel excavation in soft grounds. Bickel, 1982.

breast stoping, a method of stoping employed on veins where the dip is not sufficient for the broken ore to be remove by gravity. The ore remains close to the working face and must be loaded into cars at that point. See also '**overhand stoping**'. Fay, 1920.

breast drill, the drill that is used when the borehole is started. TNC, 1979.

breast holes, blastholes in drifting or tunnelling located next to the roof holes and with blasting action downwards. Singh and Wondrad, 1989.

breast hole helper, a blasthole located next to a breast hole and initiated before the breast holes in tunnelling.

bridge conveyor, a conveyor which is supported at one of the ends by a supporting unit and at the other end by a receiving unit in such a way as to permit changes in the position of either end without interrupting the operation of the loading unit. NEMA MB1, 1961.

bridgewire, a resistance wire which connects the ends of the leg wires inside an electric blasting cap. It is embedded in the ignition charge of the blasting cap, electric squib, or similar devices. The fine wire, sometimes a platinum wire, is heated by the passage of the electric current to ignite the ignition charge.

bridgewire detonator, contains of an incandescent bridge made of thin resistance wire, which is made to glow by the application of an electric pulse. An igniting pill is built around the wire by repeated immersion in a solution of a pyrotechnical material in a solvent, followed by drying. The igniting flash acts directly onto the detonation surface in the case of instantaneous detonators; in delayed-action detonators it is sent over a delay device onto the detonating pill so as to produce a water-tight bond with it. Non-armed bridgewire detonators have an open casing, into which a detonator may be inserted. Meyer, 1977.

bridging, a term used to indicate that the continuity of an explosive column in a borehole is broken. This may occur, either by improper placement, as in the case of slurries or poured blasting agents, or where some foreign material has plugged the hole. Konya and Walter, 1990.

Brinell hardness number, see 'Brinell hardness test'.

Brinell hardness test, a test for determining the hardness of a material by forcing a hard steel or carbide ball of specified diameter into it, under a specified load. The result is expressed as the *Brinell hardness number*, which is the value obtained by dividing the applied load in kilograms by the surface area of the resulting impression in square millimetre. A.S.M. Gloss, 1961.

brisance, the rise time and amplitude of the transient pulse caused by a certain explosive in a blasthole of a certain diameter in rock or placed free in water or air. The rise time and amplitude depend on both the characteristics of the explosive (velocity of detonation (c_d)) and its confining medium (Young's modulus). An explosive with a high c_d has a higher brisance. A rock with a high value of Young's modulus will get a larger rise time and strain amplitude. Explosives with a detonation velocity larger than 5 000 m/s are called *high brisance explosives* and those having a detonation velocity less than 2 500 m/s *low brisance explosives*. Brady, 1985.

brisance test (Hess), see 'Hess cylinder compression test'.

brisance test (Kast), see 'brisance value'.

brisance value (P_b), (MJ/s or W), in the 1920's Kast introduced the brisance value as a measure of the ability of an explosive to fragment or demolish a solid object when detonated in direct contact with it. The symbol for brisance value used here is large P (effect or power) because brisance is defined as the energy release per time unit. The *brisance value P_b* is defined as follows;

$$P_b = p_{atm} V_{eg} \frac{\frac{Q_e}{C_m} + 273}{273} \rho_e c_d \qquad (B.10)$$

where p_{atm} is the atmospheric pressure in Pa, V_{eg} is the specific volume of the gaseous explosive reaction products in m³/kg at temperature 273°K and pressure p_{atm} in Pa, Q_e is the explosive heat (heat of explosion) in J/kg and C_m is the mean specific heat of the explosion products in J/kg, ρ_e is the density of explosive in kg/m³ and c_d is the velocity of detonation in m/s. The brisance value was earlier used to estimate blasting performance but is judged to be only usable for determining the effect of boulder blasting with mud capping. Persson et al., 1993. *A new definition of brisance has to be developed where the specific volume of gaseous reaction products are determined at the actual temperature after detonation and not at 273°K.* May be then it can also be used for confined blastholes.

Another brisance value P_b proposed by Trauzl in 1959 reads as follows:

$$p_{spg} = \rho_e c_d 101 \cdot 10^{-6} V_{eg} \left(\frac{t}{273} + 1 \right) \qquad (B.11)$$

$$P_b = \rho_e c_d p_{spg} = \rho_e^2 c_d^2 101\cdot 10^{-6} v_e \left(\frac{t}{273}+1\right)$$ (B.12)

where p_{spg} is the specific gas pressure in Pa, t is the temperature of explosion in °C, ρ_e is the charged density of explosive in kg/m³ and c_d is the velocity of detonation in m/s and v_e is the specific gas volume of explosive gases per kg of explosive in m³/kg. Observe that the product of the charged density of explosive and the velocity of detonation is equal to the the *acoustic impedance of the explosive*. See also '**Kast cylinder compression test**'. Trauzl, 1959.

brittle, see '**brittle material**'.

(brittle) crack propagation (c_c), **(m/s)**, a very sudden propagation of a crack with the absorption of no energy except that stored elastically in the body. Microscopic examination may reveal some deformation even though it is not noticeable to the unaided eye. ASM Gloss, 1961.

brittle fracture, fracture with little or no plastic deformation. ASM Gloss, 1961. In fracture mechanics a fracture characterized by the lack of plastic deformation at the crack tip.

brittle material or **mineral**, a material for which the yield point and the point of rupture lies close together. Briggs, 1929. According to this definition no rock types can be be classified as brittle. It is also defined verbally as a *non-ductile material* that *fails catastrophically* under dynamic loading conditions. In material science it is generally applied to materials that fail in tension rather than shear, or when there is little or no evidence of plastic deformation before failure. This implies that the stress-strain curve is linear up to the failure of the material. Examples of brittle materials are hard rocks (but they are not perfect brittle), ceramics and glass. Fragblast, 2007. In mineralogy a brittle material is *non-flexible* or *non-ductile*, that is, that a stone will crumble under a knife or hammer, but not necessarily that it is weak or fragile. Shipley, 1945. If a mineral breaks into fragments or powder under a light blow or crumbles into a dust when cut, it is brittle (for example, calcite and quartz). Stokes, 1955.

brittleness, proneness of a solid to crack at low stress level before the onset of plastic deformation. See also '**fracture toughness**'.

brittleness (friability) test (S_{20}), **(dimensionless)**, a Swedish measure of rock resistance to crushing due to repeated mass-drop impacts (impact wear). The rock material to be tested by impact wear, 0,5 kg with a density of 2 650 kg/m³ in the fraction of 11,2–16,0 mm is placed into a thick steel cylinder where it is further crushed 20 times by a falling mass (14 kg). The percent of the total material with size below 11,2 mm square mesh size is the S_{20}-value. The mean value for a minimum of three to four parallel tests is chosen as the S-value or the rock sample. This value together with a sliding wear test, the Siewers test value (J), are used to calculate the drilling rate index (DRI), a relative drillability number developed by University of Trondheim in Norway. It is a relative reliable and accurate method for predicting rock drillability. Tamrock, 1989.

broach, to restore the diameter of a borehole by reaming it with a broaching bit. Long, 1920.

broaching, a line of closely spaced holes is drilled along the required line of breakage. The rock between the holes is knocked out with a broach and removed with the aid of wedges. A method of rock excavation employed where it is important that the adjacent rock formation should not be shattered by explosive. Ham, 1965.

broaching bit, see '**broach**'.
broken charge, a charge of explosive in a drill hole divided into two or more parts that are separated by stemming. Fay, 1920.
brow, the edge of the roof in a crosscut or longitudinal drift in sublevel caving next to the blasted ore or the roof edge in the loading drifts next to a draw bell.
Brower compressive test, quantitative measurement of the sensitivity of explosives and other energetic materials to initiation by the adiabatic heating of a gas, for example the situation in bubble compression. The apparatus developed by Kay Brower, consists of a short vertical hardened steel thick-walled cylinder, closed at its lower end, with a sapphire window, and having a close-fitting 12,7 mm diameter piston with O-rings at its upper end, sealing off a volume of the cylinder bore. The piston is restricted in its upward motion so that it cannot leave the cylinder. A drop weight, or a fast moving projectile, is allowed to impact the exposed top end of the piston, pushing the piston down into the cylinder, compressing the gas (normally argon) that is enclosed in the cylinder together with a very small volume of the explosive to be tested for adiabatic gas compression sensitivity. The piston bounces on the compressed gas and this limits the time the explosive is exposed to the high gas temperature. The amount of chemical energy released by surface burning (or consumed by endothermic decomposition) during the short time the explosive surface is exposed to the high temperature can be measured by the drop weights rebound, and a sample of the reaction products remaining in the cylinder afterwards can be drawn through a valve for chemical analysis. Optical studies of the radiation can be made through the sapphire window. Persson, 1993.
Brunton pocket compass, see '**clinometer**'.
BSV (acronym) **(MJ/kg)**, see '**blasting strength value**'.
bubble energy (W_b), **(MJ)**, energy of the expanding gas of an explosive. It is measured and quantified by underwater tests.
bubble test, method to test explosives under water with respect to the energy (amplitude of the P-wave) and gas volume produced (size of gas bubble). See '**underwater test**'.
bucket, round open container used for carrying water, coal, sand etc. Atlas Copco, 2006.
buckling, producing a bulge, bend, bow, kink or other wavy condition in sheets or plates by compressive stresses. ASM Gloss, 1961.
buffer, previously shot material not removed, resting against a face to be shot. Konya and Walter, 1990.
buffer blasting (*towards broken rock*), blasting against broken rock where the rock broken is acting as a buffer. Model and full-scale blasting tests show that when using a certain specific charge, the fragmentation may improve if the swell factor for the buffered rock is ~1,5, compared to no buffer at all (according to tests performed in the Soviet Union and in Sweden).
buffer blasting (*controlled contour blasting*), a term used for the combination of presplitting and smooth wall blasting; used in surface mining with large diameter blastholes, see Fig. 11. The charge concentration is not only reduced in the contour row, but also in the two rows (buffer rows) next to the contour row. The drill hole diameter is the same in the buffer zone as in the production blast area. Holes are located half to three fourths of the normal production blast values. Subdrilling is totally avoided for the buffer holes.

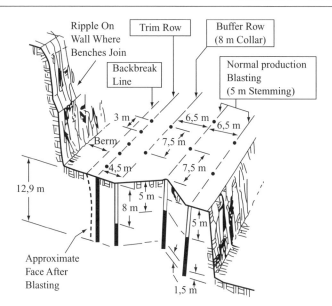

Figure 11 Buffer blasting. After Kennedy, 1990.

buffer row, the row next to the last row (perimeter or trim row) of the main blast. These holes have increased stemming to prevent cratering back at the surface through the perimeter row. Kennedy, 1990.
building stone blasting, see '**dimensional stone blasting**'.
building demolition blasting, the demolition of buildings by explosives.
bulk density *or apparent density of an explosive before charging* (ρ_e), **(kg/m³)**, the mass of the explosive divided by its volume before charging into the blasthole.
bulk density of explosive (after charging), (ρ_{ec}), **(kg/m³)**, the mass of explosive in the borehole divided by that volume of the blasthole which is charged. For cartridged explosives the bulk density is lower than the explosive density because of difficulty of getting 100% packing of the explosive and for powder explosives, like ANFO, the bulk density will be larger because the explosive needs to be compressed a little to get it to fasten in upholes.
bulk explosive, explosive material prepared for use without packing. These kind of explosives will fill the cross-section of the borehole after charging (decoupling ratio = 1).
bulkhead, a partition (wall) built-in in an underground structure or a structural lining to prevent the passage of air, water, or muck. ISRM, 1975.
bulkhead connection (machinery), hydraulic connection that can be fixed in a hole in a plate or "bulk head" in such a way that hoses can be affixed to either side. Atlas Copco, 2006.
bulking or **swelling factor** (*S*), **(dimensionless)**, the increase in volume of a rock mass when it has caved or been otherwise broken and removed from its *in situ* state is defined by the swell factor. See '**swell factor**'.
bulk mix, see '**bulk explosives**'.
bulk modulus (*K*), **(Pa)**, *modulus of volumetric expansion or inverse of compressibility* (*incompressibility*), a derived quantity characterizing the inverse compressibility of

a material. It is a quantity that expresses a materials resistance to elastic changes in volume. It can be calculated as follows;

$$K = -\frac{p}{e} = \lambda + \frac{2}{3}G = \frac{2(1+v)G}{3(1-2v)} = \frac{E}{3(1-v)} = \rho c_P^2 - \frac{4}{3}\rho c_S^2 \qquad (B.13)$$

where p is the hydrostatic pressure in (Pa) and e is the volumetric strain, λ is the Lame's constant and G the shear modulus in (Pa) valid for isotropic media, and v is Poisson's ratio, E is Young's modulus in (Pa), ρ the density of the material in (kg/m³), c_p is the longitudinal wave velocity in (m/s) and c_s is the transverse wave velocity in (m/s). $1/K$ = compressibility, (κ), (1/Pa).

bulk specific volume (v_b), (m³/kg), bulk volume per unit of mass.

bulk strength (s_b), (MJ/m³) or **volume strength (s_v), (MJ/m³)**, energy released per unit volume of explosive. Usually expressed as '**relative volume strength**' (*relative bulk strength*), when related to the '**volume strength**' (*bulk strength*) of a standard explosive such as ANFO. (ANFO is the normal reference explosive with densities between 750 and 870 kg/m³, while blasting gelatine is often quoted having a density of 1 550 kg/m³). It is recommended to use the term '**volume strength**' instead of '*bulk strength*'.

bulk underground mining, use of large scale underground mining methods like block caving, sublevel caving, longwall caving and sublevel stoping.

bulk volume (V_b), (m³), the volume of the solid material plus the volume of the sealed and open pores present. Dodd, 1964.

bull, to enlarge the bottom of a drilled hole to increase the volume for explosive charge at the bottom of the blasthole.

bull bit, a 'flat drill bit'. Fay, 1920.

bulldoze or **mud capping**, see '**adobe charge**'.

bulled hole, a blast hole which has been enlarged (chambered) to accommodate extra explosive over a portion of its length (usually at the bottom) by exploding a small charge. AS 2187.1, 1996. A borehole where portions have been washed out or fallen away due to the action of water or equipment used within the hole.

bullet-resistant, ability to resist penetration by a bullet. The US standard for magazine walls or doors (of construction) stipulates resistance to penetration by a bullet of 150-grain M2 ball ammunition having a nominal muzzle velocity of 822,96 m/s (2 700 ft/sec) fired from a 0,30-calibre rifle at a distance of 30,48 m (100 ft) perpendicular to the wall or door.

bullet sensitivity, see '**bullet test**'.

bullet test, a method to test the sensitivity of explosives with respect to impact. Bullets are shot against a sample of the explosive. The velocity of the bullet is increased until the explosive is initiated.

bullfrog, see '**barney**'.

bulling, a procedure intended to enlarge a section of a blast hole (usually the bottom) in order to accommodate extra explosive over that section. AS 2187.1, 1996. Also the firing of explosive charges in the cracks of loosened rock. Also the clay stemming is forced around the charge by a bulling bar. Nelson, 1965.

bull hole, large empty centre hole in a parallel hole cut. Olofsson, 1990. See '**parallel hole cut**'.

bulling bar, an iron bar used to pound clay into the crevices crossing a borehole, which is thus rendered gastight. Fay, 1920.

bull nose or **camel back**, a convex pillar or surface at the entry to a drawpoint.
bullnose bit, a non coring bit having a convex, half hemispherical shaped crown or face. Also called *wedge bit, wedge reaming bit, wedging bit*. Long, 1920.
bull point, synonym for 'boulder buster'. Long, 1920.
bull real, the **churn drill winch** that lifts and lowers the drill string. Also called *spudding reel*. Nichols, 1965.
bull rope, a heavy rope or cable from which a cable drill and stem are suspended dropped, or churned up and down, in drilling a borehole. Long, 1920.
bull wheel, the large winding drum on which the drill cable or bull rope of a churn or a cable-tool drill is wound. Long, 1920.
bump, see '**rock burst**'.
bunch connector, detonating cord prepared in two loops through which up to 20 shock tubes leading to detonators in blastholes can be initiated by the detonation from the cord. The cord is initiated by a detonator attached to a shock tube. This type of initiation is commonly used in drift and tunnel blasting.
burden (abbreviation), for '**burden distance**', see '**burden distance**'.
burden, *burden in bench blasting* (B), **(m)**, shortest perpendicular distance between the centre line of a charge and the free or buffered face.
burden, *burden in crater blasting* (B), **(m)**, shortest perpendicular distance between the centre of a charge and the free surface. Usually the terms 'charge depth' or 'depth of burial' are used in the literature.
burden, *blasted burden* (*true or effective burden*) (B_b), **(m)**, shortest perpendicular distance from the centre line or the centre of a charge to the free surface created by earlier charges fired in the round. The exact shape of the free surface created by earlier firing charges cannot always be determined because it depends on the real delay times and firing sequence.
burden, *critical burden in bench* or *crater blasting* (B_c), **(m)**, the minimum burden distance with no breakage and displacement.
burden distance, shortest perpendicular distance between the centre line or centre of a charge and the free surface.
burden, *drilled burden* (B_d), **(m)**, the burden distance measured in the plane containing the collars of blastholes, between the front row blastholes and the face or between one row of blastholes and the next row. This distance is only equal to practical burden when the bench is vertical. With inclined holes, for geometrical reasons, the drilled burden is always larger than the practical burden.
burden, *maximum burden* (B_{max}), **(m)**, the largest burden distance which can be used for good fragmentation if all boreholes are located at their correct positions (no borehole deviation). Because 'good fragmentation' is difficult to define in quantitative terms the use of the term 'maximum burden distance' is not recommended.
burden, *optimum burden* (B_{opt}), **(m)**, the burden distance for which the combined cost of drilling, blasting, mucking, hauling, and crushing is minimum.
burden, *optimum breakage burden* (B_{optb}), **(m)**, in bench and crater blasting the burden distance which, for a certain charge size, gives the maximum amount of rock broken. In the interest of safe breakage, the burden distance in crater blasting should usually be slightly smaller than the optimum breakage burden.
burden, *optimum fragmentation burden* (B_{optf}), **(m)**, the burden distance which gives the maximum surface area of fragmented rock.

burden, *practical burden* (B_p), (m), the planned burden used in blasting which is derived by subtracting the greatest blasthole deviation (caused by drilling) from the maximum burden. For vertical benches $B_p = B_d$.

burden, *reduced burden* (B_r), (m), the reduced burden is defined as $B_r = \sqrt{B_p S_p}$ where B_p is the practical burden and S_p is the practical spacing. If the properties of the rock and explosives are not known in detail, the following formulas, based on regression analysis of burdens and spacings used in more than one hundred surface and underground mines, can be used for rough estimates of the reduced burden. Rustan, 1990.

For surface mines with blasthole diameters of 90–380 mm;

+52% expected maximum value

$$B_r = 18.1 d^{0.689} \tag{B.14}$$

−37% expected minimum value

where d is the blasthole diameter in m. It is assumed that the decoupling ratio is equal to 1.

For underground mines with blasthole diameters of 50–165 mm;

+40% expected maximum value

$$B_r = 11.8 d^{0.630} \tag{B.15}$$

−25% expected minimum value

where d is the blasthole diameter in m. It is assumed that the decoupling ratio is equal to 1. Formulas including rock and explosive properties, see Kou and Rustan, 1992.

burden velocity (v_b), (m/s), the average velocity of the burden rock mass after detachment from the solid rock mass. The velocity depends on the strength of the rock and explosive, and the confinement (burden, blasthole inclination, etc.). By high speed photography an empirical formula has been developed. Chiappetta, 1991.

$$v_b = \frac{k_1}{e^{k_2 B}} \tag{B.16}$$

where v_b is the velocity of the burden (m/s), k_1 and k_2 are constants depending on explosive and rock properties, and B is the burden distance (m). In full scale bench blasting burden velocities measured are about 15–20 m/s. The peak velocity is normally located just in front of the blasthole and the velocity reduces further coming closer to the two side cracks. The kinetic energy in a blast can be calculated by the following formula.

$$W_k = \frac{m v_b^2}{2} \tag{B.17}$$

where W_k is the kinetic energy (also called heave energy) in Joule (J), m is the broken mass of rock in kg, and v_b is the mean velocity for the broken mass in m/s.

Bureau International des Poids et Measures (BIPM), the International Bureau for Weights and Measures, established in 1875 and located in Sèvres near Paris.

Burleigh or **Burley** (historical term), a miners term for any heavy two-man drill. The Burleigh was the first successful rock drill machine. Hess, 1968.

burn cut, a cut where all holes are drilled parallel to the tunnel axis and slightly upwards for dewatering, and where all the drill holes are of the same diameter, see Fig. 12. Some of the holes are charged heavily, while others are left without a charge to accomplish the swelling of the rock. The distances between the holes are small (~0,2 m) and usually a square pattern for the centre holes is used.

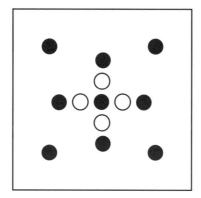

white = empty holes or relief holes   **black** = charged holes

Figure 12 Burn cut. After Atlas Copco, 1982.

burn drilling, see '**jet piercing**'. KTH, 1969.

burned bit, as a result of high speed, excessive pressure and poor water circulation, sufficient heat may be generated at the bottom of a borehole to cause diamond crown to soften, resulting in displacement of diamonds and a ruined bit. Nelson, 1965.

burning rate, the linear burning rate of a propellant is the rate at which the chemical combustion reaction is propagated by both thermal conduction and radiation. The combustion gases are flowing in the direction opposite to that of the combustion progress (unlike in detonation). Meyer, 1977.

burst, see '**rock burst**'. Also an "explosively like breaking of coal or rock" in a mine due to pressure. In coalmines they may or may not be accompanied by a copious discharge of methane, carbon dioxide, or coal dust. Also called *outburst, bounce, bump, rock burst*. USBM Bull 309., 1929.

burst current, the value of the current at which the bridge-wire or foil explodes. RISI, 1992.

burst energy release index (BERI), (W_{bi}), (**J**), the amount of energy released by the failure of rock specimens in a soft testing machine. The extent of vibrations generated in the machine is proportional to the amount of energy released during fracture. The BERI is the vector sum of the peak particle velocities of the vibrations in the vertical, longitudinal and transverse directions. Singh, 1989.

bursting charge or **primer** (Q_{pri}), (**kg**), see '**primer**'.

bursting pressure, pressure at which rock, a hose or vessel will rupture when subjected to an internal pressure. Atlas Copco, 2006.

bursting time (t_b), (**s**), the time between the application of an electric current and the setting off of the explosive charge. Nelson, 1965.

Burton A (tradename), a non gelatinous permissible explosive used in mining. Benneth 2nd, 1962.

burst-proneness index (R_{burst}), **(dimensionless)**, the ratio of the *strain energy retained* W_{ser} (J) to the *strain energy dissipated* W_{sd} (J). $R_{burst} = W_{sr}/W_{sd}$ is used as a measure of the tendency of a rock type to burst. It is determined from the elastic hysteresis loop parameter in the uniaxial compression loading and unloading tests on rock specimens up to approximately 80% of their compressive strength. Singh, 1989.

bushing, removable metal tube mounted in an opening. Atlas Copco, 2006.

bushing bearing, removable metal tube mounted in a bore to reduce friction between rotating parts. Also called a friction bearing. Atlas Copco, 2006.

bushing half, one half of a bushing manufactured in two parts. Atlas Copco, 2006.

buster, an expanding wedge used to break down coal or rock. Pryor, 1963.

bus wires, the two wires, joined to the connecting wire, to which the leg wires of the electric caps are connected in a parallel circuit, see Fig. 13. Each leg wire of each cap is connected to a different bus wire. In a series-in-parallel circuit, each end of each series is connected to a different bus wire. USBM, 1983.

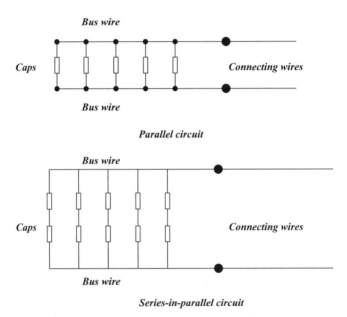

Figure 13 Definition of bus wires in an initiation circuit.

but shot, in coal mining, a charge placed so that the face or burden is nearly parallel with the bore hole. Fay, 1920.

butt, rock exposed (walls and roof) at right angles to the face, and in contrast to the face, generally having a rough surface. Also called *end* in Scotland. B.C.I., 1947.

butt, boot-leg, socket or *gun,* that portion or remainder of a shot hole in a face after the explosive has been fired incompletely after the blast and that still may contain explosives and is thus considered hazardous. A situation in which the blast fails to cause total failure of the rock because of insufficient explosives for the amount of burden, or owing to incomplete detonation of the explosives.

butt entry, an entry driven at right angles to the butt. B.C.I., 1947.

butt heading, see '**butt entry**'. Fay, 1920.

buttock, a corner formed by two coal faces more or less at right angles, such as the end of a working face, the fast side, any short piece of coal approximately at right angles to the face, a rib, the rib side. Mason, 1951.

button bit, percussive drill bit having button-like spherical, cemented tungsten carbide inserts that engage and remove the rock during drilling. They are manufactured in sizes from ~35 mm and upwards. Atlas Copco, 2006. Button bits are more wear resistant than insert bits. They offer longer grinding intervals, longer working life and improved penetration rate, and should be the first choice for most applications. They are manufactured in sizes from 35 mm and up. Sandvik, 2007.

button cutters, picking tools used on road headers for disintegration of rock. Button bits are mounted on the tools.

button integral steel, drill rod, normally having a hexagon shape, with button like tungsten carbide cutting tools soldered into one end and the other designed to fit into the rock drill chuck Atlas Copco, 2006.

buttress, a pier or projection built out from a wall to increase its strength and add to its resistance or thrust, which may arise from earth or water pressure or from an arch. Ham, 1965.

buzzer, audio warning. Atlas Copco, 2006.

BWI (acronym), for '**bit wear index**'.

c **(m/s)**, electromagnetic- and mechanical wave propagation velocity.

cab, enclosed compartment having windows and a protective roof structure and containing the operators seat and the controls to operate the equipment. Atlas Copco, 2006.

CAB (acronym), for '**computer aided blasting**'.

CABD (acronym), for '**computer aided blast design**'.

cabin, a small rough house of usually one room, such as a prospector's cabin. Hess, 1968. See also '**control cabin**'.

cabin ergonomy (drilling machines), includes aspects like superb view of the drill hole from the operators seat, facilitating accurate collaring and control of the drilling operation. The angled, laminated glass windscreen is equipped with a large wiper, as are the right hand side and roof windows, and improved air ventilation effectively clears condensation mist from all of these windows. A jumbo-sized rear view mirror gives good visibility and safer rig moving. For additional safety a back camera may be installed. Easy to read instrumentation that is simple to learn and to use. The cabin is laid out so that the operator can monitor and control the entire drilling process without changing body position. This relieves neck, shoulders and back from strain. The ergonomically-designed seat is vertically and laterally adjustable, with control levers and control panel located in, or in close proximity to the collapsible arm rests. The seat is slewable for easy entry into the cabin, and the door is fitted with a safety stop that prevents crushed fingers. Good insulation affords a noise level below 80 dB and rubber-damping of the cab mounting reduces vibration and gives greater comfort during tramming. The cabin has an efficient CFC-free air conditioning system for booth cooling and heating, and many surfaces are textile covered for greater comfort. The cab complies with the European and International safety demands for Rol-Over Protective Structure (ROPS) and Falling Object Protective structure (FOPS). Atlas Copco, 2006.

cable adaptor, threaded pin attached to the end of a cable to allow it to be secured by a nut. Atlas Copco, 2006.

cable anchor, see '**cable reinforcement**'.

cable bolt, a reinforcing element, normally made of steel wires which are layed to a strand or a rope configuration and installed un-tensioned or tensioned with cement grouting in the rock mass.

cable bolt reinforcement, a rock mass reinforcement method using cables, high tensile strength wire, mine rope, and reinforcing strand (multiple cables) that are grouted into drill holes. The method started to be used in the 1980's mainly for reinforcement of hanging walls in sublevel stoping but also in cut and fill mining where the roof in the stopes was stabilized with cable bolts. Diameters of the cables are in the order of 25 to 50 mm and lengths normally 12 to 50 m. The cables can be fully grouted cable dowels (tendons) or pre-tensioned before grouting. To improve the friction along the cable, anchors can be attached at various points along the full length of the cable. Normally cable reinforcement is mounted on 2 m square grids. Cables less than 6 m are called short. Equipment for mechanised cable bolting has been developed. One man can drill, grout and install the cable. Possible cable lengths in mechanized cable bolting are between 4–40 m.

cable bolter, a machine developed for mechanized mounting of cables in drill holes.

cable bolting, process of drilling long holes in rock and pushing cables into the holes together with cement grout with the purpose to stabilize the rock mass. In hard rock, cable lengths up to 25 m can be mounted and in soft rock up to 32 m. Normally cables with a diameter of 38 mm are being used. Atlas Copco, 2006. See also '**cable reinforcement**'.

cable churn drill, see '**cable drill**' and '**churn drill**'. TNC, 1979.

cable churn drilling, a drill bit attached to a stem and swivel socket which are suspended on a cable and allowed to fall freely against the rock. Since drilling is effected by crushing, the fall and weight of the drilling tool must be suited to the hole diameter and type of rock. The number of strokes depends on the air resistance to fall in the drill hole, and the inertia of the drilling tool. Churn drills are either wheel-mounted or, in the case of larger units, mounted on caterpillar tracks. They are driven by electric motors or internal-combustion engines. The chisel bit, which could be 0,5 to 1 m long, is screwed into the stem. The steel cutting edges are renewed either by forging or by hard surfacing. Bits with carbide inserts have not, so far, proved sufficiently durable to be economical. Data for the most commonly used sizes of churn drills are as follows: Drill hole diameter 90–300 mm, weight of drilling tool 200–2 500 kg, length of stroke 500–900 mm and number of strokes 50–60/min. Fraenkel, 1952.

cable dowel, a cable not prestressed, often with multiple strands, grouted into a drill hole with the purpose to reinforce the rock mass.

cable drill, a heavy drill rig in which a rope is used for suspending the tools in the borehole. See also '**churn drill**'. Nelson, 1965.

cable drilling, see '**cable churn drilling**'.

cable drum, drum like device or cylinder on which the electric cable, water hose and other communication lines with the full-face boring machine or other mobile machines are stored.

cable grip, device that fits around the cable to anchor it and avoid putting tension loads on the connector. Atlas Copco, 2006.

cable railway, an inclined track up and down which travel wagons fixed at equal intervals to an endless steel wire rope, located either above or below the wagons. Ham, 1965.

cable reel, a cylinder drum to store the blasting cable when not being in use. Also a powered electrical cable drum used on a drill rig to store the trailing cable during tramming.

Cabletec LC (tradename), a machine that can push a reinforcement cable often 38 mm in diameter into the drill holes from a drum. A cement slurry is used to grout the cable to the rock surface in the cable hole. Two booms are used one for drilling and one for grouting and cable insertion. The machine is a fully mechanized rig for drilling and cable bolting by a single operator. Atlas Copco, 2007.

cable tool cuttings, the rock fragments and sludge produced in drilling a borehole with a churn drill. Long, 1920.

cable-tool drill, synonymous with 'churn drill'. Long, 1920.

cable tool drilling, the *cable tool spudding machine* is capable of handling many drilling applications. The technique used must be adjusted to suit the purpose of drilling. Many techniques have been developed. Here we look at the more common techniques, Normal cable tool operations (*consolidated formations*): The spudding action is operated to allow the tools to drop freely, catch the tools before the bottom so that when the bit strikes the bottom the cable is stretched, quickly accelerate the tools upwards, allow the weight to come off the cable so that the swivel turns to rotate the bit. *Drilling in unconsolidated formations* (*churn drilling*), the mast must be fitted with an effective recoil system. The recoil provides extra strength for the rapid upward acceleration of the bit after it drops into the material at the bottom of the hole. Stein, 1997.

cable tool rig, synonymous for 'churn-drill rig'. Long, 1920.

cactus grab, a digging and unloading mechanical attachment hung from a crane or excavator. It consists of a split and hinged bucket fitted with curved jaws or teeth, which dig into the loose rock while the bucket is being dropped, and contract to lift the load while it is being raised. It is used increasingly for mechanical mucking in shaft sinking. A standard cactus double rope grab for shaft sinking has a capacity of about 0,5 m^3, it weighs about 2,5 t and can fill a 5 t capacity hopper in about 4 min. Nelson, 1965.

CAD (acronym), see '**computer aided blast design**'.

CAF (acronym), see '**cut and fill mining**'.

cage, mining term for elevator. Kentucky, 1952.

cage guides, conductors made of wood, iron or steel, or wire rope, used for the purpose of guiding the cages in the shaft and to prevent them from swinging and colliding with each other while in motion. Nelson, 1965.

CAI (acronym), for '**Cherchar abrasivity index**' see '**abrasive test**'.

caisson, a cylindrical steel section of shaft, used for sinking through running or waterlogged ground. A horizontal caisson is used for tunnelling through similar ground, perhaps with pressure locks to aid in keeping out water. Pryor, 1963.

caisson sinking, drum shaft, drop shaft, a method of sinking a shaft through wet clay, sand or mud down to firm strata by using steel cylinders to support the walls.

caliber, the inside diameter or bore of a tube, pile or cylinder. Long, 1920.

caliper, an instrument used to measure precisely the thickness or diameter of objects or the distance between two surfaces, etc. Long, 1920.

caliper log, continuous record of the variations in mean diameter or in cross-sectional area of a borehole with depth. Institute of Petroleum, 1961.

calyx, a steel tube attached to the upper end of a core barrel having the same outside diameter as the core barrel. The upper end is open except for two web members running from the inside of the tube to a ring encircling the drill rod. The calyx serves as a guide rod and also as a bucket to catch cuttings that are too heavy to be flushed out of the borehole by the circulation fluid. Also called 'bucket', 'sludge barrel' and 'sludge bucket'. Long, 1920.

calyx drill (shot drill), a rotary core drill which uses hardened steel shot for cutting rock, and will drill holes from diamond drill size up to 1,8 m or more in diameter. Drilling is slow and expensive, and holes cannot be drilled more than 35° off the vertical, as the shot tends to collect on the lower side of the hole. Lewis, 1964.

calyx drilling, a method of rotary drilling using a toothed cutting bit or chilled shot. B.S. 3618, 1963.

calyx rod, a drill rod with circular cross section used on a shot drill, usually outside coupled and of larger diameter than diamond drill rods. Long, 1920

camel or **bull nose**, a convex pillar or surface at the entry to a draw point.

Canadian Institute of Mining and Metallurgy (Can. I.M.M.), compare with 'Institute of Mining and Metallurgy' London that is organized in a similar way to this organisation.

canopy, safety roof made of steel placed at the top of the full-face-boring machine.

cantilever, a support system in which one end of a beam is fixed and the other is free. Construction sometimes used to give grizzly vibrating freedom. Pryor, 1963.

cantilever grizzly, grizzly fixed at one end only, the discharge end being overhung and free to vibrate. This vibration of the bars is caused by the impact of the material. The disadvantage of the ordinary bar grizzly is clogging due to the retarding effect of the cross rods. This has been overcome in the cantilever grizzly by eliminating the tie rods except at the head end where they are essential. The absence of these rods below the point of support also aids in preventing clogging as it permits the bars to vibrate in a horizontal plane, which keeps the material from wedging. Pit and Quarry, 1960.

cap, see '**blasting cap**'.

cap, a hardhat for the protection of the head. Also called helmet. Hanko, 1967.

capacitor blasting machine, machine to induce current to the blasting caps. A capacitor is charged by a handle and discharged at blasting.

cap crimper, a mechanical device for crimping the metallic shell of a blasting cap securely to a section of inserted safety fuse. IME, 1981.

cap nut, threaded nut with an opening in only one side. Atlas Copco, 2006.

capped fuse, safety fuse to which a blasting cap has been crimped on to one end.

capped primer, a package or cartridge of cap-sensitive explosive which is specifically designed to transmit detonation to other explosives, and which contains a blasting cap. USBM, 1983.

cappel or **capping**, a fitting at the end of the winding rope to enable the bridle chains of the cage to be connected by a pin through the clevis. Nelson, 1965.

capping station, a special location expressly used for preparing capped fuses. AS 2187.1, 1996.

cap screw, see '**bolt**'.

cap sensitivity, sensitivity of an explosive to initiation by a test blasting cap No. 8 (0,40–0,45 g of PETN) or a fraction thereof. The base charge of blasting cap No. 8 may be different in other countries. In South Africa the charge mass is 0,8 g.

cap sensitivity test, a test to determine the sensitivity of an explosive to shock from a standard detonator. It is used as one of the criteria for classifying energetic materials as explosives in the United Nations Hazards, Division 1.5. Persson et al., 1994.

cap set, a term used in square-set mining methods to designate a set of timber using caps as posts, resulting in a set of timber shorter than the normal set. USBM, 1990.

cap shot, a light shot of explosive placed on the top of a piece of shale that is too large to handle, in order to break it. B.C.I., 1947.

cap wires or *leg wires*, the two single wires or one duplex wire extending out from an electric blasting cap. IME, 1981.

carbide tool, tool usually for drilling rock, made from tungsten carbide. Atlas Copco, 2006.

carbon bit, a diamond bit in which the cutting medium is inset carbon. Long, 1920.

carbon monoxide, CO is a poisonous gas, tasteless, odourless, and a by-product of the detonation of an explosive. An inadequate amount of oxygen in the explosive causes the formation of excessive carbon monoxide content in the blast fumes.

carbon dioxide blasting, a method of blasting coal that has been undercut, top cut, or sheared. Into one end of a seamless high-grade Mo-steel cylinder 50–75 mm in diameter and from 0,9 to 1,5 m long is put a cartridge containing a mixture of potassium perchlorate and charcoal with an electric match. The other end is sealed by a metal disc weaker than the shell and held in place by a cap that has holes at 45° to the axis of the cylinder. The cylinder is filled with liquid carbon dioxide at a pressure of 6,9 MPa and inserted in the borehole with the cap hole pointing outwards. The heating mixture is lit and raises the gas pressure so that the disc is sheared; the carbon dioxide escaping through the angular holes tends to hold the cylinder in place break and push the coal forward. If the gas pressure is not enough to break the coal, the cylinder, if not properly set, will be blown from the borehole. The cylinder can be used over and over. It is claimed that a greater portion of lump coal is obtained than with ordinary explosives. Some smelters loosen slag in the same way. Hess, 1968.

Cardox (tradename), a system that uses hollow cartridges made of alloy steel and filled with liquid carbon dioxide, which, when initiated by a mixture of potassium perchlorate and charcoal, creates a pressure, 70–130 MPa, adequate to break and undercut coal. USBM, 1983.

careful blasting, see '**controlled blasting**'.

Carribel (tradename), a permitted explosive of medium strength that can be used in wet boreholes provided its immersion time does not exceed 2 to 3 hours. The maximum charge weight in British coalmines is 680 g and can be used for coal and ripping shots in conjunction with short delay detonators. Nelson, 1965.

Carrick delay detonator, a special designed detonator where the tube is made of a non-slagging bronze element, as opposed to the aluminium used in conventional detonators. Thus the possibility of the ignition of post-blast gases from the hot molten aluminium particles (called slagging) is eliminated. It is a special designed blasting cap for use in gassy coal mines. Sen, 1995.

carrier, the basic platform for mobile machines on which specialized equipment is mounted like drilling booms, scaling impact breaker, charging equipment, loading equipment for rock etc.

carrier unit, integrated mobile vehicle forming the base of a drill rigs. Atlas Copco, 2006.

carton, a lightweight inner container for explosive materials, usually encased in a substantial shipping container called a case. Atlas Powder, 1987.

cartridge count (stick count), a method of expressing the mass of an explosive cartridge by listing the number of cartridges per 23 kg (50-pound) case (in the USA). This incorrect measure should be replaced by grams or kilograms per cartridge, because other countries may have other case masses, for example South Africa (25 kg).

cartridge (explosive), a performed unit of explosive enclosed in thin or thick paper or plastics to a predetermined diameter and length.

cartridge (rock bolting), thin plastic cartridges filled with resin or cement and inserted into the drill hole that is planned for rock reinforcement. The turning of the rebar will open up the cartridges and the result will be a grouted rebar. Stillborg, 1994.

cartridge density, the ratio between the mass of an explosive cartridge and its volume.

cartridged explosive, explosive enclosed in thin or thick paper or plastics to a predetermined diameter and length.
cartridge fuse, a fuse enclosed in an insulating tube in order to confine the arc when the fuse blows. Crispin, 1964.
cartridge loader, device to blow cartridges by means of compressed air through a plastic hose into the drill hole. This results in a high bulk explosive density.
cartridge pin, a round stick of wood on which the paper tube for the blasting cartridge is formed. Fay, 1920.
cartridge punch, a wooden, plastic, or non-sparking metal device to punch an opening in an explosive cartridge to accept a detonator or a section of detonating cord. IME, 1981.
cartridge strength, see '**mass-** or **volume strength**'.
Cascade mining (CM), a room and pillar mining method developed in the 1960's for flat dipping (20°–30°) ore bodies at Mufulira mine in Zambia. Longitudinal drifts are being enlarged to stopes and pillars are left between the stopes. When the stopes have been blasted and mucked the higher laying pillar is broken by blasting and mined (100% pillar recovery is the goal).
case, a substantial outer shipping container meeting the U.S. Department of Transportation (DOT) specification for explosive materials. Atlas Powder, 1987.
case, see '**blankoff**'.
case insert, a set of printed, precautionary instructions, including the U.S. Institute of Makers of Explosives (IME) Do's and Don'ts, which is inserted into a case of explosive materials. Atlas Powder, 1987.
case liner, a plastic or paper barrier used to prevent the escape of explosive materials from a case. Atlas Powder, 1987.
case off, see '**blankoff**'.
casing or **casing tube**, a tube inserted in the drill hole to stabilize the wall of the hole.
casing bit, a diamond-set rotary bit designed to bore out an annulus slightly larger than the casing. It is withdrawn before the casing is inserted. B.S. 3618, 1963. Drill bit used to drill a hole for setting casing. Atlas Copco, 2006.
casing bowl and slips (synonyms), see '**casing spider**'. Long, 1920.
casing dog, a lifting device consisting of one or more serrated sliding wedges working inside a cone-shaped collar. Used to grip and hold casing while it is being raised or lowered into a borehole. Long, 1920.
casing drive hammer (drive hammer), a weight used to drive casing down a borehole. Also called monkey. B.S. 3618., 1963.
casing jar hammer (jar hammer), a drive hammer used to extract casing. B.S. 3618, 1963.
casing shoe, type of diamond bit specifically designed for enlarging a hole for setting casing. Atlas Copco, 2006.
casing socket, threaded tube used to connect drill rods and casing having dissimilar threads. Atlas Copco, 2006.
casing spear, a drill tool used in connection to drilling to place a liner in the hole. Stein, 1997.
casing spider, a holding device resting on the drilling floor, consisting of two or more heavy cone-shaped bowls or collars, used to suspend casing in a drill hole during makeup or breakout. Also called 'casing bowl and slips'. Long, 1920.

casing tube, threaded tubes that are installed to form a lining in a diamond drill hole to prevent the side walls from sluffing or caving into the hole. Atlas Copco, 2006.

cast bit, a bit in which the diamond set crown is formed on a bit blank by pouring molten metal into a prepared mold. Also called cast-set bit and cast-metal bit. Long, 1920.

cast blasting, see '**blast casting**'.

cast booster, a high density explosive unit designed to provide a high detonation velocity (c_d) a high detonation pressure, for adequate initiation of the main/column charge. It may be made up of a more sensitive inner core to accept initiation from a detonating cord and detonator. The inner core may be made of Pentolite (a 50%/50% mixture of PETN and TNT) or only PETN. Sandhu and Pradhan, 1991.

cast charge, a charge of solid high explosive used to detonate less sensitive explosive materials. See also '**cast booster**'.

casting with gliding form, a support method used in tunnels and shafts that needs to stand for 50–100 years. The form is moved at certain time delays to make it possible for the concrete to cure. The same form is used all the time. Hanko, 1967.

cast, extruded or **pressed booster**, a cast, extruded, or pressed solid high explosive used to detonate explosive materials of lower sensitivity. Atlas Powder, 1987.

casting of explosives, increasing the density of an explosive at the manufacturing by casting. This increases the brisance of the explosive and this technique is therefore used for military explosives but also for primers and boosters for commercial explosives.

cast-metal bit (synonym), see '**cast bit**'.

cast pile, *spoil or spoil bank*, the pile of material that is cast or moved, from its original position by an excavating machine, as opposed to that loaded into haulage equipment. It is usually composed of waste material.

cast primer, primer with hard consistency, used to initiate blasting agents. Usually an explosive with high detonation velocity (c_d). See also '**cast booster**'.

cast set, a bit produced by casting process, see '**cast bit**'.

cast set bit, synonym for '**cast bit**'.

cataclastic, a texture found in metamorphic rocks in which brittle minerals have been broken and flattened in a direction perpendicular to the pressure stress. A.G.I., 1960.

catalyzer, engine exhaust cleaning device that uses a compound to remove un-burnt particles. Atlas Copco, 2006.

catch bench, a bench in an open pit made so wide so it can catch falling stones from benches above.

catch scaffold, a platform in a shaft a few feet beneath a working scaffold to be used in case of accident. Fay, 1920.

caterpillar treads, an attachment like an apron conveyor, placed around and connecting the front and back wheels of selfpropelled machines, furnishing a broad track that allows the machine to traverse rough, uneven, soft, or sandy country. If the distance between the wheels is considerable, idlers help to aline the track. USBM, 1968.

caulking, substance or material placed in joints to seal or make waterproof. Atlas Copco, 2006.

cautious blasting or **controlled blasting**, an umbrella term for controlled contour blasting, controlled ground vibrations, controlled throw, etc.blasting with certain limitations due to the surroundings e.g. limitations concerning flyrock, ground vibrations, air shock waves. Nitro Nobel, 1993. See also '**controlled blasting**'.

cavability, the ease that a rock mass can be caved. Laubscher has developed a classification scheme for cavability, see Table 7. Laubscher, 1981.

Table 7 Cavability according to geomechanical classes 1 to 5. Laubscher, 1981. Two last rows calculated by Rustan, 2007.

Caving parameter	Geomechanical class				
	1	2	3	4	5
Cavability	Not cavable	Poor	Fair	Good	Very good
Fragment size	–	Large	Medium	Small	Very small
Secondary blasting	–	High	Medium	Small	Very small
Undercut, equivalent hydraulic radius = A/P (m)	–	<30	30–20	20–8	>8
Undercut, equivalent squared length (m)	–	120	120–80	80–32	32
Undercut area (m^2)	–	14 400	6 400–14 400	1 024–6 400	1 024

A = area of the undercut and P = perimeter of undercut.

cave, falling in of roof strata, sometimes extending to the surface and causing a depression therein. Also called '*cave-in*'. Hudson, 1932.
cave back or **crown**, the roof of the cave where fragments have been detached from the in situ rock mass.
cave hole, a depression at the surface, caused by a fall of roof in the mine. Fay, 1920.
cave-in, collapse of walls or roof of mine excavation. Pryor, 1963.
cave inducement or **induced block caving**, the process of inducing caving by some technique in addition to, or other than, undercutting, e.g. drilling and blasting or hydraulic fracturing.
cave-in-heave, the partial or complete *collapse of the walls of a borehole*. Brantly, 1961.
cave initiation, the process of the initiation of 'natural' caving by undercutting and drawing some of the broken ore.
cave propagation, the process of propagation of an initiated cave by the progressive drawing of broken ore in a planned and controlled manner.
caving, the process of the detachment of *in situ* rock from the cave back. A term also for mining methods allowing the failure of the roof above the mining area.
caving angle or **angle of caving** (α_{cav}), (°), the angle in a section of the *loosening draw body* cut through its vertical axis. The angle is defined by two points. The first point is defined as the outmost influenced point from the centre of the outlet and the second point by the outmost affected point on the higher level of interest, see Fig. 3 at angle of caving.
caving mining methods, mining methods in which the ore is broken by gravity forces due to the undercutting of the ore like in block caving or by drilling and blasting the ore against the hanging wall rock. There are five caving methods 1) *Block caving* (*BC*) where the ore is caved by gravity forces including caving chutes, branched raises or draw points like in sublevel caving. Sometimes drilling and blasting is needed for some parts of the ore and this variant is called *Induced block caving* (*IBC*). 2) *Sublevel caving* (*SLC*) is the mining method where the ore is drilled and blasted and the hanging wall is being caved. The broken ore and waste is moved by gravity towards the draw point. 3) *Sublevel block caving* (*SLBC*) where the undercut is made by a sublevel caving layout and the ore is thereafter blocked by gravity. The method was developed at LKAB

in Malmberget to mine a pillar. The method was there given a local name 'Slot blocking' 4) *Longwall mining (LM)*, is used for flat dipping coal seams which may be mined continuously by drilling and blasting or mechanical cutting of the coal. The overlaying material is allowed to cave and subside until the subsidence reaches the surface. 5) *Top slicing (TS)*, a mining method where timbered drifts are driven in ore and caved afterwards. The next slice is made below the first slice and having mats to protect the cave to penetrate downwards.

caving rate v_c **(mm/day)**, the average rate, usually expressed in mm per day, at which the cave naturally propagates upwards.

cavity monitoring system, measurement of the contour of cavities underground by a laser rangefinder *scanning head equipment* inserted into the cavity on a carbon fibre boom. It can measure distances up to 100 m to surfaces without artificial reflector material. Liu, 1994.

CBLSC (acronym), for 'circular bipolar linear shaped charge'. Rustan, 1983.

CCB (acronym), see '**controlled contour blasting**'.

CCNBD (acronym), see '**Cracked Chevron-Notched Brazilian Disc**'.

CDA (acronym), see '**continuous deformation analysis**'.

CDISK (abbreviation), see '**Chevron notch Brazilian test**'.

CEFIC (abbreviation), for '**Conseil Europeen de l'Industrie Chimique**' (in French) and 'European Council of Chemical Manufacturers Federation' (in English).

cell powder factor (CPF), (q_{cell}), **(kg/m³)**, see '**cell specific charge**'.

cell specific charge (CSC), (q_{cell}), **(kg/m³)**, in practical rock blasting it is interesting to know the specific charge in different parts of a round. This is especially important in ring drilling where the specific charge will vary a lot. The blast round is therefore divided into defined cell volumes, squares, circles etc, and the specific charge is calculated for each cell. Some examples of mining methods where ring drilling is being used are sublevel caving, sublevel shrinkage stoping and sublevel stoping. King, 1988.

cement cartridge, a cartridge containing cement to be used for grouting rock bolts into drill holes. The rock bolt is rotated through the cement cartridge to spread the cement around the rock bolt.

cemented bolt, normally a reinforced bar grouted by cement into a drill hole to reinforce the rock mass. Fraenkel, 1952.

cemented hydraulic fill or *paste fill*, **(CHF)**, in pipelines transportable fractions of waste rock, concentrator tailings or sand mixed together with cement and water and used for the stabilization of underground space.

cemented rock fill (CRF), waste rock bound together with cement or fly ash to increase the strength of the material and to be used as a fill material in mines for mined out areas. The purpose with the fill is to avoid subsidence on the surface. Rustan, 2007. CRF originally consisted of spraying cement slurry or cemented hydraulic fill on top of stopes filled with waste rock, as practiced at Geco and Mount Isa mines. Nowadays, cement slurry is added to the waste rock before the stope is filled. Where rock is quarried on surface, it is normally gravitated to the mining horizon through a *fill raise*, from the base of which trucks or conveyors are used for lateral transport underground. Advantages of CRF include a high strength to cement content ratio, and provision of a stiff fill that contributes to regional ground support. Atlas Copco, 2007.

cement fluid end, portion of a cement pump that the cement passes through. Atlas Copco, 2006.

cement grouted cable bolts, one or more steel wires laid to a strand or rope configuration and installed tensioned or un tensioned together with cement grouting in a drill hole with the aim to reinforce the rock mass, see Figure 14. Stillborg, 1994.

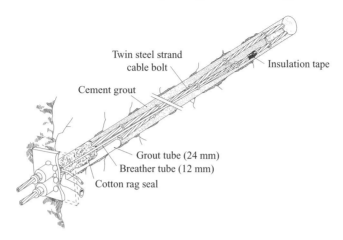

Figure 14 Cement grouted cable bolt—Twin steel strand. After Stillborg, 1994.

cement grouting, injection of cement into drillholes and intersecting cracks with the purpose to reduce water inflow and increase the friction properties of the rock mass. Sometimes fine graded aggregate needs to be added.

cement gun, see '**grout injector**'.

cement injector, see '**grout injector**'.

CEN, see '**European Standardization Committee**'. Standardization of types of explosives are included in the standard.

CENRBB (acronym), see '**chevron edge notch round bar in bending**'.

center bore, see '**set inside diameter**'.

center core method, a method of tunnelling large tunnel areas whereby the centre of the tunnel is left to the last for excavation. Sandström, 1963.

center cut, a) the boreholes, drilled to include a wedge shaped piece of rock and which are fired first in a heading, tunnel drift, or other working place. See also '*center shot*'. Fay, 1920. b) a vertical cut or groove made in coal at or near the center of a working face to facilitate blasting. Grove, 1968.

center cutter, a cutter placed in the center of the drill head for a full-face boring machine.

center flushing hole, the axial hole in the drill rod for the supply of cooling water to the drill bit. Sandvik, 2007.

centering end, ring shaped piece installed in hole or cylinder to position something in the center. Atlas Copco, 2006.

center plug, a small diamond-set circular plug, designed to be inserted into the annular opening in a core bit, thus converting it to a non coring bit. Long, 1920.

center shot, a shot in the center of the face of a room or entry. Also called '**center cut**'. Fay, 1920.

Centigrade scale, the temperature scale defined by Celsius. Freezing water defines the zero point, 0°C, and the boiling of water defines +100°C.

centralizer, drilling accessory used to make the drill string rotate on the same axis in order to reduce borehole deviation.

Central Mining Research Station (CMRS), a mining research centre located in *India*. The Institute has developed an own rock mass classification system for Indian coal mines, because the existing classification systems developed by Bieniawski at Council for Scientific and Industrial Research CSIR in South Africa and Barton at the Norwegian Geotechnical Institute NGI, were not found to be entirely suitable. Hudson, 1993.

centres, the distance between two adjacent parallel blastholes.

certified blaster, a blaster authorized through governmental agency certification to prepare, execute and supervise blasting.

CGr (abbreviation), for '**coarse gravel**' defining fragments with sizes from 20 to 63 mm.

chain, set of links fitted into one another, usually of metal and used to secure, lift or bind something. Atlas Copco, 2006.

chain coal cutter, a coal cutter that cuts a groove in the coal by an endless chain travelling round a flat plate called a jib. The chain consists of a number of pick boxes. Each box holds a cutter pick fastened into the box by a setscrew or similar device. The cutter pulls itself along the face by means of a rope at a speed varying from 0,2 to 1,5 m/min or more. Nelson, 1965.

chain curtain, a number of chains hanging from the roof and down protecting the entrance to a rock shaft at the dumping point or at the inlet to a crusher.

chain feed, a feeding mechanism by which the up-and-down movements of the drill stem are controlled by a link chain running on sprocket gears. Long, 1920. Drill feed where the force applied to move the drill forwards and back is transmitted through a roller chain. Atlas Copco, 2006.

chain feeder, is used for the linear movement of the drill machine along the feeder beam by an endless chain.

chain link mesh, steel net that is fastened to rock bolts and used to protect personnel from rock fall.

chain machine, a coal-cutting machine that cuts the coal with a series of steel bits set in an endless chain moved continuously in one direction either by an electric motor or a compressed-air motor. The chain machines may be divided into four classes, breast-, shortwall-, longwall- and over-cutting machines. Kiser, 1929.

chain tensioner, generally a threaded arrangement used to apply tension to the chain of a "chain feed". Atlas Copco, 2006.

chamber-and-pillar, see '**breast-and-pillar**'. Fay, 1920.

chamber blasting, a large-scale blast in which bulk explosives are detonated in excavated subterranean chambers. Also called coyote blast; gopher hole blast. Webster 2nd ed., 1960.

chambering (springing), the enlarging of the bottom of a quarry blasting hole by the repeated firing of small explosive charges. The enlarged hole or chamber is then loaded with the proper explosive charge, stemmed and fired to break down the quarry face. Long, 1920.

change house, the house, normally on surface, where the miners change their cloth to working cloth and vice verse.

channel effect, effect where a shock wave in air, in the gap between a pipe charge and the blasthole wall, moves faster than the detonation front in the explosive. The pressure generated from the air shock wave may be large enough to dead-press the explosive, and decelerate the detonation front completely. This effect may occur at certain decoupling ratios, e.g. when using pipe charges in controlled contour blasting. Johansson and Persson, 1970.

chap, see '**scaling**'.

Chapman-Jouguet plane (CJ-plane), the interface separating the steady and the non-steady regions at the detonation front. The boundary between the energy release which is supporting the detonation front, and hence driving the detonation velocity (c_d), and the energy which is released too late for this. In an ideal detonation, all energy supports the detonation front, and the CJ zone thus coincides with the end of the reaction zone. The extent to which the measured detonation velocity falls short of the theoretical value is an indication of the proportion of energy released behind the CJ plane.

characteristic impedance for an explosive (Z_e), (kg/m² s), see 'acoustic impedance of an explosive'.

charge (noun), 1) a quantity of explosive, 2) explosive placed in a drill hole or confined space. AS 2187.1, 1996.

charge or load (verb), to put the explosive into the borehole, arrange the fuse, squib, and tamp.

charge concentration linear or **linear charge concentration (q_l), (kg/m)**, mass of explosive per meter of drill hole. Nitro Nobel, 1993.

charge density (ρ_{ec}), (kg/m³), see 'charged explosive density'. This term is often wrongly used to denote linear charge concentration (kg/m).

charge depth **(B), (m)**, see 'burden or burden in crater blasting'.

charge factor, charge load, splitting factor, **specific splitting charge (q_a), (kg/m²)**, mass of explosive in contour holes per blasted contour area.

charge load, charge factor, splitting factor, **specific splitting charge (q_a), (kg/m²)**, mass of explosive in contour holes per blasted contour area.

charge mass, (Q), (kg), the mass of the charge.

charge mass per delay (Q_{del}), (kg), the mass of explosive charge considered to detonate in a delay with a separation smaller than 8 ms to the next delay.

charged explosive density (ρ_{ec}), (kg/m³), the ratio of the mass of the charge in the charged section of a borehole to the volume of that section of the borehole. Multiplication of the mass strength of an explosive by the charged explosive density yields the volume strength (*bulk strength*) of the explosive. For liquid explosives the density of the explosive will increase with the depth of the explosive column due to the hydrostatic pressure. For very high blastholes the explosive may be dead pressed by the increased pressure.

charge weight scaling formula, a ground vibration formula to calculate the peak particle velocity as a function of charge and distance from the blast. See 'scaled distance formula'.

charge with sticks, a non-continuous charge in a borehole separated by non-explosive material e.g. wood sticks.

charging, the act of placing explosives in the desired position for firing. AS 2187.1, 1996.

charging density **(ρ_{ec}), (kg/m³)**, *see* 'charged explosive density'. This term is often wrongly used to denote linear charge concentration (kg/m).

charging equipment, see 'charging machine'.

charging hose, plastic hose used for transportation of bulk or cartridged explosives into the blasthole.

charging machine, mechanical device to transport (sometimes also pack) the explosive into the blasthole. It is used both for cartridges and bulk explosive. Usually different machines are used for powder, cartridges, and liquid explosives like slurry and emulsion explosives.

charging platform, boom and basket arrangement attached to the front of a drill rig or other vehicle and used to lift men and materials into position for loading explosives into drill holes. Atlas Copco, 2006.

charging platform boom, rig mounted boom equipped with a man-basket that can be moved vertically and horizontally, used as platform for loading explosives into blast holes rig mounted boom equipped with a man-basket that can be moved vertically and horizontally, used as platform for loading explosives into blast holes. Atlas Copco, 2006.

charging rod, see 'tamping rod'.

Charpy test, a bending strength test of a material. A pendulum-type, single blow impact test in which the specimen, usually notched, is supported at both ends as a simple beam and broken by a falling pendulum. The energy absorbed, as determined by the subsequent rise of the pendulum, is a measure of impact strength or notch toughness. ASM Gloss, 1961. (This test is similar to the fracture toughness test only with the difference, that in the fracture toughness test, the load on the beam starts from zero and is slowly increased to breakage).

chasing, process of using a die to re-cut a damaged thread. Atlas Copco, 2006.

check valve, valve mounted in a circuit to permit fluid flow in only one direction. Atlas Copco, 2006.

chemical (grouting) injection, reinforcement method where chemicals are used instead of cement e.g. plastic or silica compounds. Synonymous with 'resin grouting'. TNC 1979. Some chemicals like "silica gel" is very poisonous and must be used with great care.

chemical drill, an exotic drilling method where chemicals are used to disintegrate rock. Successful results have been reached in sandstone, limestone and granite. The chemical used was hydrofluoric acid H_2F_2 which is a very reactive liquid and very dangerous in handling. KTH, 1969.

Cherchar abrasiveness test, the test is conducted by abrading a steel pin under a static load of 7 kg by moving the pin 1 cm on the rock in 1 s. The rock surface should be flat an oriented normal to the axis of the pin which is a cone with 90° top angle. The average diameter of the wear flat, measured in 1/10 mm units, from five tests on rock with a grain size less than 1 mm is defined as the Cerchar abrasivity index. Rocks having larger grain size will require more tests.

cherry picker, a device that lifts the empty car to let the loaded car pass underneath. The system can only be used in relatively high tunnels. Atlas Copco, 1982.

chevron blasting, blasting technique where blastholes are tied up in such a way that single holes in the front row detonate together with a lagging hole in each subsequent row, see Fig. 15. A tie-up which initiates the hole immediately behind the hole in the front is called a V0 Chevron, while one tied to the hole next to the hole in the front is called a V1, and so on. An

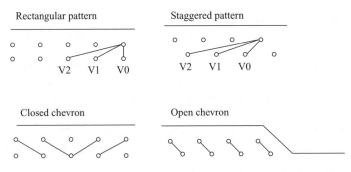

Figure 15 Chevron, open and closed initiation. After Cunningham, 1993.

'open chevron' opens at the end of the bench with two free faces. A 'closed chevron' opens in the centre of the bench and moves outwards from this position. Cunningham, 1993.

chevron edge notch round bar in bending (CENRBB), see '**fracture toughness**'.

chevron-notched Brazilian disc, see '**cracked chevron-notched Brazilian disc (CCNBD)**'.

CHF (acronym), see '**cemented hydraulic fill**'.

chilled dynamite, the condition of the dynamite, when subjected to a low temperature not sufficient to congeal it, will seriously affect the strength of the dynamite. Fay, 1920.

chilled shot drill, see '**shot drill**'.

chimney (from overbreak), overbreak can reach large heights especially if one joint set is parallel to the drift and steep and there is a joint set perpendicular to this and where the intersection line is almost horizontally and thereby parallel to the drift.

chimney caving, *piping* or *funnelling*, the progressive migration of an unsupported mining cavity through the overlying material to the surface. The surface subsidence area may be of a similar plan shape and area to the original excavation. Chimney caves may form in weak overburden materials e.g. on the Zambian copper belt, in previously caved material or in regularly jointed rock which progressively unravels. Chimney caves have been known to propagate upwards to surface though several hundreds of metres. An unexpected chimney cave occurred also at the Idkerberget mine in Sweden.

chip, small angular, and generally flat pieces of rock or other materials. Long, 1960.

chip bit, a bit in which the major portion of the inset diamonds are either diamond chips or thin, tabular-shaped, low-grade drill diamonds Long, 1920.

chip blasting, shallow blasting of ledge rock. Nichols, 1956.

chipping hammer, a handheld hammer for adjustments of surfaces. A piston is making a reciprocal movement inside a cylinder. This movement takes place when a slide valve allows compressed air to act alternately on the top and bottom of the piston. Different hammer characteristics are obtained by varying the length of stroke, the piston diameter and the mass of the piston. There are about a dozen different tools (wedge, wide chisel, narrow chisel, moil point, tamping tool, spade, asphalt cutter, digging blade, tie tamper etc.) that can be used for the chipping hammer. Atlas Copco, 1982.

chippings, crushed angular stone fragments from 3,17 mm (1/8 inch) to 25,4 (1 inch) mm.

chisel, a tool of great variety whose cutting principle is that of the wedge. Crispin, 1964.

chisel bit, a percussive-type rock-cutting bit having a single, chisel-shaped cutting edge extending across the diameter and through the center point of the bit face. Also called 'chisel-edge bit', 'chisel-point bit', 'chopping bit' and 'Swedish bit'. Long, 1920.

chisel drag bit, see '**drag bit**'.

chisel edge bit, synonymous for 'chisel bit', also 'chisel-point bit', 'chopping bit' and 'Swedish bit'.

chisel-point bit, synonymous for 'chisel bit', 'chisel-edge bit' and 'chopping bit' and 'Swedish bit'. Long, 1920.

chisel ripper, demolition tool. Atlas Copco, 2006.

chisel steel, a carbon steel containing 1% carbon. It is readily forged and used for chisel making, large punches, miner's drills, etc. Camm, 1964.

chock (historical term), a square pillar for supporting the roof constructed of prop timber laid up in alternate cross-layers, in log-cabin style, the center being filled with waste. Fay, 1920.

chocking (historical term), the supporting of undercut coal with short wedges or chocks. C.T.D. 1958.

choke blasting, blasting with insufficient expansion space. A choke blast is therefore considered a failed blast. Blasting against fill or broken rock with sufficient space is called buffer blasting. Typically, 30% expansion space is required to avoid choke blasting in underground excavations. Model and half-scale tests show that the finest fragmentation is obtained with 40 to 50% expansion space. In the cut in a development round, the minimum expansion space needed for the first blasthole is 10 to 15%.

choke mass blast, blasting in open stopes but only drawing 10% of the material with the purpose to limit caving of the stope walls. Hudson, 1993.

choke screw, threaded, adjustable flow restriction used in pneumatic and fluid circuits. Atlas Copco, 2006.

chopping, see '**slabbing**'.

chopping bit, a steel, chisel-shaped cutting-edged bit designed to be coupled to a string of drill rods and used to fragment, by impact, boulders hardpan, and lost core in a borehole. Also called 'chisel bit', 'chisel-edge bit', 'chisel-point bit', 'log-shank', 'chopping bit' and 'Swedish bit'. Long, 1920. See '**chisel bit**'.

chuck, the part of a diamond or rotary drill that grips and holds the drill rods or Kelly and by means of which longitudinal and/or rotational movements are transmitted to the drill rods of Kelly. Long, 1902.

chump, to drill a shothole by hand. Fay, 1920.

chunk, see '**boulder**'.

churn or **chump**, a long iron rod used to hand bore shot holes in soft material, such as coal. Pryor, 1963.

churn drill, a percussion drill. A portable drilling equipment usually mounted on four wheels and driven by steam-, diesel-, electric, or gasoline-powered engines or motors. The drilling is performed by a heavy string of tools tipped with a blunt-edge chisel bit suspended from a flexible manila (rope) or steel cable, to which a reciprocating motion is imparted by its suspension from an oscillating beam or sheave, causing the bit to be raised and dropped, thus striking successive blows by means of which the rock is chipped and pulverized and the borehole deepened. Also, the act or process of drilling a hole with a churn drill. Extensively used by the diamond-drilling industry to drive pipe vertically through difficult and deep overburden or fractured barren ground before coring operations with a diamond rill. Also called American system drill, cable-tool drill, rope-system drill, spudder, spud drill and well drill. Long, 1920.

churn drilling, drilling method where the drilling effect is obtained by the drill steel striking against the rock. The steel moves up and down with the percussion device, which is actuated either by force of gravity or by means of externally applied mechanical force. The cuttings are usually removed by silting in water and flushing intermittently or sometimes by continuous flushing with water or air. Fraenkel, 1952. See also '**cable churn drilling**'.

chute, a mechanical construction at the bottom of a shaft by which the flow of ore or waste rock through the shaft can be regulated to fill mine cars or trucks, for haulage to a deeper shaft.

chute caving, the method involves both overhand stoping and ore caving. The chamber is started as an overhand stope from the head of a chute and is extended up until the back weakens sufficiently to cave. The orebody is worked from the top down in thick slices, each slice being, however, attacked from the bottom and the working extending from the floor of the slice up to an intermediate point. The cover follows down upon the caved ore. Also called caving by raising or block caving in chute. Fay, 1920.

circ grip, tong used for installing and removing "**circlips**". Atlas Copco, 2006.

circlip, spring steel clip that is inserted into a grove cut in a cylinder or rod. Atlas Copco, 2006.

circuit, a closed path for conveying electrical current. See also '**series blasting circuit**', 'parallel blasting circuit' and 'series in parallel blasting circuit'.

circuit tester, see '**blasting galvanometer**'.

circular bipolar linear shaped charge (CBLSC), see '**shaped charge**'.

CJ (acronym), see '**Chapman-Jouget**'.

CJ-plane, see '**Chapman-Jouguet plane**'.

CL (acronym), for '**cutter life**' see '**abrasive test**'.

Cl (abbreviation), for '**clay**' defining particles with sizes ≤0,002 mm.

claim, the portion of land held by one claimant or association by virtue of one location and record. Nelson, 1965.

clamp, device used to affix a metal coupling to a rubber hose. Atlas Copco, 2006.

clamping bracket, bracket used to hold something in place. Atlas Copco, 2006.

clamping piece, piece used to keep something in place by exerting a force on it. Atlas Copco, 2006.

Class A explosives, explosives, as defined by the U.S. Department of Transportation, that possess detonating or otherwise maximum hazard, such as dynamite, nitroglycerin, lead azide, blasting caps, and detonating primers. Atlas Powder, 1987.

Class B explosives, explosives, as defined by the U.S. Department of Transportation, that possess flammable hazard, such as but not limited to, propellant explosives, photographic flash powders, and some special fireworks. Atlas Powder, 1987.

Class C explosives, explosives, as defined by the U.S. Department of Transportation, that contain Class A or Class B explosives, or both, as components but in restricted quantities. Atlas Powder, 1987.

classification code of explosives, a statement of the class, division and compatibility group to which an explosive has been assigned. Explosives are also classified according the strength of explosive, toxic gas produced when detonated etc.

classification of explosives, the separation of dangerous goods into classes and divisions according to their hazard characteristics in storage and transportation. AS 2187.1, 1996.

Classification as per explosives rules.
- Class 1   Black powder (gun powder)
- Class 2   Nitrate mixtures
- Class 3   Nitro compounds
    - Division I, Blasting gelatine
    - Division II, Guncotton, PETN, TNT, PRIMEX etc.
- Class 4   Chlorate mixture (not commonly used in mines)
- Class 5   Fulminate
- Class 6   Division I, Safety fuse, ignitor cord, safety electric fuse, percussion caps.
    - Division II, Plastic ignitor cord, detonating cord/fuse, electric fuse, fuse, fuse ignitors etc.
    - Division III, Detonators, delay detonators, relays etc.
- Class 7   Fireworks
- Class 8   Liquid oxygen explosive

classification of rocks, see '**rock mass classification**'.

claw coupling, type of air and water hoses connector having a "bayonet" type locking arrangement that can be connected by hand. Atlas Copco, 2006.

clay, a fine-grained, natural, earthy material composed primarily of hydrous aluminium silicates. It may be a mixture of clay minerals and small amounts of no clay materials or it may be predominantly one clay mineral. The type of clay is determined by the predominant clay mineral present (that is kaolin, montmorillonite, illite, halloysite, etc.). Bureau of Mines Staff, 1968. According to *international classification*, clay has a *grain size less than 0,002 mm.* CTD, 1958.

clay boring bit, a special coring bit used on split-inner–tube core barrels. Thickness of bit face is reduced and inside shoulder is not inset with diamonds, to allow a sharp edged extension of the inner barrel to extend through and project a short distance beyond the face of the bit. Also called clay bit, mud bit. Long, 1920.

clay dummy, a cartridge of clay used to stem a blasthole.

cleat, a main joint direction in a coal seam along which it breaks most easily. Runs in two directions, along and across the seam. Pryor, 1963.

cleavage, as originally defined, rock cleavage is any structure by virtue of which a rock has the capacity to part along certain well defined surfaces or planes more easily than along other surfaces or planes. Geologist usually employs the term for secondary structure produced by metamorphism or deformation rather than for original structures, such as bedding or flow structures. Stokes, 1955. In mineralogy, the property possessed by many minerals of being rather easily split parallel to one or more of the crystallographic planes characteristics of the mineral. Fay, 1920. A tendency in rocks to cleave or to split along definite, parallel, closely spaced planes which may be highly inclined to the bedding planes. It is a secondary structure, commonly confined to bedded rocks, developed by pressure, and ordinarily accompanied by at least some recrystallization of the rocks, Cleavage should not be confused with the fracturing of rocks, which is jointing. The property or tendency of rock to split in a brittle fashion along secondary, aligned fractures or other closely spaced, planar structures or textures, produced by deformation or metamorphism. The main types of cleavage are a) slaty cleavage, the very pervasive cleavage found in typical roofing slates b) crenulation cleavage, a planar structure produced by micro-folding, schistocity produced by parallel alignment of tabular materials, like micas in rocks which have undergone more intense metamorphic recrystallization. In quarrying the cleavage is often called the rift. Nelson, 1965.

cleavage crack, a microcrack which is located in a cleavage plane of the mineral grain or along a grain boundary.

CLI (acronym), for '**cutter life index**' in full face boring, see '**abrasive test**'.

clinometer or **inclinometer**, an instrument for measuring angular deviation from horizontal or vertical in drilling. A clinometer can be as simple as a plumb bob and a protractor. However, the *Brunton Pocket Compass* is a combined compass and clinometer and is often used by geologist. Stein, 1997.

closed porosity or **sealed porosity** (n_s), (%), the ratio of the volume of closed or sealed pores to the bulk volume expressed as a percentage. Dodd, 1964.

closure of cracks (micro), microcracks can be easily closed under stress (mainly those located approximately perpendicular to the maximum stress) but hardly pores. Miguel, 1983.

clutch house, housing that encloses the moving parts of a 'clutch'. Atlas Copco, 2006.
clutch lever, lever used to engage and disengage a clutch. Atlas Copco, 2006.
clutch rod, rod linking the clutch lever and the clutch. Atlas Copco, 2006
CM (acronym), see '**Cascade mining**'.
CMRS (acronym), see '**Central Mining Research Station**' in Dhanbad, India.
Co (abbreviation), for '**cobble**' defining fragments with sizes from 63 to 200 mm.
coal drill, usually an electric rotary drill of a light, compact design. Aluminium and its alloys usually are used to reduce the weight. Where dust is a hazard, wet drilling is employed. With a one horsepower electric drill, advance speeds up to 1,8 m/min is possible. Light percussive drills, operated by compressed air, hand-operated drills are also employed. Nelson, 1965.
coal dust explosion, a mine explosion caused by the ignition of fine coal dust. It is believed that an explosion involving coal dust alone is relatively rare. It demands the simultaneous formation of a flammable dust cloud and the means of ignition within it. The flame and force of a firedamp explosion are the commonest basic causes of coal dust explosion. The advancing wave of the explosion stirs up the dust on the roadways and thus feeds the flame with the fuel for propagation. See also '**colliery explosion**' and '**stone-dust barrier**'. Nelson, 1965.
coal dust index (I_{cd}), **(%)**, percentage of fines and dust passing the 0,3 mm sieve (48 mesh). Benneth 2nd ed., 1962.
coal mine explosion, the burning of gas and/or dust with evidence of violence from rapid expansion of gases. USBM, 1966.
coal mining explosives, the statutory requirements regarding the use of explosives in coal-mines are very stringent. In gaseous mines only permitted explosives are allowed. Nelson, 1965.
coal puncher, a coal cutter of the reciprocating type, used for undercutting and nicking coal. Also called pick machine. Fay, 1920.
coal seam, a bed or stratum of coal. Craigie, 1938–1944.
coarse strainer, screen-like device used to remove large dirt particles from a fluid or a gas. Atlas Copco, 2006.
cobbing, hand concentration in which lumps of concentrate are detached from waste, using a 1,5 kg chisel-edged hammer. Term also used for whole sorting operation. Pryor, 1963.
cobbing hammer, a special chisel type of hammer to separate the mineral in a lump from the gangue in the hand picking or ores. Nelson, 1965.
cobble (Co), (very coarse soil), rock fragments >63 to 200 mm. SS-EN ISO 14688-1:2002, 2003.
cock, simple manually operated shut-off valve. Atlas Copco, 2006.
coefficient, a number that serves as a measure of some property (as of a substance or body) or characteristic (as of a device or process) and that is commonly used as a factor in computations. Webster, 1961.
coefficient of anisotropy (R_{ani}), **(dimensionless)**, the ratio between the maximum uniaxial compressive strength to the minimum uniaxial compressive strength in a rock sample that includes joints. The coefficient is dependent on the orientation angle of the joints to the stress direction. Values up to ten have been recorded in Martinsburg slate however usually it does not exceed 4,5.
coefficient of consolidation (soil), (C), **(m²/s)**, a quantity defining the compaction of soil after drainage. It is defined as follows,

$$C \approx \frac{k}{V_c \rho_w} \qquad (C.1)$$

where k is the permeability in (m/s), V_c is the coefficient of volume compressibility in m²/kg or 1/Pa, ρ_w the unit weight of water in kg/m³. Hudson, 1993. For normal consolidated clays C varies between $0,5 \cdot 10^{-6}$ and $3 \cdot 10^{-3}$.

coefficient of curvature (C_c), (dimensionless), measure of the shape of the grading curve (size distribution) within the range from k_{10}, k_{30} to k_{60}.

$$C_c = (k_{30})^2/(k_{10} \cdot k_{60}) \qquad (C.2)$$

where k_{10}, k_{30} and k_{60} are the particle sizes corresponding to the ordinates 10%, 30% and 60% by mass of the percentage passing. SS-EN ISO 14688-2:2004, 2003.

coefficient of friction (μ), (dimensionless), the resistance to motion which is called into play when a solid body or liquid is attempted to slide on a surface over another with which it is in contact. The frictional force F opposing the motion is equal to the moving force up to a value known as limiting friction. Any increase in the moving force will then cause slipping. Static friction is the value of the limiting friction just before slipping occurs. Kinetic friction is the value of the limiting friction after slipping has occurred. This is slightly less than the static friction. The coefficient of friction is the ratio of the limiting friction force F to the normal force N between the two surfaces. It is constant for a given pair of surfaces. Walker, 1988. The relationship is controlled by many variables, such as whether is applied to limiting (holding) friction, sliding friction, rolling friction or the internal friction of a bulk material; and whether the surfaces in contact are smooth or rough, as well as the kind of material pairing surface conditions, e.g. whether it is wet or dry, lubricated or non-lubricated. ASA MH4.1, 1958.

coefficient of permeability (k), (m/s), the proportional constant in Darcy's law for gas or fluid flow in porous media. See '**hydraulic conductivity**'.

cohesion (C), (MPa), shear resistance at zero normal stress. An equivalent term in rock mechanics is '*intrinsic shear strength*'. ISRM, 1975.

collar (noun), the mouth or opening of a borehole, drill steel, or shaft.

collar (verb), see '**collaring of a hole**'.

collared shank, forged end of drill steel for the purpose of fitting into the drilling machine. The collar is transmitting the impact energy as well as rotation. Sandvik, 2007.

collaring deviation or **error in drilling (D_c), (m or %)**, error of final position of hole compared with the intended setting-out point.

collaring of a hole, the act of starting a drill hole.

collar length or **-distance (l_{co}), (m)**, the distance from the top of the explosive column in the blasthole to the collar of the blasthole. This part is usually filled with stemming. USBM, 1983.

collector, device where material from several sources come together. Atlas Copco, 2006.

colliery, a coal mine. Pryor, 1963.

colliery explosion, an undesirable explosion in the workings or roadways of a colliery as a result of the ignition of fire damp (mainly methane), coal dust or a mixture thereof.

collision blasting, by using a V-pattern of initiation in bench blasting the blasted rock can be ejected so rock pieces will collide in the air and with increased crushing as a result. Kinetic energy will be transferred to crushing energy and thereby fragmentation is improved. Langefors, 1963.

columnar charge, a continuous charge in quarry borehole. B.S. 3618, 1964.

column charge, charge placed between the bottom charge and the stemming in a blasthole. Because of less confinement of the column part of a blasthole, the linear charge concentration (kg/m) in this part of the blasthole should be ~40% less than the bottom charge.

column charge length (l_c), (m), the length of the column charge in a borehole.

column charge mass, (m_c), (kg), mass of column charge.

column charging, the charging of a drill hole with a continuous charge. AS 2187.1, 1996.

column depth or column height (l_c), (*m*), see '**column charge length**'.

column load, see '**column charge**'.

column rig, drill mounted on central pillar and can be lowered or raised on the pillar. Atlas Copco, 2006.

combination shot, a blast made by dynamite and permissibles, or permissible explosives and blasting powder in the same hole. It is bad practice and in many states in the USA it is prohibited. Fay, 1920.

combined overhand and underhand stoping, this term signifies the workings of a block of ore simultaneously from the bottom to its top and from the top to the bottom. The modification are distinguished by the support used, as open stope, stull-supported stopes (timber prop supported stopes, or pillar-supported stopes. Also known as combined stopes, combination stoping, overhand stoping and milling system, back and underhand stoping milling system. Fay, 1920.

combustion, an exothermic chemical reaction which generates products at high temperature and with light as well as heat (energy). The rate may be slow to rapid.

commercial explosives, explosives designed, produced, marketed and used for commercial or industrial applications rather than for military purposes. Atlas Powder, 1987.

comminution, the action of reducing a material volume to a smaller size. It is a term distinct from fragmentation by blasting in that it more often refers to the mechanical processes applied in mineral processing for grinding and pulverisation.

compound landslide, includes a failure surface which is partly rotational and partly translational. Hudson, 1993.

compressed air blasting or **air shooting**, a method originated in the US for breaking down coal by compressed air. Air at a pressure of 83 MPa is conveyed in a steel pipe to a tube or shell inserted in a shot hole. The air is admitted by opening a shooting valve and is released in the hole by the rupture of a shear pin or disc. The sudden expansion of the air in the confined hole breaks down the coal. Advantages: 1) a high proportion of large coal, 2) no danger of methane ignition, 3) no toxic or disagreeable fumes. Nelson, 1965.

compressed-air filling, a filling method in mines using fill in which compressed air is utilized to blow the filling material e.g. crushed rock or sand into the mined-out stope. Stoces, 1954.

compressed air shield, the outlet of compressed air behind the full-face boring head to create an overpressure and thereby avoid dust to penetrate into the working area of the full-face boring personal.

Commission of Fragmentation by Blasting, a special commission was formed in 1991 within the International Society for Rock Mechanics ISRM to deal with scientific blasting problems, standardization etc. The Commission was responsible for the arrangement of Scientific Symposia's in Blasting called 'The International Symposia's on Rock Fragmentation by Blasting' abbreviated 'Fragblast' Symposia are being held normally each third year. The First Symposium was held in Luleå, Sweden in 1983, 2nd Keystone, Colorado, USA in 1987, 3rd Brisbane, Australia in 1990, 4th Vienna, Austria in

1993 and 5th Montreal, Canada in 1996, 6th Johannesburg, South Africa in 1999, 7th Beijing, China in 2002, 8th Santiago Chile 2006 and 9th and Grenada Spain 2009.

compaction, decrease in volume of a loose material (soil or rock particles).

competent rock, a rock mass with good strength which is defined by a Young's modulus in excess of 20 GPa and a uniaxial compressive strength in excess of 50 MPa.

composition B, a mixture of RDX (cyclo-trimethylene-trinitramine) and TNT (tri-nitrotoluene) which, when cast, has a density of 1 650 kg/m³ and a detonation velocity (c_d) of 7 625 m/s. It is used as a primer for blasting agents. USBM, 1983.

compressed air gas cartridge, see 'airdox'.

compressibility (κ), (1/Pa), property of a material to decrease in volume when subjected to a load. It is defined as follows;

$$\kappa = -\frac{1}{V}\frac{dV}{dp} \tag{C.3}$$

where V is the original volume in m³, dV is the differential change in volume m³ due to the differential change in pressure dp in (Pa). Compressibility is equal to the inverted value of bulk modulus ($K = 1/\kappa$). See also '**bulk modulus**' for calculation of compressibility from the elastic parameters of the material.

compressive strength (uniaxial), (σ_c), (MPa), the resistance of a material against failure when subjected to compressive stresses. Uniaxial compressive strength tests are those most commonly used to classify rock strength. Recent research emphasises a correlation between compressive strength and fracture toughness. The compressive strength of rock ranges between 10 and 500 MPa. A classification scheme has been made by Deere and Miller, 1966, see Table 8.

Table 8 Rock mass classification based on the uniaxial compressive strength. After Deere and Miller, 1966.

Class	Description	Uniaxial compressive strength MPa	Examples of rock types
A	Very high strength	>200	Quartzite, dolerite, gabbro, basalt
B	High strength	100–200	Marble, granite, gneiss
C	Medium strength	50–100	Sandstone, slate, shale
D	Low strength	25–50	Coal, siltstone, schist
E	Very low strength	<25	Chalk, rock salt

Compressive strength can also be correlated to point load tensile strength not to specific gravity, P-wave velocity, S-wave velocity, bar velocity, modulus of rigidity Young's modulus and Poisson's ratio. D'Andrea, 1965.

compressive stress (σ), (uniaxial), (MPa), normal stress tending to shorten the body in the direction in which it acts. ISRM, 1975.

computer aided blasting (CAB), the use of information technology and digital timing systems in improving expertise, blast modelling, data banks and appropriate sensing and control systems so as to optimise blasting effects.

computer aided blast design (CABD), the use of computer to simulate rock blasting. The main fields studied are: 1) Development of cracks and fragmentation around blastholes. Harries, 1973, 1975 and 1977 and 2) Simulation of throw and muckpile shape.

compression strength, see '**compressive strength**'.

compressive spring, coil spring designed to resist compression when a load is applied. Atlas Copco, 2006.

compressor, air compressor that increases the pressure of atmospheric air by confining and reducing its volume. Atlas Copco, 2006.

compressor hood, plate metal structure surrounding and protecting the workings of a compressor. Atlas Copco, 2006.

computer code, codes developed for calculation of burden, blast geometry, fragmentation, throw etc. Persson et al., 1994.

concave bit, a tungsten carbide drill bit for percussive boring. The cutting edge is concave, while in the conventional type the edge is convex. The new bit remains sharper for a longer period before regrinding becomes necessary and gives a higher penetration speed. Also called 'saddleback tip'. Nelson, 1965.

concentrated charge, the explosive charge loaded into the enlarged chamber at the bottom of a quarry blasthole. See also '**chambering** or **springing**'. Nelson, 1965. This method is commonly used in China today.

concentric pattern, diamonds set in bit face in concentric circles so that a slight uncut ridge of rock is left between stones set in adjacent circles. Compare '**eccentric pattern**'. Long, 1920.

conceptual modelling, a technique of representing a mental image of an action or object in the form of mathematical equations and logical relationships. Anon., 1992.

conceptual model of rock blasting, see Table 9.

Table 9 A preliminary conceptual model of rock blasting. Rustan, 2008.

Fragmentation area or lines of fragmentation	Cracks	Wave types	Strength criteria overrun	Estimated contribution to fragmentation (%)
Crushing zone	Radial, tangential and different angles	Shock wave due to initial gas pressure	Dynamic compressive and shear strength	2
Two side cracks	Radial	Shock wave and initial wave, reflected wave and quasi static gas pressure	Mode I, Dynamic tensile strength crack	1
Bench front	Circular, scabbing cracks	The P-wave compression wave is reflected as P-wave tensile wave	Dynamic tensile strength	2
Whole prism affected starting from bench face and moving towards the blasthole.	Bending cracks vertical and horizontal	Quasi static gas pressure	Bending strength	40
Release of strain energy. Whole prism	Shear and tensile cracks	Quasi static gas pressure	Dynamic shear and tensile strength	55
				100

conchoidal fracture, a fracture with smooth, curved surfaces showing concentric undulations resembling the line of growth on a shell. Conchoidal fracture is well displayed in quartz and flint stone, and to a lesser extent in anthracite. Nelson, 1965.

concrete arch, reinforcement of shaft, drifts, tunnels, workshops underground by reinforced concrete.

concrete blasting, the demolition of concrete by explosives. Lauritzen and Schneider, 1992.

concrete crusher, mechanical device used to demolish concrete structures. Atlas Copco, 2006.

concrete drill, drill specifically designed for drilling in concrete. Atlas Copco, 2006.

concrete segmental lining, lining of the inside of a tunnel or roadways in coal mines. In coalmines they used to have 0,3 m wide concrete blocks to support the roadways. More than 50 segments were required for a 5 m diameter roadway. Improved performance could be achieved by separating the blocks with wood inserts or strips of old conveyor belting. The installation of such lining was however very labour intensive, and 5 concrete blocks replaced their use. Hudson, 1993.

concurrent caving, there are five main caving methods in mining, block caving, sublevel caving, sublevel block caving, longwall caving and top slicing. The term concurrent is here used to mark that it is a prerequisite that overlaying bedrock also must cave.

concussion, shock or sharp airwaves caused by an explosion or heavy blow. Nichols, 1962. The inaudible part of an air blast. Frequency below 20 Hz. Persson et al., 1994.

concussion charge or **bulldoze**, a ready-made charge put in close contact with the surface of a boulder with the purpose to demolish it. The charge is usually enclosed by a plastic envelope making it easier to obtain a close contact between the charge and the boulder surface.

condensed medium, a solid material like rock and metals.

condensed phase of explosive, see '**commercial-** or **civil- explosives**'.

condensor-discharge blasting machine, a blasting machine that uses batteries or magnets to energize one or more condensors (capacitors), whose stored energy is released into a blasting circuit. USBM, 1983.

conductivity (hydraulic) or **permeability** (c_h), (m/s), see '**hydraulic conductivity**'.

cone bit (synonym), see '**roller bit**'. Long, 1920.

cone crusher or **gyratory crusher**, a machine for reducing the size of materials by means of a truncated cone revolving on its vertical axis within an outer conical chamber, the annular space between the outer chamber and cone being tapered. To facilitate the cone crusher to accept extra-large feed, the standard gyratory crusher was redesigned with the throat extended on one side. This type of crusher is called *jaw-type gyratory crusher*. Constable, 2009.

cone cut, a cut normally used in drifting in which a number of central holes are drilled towards a focal point and, when fired, break out a conical section of strata. This cut is seldom used.

cone rock bit, a rotary drill, with two hardened knurled cones which cut the rock as they roll. Synonym for 'roller bit'.

confined detonation velocity or **confined velocity of detonation** (c_{dc}), (m/s), velocity of detonation in a defined diameter hole and confinement. Steel pipes are not recommended for these measurements, both for the sake of safety, and because steel can give misleading results.

confined shear strength (τ_l), (MPa) of brittle materials, it has been confirmed by tests that the strength of a material increases with confinement. Lundborg, 1972 developed a formula to calculate the confined shear strength τ_l dependent on the normal pressure σ_n,

$$\tau_l = \tau_0 + \frac{\mu \sigma_n}{1 + \frac{\mu \sigma_n}{\sigma_i + \sigma_0}} \tag{C.4}$$

where τ_i is the limit value of the shear strength in Pa at very large normal stress σ_n in Pa and τ_0 is the shear strength at zero normal pressure in Pa and μ is the coefficient of friction between crack surfaces. For a particular sandstone τ_i = 900 MPa, τ_0 = 20 MPa which means that under very high normal forces the shear strength will increase up to 45 times at 900 MPa confinement.

confined swell (in sublevel caving), the swelling of the blasted ring in sublevel caving against the caved rock. MassMin, 2008.

confinement, degree of confinement or **fixation factor (f), (dimensionless)**, constraining effect of the material surrounding an explosive charge. The confinement of a charge depends on the strength and density of the surrounding material; the number-, orientation-, shape-, and other characteristics of the free faces to the gravity field; the distance from the charge to the free faces; the static or confining stress field working on the material; the material properties surrounding the free faces, e.g. air, water or buffered rock; and finally the amount of rock being broken by the blast. No general system has been developed for the quantification of confinement. An estimate of the degree of confinement (f) for benches with different inclinations is obtained by the use of a 'fixation factor' which is used to multiply the practical burden for the calculation of the burden adjusted to the degree of confinement. The degree of confinement (fixation factors (f_b) are shown in Table 10.

Table 10 Degree of confinement (fixation factors) used in bench blasting. After Langefors and Kihlström, 1978.

	Blasthole inclination (f_b)			
	Vertical	3:1	2:1	Free bottom
Bench with one row of holes	1,0	0,9	0,85	0,75
Downward stoping, one row of holes	0,8	0,7	0,65	0,60

conic shaped charge, a metal cone where the outside is covered with a layer of explosives, e.g. PETN because a high detonation velocity (c_d) is necessary. A metal jet is formed upon detonation which exerts a strong penetration effect on the rock.

conjugated joints/faults, two sets of joints or faults that formed simultaneously under the action of field stress conditions (usually shear pairs).

connecting rod, rod having a fork at each end that when fastened between two parts allows a certain amount of swivel movement. Atlas Copco, 2006.

connecting wire or *firing wire*, any insulated wires (usually of lighter gauge than the firing cables) used in a blasting circuit to extend the length of the cap wires (leg or leading wires) to the shot firing cable.

connection plate, plate used to connect two or more parts together. Atlas Copco, 2006.

connection sleeve, a small tube for connection of detonating cord and fuse or electric wires leading to electric blasting caps. TNC, 1979. It may also be the end of a stiff pipe charge, which has a smaller outer diameter than the other end, and where the larger diameter end of another charge can be pushed over, about 20–30 mm. The pipe charges can therefore be added to a continuous charge with a length which is a multiple of the charge length.

connector, metallic or plastic device to connect two detonating cords in order to accomplish a delay of the detonation between the cords.

conservation equations, there are three conservation equations that always must be fulfilled in physical processes, the conservation of energy, -mass and -momentum.

consistency index (I_c), (%), numerical difference between the '*liquid limit*' and the natural '*water content*' expressed as a percentage ratio of the '*plasticity index*'.

$$I_c = (w_L - w_m)/I_P \qquad (C.5)$$

Where I_c is the consistency index in %, w_L is the liquid limit in mass-%, w_m is the water content in mass-% and I_P is the plasticity index in mass-%. SS-EN ISO 14688-2:2004, 2004.

consolidation, in geology, any or all of the processes whereby loose, soft, or liquid earth materials become firm and coherent. In soil mechanics, it refers to the adjustment of a saturated soil in response to increased load and involves the squeeezing of water from the pores and decrease in void ratio. Stokes, 1955. The gradual reduction in volume of a soil mass resulting from an increase in compressive stress. ASCE, 1958.

constitutive equation or **material law**, stress-strain relationship or force-deformation function for particular materials.

contact injection, injection of contact material between two surfaces, e.g. a rock surface and shotcrete surface. TNC, 1979.

constriction, see '**confinement**'.

continuity of detonation test (cartridge train), the ability to transmit the detonation between two coaxial cartridges having the same diameter and separated by an air gap. European test EN 13631-11. Sanchidrian, 2007.

continuous critical burden test, a single borehole is drilled inclined into a rather competent and planar rock surface representing a borehole with steadily increasing burden. By blasting a hole drilled in this way the exceeding of a maximum critical burden, at which no breakage at all will occur, is ensured. Blastability is thereby equated with the critical burden which can be defined by the study of the shape of the excavation area. Whimmer, 2008.

continuous deformation analysis (CDA), the process of simulating deformations continuously in solids with the help of computer codes.

continuous measurement of the velocity of detonation (VOD), (c_d), (m/s), there are several methods available;

1 SLIFER, see '**SLIFER method**'.
2 Continuous resistor, see '**Resistor method**'
3 Electrical time domain reflectometry method (ETDR), see '**Electrical time domain reflectometry method**'.

continuous mining, refers to mining methods where e.g. the ore is loaded on conveyors bringing the ore to surface via underground bins and hoisting shafts. Alternative method is an immediate crushing of ore at the mining front and pumping the crushed ore to surface.

continuous resistor method, see '**resistor method**'.

continuum mechanics, field of mechanics where a material is considered to form a continuum. Deals with the load capacity of solids where the local stress and displacement fields are continuous. The material modelled can not have any discontinuities like fault and joints but the material itself is damaged due to the presence of microscopic defects such as microcracks, flaws or voids.

continuum models for layered and blocky rock, the modeling of non-homogeneous materials can be done by continuum theories only if the characteristic fabric length (e.g. joint spacing) of the material is vanishing small as compared to some characteristic structural length (e.g. tunnel diameter). In all other cases either a generalized continuum theory has to be employed or one has to resort to a numerical scheme such as the distinct element method. Hudson, 1993.

contraction, a shortening of a word by using only some of the letters in the word, e.g. Dr for doctor.

continuity of detonation test (cartridge train), the ability to transmit the detonation between two coaxial cartridges having the same diameter and separated by an air gap. European test EN 13631-11. Sanchidrian, 2007.

contour blasting, see '**controlled contour blasting**'.

contour or **perimeter**, the final profile or surface planned in an excavation.

contour holes or **perimeter holes**, blastholes drilled along the final contour or perimeter of the excavation.

contour ratio, see '**half cast factor**'.

control box, movable set of valves, switches and gauges used to activate the functions of a machine. Atlas Copco, 2006.

control cabin, a room for the operator of drilling machines, hydraulic breakers, LHD-machines etc from where the machine is controlled.

control device, apparatus used to activate the functions of a machine. Atlas Copco, 2006.

controlled blasting, special blasting technique where ground vibrations, throw, and backbreak are controlled. This is often accomplished by using decked charges at large hole blasting and initiation of each deck with a separate delay, thereby reducing the maximum charge mass (weight) detonated per delay. Another technique it to use decoupled charges. In the USA this term is used to define only controlled contour blasting.

controlled caving, a (coal) mining method utilizing the advantages of longwall but at the same time without filling. In this method the working room in front of the working face is protected by close lines of props and cribs, which are portable and easily taken apart. As the face proceeds the cribs are shifted as well as the props with the face. Leaving the mined-out room to cave. This method is also called mining with self-filling. Stokes Vol. 1, 1954.

controlled contour blasting, special blasting method where special care is taken to avoid overbreak and damage of the remaining rock surfaces. The method is used in tunnels, drifts, raises, open pits and production blasts, and it is designed to minimize overbreak and leave clean-cut solid walls. The last row of holes is charged with a reduced linear charge concentration. One charging method is to use tube charges with considerable smaller diameter than the blasthole cantered in the perimeter holes by sleeves and with the aim to reduce the blast induced damage zone. This technique is used in small diameter blastholes, <45 mm, and for intermediate, and large diameter blastholes the explosive may be diluted by a non reactive- or less reactive material.

The rows next to the contour may also have reduced charge concentrations in order not to give a larger damage than the contour row. Hole spacing and burden are reduced to minimize overbreak and obtain a more stable contour. Four different methods regarding initiation are in use: 1) smooth blasting where the contour holes should be initiated with the same delay after all other holes have been initiated, 2) presplit ting, where the contour holes are initiated first, and with the same time delay, 3) cushion blasting, which may be a combination of smooth blasting and presplit ting, see '**cushion blasting**' and finally 4) micro-sequential contour blasting, with short delay time (1,5 ms) between the blastholes in the contour row. Reduction of the charge concentration can be achieved by decoupling the charges, by using pipe charges or reducing the strength of the explosive. Controlled crack initiation at the blasthole wall can be achieved by notching the borehole mechanically after drilling by a special notching tool or using linear shaped charges. The borehole pressure used in controlled contour blasting in drifts should normally be in the order of 70–150 MPa. When *linear shaped charges* are used, the borehole pressure can be reduced to 20–40 MPa. Main and sub-parameters at controlled contour blasting. Parameters affecting in controlled contour blasting are shown in Table 11.

Table 11 Main and sub-parameters at controlled contour blasting. Rustan, 1985.

Main parameters	Sub-parameters
Borehole pressure	Detonation heat (added to original text)
	Velocity of detonation
	Density of explosive
	Gas volume per kg of explosive (added to original text)
	Decoupling ratio
	Linear charge concentration
	Length of borehole
	Shape of the charge
Borehole confinement	Shape of the tunnel
	Spacing and burden and the ratio between them
	Borehole deviation and the deviation at collaring
	Number of delays of the perimeter holes
	Scatter in delay timing for the delay numbers used in the perimeter
	Misfired holes
	Rock pressure
Maximum charge per delay	The delay time must be larger than 8 ms to be regarded as separate delay

When blasting in cold regions with permafrost the controlled contour blasting design rules have to be modified due to the larger strength of frozen rock and soil compared to unfrozen. Mellor, 1975.

controlled perimeter blasting, see '**controlled contour blasting**'.
controlled trajectory blasting (CTB), a method to control the trajectory of throw in cast blasting.
control lever, lever used to engage and disengage a device such as a hydraulic valve. Atlas Copco, 2006.
control lever set, all the parts required to replace a control lever. Atlas Copco, 2006.
control panel, permanently fixed set of valves, switches and gauges used to activate the functions of a machine. Atlas Copco, 2006.
convergence-confinement method, a method to interpret the interaction between ground and support in tunnels. MassMin, 2008. P 793.
convergence meter, an appliance for measuring changes in size of openings often in a vertical direction usually at the most recently exposed region such as coal- or tunnel face. It consists of a telescopic strut set between roof and floor and carrying a pen, which records the movement on a clockwork, driven chart. Also called '**ronometer**'.
convergence recorder, see '**convergence meter**'.
conversion kit, set of parts replacing those previously installed, with the intent of improving or changing the function of the original installation. Atlas Copco, 2006.
conversion set, cylindrical connection between dissimilar parts. Atlas Copco, 2006.
conveyor, an endless band usually made of rubber and reinforced by steel cables leaning on small rollers and stretched by two end drums. The material (normally less than 300 mm in size) is transported from point A to B on the upper side of the conveyor. The conveyor could be horizontal or inclined. There are different kind of conveyors like '*bridge-*', '*flight-*', '*loading-*', '*machine-*', '*scraper-*' and '**chain conveyors**'.
conveyor haulage, see '**conveyor**'.
cooling coil, spiral wound hollow tube through which a liquid is circulated to remove heat from a fluid passing over it. Atlas Copco, 2006.
co-operating charges, simultaneously detonating charges or the time delay between charges <8 ms.
Coprod (tradename), a drilling method where a drill steel without threads are stacked on each other and used to transfer the impact energy from the drill machine to the drill bit. For rotation another set of drill steels are used consisting of tubes enclosing the other drill string. Compared to conventional top hammer drilling where 15% of the energy is lost in the threads Coprod system loses only 5% of its energy and is thereby achieving faster penetration, better hole straightness and longer service life. Hole diameters are normally from 90–127 mm. The drilling speed of a top hammer is thereby combined with the larger hole straightness of down the hole drilling DTH. Atlas Copco, 1997.
cord cutter/punch, a special designed tool that can both cut the detonating cord and also make a hole with a nail to insert the detonator into the cartridge.
cordeau detonant fuse, a kind of detonating cord, see '**detonating cord**'.
Cordtex (tradename), a detonation fuse suitable for opencast and quarry mining. It consists of an explosive core of pentaerythritol tetranitrate (PETN) contained within plastic covering. It has an average velocity of detonation of 6 500 m/s. Cortex detonating fuse is initiated by electric or a No. 6 plain detonator attached to its side with an adhesive tape. Nelson, 1965.
Cordtex relay (tradename), a device to achieve short-interval delay firing with Cordtex. A relay is an aluminium tube with a delay device, and is inserted in a line of Cordtex where required. The relays are made with two delays, 15 and 20 ms respectively. See also '**detonating relay**'.

core, the sample of rock obtained through the use of a hollow drilling bit, which cuts and retains a section of the rock penetrated. A.G.I., 1957. A tube inside a drill pipe and which is supported by a bit to receive the core in core boring. Webster, 1961.

core barrel, a hollow cylinder attached to a specially designed bit and which is used to obtain and to preserve a continuous section, core, of the rocks penetrated in drilling. A.G.I., 1957. Device consisting of two or more concentric tubes and other parts, mounted directly above the core lifter case and used to receive cores produced when drilling with a kerf bit. Atlas Copco, 2006.

core barrel bit, obsolete name for a core bit that could be coupled directly to the bottom end of a core barrel. Long, 1920.

core barrel coupling, a coupling for a core barrel. LKAB, 1964.

core barrel rod, synonymous for guide rod see '**guide rod**'. Long, 1920.

core barrel stabilizer, tubular device, having raised ribs fitting closely inside the hole, mounted immediately above a core barrel to prevent vibration when drilling. Atlas Copco, 2006.

core bit, a hollow, cylindrical boring bit for cutting a core in rock drilling or in boring unconsolidated earth material. It is the cutting end of a core drill. USBM, 1990. Tubular kerfed cutting bit with the cutting surface made up of 1) Set diamonds, 2) Diamonds in matrix or 3) Tungsten carbide chips. Atlas Copco, 2006.

core box, specially constructed wooden or steel boxes used for the storage and transportation of rock cores produced during exploratory drilling. Atlas Copco, 2006.

core drill, a mechanism designed to rotate and cause an annular-shaped rock-cutting bit to penetrate rock formations, produce cylindrical cores of the formations penetrated, and lift such cores to the surface, where they may be collected and examined. See also '**adamantine drill**', '**calyx drill**', '**diamond drill**', '**rotary drill**' and '**shot drill**'. Long, 1920.

core drilling or **diamond drilling**, process of using a hollow centered drill bit to produce cylindrical rock cores with the purpose to investigate the properties of a rock mass. Atlas Copco, 2006.

core lifter, split conical ring used to retain cores in a core barrel during extraction from the hole. Atlas Copco, 2006.

core lifter case, threaded tubular housing mounted directly above the drill bit into which the core lifter is fitted. Atlas Copco, 2006.

core load (q_l), (g/m), the explosive core of a detonating cord, characterized by the number of grains/m corresponding to linear charge concentration where grain (1 grain = 28 g) is not a metric unit. In the mertric system use g/m.

coring, see '**diamond drilling**'.

corner cut, a blast where swelling can occur in three directions, see Fig. 16.

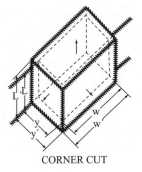

CORNER CUT

Figure 16 Corner cut. After Kennedy, 1990.

Cornish cut, a kind of *parallel hole cut*.

Coromant cut, a parallel hole cut where blastholes are located at an increasing distance from one or several uncharged holes which are used for swelling of the rock, see Fig. 17. Two rings are constructed around the empty hole or holes. A special alignment tool called a *template* can be used to get the first ring of holes in the right position and direction. The method is used in tunnelling. Langefors and Kihlström, 1978.

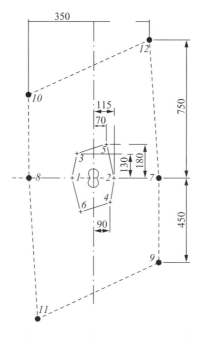

Figure 17 Coromant cut with two empty holes ɸ 57 mm. Numbers in combination with arrows shows distances in mm. After Langefors and Kihlström, 1978

correlation coefficient (R), (dimensionless), a number expressing the fitness of a curve to measurement data. R varies between 0–1 where 1 represents the case when all measurement points are located on the fitted line.

Cosserat continuum theory, continuum models for layered and blocky rock. Hudson, 1993.

cotter pin, see '**split locking pin**'.

countersink drill, a drill used to ream conically the top of a drill hole in manufacturing machine parts. Sandvik, 2007.

counter sinking, the act of conically reaming the top of a drill hole. Sandvik, 2007.

counterweight, a weight used together with a cylindrical drum to reduce the hoisting moment of a single load. Lewis, 1964. Also a weight placed to counter the instability caused by a cantilever mounted structure. Atlas Copco, 2006.

country rock, the rock traversed by or adjacent to an ore deposit. Fay, 1920.

coupled wave, a surface seismic wave of complex motion in an elastic medium. It can be described by mathematics. Also called C-wave. A.G.I., 1960.

coupling (blasting, mechanical), in blasting coupling concerns the efficiency of the transfer of energy from an explosive reaction into the surrounding rock in a borehole or on the surface of rock. Perfect coupling prevails if there are no losses due to absorption or cushioning (air between explosive and rock). The amount of energy transferred depends on the impedance ratio, see '**impedance ratio**'.

coupling (drilling), metal piece which joins the drill rods.

coupling (electrical), capacitive or inductive coupling from power lines, which may transfer electrical energy into an electric blasting circuit.

coupling (explosive in blasthole), the degree and quality of interaction (filling) of the explosive charge with the borehole volume and borehole wall. It is defined by the volume of explosive in relation to the total volume of the blasthole. There exists to kind of coupling ratios; *axial coupling ratio* R_{ca} and *radial coupling ratio* R_{cr}. The radial coupling ratio e.g. for a 50 mm diameter column charge in 100 mm blasthole is 0,25. Bulk loaded explosives yield complete coupling 1,0, in the charged parts. Untamped cartridges are decoupled. See also '**decoupling**'.

coupling half, one half of a coupling manufactured in two parts. Atlas Copco, 2006.

coupling (jointing), the act of connecting or jointing two or more distinct parts.

coupling nut, threaded nut used to connect two threaded rods together. Atlas Copco, 2006.

coupling ratio (R_c), (dimensionless), the ratio of the diameter of the charge to the diameter of the blasthole in charged parts of the blasthole.

coupling sleeve, an element of a drill string consisting of steel tube with a female thread with the aim to couple rods together with sufficient firmness to ensure that energy transmission will be effective. Atlas Copco, 1982. Tubular device used to connect two or more parts e.g. a threaded cylindrical coupling used to connect two drill rods having threaded ends. Atlas Copco, 2006. See '**connection sleeve**'.

course, the direction of the orebody in regard to the geographic north.

cover, box like structure used to cover or protect another part. Atlas Copco, 2006.

cover hole, one of a group of boreholes drilled in advance of mine workings to probe for and detect water-bearing fissures or structures. Long, 1920.

covering, protection of a round to avoid flyrock, see '**blasting mat**'.

cover plate, general term for anything used to protect something. e.g. a flat steel plate used to cover an inspection hole in a hydraulic tank. Atlas Copco, 2006.

covolume (α), (m^3/kg), the volume occupied by a gas when it is compressed to its maximum density, i.e. the effective volume of the atoms and/or molecules themselves.

coyote blasting (or shooting), coyote-hole blasting, gopher-hole blasting, a term applied to the method of blasting in which large charges are fired in small adits or tunnels driven, at the level of the floor, in the face of a quarry or slope of an open-pit mine and where the space is completely filled with explosives. See also '**chamber blasting**'. Common blasting method in China.

CPEX, a non-ideal detonation code developed by ICI. See '**non-ideal detonation**'. Cunningham, 1991.

CPF (acronym), for 'cell powder factor'. See '**cell specific charge**'.

cps (acronym), unit for rotation velocity, see '**cycles per second**'.

crack, a displacement discontinuity in a solid or at the interface between solids. In blasting the term crack is often used to characterize the blast induced extension of the fracture network.

crack aspect ratio (R_{cra}), (dimensionless), the ratio of the maximum width (crack opening) to the length of a crack.

crack bifurcation, the process of a (propagating) crack dividing into two or several cracks. This occurs when there is an excess of crack driving force in the stress field.

crack density (ς), (mm^{-1}), (on **micro** level), is defined for each grain as the sum of the intragranular and transgranular crack length in mm divided by the area of the grain. Montoto, 1982.

Cracked Chevron-Notched Brazilian Disc (CCNBD) method, method to determine the fracture toughness of ceramics for Mode I cracks developed by Shetty in mid 1980's and also started to be used for rock samples in 1989. A disc prepared for Brazilian test is notched axially by a circular disc from both sides and in the same plane to a depth that creates an open notch see Fig. 18. Fowell, 1991.

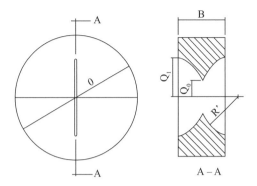

Figure 18 Geometry of cracked notched Brazilian disc specimen with basic notation. After Fowell, 1991.

crack extension force (F_c), (N), a crack will start to propagate due to a crack extension force F_c in a direction which minimizes the subsequent energy release rate. Krantz, 1979.
crack frequency or **frequency of cracks (f_c), (No./m)**, number of cracks, crossing a 1 m section of a straight line in a borehole or on the rock surface.
crack length or **length of crack (l_{cr}), (m)**, (maximum) extension of a planar crack.
crack morphology, the topography of and the markings on the fracture surfaces, studied for the determination and analysis of the state of stress (tensile, shear, etc.) and the conditions which led to the rupture.
crack mouth opening displacement (b_m), (mm), the opening of a propagating crack at the crack mouth or tip. Korzak, 1989.
crack opening displacement (b_t), (mm), the width of a propagating crack at a defined time (time t). Korzak, 1989.
crack propagation, the process of crack extension. Crack propagation is controlled by irregularities, grain boundaries, and planes of weakness. The real area of fracture surfaces is larger than the nominal area of fracture. The *critical strain energy release rate (G_c)* measured often shows considerable scatter. The process of crack propagation is fairly complex, involving various stages, such as crack initiation, sub-critical crack growth, the critical condition of instability, unstable crack propagation, crack branching, crack arrest and possibly crack reinitiation. Reinitiation of crack as well as crack branching are phenomena often observed and fundamental for fragmentation blasting.
crack propagation velocity (c_c), (m/s), speed of movement of the front of a crack. The crack velocity depends on the physical, mechanical, and other properties of the rock or rock mass, and varies from 0 up to the Rayleigh wave speed. Values up to 38% of the velocity of sound have been measured. Johansson and Persson, 1970. Larger values have been measured in limestone under special conditions. Petrosyan, 1994.
crack propagation velocity measurement, can be obtained with velocities gauges, electrical impedance measurements, conducting wires, high-speed photography and ultrasonic interference. Swan, 1975.

crack range, the volume of solid material surrounding a crack that experiences magnified strains which arise from the stress intensifying effect of the crack. Giltner, 1993.

crack resistance or *crack resistance energy* (*G*), (J/m²), see '**strain energy release rate**'

crack surface, the surface area of a crack created during crack extension. In a rock formation the surfaces of cracks may be non-planar, in case of insufficient clearance, friction (shear stresses) and interlocking (normal stresses) may be present along sections of the crack.

crack width or **width of crack** (*b*), **(mm)**, the width or opening of a crack which is subjected to tensile opening stresses or internal compression.

cradle, a movable platform or scaffold suspended by a rope from the surface, upon which repairs or other work is performed in a shaft. Fay, 1920. Generally a device partially surrounding another object to retain and support it. Atlas Copco, 2006.

Craelius core orientator, a device to measure the low side of an angled hole. A small free moving steel ball is used for this purpose. Stein, 1997.

Cranz–Schardin multiple spark camera, a high speed camera that can produce 24 "snapshot" frames of the dynamic process photographed at discrete time intervals at a framing rate of approximately 30 000–900 000 frames per second. Daehnke, 1996. The camera is described in detail, see Riley, 1966.

crater, an excavation, normally cone or bell shaped, in rock or soil created by an explosion on or below the surface. *Apparent craters* and *true craters* are formed, see '**crater blasting**'.

crater blasting, a blasting method where the blastholes are drilled perpendicular to the surface being blasted, and the holes are charged with a concentrated charge (length of charge <6 times the blasthole diameter) near the surface. After blasting a crater is formed. Nomenclature used in crater blasting is shown in Fig. 19.

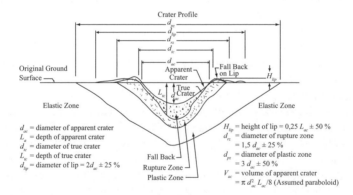

Figure 19 Schematic drawing of a crater with the important dimensions indicated and definitions of common terms used. After Nordyke, 1962.

The quantity of interest is the burden (charge depth) associated with the maximum value of rock volume broken per kg of explosive used. This burden is called the *optimum breakage burden* (B_{ob}) and can theoretical be predicted according to the formula developed by Livingstone, 1956.

$$B_{ob} = k\sqrt[3]{Q} \quad \text{Livingstone crater formula} \quad (C.6)$$

where B_{ob} is the optimum breakage burden distance in (m), k is a constant of proportionality expressing rock and explosive properties, and Q is the mass of explosive in (kg). A data analysis for more than 1 000 craters, formed in different kinds of rocks and due to different sizes of charges varying from model scale up to super scale (nuclear) resulted in the following empirical formula. Dick and al, 1990.

$$\log\sqrt{(B_{ob}^2 + r_a^2)} = 1,846 + 0,312 \log Q \quad (C.7)$$

where B_{ob} is the optimum breakage burden distance or charge depth, r_a is the apparent crater radius and Q is the equivalent TNT charge mass. The volume V of the crater in m³ can be calculated by the Livingston's crater volume formula

$$V = kR_{eun} M_i LQ(k_s)^3 \quad \text{(Livingston crater volume formula)} \quad (C.8)$$

where k is the material behaviour index (kg⁻¹ J⁻³ m³), R_{eun} is the energy utilization number or the ratio between the crater volume V divided by the volume at optimum breakage V_{ob} or $R_{eun} = V/V_{ob}$ ($R_{eun} = 1$ at optimum depth). R_{eun} is therefore dimensionless, M_i is the material index defined by the quota between the optimum crater volume V_o divided by the optimum crater depth or the optimum burden B_o cubed,

$$M_i = V_o / B_o^3 \quad (C.9)$$

and L is the stress distribution number or a shape factor for the charge (if the charge is spherical $L = 1$) L is dimensionless, Q is the charge weight in kg and k_s is the strain energy factor in (J).

crater blasting test, a test method to find the optimal depth of a charge in crater blasting. Optimal depth is defined as the depth where the maximum volume of material is broken for a certain quantity of explosive. In the crater test, a given size charge (height equal to six times the hole diameter) is buried at various depths perpendicular to the free face and fired. The achieved crater volumes and critical depths where no breakage occurs are measured. From this data, a strain energy factor can be calculated for the Livingston crater blasting formula, see '**crater blasting**'. The result can only be used for full scale crater blasting and not for bench blasting because of different confinement, geometry and blast direction.

crater charge test, the burden (charge depth) in crater blasting is varied by single hole blasting of craters until the optimum depth (optimum breakage burden) is reached. This burden is defined as that burden that gives maximum volume of broken rock. The charge depth is thereafter increased until no breakage occurs, the critical depth or critical burden. The selected burden should always be a little less than the optimum breakage burden.

crater cut, a *parallel hole cut* and the cut is achieved by *crater blasting*, see Fig. 20. Uncharged drill holes for swelling are not used. The technique to crater blast the cut is not so efficient, so the method is not used in drifting. Langefors, 1963.

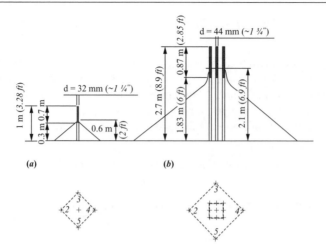

Figure 20 Crater cut, a parallel hole cut and with breakage parallel to direction of holes. After Langefors, 1963.

crater lip height (H_{lip}), (m), the vertical height from the top of the rim to the original level of the surface being blasted.
crawler, one of a pair of an endless chain of plates driven by sprockets and used instead of wheels, by certain power shovels, tractors, bulldozers, drilling machines, etc., as a means of propulsion. Also any machine mounted on such tracks. USBM Staff 1968.
crawler excavator, see '**crawler loader**'.
crawler loader, a machine for loading of rock using crawlers for movement. TNC, 1979.
crawler rig, drilling rig mounted on a crawler. Atlas Copco, 2006b.
Crawshaw-Jones fume test, a small sample of the explosive is detonated in a closed tank, and the reaction products analysed for its content of toxic fumes. The concentrations measured with these laboratory scale methods are generally quite different from the concentrations measured in full scale. The method works with a very small quantity of test explosive to be detonated and this influences the detonation performance. Many blasting agents with a large critical diameter can not be tested because they don't detonate in small sample sizes. *Ardeer tank method* and *Bichel Gauge* method uses similar layouts. Persson et al., 1994.
creep ($\acute{\varepsilon}$), (s^{-1}), a time dependent partly plastic displacement, $\acute{\varepsilon} = d\varepsilon/dt$, of solids caused by stress. Lapedes, 1978. There are two types of creep mechanisms '*diffusion creep*' and '*dislocation creep*'. In actual crystals there are always lattice defects, a lattice vacancy or an interstitial atom is a point defect. The diffusion of atoms takes place through these point defects. Dislocation creep is dominant under stress so high that the yield stress can be disregarded. Hudson, 1993. See also '**squeezing rock or ground**'.
creep law, in practical engineering a number of empirical and theoretically derived creep laws have been introduced for quantitative description of rock creep at low temperatures. Most of them can be expressed in the form

$$\varepsilon'(t) = At^{-n} (0 \le n \ge 1) \tag{C.10}$$

where $\dot{\varepsilon}(t)$ is the creep strain rate in s^{-1} and A and n are constants that depend on the material as well as on the test conditions. $n = 1$ gives the logarithmic creep law, which seems to be generally applicable at stresses lower than 2/3 of the failure stress and at temperatures lower than $0{,}2\ T_m$. Andrade (person name) type creep with $n = 2/3$ has been frequently observed at higher stresses and temperatures. Hudson, 1993.

crescent wrench, a wrench with adjustable gap. Stein, 1997.

crest, the top of the face in a bench created by a previous shot. Konya and Walter, 1990.

CRF (acronym), see '**cemented rock fill**'.

crib, a structure composed of frames of timber laid horizontally upon one another, or of timbers built-up as in the walls of a log cabin. Fay, 1920. This technique was used to support the roof especially in coal and gold mines.

crimp, the folded ends of a paper explosive cartridge, or the circumferential depression at the open end of a blasting cap which serves to secure the fuse. IME, 1981.

crimper, a special hand tool or bench mounted tool used for crimping a blasting cap onto a length of safety fuse. AS 2187.1, 1996.

crimping, the act of securing a fuse cap or igniter cord connector to a section of a safety fuse by compressing the metal shell of the cap by means of a cap crimper. Atlas Powder, 1987.

critical burden in bench blasting (B_c), **(m)**, the minimum burden distance with no breakage and displacement.

critical decoupling ratio (R_{dcc}), **(dimensionless)**, the ratio between the borehole diameter and the diameter of a cylindrical charge when there is no longer a transition zone around the blasthole. For granite this ratio is ~4,2 and for limestone ~8,4. Atchison, 1964 a and b.

critical density (ρ_c), **(kg/m³)**, the unit weight of a saturated granular material below which it will lose strength and above which it will gain strength when subjected to rapid deformation. The critical density of a given material is dependent on many factors. ASCE, 1958.

critical depth in crater blasting (B_c), **(m)**, the minimum burden (depth of burial) of an explosive charge at which it will not break a crater to a free surface. A small decrease in burden (depth) will result in breakage.

critical diameter of cartridge (d_{cc}), **(m or mm)**, the diameter for an unconfined cylindrical explosive charge (located in air), below which the detonation is unstable. In the case of poured explosives, thin cardboard tubes or plastic pipes can be used to make them cylindrical. The critical diameter of explosive charges in boreholes also depends on the decoupling factor, the density, temperature and pressure of the explosive. The density of the explosive varies with the charging method. For ANFO poured into boreholes the density will be ~840 kg/m³ and dcc ~100 mm, while for pneumatically charged ANFO the density will be ~1 000 kg/m³ and dcc ~50 mm.

critical diameter test d_{cc}**-test**, determination of the minimum diameter for stable detonation under the worst possible conditions which means placed in air (unconfined). Decreasing critical diameters are found at increased confinements. Steel pipes will enable the explosives to be fired in smaller diameters than air because of larger confinement.

critical strain energy release rate for mode I, II or III loaded cracks (G_{Ic}), (G_{IIc}) or (G_{IIIc}), **(J/m²)**, critical energy release rate associated with the three basic modes of fractures: mode I, II and III, respectively.

critical stress intensity factor (K_{Ic}), **(N/m$^{3/2}$)**, critical level of the theoretically derived and calculated stress intensity, associated with the onset of brittle fracture, compare with 'fracture toughness'.

critical void ratio (e_c), **(dimensionless),** the void ratio corresponding to the '*critical density*' of a soil. See also '**critical density**'.

cross bit or **cruciform drill bit**, drill bit for percussive drilling with four cemented carbide inserts. Sandvik, 2007.

cross-braded chisel bit, see '**cross chopping bit**'. Long, 1920.

cross chopping bit, bit with cutting edges made by two chisel edges crossing at right angles with the intersection of chisel edges at the center of the bit face. Used to chop (by impact) lost core or other obstructions in a borehole. Also called '*cross-braded chisel bit*' and '*cruciform bit*'. Long, 1920.

crosscut, a near-horizontal drift driven across the course of a vein. In general, a drift crossing the direction of the main workings or a connection between a shaft and a vein.

cross-hole seismic, radiation of seismic waves from one borehole and monitoring the signal in a borehole nearby (crosshole). Usually the boreholes are arranged parallel to each other. The method is used to explore the rock mass regarding fractures, voids and strength. Parameters measured include P-and S-wave velocities, rise time, as well as arrival time of the pulse.

cross-hole resistivity, measurement of the electric resistivity of the rock mass between electrodes placed into two parallel drill holes. Also called "cross-hole resistivity tomography". Noguchi, 1991.

cross-hole tomography, geophysical signals radiated from one borehole and monitored in a boreholes nearby (crossholes) to examine a volume of rock mass. It is a geophysical method to investigate the rock mass before excavation.

cross-linking agent, a chemical which is added to a fluid explosive in the charging operation for the improvement of the consistency and bending stiffness after the explosive has been positioned in the borehole. It is especially interesting when charging large diameter upholes. The cross-linking agent effects an almost instantaneous increase in the viscosity of the explosive.

crossover, an adapter used to connect dissimilar threads. Atlas Copco, 2006.

crossover adapter, threaded sleeve (adapter) used to connect two dissimilar threads. Atlas Copco, 2006.

crossover sub, short threaded drill tube used to connect to dissimilar threads. Atlas Copco, 2006.

cross sectional area (A), **(m^2)**, the area of a drift, tunnel or room underground perpendicular to the longitudinal axis to the drift or room. TNC, 1979.

cross stoping, see '**overhand stoping**'. Fay, 1920.

crown pillar, the upper part of an under ground stope left to support the hanging wall. Also any horizontal pillar of unmined rock left above a caved or mined-out area.

cruciform bit, see '**cross chopping bit**'.

cruciform drill bit, drill bit that has four tungsten carbide inserts placed in the form of a cross. Atlas Copco, 2006.

crude ore, fragmented ore leaving the mine and used as intake raw material for the mineral dressing (processing).

crush, see '**squeezing rock** or **ground**'.

crushed rock (used in geological descriptions), rock material smaller than 15 mm in size. In nature this occurs when the rock is heavily jointed to 'sugar cube size'.
crushing drilling, a rotary drilling method in which drilling is performed by the crushing or grinding action of a roller bit which rotates while being pressed against the rock. Fraenkel, 1952. Also called '*roller bit drilling* or rotary *crushing drilling*'.
crushing rolls, a crusher consisting of two heavy rolls between which ore, coal or other mineral is crushed. Sometimes the rolls are toothed or ribbed, but for ore their surface is generally smooth. Fay, 1920.
crushing zone index (CZI), (I_{cru}), **(dimensionless)**, the crushing zone index is defined as follows;

$$I_{cru} = \frac{p_b^3}{k\sigma_c^2} \qquad (C.11)$$

where p_b is the borehole pressure in MPa, k is the stiffness of the rock in N/m and σ_c is the uniaxial compressive strength of the rock in MPa. If the material within the crushing zone is homogenous and isotropic the stiffness k can be defined as follows;

$$k = \frac{E_d}{1+v_d} \qquad (C.12)$$

where k is the stiffness of the rock in N/m, E_d is the dynamic elastic modulus and v_d is the dynamic Poisons ratio. The crushing zone radius r_{cr} in mm can thereafter be calculated by the following empirical formula;

$$r_{cr} = 0{,}812 \; r \; (I_{cru})^{0{,}219} \quad R = 0{,}91 \qquad (C.13)$$

where r is the borehole radius in mm. Esen, 2003.
crush zone radius (in rock blasting), (r_{cr}), **(m)**, normally the zone of material adjacent to the part of the blasthole which is charged and in direct contact with the explosive. The material in the crush zone will be crushed because of stresses in excess of the dynamic compressive strength of the material. The diameter of the crush zone depends on the strength of the rock. In hard rock, and under fully confined conditions, the radius of this zone is about two blasthole diameters. The crush zone in glass is increasing with the speed of S-wave. Rossmanith, 1978. The failure mechanism in the crushing zone is due to shear. Heusinkveld, 1975.
crush zone model (CZM), a fragmentation calculation model where the KuzRam model is used to calculate the coarse and medium fragments and the fines are calculated proportional to the crushing zone volume in the vicinity of the blasthole.
crystal cleavage, process of transgranular separation within a crystal along a cleavage plane.
CSa (abbreviation), for '**coarse sand**' defining particles with sizes from 0,63 to 2,0 mm.
CSC (acronym), see '**cell specific charge**'.
C-scale, the scale of a sound level measurement instrument that only slightly filters (discriminates between) the low frequencies. Konya and Walter, 1990.
CSi (abbreviation), for 'coarse silt' defining particles with sizes from 0,02 to 0,063 mm.

CSIR (abbreviation), see '**South African Council of Scientific and Industrial Resaerch**'.
CSIRO (abbreviation), for '**Commonwealth Scientific and Industrial Research Organisation**'. A National Research Science Organisation in Australia including research in Exploration, Mining and Minerals processing industries. It was founded in 1926. 6 500 persons are employed in 57 sites. www.csiro.au
CSIRO hollow inclusion cell, a device for determining the absolute stress tensor during a single overcoring test. Hudson, 1993.
CTB (acronym), see '**controlled trajectory blasting**'.
cubical triaxial test, tests made on cubes where the stress can be regulated individually in three perpendicular directions.
cultural vibration, vibration that is commonplace and familiar to the observer. Konya and Walter, 1990.
cup seal, a seal having a cup like shape, with various applications. Atlas Copco, 2006.
current leakage, portion of the firing current by-passing part of the blasting circuit through unintended paths such as earth, rails, water tubes etc.
current-limiting device, a device used to prevent arcing in electric blasting caps by limiting the amount or duration of current flow. Also used in the blasters' galvanometer or multimeter to assure a safe current output. USBM, 1983.
curtain, something flexible to hung in front of, to remove from view. As rubber "curtain" placed in front of hydraulic hoses to remove them from view. Atlas Copco, 2006.
curvometer, an instrument used to measure the *curvature of linings*. Hudson, 1993.
cushion, general term for a device or the act of protecting something from damaging shock waves. A mechanical spring, rubber mounting or compressible gas can act as a "cushion". Atlas Copco, 2006.
cushion blasting, a method of blasting in which 1) An air space is left between the explosive charge and the stemming (see '*air deck blasting*'), or in which the blasthole is purposely drilled much larger than the diameter of the explosive cartridge to be loaded (see '*smooth blasting*'). 2) It can also be a combination of presplitting and smooth blasting technique when boreholes are drilled close to each other as in presplitting and initiation occurs after all other holes in the round have been initiated, as in smooth blasting. 3) The contour is fired after the main blast in a similar way as smooth blasting.
cushion shooting, synonymous to '**cushion blasting**'.
cushion shot, synonymous to '**cushion blast**'.
cushion stick, a cartridge of explosive loaded into the bottom of a small-diameter borehole before the primer is inserted. The purpose of the cushion stick is to reduce damage to the detonator and leg wires. The use of a cushion stick is not generally recommended because of the possible generation of bootlegs.
cut (open pit), that part of an open pit excavation with more or less specific depth and width, and continued in like manner along or through the extreme limits of the open pit, see Fig. 21.

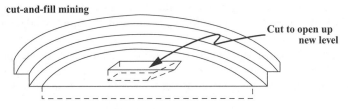

Figure 21 Example of a cut in open pit mining.

cut (road), that portion of an excavation with more or less specific depth and width and continued in similar manner along or through a hillside. A series of cuts is taken before complete removal of the excavated material is accomplished. The specific dimensions of any cut are closely related to the material's properties and required production levels. Konya and Walter, 1990.

cut and cover (CAC), a method of construction a tunnel in which an excavation is made to formation level and then filled back over the tunnel after its lining had been constructed. Ham, 1965.

cut-and-fill mining or **stoping (CAF)**, a mining method where the ore is excavated by successive flat or inclined slices working upward from the current level (undercut and fill mining), as in shrinkage stoping. After each slice is blasted, all ore broken is removed, and the stope is filled up with waste material to within half a meter up to a few meters from the back before the next slice is taken out, with just enough room being left between the waste pile and the back of the stope to provide working space.

If the strength of the orebody is low the direction of cut and fill mining must change to downward cut and fill mining where the bottom of the fill is given a higher strength so it may be used as an artificial roof when mining downwards.

cut and fill with slice method, see '**rill mining**'.

cut blasting, blasting of the first holes in tunnelling, drifting, raising or shaft sinking in order to open up a swelling room to the full depth of the round.

cut blasting initiation, see '**micro-sequential contour blasting**'.

cut easer holes, in drifting, the holes closest to the cut used to enlarge the opening formed by the cut. Olofsson, 1990. See '**relievers**'.

cut hole, one of the first holes blasted in a tunnel, drift, raise or shaft in order to create a free face for the following holes. Another meaning is the removal of a volume of rock on surface, for example in road cuts.

cut-off, a break in a path of detonation or initiation caused by extraneous interference, such as flyrock or shifting ground often caused by the firing of lower delay numbers.

cut-off hole, misfired hole resulting from the cutting-off of the initiation system or explosive column by previous firing holes, or flyrock from these.

cutter, general term used to describe a device used to cut other material. Atlas Copco, 2006.

cutter loader, a longwall machine that cuts and loads the coal onto a conveyor as it travels across the face. Cutter loaders may be grouped according to the thickness of web cut: as (1) thick web machines, such as the A.G. Meco-Moore, which cuts and loads up to 1,8 m; (2) medium web machines, which take about 0,6 m such as the Gloster-getter, Anderton shearer, and Trepanner; (3) narrow web or plough type machines, which take from 2,5 cm to 30 cm of coal during the traverse of the face. Nelson, 1965.

cutting edge, the point or edge of a diamond or other material set in a bit that comes in contact with and cuts, chips, or abrades the rock. Also called cutting point. Long, 1960.

cutting head, the rotating head at the front of the full-face boring machine.

cutting machine, general term used to describe variety of machines designed to cut other materials. Atlas Copco, 2006.

cuttings, the rock fragments created by the drill bit when it is impacting, sliding or rolling (disk cutters) over the rock surface. Sandvik, 2007.

C-wave, synonym for coupled wave. A.G.I., 1960.

cycles per second (cps), a unit for rotation velocity.

cyclic loading, successive loading and unloading of a material. Cyclic loading often causes a material to fail at a stress level lower than its determined static strength. This reduction in the strength of a material due to fluctuating or alternating stresses is called fatigue. The rock fatigues under cyclic load, if the maximum applied stress is in the 75–100% range of the compressive strength.

cyclone, device that causes air to move in a spiral motion to centrifugally remove solid particles from it. The principle is used in dust collectors on surface drilling rigs. Atlas Copco, 2006.

cyclotrimethylenetrinitramine (RDX) or **hexogen**, an explosive substance consisting of $C_3H_6N_6O_6$. The properties are colourless crystals with a heat of explosion of 6 025 kJ/kg, density of 1 800 kg/m^3, melting point at +204°C and a detonation velocity (c_d) of 8 750 m/s. It is used in the manufacture of compositions B, C-3, and C-4. Composition B is useful as a cast primer.

cylinder, general term for any part having a cylindrical form. Used to describe a complete "hydraulic cylinder" as well as the cylindrical barrel. Also the cavity or bore in which a rock drill piston moves etc. Atlas Copco, 2006.

cylinder bracket, mounting fixture for a hydraulic cylinder. Atlas Copco, 2006.

cylinder compression test, a test method for explosive which defines the 'breaking power' of the explosive which is related to the fragmentation capacity of the rock. There are two methods available, Kast and Hess methods where the latter is the most common. See '**Kast cylinder compression test**' or '**Hess cylinder compression test**'.

cylinder cut, cylindrical cut or **large hole cut**, a parallel hole cut which has one or two holes of larger diameter than the others, see Fig. 22. The large holes are used for swelling. Atlas Copco, 1982.

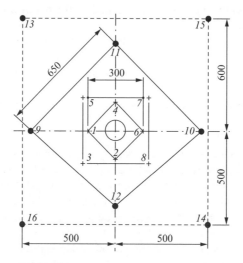

Figure 22 Cylinder cut with delay numbers. Units in mm. After Langefors and Kihlström, 1982.

cylinder expansion test, a method for testing the strength of explosives. This method was developed at the Naval Ordnance Laboratory, later renamed Naval Surface Warfare Center, Maryland. The purpose is to compare dynamic performance of explosives and to derive empirical equations of state for the detonation products. A copper tube is filled with the sample of explosive. When it is detonated, the cylinder wall will accelerate outwards due to the expansion of the detonation products. The cylinder wall motion is registered with a high-speed camera until the cylinder expands to about three times the original diameter. The pressure at this point drops to some hundred bars. Although the test gives a good measure of the expansion work, when compared to the rock blasting situation the confinement is too low to be comparable to the confined borehole in a bench blast. Persson, 1993. Sanchidrian, 2007.

cylinder link, movable device used to connect a hydraulic cylinder to another component. Atlas Copco, 2006.

cylindrical failure of slopes, one kind of failure in slopes of soil is the cylinder shape, where the failure occurs on the surface of a cylinder of soil turning around its axis.

CZI (acronym), see '**crushing zone index**'.

CZM (acronym), see '**crush zone model**'.

damage (D_{str}), **(dimensionless)**, in rock is defined by a scalar parameter D_{str},

$$0 \leq D_{str} \leq 1 \qquad (D.1)$$

such that $D_{str} = 0$ corresponds to the intact, undamaged rock, $D_{str} = 1$ corresponds to the inability to transmit tensile stress (full fragmentation), and intermediate values of D_{str} correspond to intermediate values of fracture damage. Grady, 1980 a. See also '**breakage zone**' and '**failure zone**'.

damage contour, the surface area of the 3-dimensional volume representing damaged rock mass or damaged parts of a building. A more precise term would be "*isodamage contour*".

damage criterion (volume), (D_{vol}), **(dimensionless)**, In fracture mechanics the damage D_{vol} is defined as the volume fraction of the material that has lost its load carrying ability or $D_{vol} = V_d/V_{tot}$ where V_d is damaged volume in m³ and V_{tot} is the total volume m³ affected and of interest for damage. Boade, 1980. The damage criterion characterizing rock breakage through blasting should be based on the dynamic strength properties of the material like dynamic tensile and compressive strength, dynamic shear and bending strength. Traditionally the damage criterion was linked to the static tensile strength.

damage index at blasting (D_{ib}), **(dimensionless)**, damage to remaining rock at blasting could be divided into blast induced or (mining-induced) damage caused by blasting forces and stress-induced damage caused by change in rock mass stress. A damage index for tunnelling and drifting was developed by Paventi et al., including five parameters; change in intact rock strength by increase in micro-fracturing, the amount of half cast after blasting, assessing the drumminess of the back with a scaling bar, scaling time, structure orientation compared to the tunnel or drift direction. Paventi et al., (1996).

damage parameter (based on ground vibrations), parameters used to assess the level of damage caused to a structure (rock mass, building) or machine. Ground-vibration-based damage parameters which characterize the effect of blasting on rock mass, buildings and machines depend on the structure and type of machine. For buildings the accepted damage parameter, the *peak particle velocity* (v_{max}), depends on the type of building, the wave propagation velocity (c_p) in the ground on which the building is founded, the dominating vibration frequency (f), and the number of blasts. Also for the rock mass, the damage parameter normally is based on the peak particle velocity but for computers the damage criterion is based on the *peak acceleration* (a_{max}) of ground vibration. The number of blasts is not considered a part of the damage parameter utilized in the USA. Some examples of damage levels for rock masses are shown in Table 12.

Table 12 Damage criteria in rock due to blast induced vibrations. After Bauer and Calder, 1978.

Max particle velocity (mm/s)	Effects on rock mass
<25	No fracturing of intact rock
25–635	Minor tensile slabbing will occur
635–2540	Strong tensile and some radial cracking
>2540	Complete break-up of rock mass

Damage criteria in different rock types, see '**blast induced damage zone**'.

damage volume (V_D), (m³), the volume of material that has deteriorated in strength on a microscopic or macroscopic level. Normally we are talking about damage on the macroscopic level. The computer modelling of blast damage in a rock mass, however, looks upon the start of damage in pores or flaws on a microscopic level.

damage zone, see '**blast induced damage zone**'.

dam gallery, openings created within a dam structure to allow the inspection and repair of the structure. Atlas Copco, 2006.

dampener, device used inside a rock drill machine to reduce the effect of damaging reflected shock waves travelling up the drill string. Atlas Copco, 2006.

damping, the reduction of the amplitude of a vibration of a body or of a system due to dissipation of energy, internally or by radiation. ISRM, 1975. See '**attenuation or Q-factor**'.

dangerous goods classification (United Nations), this classification is made to facilitate the safety at transportation of dangerous goods. A division into 10 classes have been done by United Nations with respect to the different general types of dangers encountered with different materials. The classes are listed in Table 13.

Table 13 Classification of dangerous goods. After United Nations, 1990.

Class	Substance
1	Explosives
2	Gases compressed, liquefied, dissolved under pressure, or deeply refrigerated
3	Flammable liquids
4	Flammable solids, substances liable to spontaneous combustion, substances (which in contact with water) emit flammable gases
5	Oxidizing substances; organic peroxides
6	Poisonous (toxic) and infectious substances
7	Radioactive materials
8	Corrosives Group I, II, and III
9	Miscellaneous
10	Substances and articles with multiple hazards

Once a material has been classified, it is included in the long list of dangerous goods, an alphabetic list of common articles and substances providing details of classification, labelling, and rules regarding its transportation. Class 1 contains all explosives, and is subdivided into six divisions with the respect to the materials behaviour in a fire, such as its propensity for mass detonation or mass explosion or for creating a mass fire hazard, see Table 14.

Table 14 Hazard divisions of class 1: Explosives. After United Nations, 1990.

Class	Hazard division of explosives	Examples
1.1	Mass detonating explosive hazard	Dynamite, cast boosters, cap sensitive emulsions, water gels and slurries
1.2	Non-mass detonating fragment-producing hazard	Ammunition with projectile hazard
1.3	Mass fire, minor blast, or fragment hazard	Non-detonating propellants and explosives
1.4	Moderate fire, no blast or fragment hazard	Class C detonators, safety fuses, other Class C explosives
1.5	Explosive substance, mass explosion, or ammunition article, unit risk	ANFO, non-cap sensitive emulsions, water gels, slurries, packaged blasting agents
1.6	Extremely insensitive substances	Currently no commercial explosive

In addition to the divisions, materials belonging to Class 1 Explosives are also divided to into 13 compatibility divisions, divisions A-N with the respect to the level of risk for explosion following flame initiation, see Table 15.

Table 15 Compatibility groups of class 1: Explosives. After United Nations, 1990.

Group	Substance
A	Primary explosive substance
B	Primary explosive substance without at least two safety features
C	Propellant explosives
D	Secondary detonating explosives without initiator or propelling charge, or article with two or more safety features
E	Secondary detonating charge without initiator, but with propelling charge
F	Secondary detonating charge with initiator and propelling charge
G	Pyrotechnics
H	Explosive and white phosphorous
J	Explosive and flammable liquid
K	Explosive and toxic chemical agent
L	Explosive with special risk (hypergolic, water activated, phosphides)
S	Package protects against all hazards
N	Extremely insensitive substances

Darcy's law, empirical equation for seepage flow in non deformable porous media. The mass flow velocity of a liquid through a porous medium due to the difference in pressure is proportional to the pressure gradient in the direction of flow. Webster 1961. The mass rate of flow or the mass flow per unit of time q_m in kg/s can be determined by the following expression that was developed by Henry Darcy in 1856.

$$q_m = -\kappa(p_i - \rho_f g_i) \quad \text{and} \quad \kappa = \frac{k}{\mu} \tag{D.2}$$

where κ is the permeability or mobility coefficient, p_i is the differential pressure in Pa, ρ_f is the density of the fluid in kg/m³ and g_i is the gravity component in the i direction in m/s² and k is the intrinsic permeability in m² and μ is the fluid dynamic viscosity in Pa·s. k is dependent on the pore geometry and in particular it is strongly dependent on the porosity n. Hudson 1993. Different formulas are used dependent on if the ground is saturated or not by water. Wikipedia, 2008, term "Darcys law".

data-shift code, a code applied by manufacturers to the outside shipping containers and, in many instances, to the immediate containers of explosive materials, to aid in their identification and tracing. IME, 1981.

D'Autriche method, a method of determining the detonation velocity (c_d) of an explosive by employing two pieces of detonating cord and a witness plate. IME, 1981.

day box, a container used at the work site for holding daily requirements of explosives.

DBA (acronym), see '**dry blasting agent**'.

DCR (acronym), for '**drift condition rating**'.

DC(T) (abbreviation), for '**disc-shaped compact test**' see '**fracture toughness test**'.

DDT (acronym), see '**deflagration to detonation transition**'.

DDZ (acronym), see '**detonation driving zone**'.

dead pressing, desensitization of an explosive caused by excessive pressure on the explosive before it is detonated, thereby causing an increase in the density of the explosive. Explosives exhibit a critical density above which detonation is inhibited. For example if the density of ANFO is increased above 1200 kg/m³, dead pressing will result.

dead pressure, see '**dead pressing**'.

dead rock or **waste rock**, the material removed in the opening of a mine which is of no value for milling purposes. Fay, 1920.

DEBM (acronym), see '**discrete element block method**' or '**discrete element block modelling**'.

debris, collection of particles such as rock fragments, sand, earth, and sometimes organic matter, in a heterogeneous mass, as found at the foot of a cliff. Fay, 1920.

decay time, the time interval required for a pulse to decay from its maximum value to some specified fraction thereof.

decibel (dB), the unit of sound pressure. The unit is commonly used when monitoring air blasts from explosions.

deck bushing, a cylinder used to stabilize the three cone roller bit at collaring the hole. Atlas Copco, 2006 b.

deck charge, an explosive charge which is separated from other charges in the blasthole by stemming or by an air cushion. Atlas Powder, 1987. The stemming between decks in dry holes should have a minimum length of 6 times the diameter of the blasthole and in wet holes 12 times the diameter of the blasthole. The stemming material that works best should have a particle size between 1/10 and 1/20 of the diameter of the blasthole. Konya, 1996.

deck charging (decking), the procedure to charge decks. Different kind of plugs are used when charging a new deck. See '**borehole plug**'.

deck charge mass, (Q_d), **(kg)**, the mass of a deck charge.

deck loading (*decking*), the loading of decks. Use '**deck charging**'.

decline, a drift driven downwards at an angle from the horizontal.

decompose, to resolve or separate into constituent parts of elements, as by means of chemical agents or by natural decay. Especially, to cause to decay or to rot. Standard, 1964.

decomposed, rock or ore altered and leached by air and water. A state of rock or ore where the original material fabric is still intact, but some or all of the mineral grains are decomposed (e.g. weathered rock).

decoupled charge, see '**decoupling**'.

decoupling, normally a separation by air between the surface of an explosive charge and the blasthole wall where it is charged or sections of the blasthole left uncharged. There are two types of decoupling; *axial (Rda) and radial (Rdr) decoupling ratio*.

decoupling factor (R_{df}), **(dimensionless)**, the ratio of the ground vibration amplitude from a decoupled detonation A_{dc} to that A_c from a fully coupled detonation of the same charge weight at a given distance. DoD, 1993.

$$R_{df} = \frac{A_{dc}}{A_c} \qquad (D.3)$$

decoupling ratio (R_{dc}), **(dimensionless)**, the ratio between the diameter of the blasthole to the diameter of the charge. Fogelson et al., 1965. The inverse of decoupling ratio is called coupling ratio.

decrease dynamic Young's modulus (E_{ddec}), **(MPa)**, is defined as the slope of the post-peak decreasing stress-strain curve. Singh, 1989.

decrease Young's modulus index (R_{dec}), **(dimensionless)**, is the ratio of Young's modulus E (modulus of rigidity) to the decrease modulus E_{dec}. The modulus of decrease is given by the slope of the post-peak decreasing stress-strain curve. Singh, 1989.

deep hardening, a hardening method for metals going into the depth of the tool material. Sandvik 1983.

Deere rock mass classification system, see '**rock quality designation**'.

deflagration, an explosive reaction associated with a moving phase change such as a rapid combustion which moves through an explosive material at a velocity smaller than the speed of sound (subsonic velocity; no shock wave produced).

deflagration to detonation transition (DDT), transition from subsonic velocity to detonation velocity (c_d) and vice versa.

deflagration to detonation transition test (DDT-test), the test substance is filled into a closed tube standing on a witness plate and ignition of the substance is made by a hot wire. The test is used to determine the tendency for the substance to undergo transition from deflagration to detonation. This test will be compulsory in future for approval of explosives according to Hazards Division 1.5 in the UN classification. Persson et al., 1994.

deflection wedge, tool used to deflect the direction of the drill bit in a hole to the desired direction. Used in prospect drilling. Sandvik, 2007.

deflectometer, an instrument for gauging any deflections of a structure. Ham, 1965.

deformation, a change in the shape or size of a solid body due to stresses, temperature, moisture, etc.

deformation lamellae, closely spaced parallel planar features which do not cross grain boundaries and frequently contain minute cavities and inclusions. Giltner, 1993.

deformation modulus, see '**modulus of deformation**'.

deformeter, an instrument measuring deformations e.g. used in scaled model analysis of a structure.

degradation, the successive breakage into smaller and smaller fragments of geological material.

degree of confinement, see '**confinement**'.

degree of freedom (in presplitting), see '**presplitting**'.

degree of packing (*P*), (%), percentage of the volume of a blasthole occupied by explosives for an intended fully charged blasthole.

delay, a distinct predetermined interval of time between the initiation of two consecutive charges in order to allow for separate firing of the explosive charges.

delay blasting, a blasting technique where delay detonators or relay connectors are used to initiate separate charges at different times.

delay cap, see '**blasting cap**'.

delay connector, short interval delay device used together with detonating cord for short delay blasting. Olofsson, 1990.

delay detonator, an electric or non-electric blasting cap which is designed to explode at a specified time after the current or ignition is started.

delay electric blasting cap, an electric blasting cap with a built-in delay that delays cap detonation at predetermined time intervals, from milliseconds up to a second or more between successive delays. USBM, 1983.

delay electric igniter, an electrical device using a fuse as the delay element. With a blasting cap on each fuse it is possible to detonate a number of charges in succession.

delay element, that part of a blasting cap which causes a delay between the initiation and the time of detonation of the base charge of the cap. The delay element usually consists of a pyrotechnical material, but micro-electronics can also be used, see '**electronic detonators**'.

delay interval, the nominal time delay between the initiation of adjacent periods in a delay series, or the nominal time between successive detonations in a blast.

delay time optimal between blastholes (t_{opt}), (ms), the time delay between two neighbouring blastholes. The purpose of using delay times is to reduce the maximum charge per delay and thereby reduce ground vibrations and to improve the fragmentation in blasting. The optimum delay time found in half scale tests according to Bergmann, 1974,

$$t_{opt} = \frac{15,6}{c_P} B \qquad (D.4)$$

t_{opt} is the optimal delay time in ms to achieve the finest fragmentation, c_p is the primary wave velocity (compressive wave velocity) in km/s and B is the burden in m.

delay period, a designation given to a delay detonator to show its relative or absolute time in a given series. Atlas Powder, 1987.

delay relay, a device incorporating a blasting cap delay arrangement to be used when delays are arranged together with detonating cord. AS 2187.1, 1996.

delay series, a series of delay detonators designed to satisfy specific blasting requirements. Basically there are two types of delay series: millisecond delay series (MS) with delay intervals in the millisecond range, and long period delay series (LP) with delay times in the order of seconds.

delay tag, a tag, band, or marker on a delay electric blasting cap or a non-electric delay blasting cap denoting the delay sequence and/or the actual delay firing time. IME, 1981.

delay time (t), (ms), time interval between the initiation and detonation of a detonator. Delay times less than 3,3 ms per meter of burden cause premature shearing between holes, and coarse fragmentation. Sandhu and Pradhan, 1991.

DEM (acronym), see '**discrete element method**' and '**discrete element modelling**'.

demec gauge, a *deformation gauge*, measuring the distance between two measurement pins. Brady, 1985.

demolition blasting, a blasting technique specifically designed for demolition of buildings, towers, bridges and industrial installations. Destruction occurs by destroying interlocking systems like beams, pillars, plates and steel girders. Sandhu and Pradhan, 1991.

demolition hammer, pneumatic or hydraulic powered hammer used to demolish buildings and other man made structures. Atlas Copco, 2006.

density index (I_D), (dimensionless), coarse soils (sands- and gravel index) is dependent upon the void ratio (e) and the void ratios corresponding to the maximum density (e_{max}) and the minimum density (e_{min}), as measured in the laboratory.

$$I_D = (e_{max} - e)/(e_{max} - e_{min}) \tag{D.5}$$

Where I_D is the density index (dimensionless), e_{max} is the void ratio at the minimum density and e_{min} is the void ratio at the maximum density. SS-EN ISO 14688-2:2004, 2004.

density of explosive (bulk) (ρ_e), (kg/m³), the mass of explosive per unit volume. The density of explosive may increase after charging, e.g. if a powdered explosive like ANFO is used, or decrease, e.g. if a cartridged explosive is used. The density of explosives normally varies between 800 and 1650 kg/m³ before charging. When liquid explosives are used in downholes, the density at the bottom will depend on the height of the explosive column, due to hydrostatic pressure.

density of explosive (after charging) (ρ_{ec}), (kg/m³), the bulk density of the explosive in the blasthole after charging may increase e.g. (ANFO) or decrease (e.g. cartridged explosives).

density of explosive (before charging) (ρ_e), (kg/m³), see '*density of explosive (bulk)*'.

density of packing (ρ_{ec}), (kg/m³), use '**density of explosive after charging**'.

density of rock (ρ_r), (kg/m³), the mass of a body divided by its volume.

density of stemming (bulk) (ρ_s), (kg/m³), the bulk density of stemming.

Department of Transportation (DOT), a USA authority regulating the transportation of explosives. Atlas Powder, 1987.

DEPS (acronym), see '**dual energy percussion system**'. Atlas Copco, 2006.

depth gauge, tough fibreglass rope covered with non-glare texture polyvinyl and printed on both sides in metres for easy reading.

depth of apparent crater (L_{ac}), (m), see '**crater blasting**'.

depth of burial (B), (m), see '**burden in crater blasting**'.

depth of pull (A_r), (m/round or %), blasted depth achieved (m/round) or the ratio between blasted depth and drilled depth multiplied by 100 (%) in driving a drift, tunnel or raise.

depth of true crater (L_{true}), **(m)**, vertical distance between visible crater bottom and the original surface.

depth per bit or **drilled depth per bit** (L), **(m)**, the length of borehole drilled with a steel bit until it must be re-sharpened. Streefkerk, 1952.

derrick, the framed wood or steel tower placed over a borehole to support the drilling tools for hoisting and pulling drill rods, casing, or pipe. Long, 1920. Perpendicular tower, on a drill rig of a particular type, that provides room for adding or removing drill rods. The vertical structure on a "crawler mounted blast hole rig" is referred to as the "mast". Atlas Copco, 2006.

derrick block, multi-sheaved pulley device that can be raised or lowered when drilling and adding or removing drill rods. Atlas Copco, 2006.

descaling, removing the thick layer of oxides formed on some metals at elevated temperatures. ASM Gloss, 1961.

desensitization, a change in the condition of an explosive rendering it impossible to detonate. Reasons for desensitization are, e.g. dead pressing, ageing, etc.

design of rock bolt systems, see 'rock bolt'.

destress blasting, a blasting method where the stresses around an underground opening are reduced. Boreholes are drilled, charged and detonated in the vicinity of the planned underground opening being excavated. The damage caused to the rock mass causes a stress redistribution.

destressed area or **destressed zone**, in strata control, a term used to describe an area where the stress is much less than would be expected after considering, the depth and type of strata.

destruction of explosive or **initiation material**, when the shelf life of an explosive or initiation material is reached it can not safely be used any longer and it has to be destructed. Every explosive or initiation material has it special technique to be destructed. Examples of destruction techniques, see Pradhan, 1996.

detachable bit, drill bit with thread or cone for simple replacement. Sandvik, 2007.

detaching hook, an appliance which releases automatically the winding rope from the cage if an overwind would occur. The hook is placed between the skip shackles, or the cage slings, and the winding rope cappel. A detaching plate is fitted to girders below the winding sheave, and in the event of an overwind, the hook becomes locked in the plate, thus suspending the skip or cage, while the winding rope it simultaneously liberated. Nelson, 1965.

detachment of rock at blasting, the separation produced between rock fragmented and later on mucked and the rock mass in place.

Detaline cord (tradename), a *detonating cord* with a linear charge concentration of 1 g/m PETN combined with a series of non-electric ms-delay detonators having 19 delay periods. The Detaline cord minimises noise and offer changing, little or no disruption to stemming or explosive charging into the blasthole. They are extremely insensitive to mechanical impact. Even a No. 8 detonator will not usually side initiate the cord. Pradhan, 2001.

detector, see '**seismometer**'.

deteriorated explosive, explosive subjected to dampness, moisture, or extreme heat during long-term storage. Deteriorated explosive materials are sometimes more sensitive to friction, and often much more hazardous to use than those in good condition. Sen, 1995.

detonating cord (detonating fuse), an initiation material or explosive consisting of a flexible line with a core of detonating explosive (a high explosive, e.g. PETN) with a small enough critical diameter to propagate a detonation. The core is wrapped with textile yarns. Usually the cord is reinforced or completely enclosed in a strong waterproof outer plastic cover. It must itself be initiated by a blasting cap. The manufactured linear charge concentration varies from 0,1 g/m (0,5 grains/foot) up to 200 g/m (940 grains/foot). The diameter will vary considerably, depending on the charge concentration of the cord. The cord may be used as part of an initiation system or to initiate explosive charges. It may also be used as an explosive itself, e.g. in controlled contour blasting. Detonating cord is mainly used on surface, but can also be used in underground blasting.

detonating cord connector, plastic hook up device for connection of detonating cord downline (in the blasthole) to the trunk line (un surface).

detonating cord downline, the section of detonating cord that extends within the blasthole from the ground surface down to the explosive charge. Atlas Powder, 1987.

detonating cord MS connector, non-electric, short-interval (millisecond) delay device for use in delaying blasts which are initiated by a detonating cord. IME, 1981.

detonating cord test, test methods where the detonating cord is tested by visual inspection for smoothness, pliability and surface blemishes. Other test methods include: diameter control, flexibility and sensitivity after flexing, water proofness and sensitivity after submerging into water, determination of velocity of detonation (c_d), test for transmission of detonation, and determination of breaking load of live detonating cord. Sandhu and Pradhan, 1991.

detonating cord trunkline, the line of a detonating cord placed on ground surface and which is used to connect and initiate other lines of detonating cord going into the blastholes, e.g. downlines (branch lines).

detonating fuse, see '**detonating cord**'.

detonating pill, see '**fuse head**'.

detonating primer, a term applied for transportation purposes to a device consisting of a detonator and an additional charge of explosives, assembled as a unit.

detonating relay connector (DRC), a unit consisting of two delay detonators with 5 to 60 ms duration mounted in end to end opposing contact. The complete unit is in the shape of a sealed plastic dog bone typically 100 mm long and colour coded according to its delay. The DRC is inserted at an appropriate position in the detonating cord line. Relay connectors are blasting delay elements primarily for use in surface mining. and quarrying operations. There are two main types: detonating relay connectors (DRC), and Nonel trunkline delays (TLD). Sen, 1995.

detonating spheres drill, an *exotic drilling method* where spheres are used as impact tool for fragmentation of the rock. The spheres are filled with two liquids separated by a membrane and at the impact the liquids are mixed and the mixture detonates. High drilling speed has been reached, 13 m/hour, at a hole diameter of 350 mm and 700 caps/hour. KTH, 1969.

detonation, an explosion in a reactive material which is characterized by a shock wave moving at greater speed than sonic velocity and accompanied by a chemical reaction whose energy release supports the propagation of the shock. The detonation is accompanied by a chemical reaction that frees a large quantity of hot, high pressure gas. The action of the wave is controlled and determined by thermo hydrodynamic equations.

detonation code, a computer program which use known chemical properties of the explosive ingredients to estimate the energy released by an explosive. The three basic parts of such a program are: list of ingredients, list of products and an equation of state. Hydrocodes vary from the empirical to the fundamental, depending largely on the quality of the equation of state. The best codes will have special equations of state for solid phase as well as for the gaseous phase. Energies calculated with different hydrocodes may be quite different, not only because of different equations of state, but also because of different conventions for the definitions of explosive energy. Some names of the detonation codes are listed in Table 16.

Table 16 Ideal detonation codes. After Sarracino, 1993.

Area of use	Empirical	Semi-empirical	Fundamental
Military	Tiger-BKW	Tiger-JCZ3	
Commercial	Explode blend	IDEX-JCZ3	IDEX-IMP

Other detonation codes like NITRODYNE can be used to model the constant volume explosion performance.

detonation energy (Q_e), (MJ/kg), see '**heat of explosion**'.
detonation heat (Q_e), (MJ/kg), see '**heat of explosion**'.
detonation impact, increase of density in front of the detonation zone in a detonating explosive and the impact of detonation products on the surrounding material.
detonation pressure (p_d), **(MPa)**, the pressure associated with the reaction zone of a detonating explosive. It is measured in the C-J plane, behind the detonation front, during propagation through an explosive column. This pressure can be estimated by the following approximation formula;

$$p_d = \frac{1}{2}\rho_e c_d^2 10^{-6} \tag{D.6}$$

where p_d is the detonation pressure in (MPa), ρ_e is the density of explosive in (kg/m³) and c_d is the velocity of detonation in (m/s). For better approximation the following formula is recommended by Atlas Powder, 1987;

$$p_d = \frac{\rho_e c_d^2}{\gamma + 1} 10^{-6} \tag{D.7}$$

where $\gamma = -\mathrm{d}\ln p/\mathrm{d}\ln V$, where p is the pressure and V the volume of the blast gases. A common value used for γ is 3,3. The detonation pressures generated by commercial explosives typically range between 2 000 and 12 000 MPa.

detonation sensitivity, a property of an explosive characterized by the minimum level of the initial impact needed to detonate the explosive. DDR, 1978.
detonation stability, properties of an explosive, under defined conditions, to keep the detonation in process.
detonation stability test, includes the measurement of both *velocity of detonation* and *critical diameter* of an explosive. The detonation stability is always measured during research and

development work and also at production control. Detonation stability tests are also used to control the ageing of an explosive. Persson et al., 1994.

detonation theory in rock blasting, the theory is based on the calculation of the expansion of the reaction products of an explosive during the chemical reaction of the explosive. It is the main feature of any method for deriving a value of the rock blasting performance of an explosive. The first attempt of such treatment was made by Wood and Kirkwood, 1954, followed by Kirby and Leiper, 1995, Leiper and Cooper, 1990 and finally further refinements were made by Kennedy and Jones, 1993. Persson et al., 1994.

detonation velocity, (c_d), (m/s), see '**velocity of detonation**'.

detonation wave, the shock wave set up by a detonation. AS 2187.1, 1996. See also '**detonation**'.

detonation zone, chemical active zone with high density and pressure behind the shock wave in the explosive.

detonate, see '**detonation**'.

detonation driving zone (DDZ), the zone after the detonation front zone which is defined by the Chapman-Jouget plane.

detonator, a blasting cap (elemented cap) assembled with safety fuse, detonating cord, shock tube, detonating gas in plastic tubes or electric wires for initiation of the cap. The initiation of the explosive can be achieved directly by the methods mentioned or via a booster. The unit, detonator and booster together, is called a primer. A booster is used when the charge cannot be initiated by a detonator only. The amount of explosive in a detonator is about 1 gram. Seismic detonators have a high degree of uniformity of firing and a reaction time usually <1 ms.

detonator factor (t), (t/detonator), amount of ton broken per detonator in a blast. Used in India.

detonator plain, see '**plain detonator**'.

detonator test, see '**blasting cap test**'.

detonator, transformer coupled, an electric detonator in which the electric initiation impulse is of high frequency. The detonator is isolated from the blasting machine by a transformer link. Leg wires from each detonator end in a coil. The lead wire is pushed through a number of these coils and initiation is started by a current passing through the lead wire. Morrey, 1982.

development, the work of driving openings to and in a proved orebody, or a coal seam, to prepare it for systematic mineral production. Gregory, 1983.

development blast, see '**development**'.

deviator stress/strain (σ_d/ε_d), (MPa), the stress/strain tensor obtained by subtracting the mean of the normal stress/strain components of a stress/strain tensor from each of the normal stress/strain components. ISRM, 1975. The difference between the major and minor principal stresses in a triaxial test. ASCE P1826, 1958.

dewatering of blastholes, the process of dewatering boreholes before charging with the help of compressed air operated pumps or electrical submergible impellent pumps.

DFEM (acronym), for '**dynamic finite element method**'.

DFPA (acronym), '**dynamic fracture process analysis**'.

DI (acronym), see '**diggability index**'.

dial, usually a circular disc having graduations on which a pointer moves to indicate measurement value.

diameter (d), (mm), a line segment which passes through the centre of a circle, and whose end points lie on the circle and the length of such a line. Lapedes, 1978.

diameter of apparent crater (d_{app}), (m), see '**crater blasting**'.
diameter of blasthole or **borehole** (d), (m or mm), see '**borehole diameter**'.
diameter of cartridge (d_c), (m or mm), see '**cartridge**'.
diameter of explosive charge (d_e), (m or mm), the mean width of the cross-section of the cartridge if it is not constant.
diameter of lip of crater (d_{lip}), (m), see '**crater blasting**'.
diameter of plastic zone at crater blasting (d_{pz}), (m), see '**crater blasting**'.
diameter of relief hole in a cut (d_r), (mm), see '**relief hole**'.
diameter of rupture zone (d_{rup}), (m), see '**crater blasting**'.
diameter of shaft, winze or **raise** (d_s), (m), see '**shaft**', '**winze**' or '**raise**'.
diameter of true crater (d_{tc}), (m), see '**crater blasting**'.
diamond bit, a rotary drilling bit studded with bort-type diamonds. A.G.I. 1960. Also called *boartbit, boart-set bit, bort bit, bort-set bit, bortz bot and bortz-set bit.* Long, 1920. The shape of the bits can be concave-, convex-, coring-, and impregnated bit. Fraenkel, 1952.
diamond core bit, drill bit encrusted with diamond material mostly used for producing core samples. Atlas Copco, 2006.
diamond cut, see '**pyramid cut**'.
diamond drill, machine used for diamond drilling.
diamond drill bit, a drill bit with diamonds as cutting material.
diamond drilling, The act or process of drilling boreholes using bits inset with diamonds as the rock-cutting tool. The bits are rotated by various types and sizes of mechanisms driven by steam, internal combustion, hydraulic, compressed air, or electric engines or motors. Long, 1920. A drilling method for *exploration purposes* where cylindrical rock cores are taken from the rock mass with the help of a hollow bit covered with diamonds.
diamond drilling machine, a machine for prospect drilling of long holes with small diamonds bits and to catch cores from the orebody and the bedrock. The machine is normally driven by electricity and consists of a *coring bit, reaming shell, core barrel, core barrel head, sediment tube, drill rod string, casing tubes, drive pipes, drill head water swivel, hoisting line, derrick sheave, derrick, hoist, prime mover for drill, pressure hose, flush pump, prime mover for pump, suction hose and drilling fluid tank.* Fraenkel, 1952 and Rustan, 2006.
diamond grinding cups for button bits, there are two types; *spherical-* and *ballistical cups*. Atlas Copco, 2006 b.
diamond grinding wheels, convex wheels used to grind button bits. Atlas Copco, 2006.
diamond indentation test, see '**hardness test**'.
diaphragm pump, pump that moves liquid by alternating air pressure applied to either side of a diaphragm. Atlas Copco, 2006.
DIAT (acronym), for digital image analysis technique see '**digital processing of images**'.
diatomaceous earth, **diatomite** or **kieselguhr**, an accumulation of fossils, usually diatoms, with some radiolaria and a smaller amount of foraminifera. Diatomite is essentially an amorphous, hydrated, or opaline silica with various contaminants, such as silica sand, clay minerals, iron, alkalis and alkaline earths. BuMines Bull. 630, 1965. Kieselguhr is used to absorb nitroglycerin in dynamite to make the nitroglycerin less sensitive to impacts. Gregory, 1980.
diatomate, see '**diatomaceous earth**'.

DIBS (acronym), for '**discrete interacting block system**', a discrete element code developed at Lawrence Livermore Laboratory. The code is capable to simulate crater blasting in hard rocks with illustration of the shear surface development and the final breakage and throw of the blocks. Hudson, 1993.

diesel hydraulic drilling, in remote areas in mines it may not be possible to install cabling for hydraulic drifters and in these cases a diesel engine coupled to an electric generator can produce the necessary electric energy for hydraulic drilling. Tamrock, 1983.

differential strain analysis ($\Delta\varepsilon$), (DSA), the measurement of strain of a material with high precision, ($\pm 2 \times 10^{-6}$), versus hydrostatic pressure supplemented with observations using the scanning electron microscope (SEM). The effects of hydrostatic pressure cycling and uniaxial stress cycling are also examined. The difference between the strain and a reference like *fused silica* is calculated and plotted in a pressure-strain diagram. The method is used to characterize *strain induced cracks* (SIC) in the rock.

differential topographic isovibration mapping, the use of hundred or more miniature seismographs randomly placed around a surface blast. After measuring the vibration the isovibration curves can be calculated. Also the difference in vibration between two blasts can be calculated. Berger, 1989.

differential volumetric variability (due to thermal stresses), if minerals are heated they normally expand different in different directions. It give rise to intra crystalline defects and significant textural discontinuities such as intergranular cracks, which modify the mechanical properties of intact rock. Houpert, 1979.

diffraction, the process that allows sound waves to bend around obstacles that are in their path. H&G 1965. When seismic waves strike the corner of end of a reflecting or refracting surface the corner will in itself serve as a point source for radiating waves back to the surface. This radiation is known as diffraction. Dobrin, 1960.

diggabiltiy index (DI), The diggability can be quantified by using microprocessor-based performance monitoring of loading equipment to calculate a diggability index (DI) based on motor torques, velocities and digging trajectories. One can thus derive a DI which relates to fragmentation size and looseness. No international standard is available to quantify diggability. A diggability index (Throw × Swell index/(height of muckpile × fragment size (80% passing)) was proposed by Mohanty, 1991.

diggability of rock, the diggability of rock is a measure of the digging capacity. The diggability of rock depends on several parameters: type of digging machine, the skill of the operator, the swelling of the rock, the size distribution of the rock, the shape of the rock and the height of the muckpile.

digging arm loader, a loading machine for rock with the working principle two gathering arms feeding a conveyor that dumps the material into mine cars. Also called gathering arm loader. Atlas Copco, 1982.

digging depth (x), (m), the optimum digging depth for loading of caved ore and waste in crosscuts at sublevel caving can be determined by the *Rankine earth pressure theory*.

$$x = H_d \left[\cot \delta_d - \tan(\frac{90° - \delta_d}{2}) \right] \quad (D.8)$$

x is the optimum digging depth in m, H_d is the height of the cross cut in m, δ_{dump} is the slope of caved rock flowing out into the drift in degrees (*angle of deposition*). The digging depths used in practice are usually less than those given by the equation above. Janelid, 1966.

digging resistance, the resistance of a material against digging. This resistance is largely controlled by the hardness, coarseness, friction, adhesion, cohesion, and mass of the material. Nichols, 1956.

digital image analysis software, is using computer program to facilitate the determination of geometric parameters of the fragmented rock mass. Example of systems developed are CIAS, Goldsize and Split in USA, WipFrag in Canada, Power Sieve Australia, IPAC and KTH in Sweden, Fragscan in France, Tupics in Germany and Fragalyst in India.

digital image analysis technique (DIAT), see '**digital processing of images**'.

digital processing of images, the use of computer technology to convert photographs, video or TV-pictures from an analogue state to a binary state for further processing and evaluation of the fragment size and shape. The developed procedure uses the following stages; 1) Image input, 2) Scaling, 3) Image enhancement, 4) Image segmentation 5) Binary image manipulation, 6) Measurement and 7) Stereo metric interpretation. Franklin and Katsabanis, 1996. The geometric aspects of images can be determined in two or three dimensions such as area, number of fragments, perimeter shape, size and orientation.

dike, see '**dyke**'.

dilatancy, property of volume increase of a material under loading. ISRM, 1975. The property of granular masses of expanding in bulk with change of shape. It is caused by the increase of space between the individually rigid particles as they change their relative positions. Fay, 1920. The quantification of dilatancy see '**dilation**'.

dilatation or **volumetric strain (e), (dimensionless)**, the ratio of the change in volume to the original volume of an element of material under stress.

dilatational-, **longitudinal-**, **compression-** or **primary (P-wave) velocity (c_p), (m/s)**, the velocity of the dilatational-, longitudinal-, compression- or primary (P-wave) wave caused by a mechanical vibration of the ground, see also '**seismic waves**'.

dilation, 1) term used in image analysis for the method of increasing the size of the fragments after erosion until the fragments just overlap. 2) deformation that is a change in volume, but not in shape. Synonym for dilatation. Billings, 1954.

dilatometer, an instrument for measuring the expansion or contraction in a metal resulting from changes in such factors as *temperature, stresses* or *allotropy* (phase changes). ASM Gloss, 1961. The dilatometer can be a 1,8 m long cylindrical borehole probe with a radially expandable dilatable membrane. It operates in 76 mm boreholes and the maximum working pressure is about 30 MPa. It is used to measure the short term deformability of the in situ rock. Liu, 1994.

dilution, contamination or mixing of worthless materials with the more valuable minerals or ore.

dimensional analysis, a technique that involves the study of dimensions of physical quantities, used primarily as a tool for obtaining information about physical systems too complicated for full mathematical solutions to be feasible. Lapedes, 1978.

dimensional stone blasting (*ornamental or building stone blasting*), the controlled blasting of stone blocks for buildings or ornamental purposes.

DIN (acronym), for '**Deutsche Industrie Norm**'.

DIN 4150, German ground vibration standard for community housing.

dint, to cut into the floor of a roadway to obtain more headroom. Fraenkel, 1952. Enlarging a drift or stope by drilling and blasting some holes parallel to the free surface.

dinter or **bottom cutter**, a coal cutter for making floor cuts. Nelson, 1965.

diode method, a method for the measurement of the velocity of detonation (VOD) of an explosive. The measurement principle is based on the insertion of a series of connected diodes in the explosive at known distances and when the detonation front passes the diodes the voltage of the series of reverse biased diodes decreases as the ionisation in the detonation front passes. The measured voltage changes are logged with time and the velocity is estimated from the slope of the distance versus time curve.

dip, (α), (°), the angle at which the strata, beds, or veins are inclined from the horizontal plane or angle formed by blasthole inclination with the vertical. Regarding the selection of mining methods, the ore bodies are divided into shallow (<15° dip), intermediate (15°–45° dip) and steep (>45° to 90° dip).

dipper, a digging bucket rigidly attached to a stick or arm on an excavating machine. Also the machine itself. USBM, 1968.

directed hydraulic impulse fracturing, two high pressure water jets (150 MPa) launched from a nozzle which is inserted into a borehole are used to create two opposite slots. The borehole and the slot are filled with water. Then, another nozzle is inserted into the hole with a membrane which can withstand pressures from 25, 50, 75 or 100 MPa. The tube to the nozzle is tightened at the collar of the borehole. The pressure active on the membrane is increased by the high pressure water jet pump until the membrane suddenly breaks and an impulse pressure is delivered to the water in the borehole and the slot. The force needed to break the rock is hereby reduced, as the area over which the pressure is acting has been increased by the slotting. Hammelmann, 1995.

directional blasting or **cast blasting (coal)**, the use of the kinetic energy of blasting to move the rock larger distances. The method is used both in mining and civil engineering. The ballistics of directional blasting has been described in detail by Chernigovskii, 1976.

directional drilling, techniques to drill more precisely towards the intended target.

direction histogram or orientation histogram, a diagram illustrating e.g. the number of cracks in different directions in intervals of e.g. 5°. Commonly used in micro fractography.

direct priming, see '**priming**'.

disc, general term used to describe a flat circular part such as found in disc brakes. Atlas Copco, 2006.

disc bit, synonym with '**disc cutter**', see '**disc cutter**'.

disc break, device usually mounted on the wheel of a vehicle whereby the deceleration and stopping result from a caliper exerting a frictional force on a disc. Atlas Copco, 2006.

disc crusher, a crusher using to parallel axes with several rotating discs on each. This crushing principle can only be used for low strength rock. LKAB, 1964.

disc cutter, a disc usually made of hardened tough steel mounted on an axle and sometimes with hard metal buttons around the periphery. Several discs are pushed against the rock face while rolling on the rock surface and breaking hand-size pieces of rock. The technique is used in full face boring of tunnels.

disc fracture, generally a planar disc-shaped crack. A disc crack can be created perpendicular to the borehole axis at the bottom of a blasthole in drifting or tunnelling by a special designed shaped charge. Occurrence of this type of crack at the bottom of a blasthole may reduce the confinement. Bjarnholt, 1987.

discharge/diverter head, device used to divert the cuttings ejected from a hole when drilling using the Odex method. Atlas Copco, 2006.

discontinuity, an interruption in the physical state of a structure or configuration, such as induced by joints, cracks, bedding planes, faults, laps, seams, inclusions, or porosity's.

A discontinuity may or may not affect the usefulness of a part. *A discontinuity has zero or relatively low tensile strength*. To a large extent, the discontinuous character of rock has a significant effect on cohesion and transfer of stress across the rock interfaces, and hence on its strength and deformational behaviour. Brown, 1981. A discontinuity is classified according to its *orientation, spacing, persistence, roughness, wall strength, aperture, filling, seepage, number of sets and block size*. Hudson, 1993.

discontinuity frequency (f_d), **(No./m)**, the number of discontinuities crossing 1 m of a straight line. The discontinuity frequency f_d (No./m) follows often a negative exponential function and e. g. for sedimentary rock masses in UK the following distribution was found,

$$f_d = f_{dm} e^{-f_{dm} x} \tag{D.9}$$

where $f_{dm} \approx 1/x_m$ is the mean discontinuity frequency in No./m of a large discontinuity population and x_m is the mean spacing in m and x is the discontinuity spacing in m. Priest, 1976. See also '**discontinuity spacing**'.

discontinuity spacing (S_{dis}), **(mm)**, the distance between adjacent discontinuities in the rock mass and is usually expressed as the mean spacing of a particular set of joints. Brady, 1985. A classification scheme is given in Table 17.

Table 17 Classification of discontinuity spacing. After Brady and Brown, (1993).

Description	Spacing S_{dis} (mm)
Extremely close spacing	<20
Very close spacing	20–60
Close spacing	60–200
Moderate spacing	200–600
Wide spacing	600–2000
Very wide spacing	2000–6000
Extremely wide spacing	>6000

Source: ISRM, 1981.

discontinuity set, the discontinuities can be grouped together in certain directions in space and such a set is called a discontinuity set. Table 18 shows how many discontinuities sets that can be found in different kind of rocks.

Table 18 Classification of discontinuities sets.

Class	Description
1	Massive, occasional random joints
2	One joint set
3	One joint set plus random joints
4	Two joint sets
5	Two joint sets plus random joints
6	Three joint sets
7	Three joint sets plus random joints
8	Four or more joint sets
9	Crushed rock, earth-like

Source: ISRM, 1981.

discontinuous deformation analysis (DDA), is a displacement-based method for modelling the dynamics, kinematics and deformation of an aggregation of blocks. Shi G-H, 1990.

discontinuous subsidence, the downwards movement of the surface above a mined area producing a surface profile that is not smooth or even but contains breaks or steps involving vertical or sub-vertical surfaces.

discrete diode method, see '**diode method**'.

discrete electrical make/breaks targets, see '**electrical make/break targets**'.

discrete element block modelling (DEBM), see '**discrete element method**'.

discrete element method (DEM), numerical technique where the continuous or discontinuous rock mass is modelled by distinct elements. A method that can be used to model movements of individual blocks in blasting and gravity flow. A modelling method where the rock mass is divided into separate blocks and the forces from the explosive can be calculated on each block. The throw of the material and e.g. the crater developed in crater blasting can be determined but not the fragmentation of the individual blocks. Heuzé, 1990. There are *four classes of discrete element methods* and the *limit equilibrium method*, see Table 19.

Table 19 Attributes of the four classes of discrete element method and the limit equilibrium method. After Cundall 1993.

Property	Contact Property	Class 1 Distinct element method	Class 2 Modal method	Class 3 Discontinuous deformation analysis	Class 4 Momentum exchange method	Limit equilibrium, method
Contacts	Rigid	–	–	xxx	xxx	xxx
	Deformable	xxx	xxx	–	–	–
Bodies	Rigid	xxx	xxx	xxx	xxx	xxx
	Deformable	xxx	xxx	xxx	–	–
Strain	Small	xxx	xxx	xxx	–	–
	Large	xxx	xxx	xxx	–	–
No. of bodies	Few	xxx	xxx	xxx	xxx	xxx
	Many	xxx	xxx	x	xxx	x
Material	Linear	xxx	xxx	xxx	–	–
	Nonlinear	xxx	x	–	–	–
No fracture		xxx	xxx	xxx	–	–
Fracture		x	xxx	–	–	–
Packing	Loose	xxx	xxx	–	xxx	–
	Dense	xxx	x	xxx	x	xxx
Static		xxx	xxx	xxx	–	xxx
Dynamic		xxx	xxx	x	xxx	–
Forces only		–	–	–	–	xxx
Forces and displacements		xxx	xxx	xxx	xxx	–

The method has been used to model movements of individual blocks around an underground opening, drift or tunnel. In the case of drift, the resulting influence of thermal loads, internal tunnel pressures, pore pressure and fracture flow resulting from the hydraulic ground water pressure has been modelled. Hökmark, 1990.

It is a computer calculation method for a two-dimensional representation of a jointed rock mass, but has been extended to application in particle flow research, studies on microscopic mechanisms in granular material and crack development in rock and concrete. The most recent two-dimensional program UDEC was first develop in 1980 by ITASCA to combine, into one code, formulations to represent both rigid and deformable bodies (blocks) separated by discontinuities. This code can also perform, alternately, static and dynamic analysis. In 1983, work begun on the development of a three dimensional version of the method entitled 3DEC. This computer program has been used primarily to study rock bursting phenomena in deep underground mines.

discrete interacting block system (DIBS) model, is a two-dimensional polygonal particle model that was originally applied to the flow behaviour of granular solids and contained physics very similar to that of earlier discrete element models. Basically the model traces the motion of each individual particle (or element) in a system of many, as it interacts with other particles and boundaries, under applied loads and gravity. Several simplifications are made concerning the interactions between elements and the properties of the elements themselves, in order to calculate efficiently the forces and motions of large numbers of distinct blocks. Simulations of crater blasting have been made using this technique. See also '**discrete element method** or **discrete element modelling**'. Hudson, 1993.

discrete optical detector method, see '**optical detector method**'.

discrete resistor method, see '**resistor method**'.

disc-shaped compact test DC(T), see '**fracture toughness test**'.

disc spring, slightly conical perforated disc like spring. Atlas Copco, 2006.

disintegrated, a condition of rock in which it is weathered to a soil, where the original material fabric is still intact. The rock is friable, but the mineral grains are not decomposed. Compare with 'decomposed'.

disk bit, synonym for '*disc bit*', see '**disc cutter**'.

disk-shaped compact test DC(T), see '**fracture toughness test**'.

dislocation, the displacement of rocks on opposite sides of fracture. Pryor, 1963. Also a general term to describe the break in strata, for example, a fault. Nelson, 1965.

dispersion, dependency of wave velocity on wave frequency.

dispersiveness (L_{dis}), (dimensionless), a shape factor used in micro fractography defined as $P^2/4\pi A$ where P is the perimeter of the grain and A the area. Montolo, 1982. General, see also '**shape factor**'.

displacement (u), (mm or m), the linear distance from the initial to the final position of an body or a particle in a solid, regardless of the length of path followed.

disruptive strength (σ_{thyd}), (MPa), the failure stress under hydrostatic tension. ASM Gloss, 1961. Hydrostatic tension is defined as three equal and mutually perpendicular tensile stresses. ASM Gloss, 1961.

disseminated ore, ore in which the valuable mineral is fairly evenly distributed through the gangue as crystals or aggregates of regular size. Pryor, 1965.

dissolution, the taking up of a substance by a liquid with the formation of a homogeneous solution. C.T.D., 1958.

distance between levels underground (H_l), **(m)**, the vertical distance between two underground levels.

distance between main levels (H_{lm}), **(m)**, the vertical distance between main levels defined as the levels where ore is transported by train, truck, conveyor belt or loader to the main hoisting shaft. With the development of technology this distance has increased from 50 m in the beginning of the 20-century to ~250 m in the beginning of the 21-century.

distance of throw (*R*), **(m)**, distance of the dynamic displacement of a body or part thereof. In blasting the length of throw has three meanings: 1) the *maximum throw* of one *single fragment*. 2) the *maximum throw* of the *main part* of the round (where the muckpile ends) and 3) the distance of *movement of the centre point of gravity* of a rock mass from its origin in the bench to the centre of the muck pile formed after blasting. All distances are measured along the horizontal plane.

distinct element method (DEM) (synonym), see '**discrete element method**'.

distortion, a change in shape of a solid body due to stress.

distortional wave, **transverse wave**, **shear wave** or **S-wave**, see '**seismic waves**', use **S-wave**.

disturbed zone, zone around an excavation in rock characterized by stress redistribution (stresses have changed their magnitude or direction due to the excavation).

ditch blasting, the formation of a ditch in soft (wet) earth by the detonation of a series of explosive charges. IME, 1981.

ditching dynamite, a dynamite which is characterized by a large gap distance (propagation distance) from one charge to another, still keeping the detonation travelling through the line of cartridges and air gaps in ditch blasting. Usually a nitroglycerin based dynamite is used.

DNT (acronym), for '**ditnitrotoluen**' see '**phlegmatization**'.

dobie, see '**mud cap**'.

dome, a special type of anticlinal pericline where the dip is radial, i.e. in plan (horizontal) view, and the structure is close to circular. The synclinal counterpart of a dome is termed basin.

domed face plate, see '**face plate**'.

domed reamer, the reaming head has a half spherical shape on which the individual disc cutters are fastened.

dome reaming bit, button bit with convex front (dome shaped). Atlas Copco, 2006 b.

donkey, see '**barney**'.

donor, an exploding charge producing an impulse that impinges upon an explosive 'acceptor' charge. IME, 1981.

door-stopper cell, an instrument to measure rock stresses in boreholes.

dope, a mixture of fuels and oxidizers and other non-explosive ingredients. IME, 1981. Individual dry, non-explosive ingredients that comprise a portion of an explosive formulation. Atlas Powder, 1987.

doped emulsion explosive, see '**dope**'.

do's and don'ts, a list of safety precautions (IME Safety Library Publication No. 4) printed by the Institute of Makers of Explosives pertaining to the transportation, storage, handling, and use of explosive materials, and inserted in cases of explosive materials and cartons of detonators. Atlas Powder, 1987.

DOT (abbreviation), for USA '**Department of Transportation**'.

double check valve, valve that prevents flow in both pressure and return hydraulic oil lines if a loss of pressure occurs such as when a hose ruptures. Atlas Copco, 2006.

double damping system, a system in rock drill machines to damp the energy in the recoil. Rock drill machines can have one or two dampers. The double damping system in combination with the '*rig control system*' transmits maximum power through the drill string. The long and slender shaped piston in the rock drill machine, which is matched to the drill steel, permits high impact energy at high frequencies resulting in long service life of drilling consumables and efficient drilling. The drill machine is fully adjustable for various rock conditions. Atlas Copco, 2007.

double impeller breaker, see '**impact breaker**'.

double load or **double charge**, a charge in a borehole separated by a quantity of inert material (drill cuttings, clay, sand) for the purpose of distributing the blast effect, or for preventing part of the charge blowing out at a seam or fissure, in which case the inert material is placed so as to include the seam. Fay, 1920.V

double packer, see '**packer**'.

double packing, a form of strip packing which removes the localized high roof pressure from the vicinity of a roadway into a region in the goaf. It consists of two parallel packs adjacent to, and on each side of, the roadway, with the packs immediately at the roadsides built of such a width as to offer less resistance than wider and stronger packs (called buttress packs) more remote from the roadway. The principle of double packing was developed by Dr. D.W. Philips in Great Britain. Nelson, 1965.

double pipe test, a steel pipe is filled with explosive, and placed on top of a witness pipe, which in turn sits on a heavy steel anvil. The witness pipe is a small-diameter, thick-walled pipe whose deformation is used to determine the energy output of the explosive. The steel anvil provides a solid fixed base for the deformation action of the explosive to work against. The deformation of the pipe is a function of the distance along the pipe, with higher values of deformation representing greater energy release. Lownds and du Plessis, 1984.

double priming, initiation technique where the individual charge contains two primers, usually one of them placed near the top and the other near the bottom of the blasthole, and often initiated simultaneously. Safety reasons call for double priming, to make sure that at least one of the primers is initiated if one detonator fails, or in case of partial displacement of the explosive column the charge is divided into two parts. Large diameter holes underground (more than 100 mm in diameter) often have two primers. In drifting top priming is prohibited.

double spiral cut, see '**Coromant cut**'.

double thread or **multiple thread**, an elongated thread on drill rods and sometimes used on the bit end of shank rods. The thread bottoms in the drill bit. Rods with double thread are always carburized. Atlas Copco, 1982. The purpose of double thread is to increase the life length of the thread.

double V-cut, cut in tunnelling where the boreholes are drilled at an angle to each other like in a V-form and at different heights. 'Double' means two adjacent V's after each other. See '**V-cut**. Fig 24.

dowel, cylindrical pin used to ensure the correct position of a part. Atlas Copco, 2006.

dowel (a rock bolt not pre-tensioned), may be of similar materials to rock bolts and is grouted along its full length. Materials used are reinforced bars, wood, fibre glass or may take the form of a cable. Recent developments in dowel systems include the proprietary friction anchor or "**split-set**" system developed by Scott, 1977, and the *"Swellex system"* developed by the Atlas Copco Co. Brady, 1985.

dowel holder, device used to retain a dowel in position. Atlas Copco, 2006.

dowel pin, pin used to align mating parts. Atlas Copco, 2006.

downcast shaft, a shaft used for transferring fresh air from the surface to the underground workings.

down hole drill, drilling machine used for down the hole drilling where the drilling hammer (machine) is working inside the hole and the rotational force of the drill bit is caused from surface. Rustan, 2006.

down hole drill pipe, drill tube used in down-the-hole drilling.

down hole equipment, see '**drill fittings**'.

down hole inspection, see '**borehole inspection methods**'.

downline, a line of detonating cord or non-electric plastic tubing in a blasthole transmitting the detonation from the trunk line or surface delay system down the hole to the primer. The primer is attached to the bottom end and additional primers may be slid down the cord in the case of decking and/or multiple priming.

downstream process, all handling and processing of a product after a certain operation in a process.

down-the-hole (shorten form), for '**down-the-hole-drilling**'.

down-the-hole bit (DTH –bit), a bit joined together with the hammer during drilling. Sandvik, 2007.

down-the-hole drill (DTH) or **in the hole drill (ITH)**, cylindrical pneumatic percussive rock drill fitted with a drill bit and attached directly to the bottom of the drill string so that it can be inserted into the hole as it drills. Atlas Copco, 2006. See also '**down-the-hole hammer**'.

down-the-hole drilling, process of drilling blast holes, water wells or energy wells, using a down-the-hole-drill. Atlas Copco, 2006.

down the hole equipment, see '**drill fittings**'.

down-the-hole hammer (DTH), a percussive or hammer drill where the bit-driving mechanisms (percussive and rotating) are located immediately behind the drill bit, and is small enough in diameter to permit it to enter and follow the bit down into the hole drilled. Cylindrical pneumatic percussive rock drill fitted with a drill bit and attached directly to the bottom of the drill string so that it can be inserted into the hole as it drills blast holes. Atlas Copco, 2006.

down-the-hole rock drill, see '**down-the-hole hammer**'.

drag bit or **bit drag**, rotary reamer shaped drill bit with cutting action and hard metal inserts along the edges. A non-coring or full-hole boring bit, which scrapes its way through the strata which must not be too hard. It may be a two-, three-, or four bladed pattern with various curves and cutaways. The drilling fluid passes down through the hollow drill stem to the cutting point. Nelson, 1965. The blade is flattened and divided, the divided ends curving away from the direction of rotation. It resembles a fishtail. A.G.I., 1957. The cutting part is made by cemented carbide. The bit is mounted on a bit adapter and can be combined with one or more reamers. Atlas Copco, 1982.

drag cut, (*a type of fan cut*), a tunnelling method where the initial cut is performed by taking a thin slice by downward directed and inclined blastholes along the whole width of the floor level, see Fig. 23. A blasting method employed in narrow underground openings. Sen, 1995.

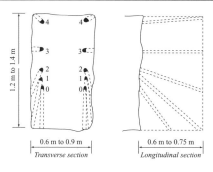

Figure 23 Drag cut. After Sen, 1995.

drag line bucket loader, see '**dragline excavator**'. Atlas Copco, 2006 b.

dragline excavator, a mechanical excavating appliance consisting of a steel scoop bucket which is suspended from a movable jib. After biting into the material to be excavated, the bucket is dragged towards the machine by means of a rope to fill it completely. The bucket is lifted and after a swing the material is dumped into a new heap. Commonly used in surface coal mining.

drag scraper bucket, a scraper bucket operated by ropes but without hinges. LKAB, 1964.

drain, orifice, tube or other opening at the bottom of a vessel used to empty the liquid from the vessel.

drainage, drawing off a liquid from a vessel or system. Atlas Copco, 2006. Also the removement of water from the surface.

drainage block, block designed as a conduit for drainage. Such as a manifold block in a hydraulic system designed to collect the drainage from several components and return it to the oil reservoir. Atlas Copco, 2006.

drainage cock, hand operated valve used to release a liquid or gas from an enclosed space. Atlas Copco, 2006.

drain pipe, pipe placed to provide an outlet for unwanted fluid. Atlas Copco, 2006.

draw, the process of extracting caved or broken ore by gravity forces from a caved area or stope.

draw angle or *limit angle* (α_{lim}), (°), see '**limit angle**'.

draw bell, the excavated channel used to transport broken ore to the draw point on the loading level. Normally the draw bell is constructed in ore, ideally having the shape of an inverted multilated cone.

draw body, term used in gravity flow to define the *shape of the volume inside the cave* at a certain volume extracted in sublevel caving or block caving. Its shape is like a drop turned upside down for a *homogenous coarse material*. A coarse and interlocked material the shape can deviate a lot from the drop shape. Algorithms have been derived from basic natural laws for the drop shape, see Bergmark-Roos formula. Rustan, 2000. The Bergmark-Roos formula was adjusted to fit drawing from wide drifts, see Kuchta, 2004.

draw column, see '**draw body**'.

draw cone, a concave cone turned upside down. This shape can be illustrated if a box is filled with black crushed rock up to half the height and the rest with white crushed rock and a hole is opened at the center of the bottom and the black material is started to be discharged the flow of the white material towards the bottom will form a trumpet or roughly a cone turned upside down. This is called the draw cone in sublevel caving and block caving.

draw control, the process of controlling the amounts of ore drawn from individual drawpoints in order to achieve a number of mining objectives.

draw control factor, an index, developed by Laubscher, varying between 0 and 1 which is a linear function of the coefficient of variance of tonnages mined between a draw point and its neighbours in a period of time of a week. MassMin, 2008,

draw height (h), (m), is defined as the maximum length of the drawbody projected on a vertical line. The drawbody may be vertical or inclined dependent e.g. on the inclination of the rings in sublevel caving. If front inclination to the horizontal is α degrees and the length of drawbody is l m the projected length of the drawbody to the vertical or the height of the drawbody $h = l \cdot \sin(\alpha - 3°)$. The deduction of 3° is due to the fact that the center axis of the drawbody is deviating from the front inclination by about 3°.

draw point, the excavated structure on the extraction or production level through which the caved or broken ore is loaded and removed from the cave. It could also be the bottom of a shaft below the caved area.

draw point drift, an opening or drift through which a drawpoint is accessed from a production drift on the extraction or production level.

draw point spacing, the spacing in the horizontal plan between the center of a draw point to the center of adjacent draw point.

draw rate (m_d), (t/day or t/shift), the rate at which caved or broken ore is drawn from individual draw points or a group of adjacent draw points.

draw zone, the surface area of the caved or broken material that will eventually report to a particular draw point during progressive draw. See also '**draw body**'.

DRC (acronym), see '**detonating relay connector**'.

DRI (acronym), for drilling rate index, see '**bit wear index** and **drillability**'.

drift or **drive**, a horizontal or slightly inclined (for drainage) underground passage. A drift follows the vein, as distinguished from a crosscut, which intersects it, and a level or gallery, which may do either.

drift and pillar, a system of mining coal similar to the room and pillar system. See '**room and pillar**'. Fay, 1920.

drift condition rating (DCR), a method to quantify blast-induced rock damage. The drift condition rating comprises two components 1) Drift roof condition related to the rock mass integrity and the percentage of half cast visible and 2) Amount of overbreak. The empirical rating varies from 0–9. Forsyth and Moss, 1990.

drift confinement (f_l), (m^{-1}), is defined as the blast hole depth divided by the drift area.

drifter, see '**hammer drill**'.

drifter drilling, drilling of boreholes in tunnels or drifts. Sandvik, 1982.

drifter or **drilling jumbo**, rock drill rig used for drilling holes in tunnelling, drifting and for winning or ore. The drilling action is normally based on the percussive drilling. The machine can be operated by pneumatics or hydraulics.

drifter rod, drill steel for drilling the *full depth of the borehole*. Sandvik, 2007. Robust hexagon shaped drill rods used for drilling blast holes in tunnels and mining drifts. Atlas Copco, 2006.

drifting, drilling, blasting and excavating rock to create transportation and access openings to ore bodies in an underground mining operation. Atlas Copco, 2006.

drifting drill string, arrangement of bit, rod and adapter used to drill holes when excavating horizontal transport openings or drifts in underground mining. Atlas Copco, 2006.

drift profile (A), (m^2), the area of the drift in a plane perpendicular to the length axis of the drift. TNC, 1979.

drift section (A), (m^2), the area of the drift in a plane perpendicular to the length axis of the drift. TNC, 1979.

drill, any cutting tool or form of apparatus using energy in any one of several forms to produce a circular hole in rock, metal, wood, or other material. See also '*calyx drill, churn drill, core drill, diamond drill, rock drill, rotary drill and shot drill*'. Long, 1920.

drillability, the resistance of rock or a mineral compound against drilling is controlled by the following engineering properties of rock; hardness, abrasiveness, texture, structure and breaking characteristics. It is function of rock properties such as mineral composition (quartz content is very important), grain size, various degree of weathering, etc. Several empirical methods have been developed for predicting drillability in different rocks; and some typical classifications schemes are: a) Drilling rate index DRI, see '**drilling rate index**', b) Mohr's hardness test scale and c) Protodyakonov classification according to the compressive strength of the material. Compare with '*bit wear index*'. Tamrock, 1989. See also '**drilling rate index**'.

drill abort, see '**drill diamonds**'.

drill and blast method, is the most common used method for drifting in mines. In tunnelling, full-face boring or roadheaders may also be used. The drill and blast method can be divided into 7 unit operations; 1) surveying and marking of drillholes, 2) drilling, 3) charging, 4) blasting, 5) ventilation, 6) loading, 7) scaling and 8) rock reinforcement.

drill and notching tool, a drill bit combined with a notching tool having two diametrically placed carbide steel cutters mounted on bearings and therefore the drill bit rotates independently to the notching tool. Young, 1983.

drill bit, a detachable cutting tool used to cut circular holes in rock, wood, metal, etc. It may be the part of an integral steel or the corresponding loose part in extension drilling. Also called *drill crown, drill bort* or *drilling bort*. Long, 1920. It is a drilling tool that transmits thrust and rotation from the drill machine, via the drill bit, to the rock. The drill bit shape is cylindrical. Normally, inserts or button bits are used in percussive drilling. The inserts and buttons are made by hard metal (tungsten carbide) but also diamonds can be used to give bits a long life length.

drill boom area coverage (A_b), (m^2), the area that can drilled by the drill boom.

drill bort, see '**drill bit**'.

drill carriage or **jumbo**, see '**drill jumbo**'.

drill collar or **drilling collar**, a length of *extra heavy wall drill rod or pipe* connected to a drill string directly above the core barrel or bit, the weight of which is used to impose the major part of the load required to make the bit cut properly. A drill collar is usually of nearly the same outside diameter as the bit or core barrel on which it is used. Not to be confused with guide rod. Long, 1920

drill core, a solid, cylindrical sample rock produced by an annular drill bit, generally driven by rotation but sometimes cut by percussive methods. See also '**core**'. Long, 1920.

drill cradle, the metal beam on which a heavy drill is fed forward as drilling proceeds. B.S. 3618, 1964.

drill crown or **drilling crown**, see '**drill bit**'.

drill cuttings, the particles formed by frictional, cutting and or impact forces on the drill bit at the hole bottom during drilling and usually ejected at the hole collar.

drill diamonds, industrial diamonds used in diamond-drill bits and reaming shells for *coring*, *cutting* or *reaming* rock. Drill diamonds usually contain obvious imperfections and inclusions, although the finer grades approach tool stones in quality. Also called *drill abort*.

drill dust, particles of rock produced when drilling boreholes.

drilled burden, addition of a formula in existing dictionary how to calculate drilled burden B_d from practical burden B_p if the angle of the blastholes is α from the vertical. $B_d = B_p / \cos\alpha$.

drilled burden (B_d), (m), the burden distance measured in the plane containing the collars of blastholes, between the front row blastholes and the face (assuming the face is undamaged) or between one row of blastholes and the next row. This distance is only equal to practical burden when the bench is vertical. With inclined holes, for geometrical reasons, the drilled burden is always larger than the practical burden.

drilled depth per bit (L), (m), the length of borehole drilled with a steel bit until it must be re-sharpened. Streefkerk, 1952.

drilled length per cubic meter of the estimated yield (b_s), (1/m² or t/m), see '**specific drilling**'.

drill extractor, tool for retrieving broken piece of drill from borehole. Pryor, 1963.

drill feed, the mechanism for advancing the drill bit during boring. Nelson, 1965.

drill fittings, devices, parts, and pieces of equipment used down hole in drilling a borehole. Also called *down the hole equipment*. Long, 1920.

drill gauge, the width across the cutting bit or diameter of the drilled hole. It is continuously reduced at boring due to wear. With tungsten-carbide bits it is possible to drill long holes without the loss of gauge.

drill guide, drilling equipment for the drilling of parallel holes in normally vertical and horizontal planes in a dimensional stone quarry. Blocks up to 4 000 m³ has been detached by this equipment. Smith, 1987.

drill head, **boring head**, or *swivel head*, the assembly which applies the drilling pressure and rotation to the drill rods. B.S. 3618, 1963.

drill hole, a hole drilled in the material for the purpose of containing an explosive charge to be blasted; also called blasthole or borehole.

drill hole deflection measurement, in certain formations, particularly in schistose rock, the drill hole frequently shows a tendency to deviate from the originally planned bearing. Several instruments allow the checking of the bearing of the hole. These instruments are enclosed in a watertight case and lowered into the drill hole on a wire rope or a rod string. The instruments are called clinometers and are based upon various principles.
1 **Glass tube** in which the horizontal level of a liquid is etched by an acid.
2 **Pendulum clinometer**, in which the weight of the pendulum at different points in the drill hole is measured by means of an electromagnet.
3 **Compass** mounted in gimbals, which can be clamped by a timed clockwork mechanism.
4 **Compass** mounted in gimbals, the position of which is recorded by a camera.
5 **Concentric rings with different diameters in a series** and placed after each other in flexible tube that is inserted into the drill hole and photographed at different depths.

The deviation from concentricity of the rings is proportional to the deflection of the drill hole. This method is called **Photobor**.
6 **Three accelerometers** mounted perpendicular to each other can measure the movement in space. Fraenkel, 1952 and Rustan, 2006.

drill hole diameter (d), (mm), see '**borehole diameter**'.

drill-hole pattern, the number, position depth and angle of the blast holes forming the complete round in the face of a tunnel, stope or open pit.

drilling, the process of making circular holes in the rock for the purpose of exploration of the rock mass, charging with explosives, drainage, cabling etc. The maximum drillhole diameter for explosive charging is 0,5 m in 1996. The drilling was in the beginning done by hand but later on replaced by drilling machines driven by Generation 1: air, Generation 2: hydraulic oil and Generation 3: water.

drilling accuracy total (D_T), (degrees, m or %), precision quantity normally defined as the maximum deviation from the target position (m) in drilling divided by the length of the drill hole (m), and given in (%). In practical work the standard deviation of the drilling accuracy is often given instead of the maximum drilling accuracy. The total borehole (drill) deviation is composed by four deviations; 1) *Setting-out deviation (D_s)*, 2) *Collaring deviation (D_c)*, 3) *Alignment deviation (D_a)* and 4) *Bending or in-the-hole deviation (D_b)*.

drilling bort, see '**drill diamonds**' and '**drill bit**'.

drilling collar, see '**drill collar**'.

drilling column, the column of drill rods. At the end of the drilling column the bit is attached. B.S. 3618, 1963.

drilling deviation total, (D_T), (°), total deviation of the axis and/or slope of the borehole after drilling from the specified geotechnical layout of the hole. See also '**drilling accuracy**'.

drilling diamonds, see '**drill diamonds**'.

drilling error, deviation from the planned blasthole pattern.

drilling fluid, the thick fluid kept circulating in a borehole to clear the chippings and cool the chisel, etc. See also '**mud flush**'. Nelson, 1965.

drilling history (brief),
- 4000 year ago wells were drilled in China.
- 1126 first well drilled in Europe at the monastery Artois.
- 1635 the first use of gun powder for blasting at Nasafjäll in Sweden.
- 1700-century drilling of holes for blasting and using gun powder.
- 1800-century beginning churn drilling in Skåne, Sweden.
- 1860 churn drilling in south Sweden.
- 1863 first trial with machine drilling in Sweden at Persberg mine.
- 1864 core drilling for the Mont Cenis tunnel in Switzerland.
- 1870 rotational drilling in Sweden for wells.
- 1910 turn over from hand drilling to machine drilling in Sweden (Hand drilling 2 man, 0,6 m/hour).
- 1940's one man and one machine. (1 machine, drill rate 5 m/hour).
- 1950 (1 pneumatic drill machine, drill rate 10 m/hour).
- 1955 (1 pneumatic drill machine drill rate, 18 m/hour).
- 1963 (1 rig 2 pneumatic drill machines, drill rate 40 m/hour).
- 1970 (1 rig 2 pneumatic drill machines, drill rate 70 m/hour).

- 1970's hydraulic top hammer drilling is being introduced.
- 1990 (1 rig 2 hydraulic machines, drill rate 90 m/hour).
- 1990's Coprod system.

drilling jumbo, rock drill rig used for drilling holes in tunnelling, drifting and for winning of ore. The drilling action normally based on the percussive drilling. The machine can be operated by pneumatics or hydraulics.

drilling log or **drill log**, the record of the events and the type and characteristics to the formations penetrated in drilling a borehole. Also called *boring log*. Long, 1920.

drilling machine, a machine that can produce either rotation in *rotary drilling* or impact in *percussion drilling* and finally rotation and impact in *percussive drilling*.

drilling methods, can be divided into four main groups 1) *Percussion drilling*, 2) *Percussive or hammer drilling*, 3) *Rotary drilling* and 4) *Thermal Piercing*. These main methods can be divided into the following sub-methods.

   1 **Percussion drilling**
      Churn drilling
      a Cable churn drilling
      b Churn drilling (pneumatic, electric, etc.)
   2 **Percussive drilling** or **hammer drilling**
      a Pneumatic hammer drilling or rotary percussive drilling
      b Motor hammer drilling
   3 **Rotary drilling**
      a Drilling with cutting tools
      b Abrasive drilling
      c Drilling by crushing (with roller bits)
   4 **Thermal piercing**
      a Fusion piercing
      b Jet piercing

drilling mud or **drill mud**, a suspension, generally aqueous, used in rotary drilling and pumped down though the drill pipe to seal off porous zones and to counterbalance the pressure of oil and gas; consist of various substances in a finely divided state among which bentonite and barite are most common. Oil may be used as a base of water. A.G.I., 1957 and 1960.

drilling pattern, geometric plan for placing of blastholes in a blast in relation to the free face, if any. It also includes their length, diameter, direction and number of drill rods.

drilling platform, movable platform with one or two decks from which drilling and charging can take place. Used for driving headings. Fraenkel, 1952.

(*drilling pressure*) or **drilling thrust** (F_{bi}), **(N)**, the pressure applied to a bit in drilling operation expressed as the number of tons applied. Also called **bit load**, **bit pressure**, *bit weight*, *drilling weight*, **drill pressure**. Long, 1920.

drilling rate (v_d), **(m/min)**, the depth of penetration achieved per unit of time with a given type of rock, drill, bit diameter, and air or water pressure, etc.

drilling rate index (DRI), the drilling rate index is a relative measure of the penetration rate of the drill bit and it is therefore not an absolute value of the drilling rate in the field. The DRI is determined, see Fig. 24, on the basis of two parameters, the brittleness (friability) value S_{20} see '**brittleness value**' and the Siewers *J*-value (*SJ* value) see '**Siewers J-value**'.

The following classes for drillability have been suggested, see Table 20.

Some typical drilling rate indices are shown in Fig. 25. Lien, 1961.

drilling rate index

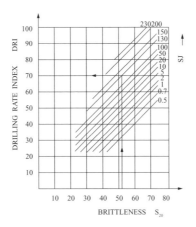

Figure 24 Determination of drilling rate index (relative drillability) from the brittleness S_{20} value (impact wear) and the *SJ* value (sliding wear). After Tamrock, 1989.

Table 20 Drillability expressed by drilling rate index. After Tamrock, 1989.

Description	Drilling rate index (DRI)
Extremely low drillability	21
Very low drillability	28
Low drillability	37
Medium drillability	49
High drillability	65
Very high drillability	86
Extremely high drillability	114

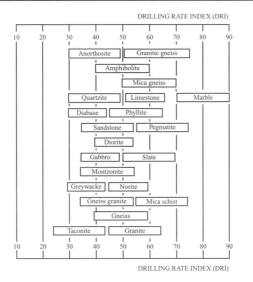

Figure 25 Drilling rate index. After Tamrock, 1989.

drilling rate measured (DRM), (v_d), (m/min), drilling rate measured in the field.
drilling rig, see '**drill rig**'.
drillings, see '**drill diamonds**'.
drilling site, the area where drilling is taking place. Sandvik, 2007.
drilling speed (v_d), **(m/min)**, see '**drilling rate**'. Bickel, 1982.
drilling thrust (F_{bi}), (N), force on the drill bit at drilling.
drilling weight or better **bit load (F_{bi}), (N)**, the force applied to a bit in drilling operation expressed as the number of tons applied. Also called *bit pressure, bit weight, drilling pressure, drill pressure*. Long, 1920.
drilling wireline hole sizes A-N, process of drilling holes with a core bit of sizes A to N using a rod string and core barrel that allows the cores to be recovered by running a tool attached to a wire line down the inside of the rods and pulling the inner barrel to surface. Atlas Copco, 2006.
drill jumbo, a drilling platform which is moved without being dismounted. Fraenkel, 1954.
drill log or **drilling log**, the record of the events and the type and characteristics to the formations penetrated in drilling a borehole. Also called '*boring log*'. Long, 1920.
drill machine, see '**drilling machine**'.
drill pattern, see '**drilling pattern**'.
drill pipe, in rotary drilling, the heavy steel pipe rotated to give motion to the drilling bit, and through which circulation of drilling fluid is maintained. A.G.I., 1957.
drill plan, see '**drilling pattern**'.
drill plane generator, a drill rig for drifting with an inbuilt computer for generation of a drill plan. From a reference point x, y, z and known direction the operator points the drill feeds at the four corners of the face, in line with geologists marks. When all adjustments have been done, the rig control system RCS will develop the most efficient round compatible with the new parameters. Atlas Copco, 2007.
drill pressure or better **drill thrust (F_{bi}), (N)**, the force applied to a bit in drilling operation expressed as the number of tons applied. Also called **bit load**, *bit pressure, bit weight, drilling pressure, drilling weight*. Long, 1920.
drill rig, equipment for mechanized drilling including one or more drill machines inclusive feeder and carrier for moving the rig. The rig can be operated by compressed air or hydraulic. The latter is most common today. Sandvik, 2007.
drill rod, steel rod (bar) designed to transfer flushing media, rotation and percussion energy from rock drill to the drill bit. Can have either threaded or tapered ends or integral with a drill bit at one end and a striking surface at the other end. Atlas Copco, 2006.
drill rod wrench, device having jaws that are able to grip the drill rod to couple and uncouple them. Atlas Copco, 2006.
drill round, complete set of holes drilled to a specific pattern to form a single blast in any rock excavation. Atlas Copco, 2006.
drill series, a series of integral drill steels in which the diameter of the drill bit decreases about 1 mm for every increase of 0,8 m in the length of the drill steel. Olofsson, 1990.
drill set, a set of *integral steels* with different lengths. Normally used in small operations and at hole diameters less than 38 mm in diameter.
drill shank, the end of the drill steel that is designed to fit into the drill machine. LKAB, 1964.
drill sleeve, part of the drill system to couple two rods together. It normally consists of a steel tube with length about 200 mm with inner thread at both ends. Sandvik, 2007.

drill steel, a long piece of steel with one small diameter hole in the centre for the flushing media, normally water, to cool the bit and remove the drill cuttings from the borehole.
drill steel break detection system, an inbuilt system on a drill rig that shuts down the drilling operation if a breakage is detected. Atlas Copco, 2006 b.
drill steel grinder, grinding machine used to restore the cutting edges to drill bits. Atlas Copco, 2006.
drill steel guide, complete drill steel guide including all its parts. Atlas Copco, 2006.
drill steel set, a set of *integral steels* with different lengths. Normally used in small operations and at hole diameters less than 38 mm in diameter. Sandvik, 2007. See also '**integral drill steel and drill set**'.
drill steel shank, the drilling accessory located between the hammer piston and the drill string at drilling.
drill steel support, device mounted at the front end of the feed to confine and guide the drill steel when beginning to drill a hole. Atlas Copco, 2006.
drill steel support arm, device used to position the drill steel support. Atlas Copco, 2006.
drill steel support cylinder, hydraulic cylinder used to position the drill steel support. Atlas Copco, 2006.
drill steel support half, half of the drill steel support bushing, manufactured in two pieces. Atlas Copco, 2006.
drill steel support holder, device used to position the drill steel support. Atlas Copco, 2006.
drill stem, see '**drill string**'.
drill string, **drilling string** or *drill stem*, all components such as drill rods, bit, stabilizers etc. that are coupled together and inserted into the hole when drilling. Atlas Copco, 2006 b.
drill thrust (F_{bi}), (*N*), force on the drill bit at drilling. Also called **bit load**, *bit pressure, bit weight, drill pressure, drilling pressure, drilling weight*. Long, 1920.
drill tube, steel tube that transfer feed force, flushing media and rotation from the drill head to the drill bit. Atlas Copco, 2006. Also drill rod used for drilling through soil. Sandvik, 2007.
drill wagon rig, semi-mechanized drilling unit, consisting of a pneumatic rock drill and feed with a positioning arrangement and a pneumatic powered hydraulic pump system with controls. The drill wagon rig is mounted on a simple three wheel undercarriage. Atlas Copco, 2006.
drill wagon underground, simple drill rig used for the underground drilling of production holes. Atlas Copco, 2006.
drivage, underground drift, tunnel, heading, crosscut *in course of construction*. It may be horizontal or inclined to a maximum of about 30°.
drivage confinement (f_t), (m^{-1}), is defined as the blast hole depth divided by the drift area.
drive (verb), to excavate horizontally, or at an inclination, as in a drift, adit, or entry. Distinguished from sink and raise. Fay, 1920.
drive belt, band passing around two wheels and transferring motion from one to the other. V-belts are a type of drive belt that may be used to drive the cooling fan, water pump and generator on an engine. Atlas Copco, 2006.
drive clamp, a collar fitted on a churn drill string to enable it to be used as a hammer to drive casing pipe. Nichols, 1965.
drive collar or **jar collar**, extra thick walled pipe or casing coupling against which the blow of a drive block is delivered when driving or sinking drive-pipe or casing. Long, 1920.

drive hammer, see '**casing drive hammer**'.
drive kelly bushing, part of a drill rig that transforms the rotation energy to the drill steel. The cross section shape of the drill rod must be irregular to make it possible to transform the rotation energy. Stein, 1977.
driven pile, a timber-, reinforced concrete-, or steel pile driven to a specified set, by drop hammer, a steam hammer, or a diesel hammer. Ham, 1965.
drive piston, piston used to provide force or movement to another object. Atlas Copco, 2006.
driver, percussive rock drill part used to transfer rotational force to the drill string. Atlas Copco, 2006.
driver blank, semi-completed piece from which a rock drill driver is manufactured. Atlas Copco, 2006.
driver kelly, is a hollow cutting tool which allows the material to be sucked up inside the hollow kelly at high velocity, ensuring all the cuttings are taken to the surface. Stein, 1997.
drive rod, an extra heavy wall drill rod to which drive collars may be attached when using a drive hammer to jar or loosen drill string equipment stuck in a borehole. Also called '*jar rod*', '*jar length*', '*jar piece*'. Long, 1902.
drive screw, rod having a helical thread and a means of rotating it to provide a force or linier movement to an object. Atlas Copco, 2006.
drive shaft, axel or shaft used to transfer power from a primary to a secondary component. Atlas Copco, 2006.
drive unit, component with all its assembled parts used to power one or more secondary components. Atlas Copco, 2006.
driving, the excavation of drift and rooms underground.
DRM (acronym), **(v_d)**, **(m/min)**, for 'drilling rate measured' in the field.
drop ball, an iron or steel mass held on a wire, rope or cable dropped from a height (by a crane) onto large boulders for the purpose of breaking them by impact energy into smaller fragments. The drop height is ~8–10 m. By using this method, secondary blasting is avoided. The method is only used in surface mining.
drop body, see '**draw body**'.
drop center, a drop center drill bit with a countersunk area in the centre of the bit face. Atlas Copco, 2006.
drop center bit, drill bit with a countersunk area in the centre of the bit face. This design is often used to achieve greater hole straightness. Atlas Copco, 2006.
drop cut, the *initial cut* made in the floor of an open pit or quarry for the purpose of developing a bench at a level below the floor. USBM, 1990. Also called '**sinking cut**' or '*lift shot*'.
drop hammer, a *pile driving hammer* that is lifted by a cable and that obtains striking power by falling freely. Nichols, 1965.
drop of motion, if a certain volume of material is drawn from the bottom of an infinite large bin with fine loose material it causes movements in the material within a volume that can been approximated by an '*ellipsoid of rotation*' also called '*ellipsoid of motion*' with its long axis vertical and placed in the center of the draw point. Janelid and Kvapil, 1966. Scaled model tests of fragmentation in mining at the Royal Institute of Technology in Sweden revealed that the shape is more similar to a **drop turned upside down** rather than an ellipsoid. Fröström, 1965. In the 1970's

Bergmark at LKAB in Kiruna was able to derive a theory for the calculation of the shape of the drop of motion from Newton's laws. Rustan, 2000. If loading is made from a crosscut over its whole width the drop shape should be modified, see Kuchta, 2002. See also '**angle of caving**'.

drop (weight) **mass impact sensitivity test**, a method to test the sensitivity of explosives. A drop mass impacts the explosive located on an anvil. Johansson and Persson, 1970.

drop (weight) **mass test**, see '**drop (weight) mass impact sensitivity test**'.

dropped core, pieces of core not picked up or those pieces that slip out of the core barrel as the barrel is withdrawn from the borehole. Long, 1920.

drop penetration test, see '**penetration test**'. Ham, 1965.

drop raise, a raise where parallel holes are drilled from an upper level to a lower level and a series of vertical crater retreat blasts are made to break the rock from the lower to the upper level. A common length of these raises is about 50 m.

drop shaft, caisson sinking or drum shaft, a method of sinking a shaft through wet clay, sand, mud and rock. When sinking shaft in rock the full area can be blasted or a part of the area by the spiral method which resembles a down going ramp similar to the spiral drill and blast concept used in drifting.

drop ways, openings connecting parallel passages that lie at different levels. A.G.I., 1957.

drop weight, synonym for 'casing drive hammer'. Long, 1920.

Drucker-Prager criterion, a failure criteria similar to the Mohr-Coulomb 2 D criteria. The advantage is, that it can be used for triaxial stress conditions. The yield surface will have a conical shape compared to the Mohr Coulomb criterion that will have a hexagonal pyramid form. The Drucker-Prager criterion can easily be programmed for finite element analysis.

drum break, brake arrangement where the braking action is the result of generating a friction between a drum and shoes being pressed against the drum. Atlas Copco, 2006.

drum shaft, synonymous to '**drop shaft**'.

dry blasting agent, an explosive that is blowable at charging but not pumpable. One example of dry blasting agent is ANFO. Slurry that is a wet blasting agent.

dry density (ρ_{dry}), (kg/m³), the weight of a unit volume of a dry sample of soil, after the latter has had water driven of by heating the sample to a temperature above 100°C.

dry drilling, drilling operations in which the cuttings are lifted away from the bit and transported out of a borehole by a strong current of air or gas instead of a fluid. Long, 1920.

DSA (acronym), see '**differential strain analysis**'.

DTH (acronym), see '**down the hole**' hammer or '**down hole drills**'.

DTH-hammer (acronym), see '**down the hole**' drill.

dual energy percussion system (DEPS), drilling system consisting of concentric drill strings. The outer string is fitted with a hollow or kerfing bit and receives percussive energy from a so called "top hammer" The inner string is fitted with a "down-the-hole drill" which generates percussive energy directly to the drill bit. Atlas Copco, 2006.

dual-pipe drill down method, drill rigs equipped with dual-pipe reverse circulation equipment may use the dual-pipe facility made up to a venture tube in the wash down shoe instead of using a standard jetting tube. When using dual-pipe techniques, care must be exercised to avoid allowing low pressures to develop at the shoe and "suck" the aquifer sands out of the hole. Stein, 1997.

dual-tube drill down method, drill rigs equipped with dual-tube reverse circulation equipment may use the dual-pipe facility made up to a venture tube in the wash down shoe instead of using a standard jetting tube. When using dual-tube techniques, care must be exercised to avoid allowing low pressures to develop at the shoe and "suck" the aquifer sands out of the hole. Stein, 1997.

dual-wall top hammer system, the dual-wall top hammer method uses an open-faced bit flush-threaded to dual-wall drill pipe. The drill string is hammered rather than rotated into the ground. The drill pipe is driven by the pile hammer, with impact applied to the outer pipe only. Stein, 1997.

dual tube reverse circulation drilling, see 'reverse circulation drilling'.

Dutch cone test, see 'soil penetration tests'.

duckbill loader, see 'shaker-shovel loader'.

ductile deformation, occurs when the rock can sustain further permanent deformation without loosing load-carrying capacity. Brady, 1985.

ductility, the ability of a material to deform plastically without formation of brittle zones or fractures. The material can sustain permanent deformation without losing its ability to resist load.

dummy, generally a stunt replacement device. In blasting it is defined as a paper bag filled with sand, clay, etc. for tamping or separating two charges in a borehole.

dummy road, a road driven forward in the waste of a conveyor face for the sole purpose of securing stone for packing purposes. Nelson, 1965.

dump, a place where the ore or waste is tipped.

dump wagon, a large-capacity side-, bottom- or end- discharged wagon on tyre wheels or (crawler track), usually tractor towed. Nelson, 1965.

durability of rock, the capacity or rock to withstand the weather and especially water and ice filling its pores. Coarse grained rocks with big voids for monumental use can resist better the effect of ice than the fine grained with small voids. Miguel, 1983.

dust catcher (drilling), see '**dust collector**'.

dust collector or *dust catcher* (drilling), a device attached to the collar of a borehole to catch or collect dry, dust like rock particles produced in dry-drilling of a borehole. Long, 1920. A vacuum cleaner is used for this purpose. Normally it sucks in both the drill cuttings as well as the dusted air from a tight rubber sleeve around the collar of the borehole.

dust collector or *dust catcher* (general), an apparatus for separating solid particles from air or gas and accumulating them in a form convenient of handling B.S. 3552, 1962. Device used to limit the amount of air born dust entering the atmosphere during rock drilling. Dust and cuttings laden flushing air is collected as it exits the drill hole and passed through a series of centrifuges and filters. The solid material is deposited in a hopper while the clean air is exhausted to atmosphere. Atlas Copco, 2006.

dust exhaust system (drilling), in drilling it might be a vacuum suction of dust from the boring head to a container.

dust explosion, an explosion associated with a sudden pressure rise caused by the very rapid combustion of airborne dust. The ignition of combustible dusts may occur in the following ways: 1) *Flame or spark,* 2) *Gas explosion or blasting* and 3) *Spontaneous combustion.* There is little known about the third mechanism, which is relatively rare in mines. The most frequent causes of major coal mine explosions are electric arcs, open flames, and explosives. Spontaneous combustion occurs in sulphide ores due to the oxidation of sulphur to sulphur dioxide. Hartman, 1961. In metalliferous mines sulphide-bearing ores may be susceptible to sulphide dust explosions.

dust extractor, see 'dust exhaust system'.

dust guard, type of cover used to protect something from the harmful effects of dust. Atlas Copco, 2006.

dyke, a minor intrusion into the crust of the earth discordant to the general structure of the rock compared with sills, which are concordant with the rock structure. They are both essentially parallel-sided, sheet-like bodies. Park, 1989.

dyna-drill, is a helical rotor pump operated in reverse. The circulation fluid pressing against the rotor forces it to rotate to allow the fluid to pass. Dyna-drills can be used in holes from 40 to 51 mm and are used for diamond core drilling and medium rotary drilling. The dyna-drill can be operated by most drilling fluids, including mud, water and air. Stein, 1997.

dynamic bulk modulus (K_d), **(MPa)**, see 'bulk modulus'.

dynamic compressive strength, (σ_{cd}), **(MPa)**, the compressive strength of a material under dynamic loading rates. The dynamic compressive strength of rock is larger than the static compressive strength, and it is directly proportional to the cubic root of the strain rate. Grady et al., 1980.

dynamic creep, creep that occurs under conditions of fluctuating load or fluctuating temperature. A.S.M., 1961.

dynamic (quasi dynamic) elastic properties of rock, Young's modulus (E_{qd}) and Poisson's ratio (v_{qd}), these properties can be calculated if the longitudinal wave velocity (c_P) and shear wave velocity (c_S) have been determined in drill cores (at strain rates in the order of 10^2 to 10^3 s^{-1}) by the Hopkinson's bar method. The quasi dynamic Young's modulus E_{qd} and quasi dynamic Poisson's v_d ratio can be calculated by the following two equations.

$$E_{qd} = \frac{c_s^2 \rho_r \left[3(c_P/c_S)^2 - 4\right]}{\left[(c_P/c_S)^2 - 1\right]} \quad \text{and} \quad v_{qd} = \frac{0.5\left[3(c_P/c_S)^2 - 2\right]}{\left[(c_P/c_S)^2 - 1\right]} \tag{D.10}$$

where c_S is shear wave velocity in m/s, ρ_r is density of the rock in kg/m^3, c_P is the longitudinal wave velocity in m/s. The values of dynamic Young's modulus and Poisson's ratio determined from these equations are usually significantly greater than those values obtained from static tests. For 5 different tested rocks the ration E_{qd}/E_s varied from 1,2–2,8 and v_{qd}/v_s from 1,2–3,8 where E_s is the static Young's modulus in Pa and v_s is the static Posson's ratio (dimensionless). The strain rate in the Hopkinsson bar test corresponds more to a *quasi dynamic test* because the strain rate is too low to simulate the conditions at the boundary of a blasthole where the strain rates would be in the order of 10^4 to 10^5 s^{-1}.

dynamic penetration test, see 'penetration test'.

dynamic pressure desensitization, the process of rendering explosives less sensitive at a higher dynamic pressure due to the increase of explosives density. See also '**dead pressing**'.

dynamic rock mechanics, the technique for describing, understanding and predicting events at high loads and strain rates in rocks for example in the blasting processes, slope failures, gravity flow etc.

dynamic strength, the strength of a material at a certain loading rate, see also '**dynamic compressive strength**'. Traditionally, the strength criterion used in rock blasting has relied only on static tensile strength, although it is known that a more proper strength criterion should be based on dynamic tensile strength. The breakage of rock is, however, also controlled by the dynamic compressive and shear strengths. In fracture mechanics the strength is only controlled by fracture toughness.

dynamic stress intensity factor, see '**fracture toughness**'.

dynamic tensile strength (σ_{td}) (MPa), the tensile strength of a material under dynamic loading. See also '**dynamic compressive strength**'. The dynamic tensile strength is dependent on the strain rate ε (s^{-1}). The *dynamic tensile strength* of US oil shale has been determined at different *strain rates* using a tensile bar, Hopkinson's bar, capacitor and gas gun. The dynamic fracture strength (tensile strength) versus strain rate is shown in Fig. 26. There is about 5 times increase in strength. Grady, 1980 a.

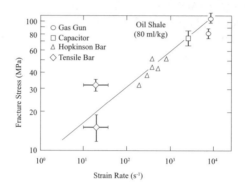

Figure 26 Dynamic fracture strength (tensile strength) versus strain rate in US oil shale. After Grady, 1980 a.

dynamic Young's modulus (E_d), (MPa), the ratio between stress and strain for a body under a minimum strain rate. *Dynamic values must always be accompanied by the strain rate used at testing.*

dynamite, an industrial explosive whose principal ingredient is nitroglycerin, $C_3H_5(NO_3)$, or specially sensitized ammonium nitrate, mixed with a material e.g. silica guhr making the mixture less sensitive to impact. Diethyleneglycol dinitrate, which also is an explosive, is often added to dynamite as a freezing point depressant. A dope, such as good pulp, and an antacid (e.g. calcium carbonate) are essential for the fabrication of dynamite. Dynamites are classified as high explosives (can be detonated by a blasting cap). See also '**blasting gelatin**'. Dynamite was the first trade name used for a commercial explosive and it was invented by the Swede *Alfred Nobel*. The key idea was to add silica guhr to make the nitroglycerin less sensitive in transport and handling. The mixture was patented on September 19 in 1867 and in the same year Nobel also promoted the manufacture of dynamite in the United States through the Giant Powder Company, which had the exclusive right of use of his patent in the USA. Forty-eight years later, in 1915, the Atlas Powder Company acquired the Giant Powder Company.

dynamite gelatin, dynamite made by gelatinising the nitroglycerin with collodial cotton before the addition of the absorbent. Bennett 2nd, 1962.

earth current, a light electric current apparently traversing the earths surface but which in reality exists in a wire grounded at both end, due to small potential differences between the two points at which the wire is grounded. Standard, 1964.

earth drill, see '**auger drill**'. Nichols, 1956.

earth drilling, drilling through loose overburden. To prevent loose material to fall into the hole a casing tube is installed down to the bedrock. Drilling methods, see '**ODEX method**'.

earth fault, electrical short circuit from live conductor to earth. Pryor, 1963.

earth fault meter, an instrument for measuring the insulation fault at low voltage without polarization. This instrument is more informative in checking detonators in loaded holes than the *insulation meter*. It is used to find and prevent or reduce current leakage to the ground when blasting in conducting ore bodies, in wet shale or clay, and in underwater blasting especially in salt water. The apparatus has no battery and can be used when loading the hole to check if the conducting wires have become damaged during this operation Langefors, 1963.

earth fault protection, a system designed to cause the supply to a circuit or system to be interrupted when the leakage current to earth exceeds a predetermined value. Also called earth leakage protection. B.S. 3618, 1965.

earth fault relay, type of switch used to automatically shut off the electric current in a system if current is leaking into the surrounding structure. Atlas Copco, 2006.

earth fault tester, see '**earth fault meter**'.

earth leakage, the loss of current to earth from any electrical circuit by unintended means.

earth wave, any elastic motion of the earth, either from natural causes such as earthquakes and storms, or created artificially by traffic, blasting, seismic exploration, etc. Seismologists recognize two main groups of earth waves, 1) *body waves* propagated in all directions though the elastic body of the earth, and 2) *surface waves* which propagate at the rock-air interface at the earths surface. Body waves are of two types, P (for primary waves) compression or longitudinal waves and S (for secondary) transverse or shear waves. Among the surface waves are a variety of transverse and rotational types, such as Rayleigh (R), Love (Q), hydrodynamic and coupled waves. Stokes, 1955.

easer, one of a number of holes surrounding the cut, in tunnelling or drifting, and fired immediately after the cut. B.S. 3618, 1964.

easer holes, holes drilled around the cut to enlarge the cut area so that the trimmers (perimeter holes) may break out the ground to the required dimensions. The positioning and number of the easer holes will depend upon the pattern of the cut shots. McAdam II,

1958. In drifting, the holes closest to the cut used to enlarge the opening formed by the cut. Olofsson, 1990.

EBC (acronym), for '**electric blasting cap**', see '**detonator**'.

EBLSC (acronym), for '**elliptical bipolar linear shaped charge**', see '**linear shaped charge**'. Qin J.F., 2009.

E-box (abbreviation), for '**electrical box**'. Atlas Copco, 2006.

EBW (acronym), see '**exploding bridge wire**' detonator.

EC (acronym), for '**explosive compaction**' see '**explosive compaction of soils**'.

eccentric bit, see '**eccentric reamer**'. Fay, 1920.

eccentric drill bit, drill bit where, the cutting surfaces are positioned in a non-symmetrical pattern, over the bit face. Atlas Copco, 2006.

eccentric lock, frictional locking device where the contact or locking surface is eccentric to the center of the pivot arm. Atlas Copco, 2006.

eccentric pattern, a mode of arranging diamonds set in the face of a bit in such a manner as to have rows of diamonds forming eccentric circles so that the path cut by each diamond slightly overlaps that of the adjacent stone. Compare 'concentric pattern'. Long, 1920.

eccentric reamer, a modified form of chisel used in drilling, in which one end of the cutting edge is extended further from the center of the bit than the other. The eccentric bit renders under reaming unnecessary. It is very useful in hard rock.

echelon cut, see '**en-echelon cut**'.

echelon pattern, see '**en-echelon pattern**'.

EDD (acronym), see '**electronic delay detonator**'.

edge detection, method for detection of the perimeter of projected surfaces of fragments. It is based mainly on thresholding, and often requires pencil tracing of photo image before digitization.

edge runner mill, a crushing and grinding unit depending for its action on heavy mullers (heavy grinding wheels which are the crushing and mixing members in a dry or wet pan), usually two in number, that rotate relative to a shallow pan which forms the base. The pan bottom may be solid or perforated. Dodd, 1964.

EDZ (acronym), see '**excavation disturbed zone**'.

EFEE (acronym), for '**European Federation of Explosives Engineers**'. The organisation was founded in 1988. Web address 'www.efee.eu'.

effective borehole pressure or **equilibrium borehole pressure** (p_{be}), **(MPa)**, see '**borehole pressure**'.

effective burden or true burden (B_b), (m), see '**burden, blasted burden**'.

effective modulus of elasticity (E_e), **(MPa)**, see '**effective Young's modulus**'.

effective porosity (n_e), **(dimensionless),** the amount of connected pore spaces that are used in relation to the total space available for fluid penetration. Brantly, 1961.

effective spacing or true spacing (S_b), (m), see '**spacing, blasted spacing**'.

effective stress (σ_e), **(MPa)**, is defined in general terms as the stress which governs the gross mechanical response of a porous material. The effective stress is the sum of the total or applied stress σ_a in MPa and the stress caused by the pressure of the fluid in the pores of the material σ_p in MPa, known as the pore pressure or pore-water pressure.

$$\sigma_e = \sigma_a + \sigma_p \qquad (E.1)$$

The concept of effective stress was first developed by Terzaghi who used it to provide a rational basis for the understanding the behaviour of soils. Brady, 1985.

effective surface energy or **specific surface energy (γ_s), (J/m²),** amount of energy required to generate a unit of new fracture surface area in a material.

effective Young's modulus or **effective modulus of elasticity (E_e), (MPa),** a fictitious elastic modulus for materials which do not exhibit a perfectly linear relationship between stress and strain. The total strain energy, i.e. the area below the stress-strain curve, is used for the calculation of the effective Young's modulus.

efficiency (η), (%), see '**blasting efficiency**'.

efficiency of detonation reaction (η_d), (dimensionless), the efficiency of detonation reaction η_d, is defined as follows;

$$\eta_d = \left(\frac{c_{dni}}{c_{di}}\right)^2 \tag{E.2}$$

where c_{dni} is the non-ideal detonation velocity measured in the blasthole in m/s and c_{di} is the ideal detonation velocity in m/s. Cameron, 1996. Not a common used measure.

effusive rock, see '**extrusive rock**'.

EFI (acronym), see '**exploding foil initiator**'.

EGDN (acronym), for '**ethyleneglycoldinitrate**', see '**NG explosive**'.

eject and dump (EAD), a special designed load haul dumper bucket for low heights underground ~1,5 m. The bucket can not be raised and therefore the rock is pushed out from the **EAD** bucket by a push plate onto feeders that transfer it to the conveyor system for transportation to the surface. Atlas Copco, 2007.

ejection of stemming or **rifling,** the process of throwing out the stemming of the blasthole as a result of blasting.

ejector drilling, drilling technique where cutting removal is performed by a media injected against rock formation. Sandvik, 2007.

elasticity, the property of a material to regain its original size and shape upon complete unloading after it has been deformed.

elastic limit, the upper limit of the elastic part of a material's stress-strain behaviour curve. Brittle rock breakage occurs at the elastic limit.

elastic moduli or **moduli of elasticity,** moduli characterizing the elastic properties of a material, e.g. *Young's modulus (E),* (GPa), *shear modulus (G),* (GPa), and *bulk modulus (K),* (Pa).

elastic ratio (R_{elast}), (dimensionless), the ratio of the elastic limit to the ultimate strength. Roark, 1954.

elastic rebound, the recovery of elastic strain. A.G.I., 1957.

elastic rebound theory, faulting arises from the sudden release of elastic energy which has slowly accumulated in the earth. Just before the rupture, the energy released by the faulting is entirely potential energy stored as elastic strain in the rocks. At the time of rupture the rocks on either side of the fault spring back to a position of little or no strain. This theory was proposed by Harry Fielding Reid. A.G.I., 1957.

elastic shock, see '**acoustic emission**'.

elastic strain (ε), (dimensionless), deformation per unit of length produced by a load on a material, which vanishes with removal of the load . Ham, 1965.

elastic strain energy (W_{se}), (J), potential energy stored in the linear part of a strained elastic solid. In the energy balance equation it is equal to the work done by deforming the solid from its unstrained state to its final state of deformation less the amount of energy dissipated by inelastic deformation and the fracture process.

elastic strain energy density (w_{se}), (J/m³), strain energy per unit of volume. A specific energy density associated with the elastic deformation of a solid. It is composed of two parts: the volumetric energy density, (J/m³), which characterizes the change in volume, and the distortion energy density, (J/m³), which takes into account a change in shape. In the case of a linearly elastic body subjected to uniaxial loading conditions, the elastic strain energy density *wse* is given by the following equation;

$$w_{se} = \frac{(\sigma_c)^2}{2E} \qquad (E.3)$$

where w_{se} is the elastic strain energy density in (J/m³), σ_c is the uniaxial compressive stress in (Pa), and E is Young's modulus in (Pa).

elastic surface waves, waves which travel only on a free surface where the solid elastic materials transmitting them are bounded by air or water. Leet, 1960. See also '**earth waves**'.

elastic waves, mechanical vibrations in an elastic medium within the elastic range. ASM Gloss, 1961.

elastic zone (in rock blasting), the zone is starting where the plastic zone ends and continues to infinity and is defined as the zone that only undergoes elastic deformation.

elastoporosity, the description of the *mechanical properties of a material including pores* with and without a fluid in the pores and under different conditions for the fluid in the pores like drained or not drained pores. The earliest theory to account for the influence of pore fluid on the quasi static deformation of soils was developed in 1923 by Terzagi who proposed a model of one dimensional consolidation. This theory was generalized to three dimensions by Rendulic in 1936. Biot developed in 1935 a linear theory of poroelasticity that is consistent with the two basic mechanisms. Hudson, 1993.

elbow coupling, fitting used to connect two pipes, tubes or hoses to change the direction of the fluid.

electrical arc drill, an *exotic method* using an electrical arc with a temperature of 5 000–15,000°C to melt the rock continuously. Calculations indicate that 100–1 000 more energy is needed compared to conventional mechanical drilling and therefore the method is not very interesting. KTH, 1969.

electrical cabinet, box like structure having an access door, used to cover and protect electrical components. Atlas Copco, 2006.

electrical current (*I*), (A), 1 ampere equals the amount of electrons moving in a conductor with the resistance 1 ohm and a potential of 1 volt over the resistance.

electrical heated drill, an *exotic drilling method* where a high voltage source is feeding an electrical current through the rock, that is heated up. Most rocks have, however, a low conductivity and therefore the method can only be used in ore or porous rock saturated with water. KTH, 1969.

electrical high frequency drill, an *exotic drilling method* where a high frequency current, 2–30 MHz and voltage 10–20 kV is heating up the rock by dielectric as well as resistive heating. The former is proportional to the frequency, but the latter is independent on frequency. The method has been used in boulder fragmentation. KTH, 1969.

electrical make/break targets, a method to measure the detonation velocity (c_d) of an explosive. Electrical make/break targets are inserted in the explosive at known distances. When the detonation front passes the electrical make/break, a contact is reached and a current can pass. The distance versus time information is interpreted to obtain the velocity of detonation.

electrical plug, device used to plug in and connect something to an electric supply. Atlas Copco, 2006.

electrical resonance method, see '**SLIFER**'.

electrical system, all the electrical components, wires and switches used to provide the various functions required on a drill rig or other structure. Atlas Copco, 2006.

electric blasting, the firing of one or more charges by electric current e.g. electric squibs, electric or electronic detonators or other electric igniting or exploding devices. This is distinct to the firing of fuse, detonating cord or NONEL.

electric blasting cap (EBC), see '**blasting cap**'.

electric blasting circuit, an electric circuit containing electric detonators and associated wiring. See also '**series blasting circuit**', '**parallel blasting circuit**', and '**series in parallel blasting circuit**'. Atlas Powder, 1987.

electric cable, coated wire made from copper or other suitable material, used for conveying electricity to a desired location or device. Atlas Copco, 2006.

electric detonator, a detonator designed for and capable of initiation by means of an electric current. Atlas Powder, 1987.

electric drill, a mechanically operated drill driven by an electric motor. Used mainly in mines.

electric exploder, see '**electric detonator**'.

electric firing, the initiation of explosive charges by means of electric blasting caps.

electric fuse, a metallic cap, usually containing fulminating mercury, in which are fixed two insulated conducting wires held by a plug, the latter holding the ends of the wires near to but not touching each other. At this plug is a small amount of sensitive *priming*. When an electric current is sent from the battery though these conductors, the resulting spark fires the priming, followed by the fulminate ($HgC_2N_2O_2$) and the charge of the explosive used. Stauffer, 1906.

electric igniter, a device containing a pyrotechnic composition for attachment to the igniter cord, safety fuse or quick-match and which are ignited electrically to initiate such cord. AS 2187.1, 1996.

electric meter, device that measures the amount of electricity consumed over a given period of time. Atlas Copco, 2006.

electric motor, motor powered by electricity, consisting of a enclosed coil and rotor, out put shaft and coupling box. An electro-hydraulic drill rig may be fitted with a number of electric motors to power hydraulic and water pumps. Atlas Copco, 2006.

electric powder fuse, this fuse was designed so that electrical shotfiring methods could be used for initiating blasting powder. The powder fuse consists of a thick paper tube containing a small charge of blasting powder, with an ordinary low tension fusehead fixed at one end. On passing electric current through the fusehead it flashes and sets off the blasting powder in the tube, which can then initiate the main charge of blasting powder in the shot hole. McAdam II, 1958.

electric rotary drill (in coal mining), a rotary drill, driven by an electric motor and may be used in coal. It may be of fan-cooled design with several rod speeds to suit different rocks. The use of aluminium or aluminium alloys is favoured where methane is liable to be present. This drill produces considerably less dust than the percussive drill Nelson, 1965.

electric socket, general term for an electrical receptacle or wall socket. Atlas Copco, 2006.

electric squib, a small shell containing an explosive compound that is ignited by the electric current brought in through the lead wires. Used for firing single small holes loaded with black powder. Lewis, 1964. More recently electric squib have been applied to commercial fireworks. Lusk, 2007.

electric storm, an atmospheric disturbance characterized by intense electrical activity, producing streaks lightning and strong electric and magnetic fields. Electric storms represent a hazard to all blasting activities.

electric time domain reflectory (ETDR), a method to measure the detonation velocity of an explosive (c_d). It determines the change in length of a coaxial cable embedded in the explosive by regularly timing the period for an electrical signal to travel along the cable, reflect off the free end and return. Spathis, 2007.

electric voltage, see '**Volt**'.

electrode ground rod driver, tool used with a breaker to install electrical grounding rods. Atlas Copco, 2006.

electro hydraulic fragmentation, an exotic fragmentation method using an electric spark to fragment rock under water. Kutter, 1969.

electromagnetic emission (EME) during rock blasting (radio waves), it was found that the EME depends on three different mechanisms 1) Rock fracturing at the time of explosion. This is believed to be the *major source for EME-emission* and probably it is due to the piezoelectric effect of minerals, 2) Electrical charged rock fragments being discharged when impacting with the pit floor. This gives a minor amplitude and finally 3) Micro fracturing of the remaining rock wall, due to pressure adjustment of the bench behind the blast. This also creates an minor amplitude. The electromagnetic emission was measured in the frequency range 20–20,000 Hz corresponding to wave lengths of 150,000 m to 150 m. This range is called *long waves*. O'Keefe and Thiel, 1991. Generally the radio frequency range is defined by wave lengths from ~1 mm to ~10,000 m corresponding to frequencies 3 GHz to 300 Hz. Ingelstam, ELFYMA.

electromagnetic geophone, see '**geophone**'.

electromagnetic induction initiation, an initiation system where by induction the electric energy from the lead wire is transferred to each detonator circuit.

electromagnetic initiation, the use of electromagnetic energy to initiate detonators. The same technique can be used as being used for radio communication. Usually, ultra low frequency systems are in use and in 1996 these systems were operational in 33 mines in the US, Canada, Australia and South Africa. Anon., 1996.

electronic blasting cap, see '**electronic detonator**'.

electronic (delay) detonator (EDD), a detonator where the time delay and other logic functions is electronically controlled, by a built-in silicon chip (integrated circuit) which feeds the main ignition current, usually from an internal capacitor. The delay time can be set with an accuracy of about 1 ms. This type of detonator has the intrinsic advantage of yielding reduced scatter in delay time and offers the possibility of selecting intervals in a wide range. Special blasting machines are required for the initiation of electronic detonators. Various kinds and degrees of sophistication exist, from the factory types with pre-set delays to those where the delays are programmable in the blasthole. Because of the greatly enhanced precision versus pyrotechnically controlled delays, and because of programmability, these devices are very important for controlled contour blasting and production blasting.

electron ray drill, an *exotic drilling method* using a source of 5–150 kV to accelerate a ray of electrons, that are focused by an electromagnetic lens into a small diameter ray that hits the rock. KTH, 1969.

elementary representative volume (ERV), when the rock is assimilated to a continuum material the ERV is the smallest volume for which there is equivalence between the idealized continuum material and the real rock. Hudson, 1993.

elemented cap, see '**blasting cap**'.

Elitz and Zimmerman test, see '**fume classification test**'.

ellipsoid of motion, see '**drop of motion**'.

elliptical bipolar linear shaped charge, see '**shaped charge**'.

elongation ratio (R_{elo}), (dimensionless), the ratio between the width and the length of a particle or fragment,

$$R_{elo} = W/L \qquad (E.4)$$

where W is the width of the fragment in m and L is the length of the fragment in m.

embankment, artificial ridge of earth and broken rocks, such as a dike or railroad grade across a valley. A.G.I., 1960.

EME (acronym), see '**electromagnetic emission**'.

emergency procedure card, a list of instructions carried on trucks transporting explosive materials and giving specific procedures in case of emergency. The list stored in the driver's cabin.

emission of light at rock fracturing, see '**exoelectron emission**' also called '**fracto-emission**'.

emissivity, the ratio of radiant energy emitted by a body to that emitted by a perfect black body. The latter has an emissivity of 1 and a perfect reflector has an emissivity of 0. Strock, 1948.

emulsion, a liquid mixture in which a fatty or resinous substance is suspended in minute particles, almost equivalent to molecular dispersion. Fay, 1920.

emulsifying agent (emulsifier), a material in small quantities that increases the stability of a dispersion of one liquid in another. ASM Gloss., 1961.

emulsion explosive, a water-resistant explosive material containing substantial amounts of oxidizers, often ammonium nitrate, dissolved in water and forming droplets, surrounded by an immiscible (unable to mix) fuel for example fuel oil. The droplets of the oxidizer solution are surrounded by a thin layer of oil and are stabilized by various emulsifiers. Sensitivity is achieved by adding voids such as small nitrogen bubbles or micro-spheres made of glass. The resulting explosive is water-resistant. The intimate mix of fuel and oil results in a high detonation velocity (c_d) and low fume characteristics for oxygen-balanced mixtures. The characteristics of the explosive may be varied by doping with solids such as AN prills. If prills make up more than 50% of the explosive, then it is known as '*heavy ANFO*'. In the USA, a non-cap-sensitive emulsion is called a blasting agent.

emulsion slurry, see '**emulsion explosive**'.

end clinometer, an instrument for measuring angular deviation from horizontal or vertical in drilling. A clinometer designed to be fitted only to the bottom end of a drill-rod string as contrasted with a *line clinometer* that can be coupled into the drill-rod string at any point between two rods. Long, 1920.

end initiated charge, the initiation of the charge is made at any of the two ends of the charge or at both ends. When using the latter method the detonation front will meet at the middle of the charge and thereby cause twice the amplitude compared to initiation at only one end. Duvall, 1965.

end piece, part forming the extremity of a structure, often functioning as a cap or protection. Atlas Copco, 2006.

end plate, a plate like an '*end piece*'. Atlas Copco, 2006.

end support, device used to support one end of a structure. Atlas Copco, 2006.

endurance limit, limiting stress below which specimens can withstand infinite cycles of stress without fracturing. The limit is considerably lower than the rupture strength. Also called *fatigue limit*.

endurance ratio (R_{end}), **(dimensionless)**, ratio of the endurance limit to the ultimate static tensile strength. Roark, 1954.

endurance strength, the highest stress repeated application or reversal of which a material can withstand without rupture for a given number of cycles (K) is the endurance strength of that material for that number of cycles. Unless otherwise specified reversed bending is usually implied. Roark, 1954.

en-echelon cut, scheme of staggered blasthole positions. The blastholes are fired in a step like fashion rather than row by row.

en echelon crack, a crack that develops between two static, not propagating cracks not laying in the same plane if uniaxial stress is applied. The connection can be perpendicular or at an angle. These kind of crack is commonly seen. Krantz, 1979.

en-echelon pattern, a) a delay pattern for detonators that causes the blasted burden (*true burden*), at the time of detonation, to be at an oblique angle from the original free face. Two variants of the en-echelon pattern are the '*chevron pattern*' and the '*open chevron pattern*'. b) in fracture mechanics/structural geology a staggered formation of cracks or delaminations.

energetic material, a material that can undergo exothermal chemical reaction releasing a considerable amount of thermal energy. This definition is most appropriate when working with explosives. It has also a wider definition to include also inert materials which can store large amount of energy at high pressure and/or high temperatures.

energy density, (*w*), (J/m³), or '**heat of explosion**'. The ratio of energy per unit volume of explosive. The method by which energy has been calculated for an explosive must be specified.

energy of conservation law, see '**conservation laws**'.

energy level (*w*), (J/m³), see "**energy density**".

energy of rupture (*G*), (J/m²), see '**strain energy release rate**'.

energy of rupture, specific energy of rupture or **specific fracture energy** (w_f), (J/m³), the work done per unit volume in producing fracture. It is not practicable to establish a definite energy of rupture value for a given material, because the result obtained depends upon the form and proportions of the test specimen and the manner of loading. As determined by similar tests on similar specimens, the energy of rupture affords a criterion for comparing the toughness of different materials. Roark, 1954.

energy or **work** (*W*), (**J**), a physical quantity characterizing the effort, gain and loss of work in a physical or chemical process. It is quantified by the scalar product of force and distance, and its unit is Joule (J).

1 J = 1 Nm = 1 Ws
1 Btu = 1,06 kJ
1 kcal = 4,19 kJ
1 kWh = 3,60 MJ

energy partitioning, the apportioning of gross explosive energy into 'shock' (stress waves) and 'gas' (heave) components. Shock energy is defined by the energy released between detonation and the point at which the expansion of the blasthole is at its maximum. Gas energy released beyond this point is used for heave. Different blasting mechanisms are activated by these components. The partitioning depends on the rock properties as much as on explosive properties. Higher detonation velocities and lower strength rocks result in higher shock energies. Note also that in this context the shock energy is entirely different from the shock energy calculation in a *pond test*. Cunningham, 1993.

energy ratio (R_{ene}), (m²/s²), a criteria for damage of rock, proposed by Crandall. A measure of the limits of safe blasting without damage to structures. It is defined as

$$R_{ene} = \frac{a^2}{f^2} = 16\pi^4 f^2 A^2 \tag{E.5}$$

where R_{ene} is energy ratio in m²/s², a is the acceleration in m/s² and f the vibration frequency Hz in s⁻¹. Energy ratios less than 3 are safe for structures, and those in excess of 6 are hazardous. Bickel, 1982.

energy ratio (ER), (R_{ener}), (m/s), the ratio of a/f where a is acceleration in (m/s²) and f is vibration frequency in (Hz). A measure of vibration level used in connection with damage from blast vibrations. Konya and Walter, 1990. Compare with Bichels energy ratio R_{ene}.

energy release, e. g. the strain energy accumulated under stress in a rock mass will be gradually released when the stress is reduced.

engineering geology, a branch of geological science forming a link between geology and engineering, particularly in civil and mining. It provides a basis of theory to guide engineering practice where earth or rock materials are directly or indirectly involved. See also mining geology and soil mechanics. Nelson, 1965.

engineering heuristics, anything that provides a plausible aid or direction in the solution of a problem, but is not necessarily the only way to do so. A heuristic is essentially a 'rule of thumb' according to Koen. Hudson, 1993.

enhancement, the improvement of an image (making it clearer) by means of image processing.

enlarging sleeve, tubular device used to transition from a smaller size to a larger one. Atlas Copco, 2006.

en passant crack, is developed when two cracks approach each other on different planes. Of course one may be stationary. It is usually not possible to tell the order of events during an interaction when looking at a photomicrograph. At first, both cracks are affected only by the distantly applied stress field. As they approach and enter each other zone of influence, additionally stresses are applied to each crack. Apparently, depending on the angle of approach, proximity and direction to the applied stress field, shear stresses within the zone of influence deflect the crack paths first away, then towards, each other. The value of fracture toughness K_{II}/K_I is first positive, then negative. They link when the tip of one crack runs into the side of the other. This *linking crack* is called a *en passant crack*. Kranz, 1979.

entropy, a non-decreasing state variable. In the theory of classical thermodynamics, the change of entropy in a given process is equal to the change of heat energy divided by the temperature. The entropy of a material increases under the following conditions: 1) In processes (purely internal or external with an interaction between the material and its surroundings) which lead to an increase in the material's heat energy, e.g. heat influx from the surroundings. 2) In exothermic chemical reactions associated with an increase in heat energy.

entry, an adit to a coal mine or a gently inclined shaft. See '**adit**'. Stout, 1980.

epigenetic, a term for mineral deposits of later origin than the enclosing rock and produced close or near to the earths surface.

EPT (acronym), see '**explosive performance term**'.

equal angle projection, a method to represent joints on an imaginary hemisphere positioned below the plane of projection so that its circular face forms the projection circle. Hudson, 1993.

equation of motion, the Newtonian law of motion states that the product of mass and acceleration equals the vector sum of the forces. $\Sigma F_x + F_y + F_z = a \cdot m$ where F_i is the force in the ith direction in N, a is acceleration in m/s² and m is the mass in kg.

equation of state, an equation describing the mutual relation between state variables of matter. One example is the equation of state of the ideal gases that gives the physical relation between pressure (p), volume (V) and temperature (T) for a substance. For an ideal gas the equation reads.

$$pV = nRT \qquad (E.6)$$

where p is the pressure in Pa, V is the volume in m³, n is denoting the mass of the amount of gas in volume V in mole, R is the gas constant in Pa·m³/mol· degree K and T the temperature in degree K. Thum, 1978.

For *blast gases* the exponential Hugoniot is describing the relation in a more precise manner.

$$pV^\gamma = nRT \qquad (E.7)$$

where γ is a *material constant* depending on the type of gas mixture.

Equations of state may involve other variables, such as internal energy W. For example, for a polytropic gas the equation of state can be formulated as:

$$W = pV/(\gamma - 1) \qquad (E.8)$$

equation of state for the detonation gases, the equation of state of the reaction products is a complex function of pressure, density, and temperature. At low density the ideal gas equation is a good approximation. At higher density, when the volume of the molecules, i.e. the covolume (α), is an appreciable part of the total volume (V), the pressure rises approximately inversely proportional to the free volume (the product of V and α). This is the main cause of the deviation from the ideal gas equation in this density regime. At very high pressure, when the molecules are in permanent contact, the pressure is mainly due to contact forces between the molecules. Several equations of state have been derived for example by Abel, Van der Waal, Cook, Maxwell, Boltzmann, and Becker. Johansson and Persson, 1970.

equilibrium borehole pressure (p_{be}), **(MPa)**, the borehole pressure at the maximum expansion of the blasthole diameter.
equilibrium pressure in blasthole (p_{be}), **(MPa)**, see '**equilibrium borehole pressure**'.
equivalent support dimension (D_e), **(m)**, of the span, diameter or wall height of the excavation, to a time dependent and safety dependant *excavation support ratio* ESR. This value is used to relate the Nick Barton Q-value to the behaviour and support requirements of and underground excavation according to the following formula

$$D_e = 2R_{ESR}Q^{0,4} \qquad (E.9)$$

Barton has given the following values for the excavation support ratio R_{ESR} (ESR).

Table 21 Excavation support ratio ESR defined by Barton 1976.

Category	ESR (R_{ESR})
A. Temporary mine openings.	3–5
B. Permanent mine openings, water tunnels for hydropower (excluding high pressure penstocks), pilot tunnels, drifts and heading for large excavations.	1,6
C. Storage rooms, water treatment plants, minor road and railway tunnels, surge chambers, access tunnels.	1,3
D. Power stations, major road and railway tunnels, civil defence chambers, portals intersections.	1,0
E. Underground nuclear power stations, railway stations, sport and public facilities, factories.	0.8

equivalent sheathed explosives, ordinary permitted explosives to which extra common salt has been added and which appear to have the good effects of actual sheathed explosives. Cooper, 1963.
ER (acronym), see '**energy ratio**'.
ergonomy, the science of constructing machines and working places so the use of the of machines and other equipment does not cause physical damage do the operator.
ergonomic hydraulic breaker, pneumatic breaker with special designed feature that make it more safe and comfortable for the operator to use. Atlas Copco, 2006.
erosion (geology), the group of physical and chemical processes by which earth or rock material is loosened or dissolved and removed from any part of the earth's surface. In includes the processes of weathering, solution corrosion, and transportation. The mechanical wear and transportation are effected by rain, running water, waves, moving ice, or winds, which use rock fragments to pound or to grind other rocks to powder or sand. Fay, 1920.
erosion (image analysis), the process of reducing the size of overlapping fragments in image analysis until the fragments do not overlap. As a result of this operation, the undisturbed projection of each fragment appears on the screen and is then used for digitizing.
erosion drill, see '**water jet drill**'.
ERV (acronym), see '**elementary representative volume**'.
ESDB (acronym), see '**European stress data base**'.
ESR (acronym), see '**excavation support ratio**'.
ETDR (acronym), see '**Electrical time domain reflectory**'.

European Federation of Explosive Engineers (EFEE), an European organisation for explosive manufacturers established in Aachen, Germany, in October 1988. The council of EFEE has developed plans for the unification of the legislation and the education of European blasters. Jonsson, 1992. EFEE offers country, company or individual membership.

European Standardization Committee (CEN), includes all EU countries and associated states (Switzerland, Norway, Iceland).

European Standards (explosives), the harmonized standards issued by the European Standardization Committee (CEN), standing in all the EU countries and associated states (Switzerland, Norway, Iceland). There are five series of such standards:
EN 13857. General.
EN 13631. High explosives.
EN 13630. Detonating cords and fuses.
EN 13763. Detonators.
EN 13938. Propellants.

European stress data base (ESDB), a data base consisting of the result from measurements in boreholes of the *horizontal minor* and *major principal stresses*. ESDB forms a subset of the world stress map (WSM). ESDB contains 1 400 data points, including other existing compilations, e.g. the Fennoscandian rock Stress Data Base. Stephansson, 1993.

excavatability, see '**diggability**'.

excavation, digging and removing soil, blasting, breaking, and loading of coal, ore, or rock in mines. Continued excavation implies continued loading and clearing away. Nelson, 1965.

excavation disturbed zone (EZD), (R_{ezd}), (m), the zone affected by an underground excavation regarding changes in any property, i.e. stresses, deformation, chemical environment, water drainage, etc. This zone is generally larger than the calculated blast induced damage zone. The changes in rock properties are due to the redistribution of stresses in the rock because of the excavation. Pusch and Stanfors, 1992.

excavation support ratio (R_{ESR}), (ESR), see '**equivalent support dimension**'.

excavator, normally a wheel or crawler mounted digging machine that can turn the hydraulic or cable operated digging arm 360° and has a shovel at the front oriented downwards for digging trenches and foundation for buildings.

excavator turn table, structure that connects the under carriage to the main body and boom of a excavator in such a way that the main body and boom can rotate 360 degrees in relation to the under carriage. Atlas Copco, 2006.

exhaust scrubber, device used to remove unburned particles from the exhaust fumes of an internal combustion engine. Atlas Copco, 2006.

exoelectron emission (at rock fracturing), see '**fracto-emission**'.

exothermic reaction, a chemical reaction which is accompanied by liberation of heat.

exotic drilling methods, several proposals for new drilling methods have been proposed and especially for large deep holes like in oil drilling; 1) turbo driven wear drill, 2) machine gun drill, 3) implosion drill, 4) ultra sound drill, 5) spark drill, 6) detonating sphere drill, 7) electrical heated drill, 8) electrical heated drill, 9) electrical high frequency drill, 10) micro wave drill, 11) induction drill, 12) nuclear reactor drill, 13) melting drill, 14) electrical arc drill, 15) plasma drill, 16) electron ray drill, 17) laser drill and 18) chemical drill. See more details under respective entry.

expanding agent, a chemical compound of which the swelling properties are achieved by mixing, e.g. water and lime (CaO) to form $Ca(OH)_2$. The compound is poured into small boreholes for the purpose of breaking small volumes of rock where restriction on ground vibrations and flyrock are large. The expanding agent breaks the rock silently without causing ground vibration problems. Jana, 1991.

expanding bushing, tubular device either cylindrical or conical that can be made to increase in diameter when a force is applied. Atlas Copco, 2006.

expanding shaft, shaft that can be made to increase in size when a force is applied to it. Usually forced by a conical bushing and threaded screw arrangement. Atlas Copco, 2006. Expanding bolt is one example of expanding shaft.

expansion agent, see '**expanding agent**'.

expansion bit, a drill bit that may be adjusted to cut various sizes of holes. The adjustment of some types may be accomplished by mechanical means while the bit is inside the borehole. Also called '*paddy*' or '*paddy bit*'. Long, 1920.

expansion cutter or **expanding cutter**, a borehole drill bit having cutters that may be expanded to cut a larger size hole than the size of the bit in its unexpanded state, also a device equipped with cutter that may be expanded inside casing or pipe to sever, or cut slits or holes, in the casing or pipe. Compare expansion bit. Long, 1920.

expansion shell anchor, see '**expansion shell bolt**'.

expansion shell bolt, a rock bolt used for reinforcement of rock using a bar and an expansion shell at the bottom of the hole with the purpose to anchor the bolt into the rock mass.

expansion reamer, see '**under reamer**'.

expansion volume, the volume of air or liquid (water) where the blasted rock can expand. If the void where the material expands is less than 15% of the volume being blasted, the breakage mechanisms will be adversely affected and the rock fragments will tend to interlace, which results in compressed material. In underground blasts of great size, it is recommended that the available expansion volume be larger than 25% to achieve an adequate flow of rocks towards the loading points and avoid the formation of hang-ups. In tunnel and drift driving, if the volume of the void in the cut is too small, the phenomenon of sinterization or plastic deformation of the finely broken material will occur. Whenever possible, it is recommended that the available expansion volume be more than 15% of the actual volume of the cut. Jimeno et al., 1995.

expansive demolition agent, see '**expanding agent**'.

explode (in rock blasting), the process of undergoing a rapid combustion (not detonation) with sudden release of energy in the form of heat that causes violent expansion of the gases formed and consequent production of great disruptive pressure and a loud noise e.g. when gun powder explodes. If the reaction velocity is lower than the sound velocity in the explosive the term used should be *explode* and if the reaction velocity is equal or larger than the sound velocity in the explosive the term used should be *detonate*.

exploder or **blasting machine,** the term has two definitions; 1) A portable apparatus *(blasting machine)* for generating the energy for firing detonators. The term is used mainly in the field and a better term to use is *blasting machine* 2) *Detonator, blasting cap, or a fulminating cartridge* placed in a charge of gunpowder or other explosive, and exploded by electricity or by a fuse. Fay, 1920.

exploding bridge wire (EBW), a wire that explodes upon application of current. It takes the place of the primary explosive in an electric detonator. USBM, 1983.

exploding bridge wire detonator, an electric detonator that employs an exploding bridge wire rather than a primary explosive. An exploding bridge wire detonator functions instantaneously. USBM, 1983.

exploding foil initiator (EFI), a foil that explodes upon application of a current. Also called a slapper. RISI, 1992.

exploding wire, if a capacitor having a high voltage is discharged over a short electrical wire (about 5 cm) the wire will break under high pressure and this will be similar to the effect of an explosive. Exploding wire can be used in *model blasting* instead of using very sensitive primary explosives PETN or lead azide.

explosion, a violent and rapid exothermic chemical reaction which generates heat, light, and noise, whereby mixtures of gases, solids or liquids react with the almost instantaneous formation of gaseous pressures near sudden heat release. It includes both rapid deflagration, detonation of chemical explosives and the process of nuclear explosion. There must always be a source of ignition and the proper temperature limit must be reached to initiate the reaction. Also a term for release of energy in a limited space at a very fast rate. Examples of explosions are the gun powder for rock blasting, the sudden rupture of high pressure vessel, tyre, balloon etc. When using explosives for rock blasting, the term *explosion* should only be used for *reaction velocities less than the sound velocity* in the explosive "sub sonic velocities". For reaction velocities *equal or larger than the sound velocity* in the explosive "super sonic velocities" the correct term is *detonation*. Rustan, 2007.

explosion energy (Q_e), (MJ/kg), see '**heat of explosion**'.

explosion heat (Q_e), (MJ/kg), see '**heat of explosion**'.

explosion pressure or **borehole pressure** (p_b), **(MPa)**, the pressure developed at the instant of an explosion. Streefkerk, 1952 and Lownds, 1986. The pressure of the gaseous products of an explosion when they occupy the volume previously occupied by the solid explosive which means no expansion of the blasthole. Usually the explosion pressure is about half of the detonation pressure. Calculation of detonation pressure, see '**detonation pressure**'.

explosive magazine, *explosive store*, or *powder house*, a surface building at a mine, quarry or construction site where explosives and detonators may be kept. It must be at a certain minimum distance from other buildings and the maximum quantity of explosives that may be kept is fixed.

explosive pressure (p_b), (MPa), see '**borehole pressure**'.

explosion proof casing, a casing that is so constructed and maintained as to prevent the ignition of gas surrounding it by any sparks, flashes or explosions of gas that may occur within such casing. Fay, 1920. The use of these casing are needed in coal mines because of the emission of methane (CH_4).

explosion state, the physical conditions behind the detonation zone in an explosive regarding, e.g. pressure and temperature.

explosion state pressure (p_d), **(MPA)**, see '**detonation pressure**'. Lownds, 1986.

explosion temperature (T_e), **(K)**, calculated temperature of the fumes of an explosive material under the assumption of detonation without any expansion of the fumes. As in any combustion, the absolute temperature is given by;

$$T_e = \frac{Q_{cv}}{\Sigma(m_r C_T)} \qquad (E.10)$$

where Q_{cv} is total heat released at constant volume in J, m_r is total mass of the reaction products in kg and C_T is the specific heat at temperature T in J/kgK. Jimeno et al., 1995.

Explosion temperature can also be defined as the calculated temperature of the fumes of an explosive material which is supposed to have been detonated while confined in a shell assumed to be indestructible and impermeable to heat; the calculation is based on the heat of explosion and on the decomposition reaction, with allowance for the dissociation equilibria and the relevant gas reaction (thermodynamic calculation of decomposition reactions). The real detonation temperature in the front of the shock wave of a detonating explosive can be estimated on the strength of the hydrodynamic shock wave theory, and is higher than the calculated explosion temperature. Meyer, 1977.

explosive, any chemical compound or mixture that reacts at high speed to liberate gas and heat and thus cause high pressures. High explosives detonate, low explosives deflagrate. Both high and low explosives can be initiated by a single No. 8 blasting cap as opposed to blasting agents which cannot be so initiated. The term includes slow burning compositions used for initiation systems, such as detonators, detonating relays, fuse heads, etc., which are covered by explosives regulations. Explosives can be classified into four main groups: high explosives, blasting agents, low explosives and '*special explosives*', see Table 22. In some cases '*explosive substitutes*' may be used.

Table 22 Classification of explosives.

High explosives		Low explosives	Special explosives	Explosive substitutes
	Blasting agents			
TNT	ANFO	Black powder	Seismic	Expanding agents
Dynamite	Slurries		Trimming	Mechanical methods
Gelatins	Emulsions		Permissible	Water jet
Hybrid charges	ANFO-slurry		Shaped charge	Jet piercing
			Binary	Compressed air/gas
			LOX	
			Liquid	

explosive accident, accident caused by the civil or military use of explosives. The frequency of accidents depends at rock blasting on the rock condition, the awareness of risks, the human factor, and the level of compliance with regulations.

explosive burning or **combustion**, oxidation at which chemical energy is transformed to heat, often during emission of light and forming new gaseous products. Burning with a speed normally of 0,01–1 m/s perpendicular to the burning front combined with a fast ignition along the surface of the material. TNC, 1979.

explosive casting, see '**cast blasting**'.

explosive characterization, explosives are characterized for blasting purposes by many parameters, like density, detonation velocity (c_d), explosive heat, mass strength, critical diameter, water resistance, etc.

explosive charge in a blasthole (Q_{bh}), (**kg**), the amount of explosive charge in a blasthole.

explosive charge in a blast (Q_T), (**kg**), the total amount of explosive charge in a blast or coyote tunnel.

explosive charger, a person filling explosive into a blasthole.
explosive compaction (EC), see '**explosive compaction of soils**'.
explosive compaction of soils, vertical drillholes are drilled through the water saturated soil and charged with explosive decks. A typical 20 m long blasthole could have 4 charges fired with the following masses and firing times starting from the bottom with 14 kg (2 500 ms), 10 kg (4 500 ms), 8 kg (6 500 ms) and 5 kg (8 500 ms). The delay time between blastholes may be in the order of 25 ms. The vibration and pressure causes the pore water to flow to the surface. For 20 long blastholes a mean settlement has been measured to 660 mm which corresponds to 3% compaction in height. Clarke, 2010.
explosive consumption total (m_T), **(kg)**, total amount of explosive used for a specific task.
explosive decks, see '**deck**'.
explosive desensitization, see '**desensitization**'.
explosive drilling, a technique developed for deep-hole drilling in especially strong and abrasive rocks. In this method, a series of small underwater explosions are used to break the rock at the bottom of the hole, the fragments from each explosion being washed away by the flushing water. The explosive used is in the form of a liquid which is transported down the hole unmixed in non explosive chemical components which are then automatically mixed in correct proportions for maximum sensitivity at the drill head. Since the energy is liberated at the bottom of the hole there is no energy loss with depth and since the explosion is underwater, the shock wave is transmitted with maximum efficiency. Min. and Minerals Eng., 1965.
explosive factor or *powder factor* (*q*), **(kg/m³)**, see '**specific charge**'.
explosive energy (Q_e), (MJ/kg), see '**heat of explosion**'.
explosive engraving, method to engrave a pictures on a metal sheets with the help of explosives. The space between the sheet metal to be formed and the matrix is evacuated to remove the resistance of the air during the forming process. Sometimes water is used as a medium to transfer the impulse from the explosive to the metal. Most explosive forming operations are carried out under water, in a tank or a pond, using heavy steel matrixes. Persson et al., 1994.
explosive forming, use of explosives to form metals, see also '**explosive engraving**' or join different metal sheets, see '**explosive welding**'.
explosive heat (Q_e), (MJ/kg), see '**heat of explosion**'.
explosive impedance (Z_e), **(kg/m² s)**, is defined by the product of explosive density ρ_e in kg/m³ and velocity of detonation, c_d in m/s.
explosive load (q_l), **(kg/m)**, see '**linear charge concentration**'.
explosive loading factor or *powder factor* (*q*), **(kg/m³)**, see '**specific charge**'.
explosive magazine, *explosive store*, or *powder house*, a surface building at a mine, quarry or construction site where explosives and detonators may be kept. It must be at a certain minimum distance from other buildings and the maximum quantity of explosives that may be kept is fixed.
explosive materials, a set of indispensable requisites for the successful performance of a blast operation. This includes (but is not necessarily limited to) dynamite and other high explosives, slurries, water gels, emulsions, blasting agents, black powder, pellet powder, initiating explosives, detonators, safety fuses, squibs, detonating cord, igniter cord, and igniters.
explosive oil, see '**Nitroglycerin**'. CCD 6th ed., 1961.

explosive oils, liquid sensitizers for explosives such as nitroglycerin, ethyleneglycoldinitrate, and metrioltrinitrate. Atlas Powder, 1987.

explosive performance, see '**explosive test**'.

explosive performance term (EPT), an empirical term which allows comparison of the relative fragmentation performance of equal volumes of different explosives at the same burden distance and borehole diameter. The EPT term was derived on half scale tests in three different rock types.

$$EPT = (0{,}36 + \rho_e) \frac{c_d^2 \rho_e Q_e}{R_{dcv}\left(1 + \frac{c_d^2}{c_p^2} - \frac{c_d}{c_p}\right)^{1{,}33}} \quad \text{(E.11)}$$

EPT = explosive performance term
ρ_e = density of explosive (kg/m³)
c_d = velocity of detonation (m/s)
Q_e = calculated maximum expansion work of explosive (kcal/g)
c_p = P-wave velocity (m/s)
R_{dcv} = volume decoupling ratio

If the numerical value of the EPT is higher than that of a standard explosive, a better fragmentation performance, i.e. smaller average fragment size, is inferred. Likewise, a smaller number than that of the standard explosive indicates inferior performance. Bergman and Riggle, 1973.

explosive permissible, see '**permissible** or **permitted explosives**'.

explosive pressure (p_b), **(MPa)**, see '**borehole pressure**'.

explosive, primary, secondary, and tertiary, a classification of explosives where the explosives are arranged in descending order of initiation sensitivity (as a result of a particular type of stimulus). One obtains an almost continuously graded sequence. Initiating explosives are called primary explosives and high explosives are called secondary explosives. Tertiary explosives contain such substances as ammonium nitrate, and fall close to the low end of the sensitivity scale.

explosive ratio (*q*), **(kg/m³)**, see '**specific charge**'.

explosive sensitivity, the ease with which an explosive will detonate or explode. An explosive must be sufficiently insensitive to withstand any shocks that may occur in handling and transporting, but, at the same time, it must detonate when initiated by an ordinary detonator, and also transmit detonation wave from one cartridge to another in a shot hole. Nelson, 1965.

explosive shattering, this method consists in soaking the ore thoroughly in water and then heating to 180°C under a pressure of 1 MPa. The pressure is then suddenly released, and the absorbed water is converted to steam that disrupts the ore. Explosive shattering is said to be more effective than ordinary crushing and grinding in liberating mineral particles without harmful over grinding. Newton, 1959.

explosive specific (*q*), **(kg/m³)**, see '**specific charge**'.

explosive store, **explosive magazine** or *powder house*, a surface building at a mine, quarry or construction site where explosives and detonators may be kept. It must be at a certain minimum distance from other buildings and the maximum quantity of explosives that may be kept is fixed.

explosive strength (Q_e), **(MJ/kg)**, can be measured by different methods in the laboratory and in the field. One of the major parameters for explosive strength is the energy content in the explosive also called *explosive heat*. When we look upon the practical use of explosives for rock blasting, it is necessary to know what result properties are the most important when using the explosive in the field, the ability to *fragment and move the rock mass* or *to reduce backbreak* and *ground vibration* in the rock mass. We must know which result parameters of the blast are the most interesting in a certain case. It is not enough to give only a strength value when explosives are classified, as for example the *Langefors weight strength* that is based on an empirical weighting of *explosive energy* plus the *gas volume* produced by the explosive. Instead we need a *system thinking* that the blast result is not only dependent on explosive strength because it depends also highly on the rock properties, the geometry and initiation design of the round. To the best knowledge today the following parameters are the most important when using explosives for rock fragmentation: 1) *velocity of detonation,* 2) *charged explosive density,* 3) *volume of gas produced per kg of explosive,* 4) *heat of explosion and* 5) *temperature of the blast gases after detonation,* 6) *density of rock* and 7) *P-wave velocity in rock*. The specific charge and joint conditions are also of very great importance for the blast result. Because the values of several of these parameters e. g. the detonation velocity (c_d) is dependent on the blasthole diameter and the density of liquid explosives are dependent on the height of the blasthole, the *single hole blasting method* in full scale (the **SHB** method), is therefore believed to be the only way to determine the total effect of a detonation in a certain kind of rock mass. *Half scale- and laboratory test methods* for determining explosives strength can only be used for a relative comparison between explosives and not for prediction of blastability in different rock types.

Full scale blasts
1. Single hole blasting tests (SHB-tests) in the rock material and with the explosive of interest and with that hole diameter and bench height which is going to be used, see Rustan and Nie, 1987. *This test method gives a comprehensive answer regarding explosive strength because it takes into account the rock mass properties and the actual rock confinement and the in situ detonation velocity (c_d) that depends on both blasthole diameter and the confinement.*

Half scale blasts
2. Underwater detonation tests for calculation of breakage index. In these test the size of the charge can be increased compared to laboratory, but the use of water still gives a different confinement compared to rock.

Laboratory tests (Only for relative classification of explosives)
3. Ballistic mortar
4. Grade strength (Strength in comparison to 60% dynamite)
5. Brisance (Only usable for evaluation for boulder blasting)
6. Trauzl lead block test
7. Plate dent test (Reflects the detonation pressure)
8. Cylinder expansion test
9. Crater test (Only useful for crater blasting)
10. Langefors weight strength. Cannot be used for modern explosives.

Regarding the energy transfer, from the explosive to the rock mass, the velocity of detonation, the P-wave velocity in the rock mass, the density of explosive and the density of rock are the most important parameters known today. Recent research indicates that the strength of the explosive is better expressed by the detonation pressure p_d than the formula developed by Langefors.

$$p_d = \rho_e c_d^2 \qquad (E.12)$$

where ρ_e *is the density of explosive in kg/m³ and* c_d is the detonation velocity of the explosive in m/s. To classify an *explosive regarding its full strength,* it is necessary to include both *explosive-* and *rock mass properties* and the *blast and initiation geometry*.

explosive stripping, the use of explosives in removing overburden in surface mining.

explosive substitutes, rock breaking methods replacing explosives such as compressed air/gas, expanding agent, mechanical methods, water jet and jet piercing.

explosive test, method or technique to assess/determine the properties of explosives. There are many different tests, e.g. 1) *Abel heat test,* 2) *Freezing and thawing test,* 3) *Liquefaction test,* 4) *Impact sensitivity test,* 5) *Friction impact sensitivity test,* 6) *Air gap sensitivity test,* 7) *Velocity of detonation test,* 8) *Water proofness test,* 9) *Strength of explosive test* and others. The UN have developed international standards for testing of explosives. UN, 1990. The Nordic countries recently developed a standard, 'Nordtest Scheme for Testing Commercial Explosives', based mainly on the UN standard. Anon., 1990.

explosive welding, two plates can be welded together if the plates are placed horizontally in close contact and on the top is placed an even layer of explosive. At the detonation of the explosive the impact from the upper plate will be so high that it is cold-welded to the lower sheet. Stainless steel can for example be welded against ordinary steel by this method.

explosively actuated device, any tool or special mechanized device that is actuated by explosives, such as jet tappers and jet perforators. The term does not refer to propellant-actuated devices. Atlas Powder, 1987.

explosively anchored rock bolt, see '**rock bolt explosively anchored**'.

exposure of the rock, removing the overburden from the bedrock.

extended charges, explosive charges spaced at intervals in a quarry or opencast blast hole. See '**deck charging**'.

external detonation test, detonation of an explosive under low confinement (air).

extension, any device used to increase the length of a structure. Atlas Copco, 2006.

extension coupling, coupling consisting of a threaded tabular section around which a loose- or tight-fitting ring is placed. The coupling connects the core barrel to the first drill rod. Also called guide-ring coupling. Long, 1920.

extension drilling, drilling operation where extension rods or tubes can be coupled together with threads to the total desired length of drilling. In underground operations the length of the drill rods are normally 1,5–2 m in length because of limited space. On surface the length of each extension rod or tube can be up to 10 m. Atlas Copco, 1983.

extension drill rod, threaded steel rod that transfers flushing media, rotation and percussion energy between rock drill and the drill bit. Atlas Copco, 2006.

extension drill steel, see '**extension drill rod**'.

extension drill string, drill string comprised of several rods having threads and connected together. Atlas Copco, 2006.

extension rod, see '**extension drill rod**'.

extension tube (drilling), drill steel made in the shape of a tube where the water can be supplied to the bit through the tube. The tubes can be coupled together with the help of threads.

extension tube (hydraulic), large tube with a slightly smaller one fitted inside in such a way that the device can be lengthened or shortened by moving the tubes. Atlas Copco, 2006. Used in hydraulic cylinders.

extensometer (MPBX), instrument used for measuring small deformations, deflection, or displacements. USBM Bull. 587. 1960. Normally measurements are done in boreholes. The distance change can be transferred to the measuring point by rods, tape or wire. The *multipoint borehole extensometer* MPBX is one of the most useful measurement methods of displacements in rock masses, see Fig. 27.

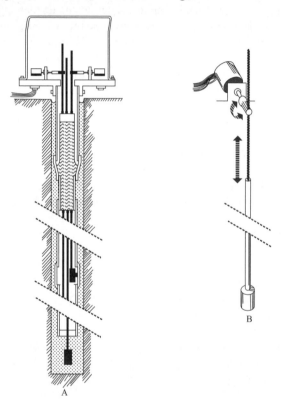

Figure 27 Typical construction of a remote read multiple point borehole rod extensometer. A) complete instrument B) Details of potentiometer read-out system. After Stillborg, 1994.

external diameter (d_{ex}), (mm or m), the outer diameter of a rod or cylinder.

external fire test, a test to determine how fire affects the explosive in the transport packaging and whether the substance explodes when it is subjected to fire. This test method is needed for the United Nation Hazard Division 1.5. Persson et al., 1994.

EXTEST (abbreviation), for '**European Commission for the Standardization of Explosive Tests**'. Student-Bilharz, 1988.

extraction, the process of mining minerals.

extraction of stuck drill steel, special fishing tools have been developed for the extraction of stuck drill steel. TNC, 1979.

extraction- or **production level**, the level in a mine where the minerals are mined and transported to the main haulage system.

extractor, device used to free stuck drill rods. Atlas Copco, 2006.

extra (ammonia) dynamite, a dynamite that derives the major portion of its energy from ammonium nitrate, also called ammonia dynamite. IME, 1981.

extraneous electricity, flows of current that circulate outside a normal conductor. In blasting it is the electrical energy, other than the actual firing current, which may be a hazard in connection with electric blasting caps. It includes stray current, static electricity, lightning, radio-frequency energy, and capacitive or inductive coupling.

extrusive rock, effusive rock or **volcanic rock**, any igneous rock derived from a magma or from magmatic materials that was poured out or ejected at the earths surface, as distinct from an intrusive or plutonic igneous rock which has solidified from a magma that has been ejected into older rocks at depth without reaching the surface.

exudation, a separation of oily ingredients out of explosives during prolonged storage, especially at elevated temperatures. For nitroglycerin based cartridged explosives a condition when free nitroglycerin can be traced. This condition is very dangerous.

eye bolt, bolt with a loop or eyelike opening at one end for receiving a hook or rope. Atlas Copco, 2006.

EZD (acronym), for '**excavation zone of disturbance**'.

fabric, orientation in space of the elements composing the rock structure. ISRM, 1975. The special arrangement and orientation of rock components, whether crystals, grains or sedimentary particles, as determined by their sizes, shapes etc. A.G.I. Suppl 1960.

fabric element, a rock component, ranging from an atom or an ion to a mineral grain or a group of mineral grains in pebbles, lenses, layers, etc., that acts as a unit in response to deformative forces. A.G.I. Supp., 1960.

fabric habit, the relations between the shape of a mineral grain and its lattice (screen size) structure. A.G.I. Supp., 1960.

face, free surface in a blast. It is any rock surface exposed to air, water or buffered rock in blasting. A face provides the rock during blasting with expansion volume. The term is often used for the end of any excavation, where work is progressing.

face-discharge bit, a bit designed for drilling in soft formations and for the use on a double-tube core barrel, the inner tube of which fits snugly into a recess cut into the inside wall of the bit directly above the inside reaming stones. The bit is provided with a number of holes drilled longitudinally through the wall of the bit through which the circulation liquid flows and is ejected at the cutting face of the bit. Also called *bottom-discharge bit*, *face ejection bit*. Long, 1920.

face drilling, the operation used underground to drill blastholes in the drift-, or tunnel face for the next round to be blasted. Atlas Copco, 2009a.

face-ejection bit, see '**face-discharge bit**'.

face hammer, used for rough dressing of stones. It has one blunt end and one cutting end. Crispin, 1964.

face of hole, the bottom of a borehole. Long, 1920.

face plate, a plate made of steel and has normally a square, circular or triangular shape and its purpose is to distribute the load at the rock bolt head uniformly into the surrounding rock, see Fig. 28. Some common types are *flat plate*, *domed plate* and *triangular bell plate*. To maintain the elasticity of the rock bolt system, the choice of the face plate is crucial. Stillborg, 1994.

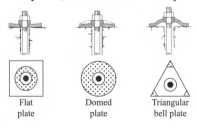

Figure 28 Some commonly used face plates and their respective effective contact surface. After Stillborg, 1994.

face shape measurement, the shape of the bench to be blasted can be measured by a pulsed laser beam from one set up, hitting the bench face. The print out for e.g. Quarryman (tradename) allows the engineer to determine, *toe burden, drill angle, face height, crest burden* and *minimum point of burden*.

facies, the aspect belonging to a geologic unit of sedimentation, including mineral composition, type of bedding, fossil content, etc. Sedimentary facies are really segregated parts of differing nature belonging to a genetically related body of sedimentary deposit. G.S.A., 1949.

Fagersta cut, a parallel hole cut of cylinder type with one center hole 75 mm in diameter, see Fig. 29. The empty hole is drilled in two steps, the first as an ordinary hole and the second as an enlargement of this pilot hole. The cut consists of two rhombic rows of blastholes with four holes in the first row and six holes in the second row. Langefors, 1963.

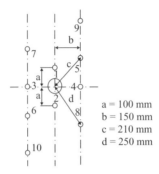

Figure 29 Fagersta cut with one uncharged opening hole with diameter 75 mm. After Langefors, 1963.

failure, the process taking place when the maximum strength of a material or specific design is achieved and visible cracks are developed. See also definition of 'damage'.

Applied it is when a structure element can not longer fulfil the purpose for which it was designed. Any structure will probably give evidence of being unsafe before actual collapse. Ham, 1965. Failure is not a unique material property. It is a *system property* because it depends not only on *different kind of loads* e.g. compression, shear, tensile, bending or torsion. It depends also on the *confinement, deviatoric stress* and the *strain rate* on the specimen tested. Rustan, 2008.

failure-induced deviatoric strain (ε_f), (dimensionless), see '**deviator strain**'. This quantity can be used to quantify damage in rock. Hearst 1976. $\varepsilon_f = 0,1$ is associated with *intense fracture*, $\varepsilon_f = 0,01$ with *macroscopic fracture*; and $\varepsilon_f = 0,001$ with *enough fracture* to produce chimneys above nuclear explosions. Schatz, 1974.

failure of rock mass, loosening or fall-out of rock from the rock mass. In reality, failure of a rock mass is almost always dominated by pre-existing or newly created weaknesses and can be grouped into two classes *brittle-* and *ductile fracture*.

failure-, rupture or **breakage wave**, when the rock mass breaks, seismic waves are created. The amplitude of these waves can be measured by accelerometers.

failure or **rupture stress (σ_{fail}), (MPa)**, stress level reached when the material breaks into pieces loosing its tensile strength.

failure zone or **breakage zone**, the zone where the rock mass has zero tensile strength. The rock mass can still take compression loads at an angle to the breakage zone but not parallel to the zone. A failure zone has still some shear strength.

fall hammer, an instrument used to determine the impact sensitivity of an explosive.

fall hammer test, the BAM fall hammer test determines the impact energy needed for initiation by dropping a mass at increasing heights onto the explosive until it detonates. Persson et al., 1994.

falling head test, a soil and rock permeability test in which the borehole is filled up with water and sealed at the end or between two points, The sealed section is set under pressure and the rate at which the water head falls is observed. The method can also be used to examine if there are open joints, faults or cracks between boreholes before blasting.

falling object protective structure (FOPS), demands on a cabin attached to a movable underground machine to make the cabin mechanically safe for falling objects according to European and International safety demands. FOPS is especially important for mechanized scaling and bolting equipment and also LHD-loaders used underground. Atlas Copco, 2006.

falling pin seismometer (old seismometer type), an instrument to quantify blast vibrations. It consists essentially of a level glass base on which a number of 6 mm diameter pins of length ranging from 152 mm to 381 mm are standing upright. The pins stand inside hollow steel rods so that each pin can fall over independently of the others. The longer the pin the less energy required to topple it. In practice it has been accepted that if the shorter pins (up to 254 mm) remain standing, then there is no possibility of structural damage to a building. See also '**vibrograph**'. Nelson, 1965.

fall of ground, the falling of stones from the back of tunnels and stopes in an underground excavation.

fallout, occurs when a rock slab or several rock slabs detach completely from the rock mass in and underground excavation, bench face, drilled hole or full-face bored or reamed hole due to the rock mass stresses and (or) gravity forces acting on the rock mass.

fan blasting, diverging blastholes normally drilled upwards for production blasting in sublevel caving and sublevel stoping. The fan is drilled more ore less vertically and perpendicular to the drift but could also be drilled inclined when the holes are used for pre-grouting.

fan cut, a cut where part of the drift area is opened by blastholes with successively larger inclination to the free face. See Fig. 30. Sen, 1995.

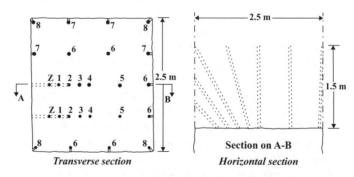

Figure 30 Fan cut. After Sen, 1995.

fan drilling, drilling pattern used in underground mining, where holes are drilled in the same plane but at different angles to form a fan or sun burst pattern.

fan drilling rig, drilling rig in which the feed mechanism can be rotated to allow holes to be drilled in a fan shaped or sun burst pattern. Atlas Copco, 2006.

far field, the distance from a point of interest in relation to a geometric quantity e.g. borehole diameter or hydraulic radius of a drift or tunnel. The limit between far, intermediate and near field should be defined for each special use of the term. For example; in large diameter hole blasting on the surface, the region greater than ~100 m from the source of the blast. In blasting underground, the far field starts at about two hydraulic diameters from the excavation drift or tunnel.

far-field stress, a stress state which is not perturbed by a heterogeneity. Hyett, 1986.

fast (igniter) **cord**, it consists of three central paper strings coated with a black powder composition and held together with cotton countering. These are then enclosed in an extruded layer of plastic incendiary composition and finished with an outer plastic covering. The overall diameter of fast igniter cord is approximately 2,5 mm. McAdam II, 1958.

fatigue, failure of a material under repeated stress. Under the repetitive loading of a material a small crack is developed and the crack is extended in small steps until the material breaks.

fatigue life, the number of cycles (K) of stress that can be sustained prior to failure for a stated test condition. A.S.M. Gloss., 1961.

fatigue limit, there are two limits for fatigue; the lower limit is characterized by the threshold value ΔK_o (1 loading cycle) and the upper limit (infinite number of loading cycles) is given by K.

fault, in geology a geological zone of disturbance such as a fracture or a fracture zone along which there has been shear displacement of the two sides relative to one another and parallel to the fracture. (This displacement may be as small as a few centimetres or as large as many kilometres). Both *axial* and *shear microcracks* play significant roles in the final stage of the *micro faulting process*.

feathered plastic collar, a device made of plastic to centralize pipe charges at their connections in contour blasting. When the pipe charges are connected, the feather is activated and its diameter is a little larger than the hole diameter. When the pipe charges are pushed into the borehole, the pipe charges are automatically locked.

feather joints, occurs preferentially in the immediate vicinity of a fault plan and intersect the fault in an acute angel pointing in the direction of the relative movement of the block containing the '*pinnate fractures*'. Hudson, 1993.

feather pair, wedge shaped steel pieces with a semicircular shape. They are placed with the flat inner surface into a drill hole and a wedged is placed between the pieces and a hammer is used to push the wedge deeper and deeper between the pieces to split rock. Atlas Copco, 2006.

Federation of European Explosive Manufacturers (FEEM), an organization for European explosive manufacturers. In 1975 FEEM was formed as a sector group of the European Council of Chemical Manufacturers Federation CEFIC. The latter organization is located to Brussels. The objective of the Federation is the improvement and development in methods of manufacture of explosives, to improve the safety and security and working conditions in the manufacture of explosives and in the transport, handling, storage and use of explosives. The number of members in 1994 were 26.

feed (noun), component of rotary or percussive rock drills upon which the pneumatic or hydraulic hammers move back and forth, also supplying the necessary thrust load for the advance. Mechanical equipment to move the drilling machine along a jig in optimal speed regarding the drillability of the rock. Specially formed beam, with all its attachments and parts and fitted with a rock drill cradle that is connected to a hydraulic cylinder or motor by pulleys and cable or chains and sprockets. The feed is fastened to the boom and is used to advance the rock drill when holes is being drilled and retract it when the hole is completed. With today's hydraulic percussive drilling machines speeds up to 3–4 m/min can be reached in hard granite for a 45 mm hole in diameter.

feed attachment, means by which the feed is attached to the boom. Atlas Copco, 2006.

feed beam, metal beam along which the rock drill is moved when drilling a hole. Atlas Copco, 2006.

feed control, device used to control speed and force that is applied to the rock drill as it moves along the feed beam. Atlas Copco, 2006.

feed cylinder, a hydraulic cylinder and piston mechanism, such as that on a diamond drill swivel head to transmit longitudinal movements to the drive rod and chuck to which the drilling stem is attached. Also called '*hydraulic cylinder*'. Long, 1920.

feeder, the mechanical part of a drilling jumbo with the purpose to create thrust on the drill bit. Sandvik, 1981.

feed force (F_f), (N), the thrust on the drill bit at drilling.

feed holder, device on which the feed is mounted that allows it to move forwards and back to position it against the rock. Atlas Copco, 2006.

feed piston bore, perforation in the center of a feed cylinder piston. Atlas Copco, 2006.

feed pressure (p_{feed}), (kPa), the pressure in the hydraulic system used for creating thrust on the drill bit.

feed rate (v_f), (m/min), the velocity of the feeder of the drill machine.

feed rate control, device used to control the rate at which the drill is moved forward when drilling. Atlas Copco, 2006.

FEEM (acronym), for '**Federation of European Explosive Manufacturers**'.

felicity ratio (R_f), (dimensionless), a term used in acoustic emission technique to quantify that acoustic emission starts a little earlier than before the maximum stress level used in the material earlier. It is defined as follows,

$$R_f = \sigma_{onset}/\sigma_{max} \tag{F.1}$$

where R_f is the felicity ratio (dimensionless), σ_{onset} is the stress level in MPa when acoustic emission starts and σ_{max} is the maximum earlier stress level applied earlier in the material in MPa. Seto, 1995.

FEM (acronym), for '**finite element method**'.

female thread or **box thread**, the thread on the inside surface of a coupling or tubular connector. Long, 1920. Thread cut on the inside of a perforation. The term *box thread* is used when referring to drill tubes. Atlas Copco, 2006.

ferrite, pure or nearly pure metallic iron as a crystalline constituent of manufactured iron and steel. English, 1939.

fertilizer grade ammonium nitrate, a grade of ammonium nitrate as defined by The Fertilizer Institute (USA). IME, 1981.

FF (acronym), see '**fracture frequency**'.

FGr (abbreviation), for '**fine gravel**' defining particles with sizes from 2,0 to 6,3 mm.

fibreglass rope, a measuring-tape covered with non-glare texture polyvinyl and printed on both sides for easy reading.

fibrecrete, shotcrete mixed with steel fibres and with the purpose to reinforce the shotcrete.

Fick's law, states that the *mass transport* or mass flux is a function of the product of the *concentration gradient* and *diffusion coefficient*. Hudson, 1993.

fill, tailings, waste, etc. used to fill underground space left after extraction of ore and with the purpose to support the hanging wall from caving and to prevent subsidence of ground above the mined orebody. The fill should be stable enough to work on with men and machines. The fill is termed '*hydraulic fill*', if flushed into place by water.

filler cap, a seal, normally circular, that can easily be removed from a tank to allow it to be filled with fluid. Atlas Copco, 2006.

filler pump, a pump used to fill fluids such as diesel fuel into the tank on a drill rig. Atlas Copco, 2006.

fill factor, the approximate load the dipper (digging bucket) actually is carrying expressed as a percentage of the rated capacity. The fill factor is commonly called the *dipper factor* for shovels or the *bucket factor* for draglines. Woodruff, 1966.

filling material, material such as waste, sand, ashes, and other refuse used to fill in worked-out areas of excavations. Stoces, 1954.

filter, device filled with fiber like or porous material, through which a gas or liquid is passed to remove impurities. Atlas Copco, 2006.

filter set, number of filters making up a complete set or a filter and related parts needed to replace a filter. Atlas Copco, 2006.

fines, particles less than a defined maximum size, that depends on the type of operation. In a limestone quarry <25 mm could be defined as fines and in an iron ore mine e g LKAB <10 mm will be defined as fines. Rustan, 2008. In a coal mine <0,5 mm can be regarded as fines. Measured fines, <1 mm, in six different mines varied from 2–18%. Onederra, 2004.

fine sand, all grains between 0,125 and 0,25 mm in diameter. AGI, 1957. In Sweden fine sand is defined as grains between 0,02 to 0,2 mm. Magnusson, 1963. International definition is lacking.

finger bit, a steel rock-cutting bit having finger like, fixed, or replaceable, steel-cutting points affixed. Long, 1920.

finger lifter, a basket-type core lifter. Long, 1920.

finish grinding, the final grinding action on a work piece where the objectives are surface finish and dimensional accuracy. ASM Gloss, 1961.

finish machining, see '**finish grinding**'.

fire, (imperative) in blasting the act of initiating an explosive.

firedamp, a combustible gas that is formed in mines due to the decomposition of coal or other carbonaceous matter, and which consists chiefly of methane; also the explosive mixture formed by this gas with air. Webster, 1961.

fire-resistance, ability to resist fire or offer reasonable protection against fire.

fire setting, an ancient method for tunnelling or mining by heating the rock with a wooden fire against the face of the mineral, which was then quenched with water, thus causing cracking. Pryor, 1963.

fireworks, combustible or explosive compositions or manufactured articles designed and prepared for the purpose of producing audible or visible effects. IME, 1981.

firing, the process of initiating an explosive charge or the operation of a mechanism which results in a blasting action. B.S. 3618, 1964.

firing cable, see '**shotfiring cable**'.

firing circuit, electrical circuit, which may be simple series, simple parallel, reverse parallel, series-in-parallel, or parallel-in-series, and obtained by connecting detonator lead wires in the appropriate ways.

firing current, electric current purposely introduced into a blasting circuit for the purpose of initiation. Also, the amount of current required to activate an electric blasting cap. USBM, 1983.

firing line, lead wire or **shot firing cable**, a line, often permanent, extending from the firing location via a connecting wire to the electric blasting cap circuit.

firing pattern, plan showing the intervals and/or firing sequence of a round. Nitro Nobel, 1993.

firing sequence, the firing of charges in predetermined order. AS 2187.1, 1996.

firing wire or **connecting wire**, insulated wires used to connect the cap wires of detonators (leg or lead wires) to the source of energy to be used for firing. A new wire is used at each blast.

firing machine, see '**blasting machine**'.

firing time, the time for each separate charge to be initiated for detonation in a blast round.

first degree of freedom, see '**presplitting**'.

Fisher constant (K), (**dimensionless**), see '**Fisher distribution**'.

Fisher distribution, a distribution used to quantify the joint orientation in a rock mass. It is based on analysis of orientation statistics assuming that a population of orientation values is distributed about some "true" value. This is directly equivalent to the idea of discontinuity normals being distributed about some true value within a set. Fisher assumed that in the probability distribution, $P(\theta)$, an orientation value selected at random from the population makes an angle of between $\theta + d\theta$ with the true orientation.

$$P(\theta) = \frac{K \sin\theta \cdot e^{K\cos\theta}}{e^K - e^{-K}} \qquad (F.2)$$

where K is the Fisher constant controlling the shape of the distribution. Fischer, 1953.

fishing pike extension, extension piece for a pointed tool used for recovering lost drill rods or other equipment from a drill hole. Atlas Copco, 2006.

fishing rod string, string of rods put together specifically for recovering rod or tubes lost down the hole. Atlas Copco, 2006.

fishing sleeve, a tool to grab the lost rods in the drill hole and remove them. Sandvik, 1983.

fishing tap, threaded conical device lowered into a drill hole and threaded into a detached drill string as a means of recovering it. Atlas Copco, 2006.

fishing tool, a tool to recover or overcome broken bits or other harmful objects from the bottom of a borehole. Pieces of metal are sometimes recovered by the use of a strong magnet attached to the drill string. Nelson, 1965.

fishtail bit, a rotary bit used to drill soft formations. The blade is flattened and divided, the divided ends curving away from the direction of rotation. It resembles a fishtail. Also called a 'drag bit'. A.G.I., 1957.

fissile, capable of being split, as schist, slate, and shale. Fay, 1920.

fissility, the property of rocks characterized by separation into parallel laminae as slate, schist, etc. Webster, 1960.
fissure, see '**discontinuity**'. A mechanical open or filled discontinuity of a width up to 0.1 m. Norwegian Rock Mechanic Group, 1985. An extensive crack, break, or fracture in the rock. A mere joint or crack persisting only for 0,05–1 m is not usually termed a fissure by geologists or miners. Fay, 1920.
fissure growth velocity (c_c), **(m/s)**, see '**crack velocity**'.
fissure propagation velocity (c_c), **(m/s)**, see '**crack velocity**'.
fittings, see '**connection**'.
fixation factor, see '**confinement**'.
FLACK, a 2-D (two dimensional) finite difference code. Hudson, 1993.
flags-danger, flags, usually red, that may or may not be imprinted with a warning and are used to caution personnel around explosives operations, or displayed on trucks transporting explosives. Atlas Powder, 1987.
flakiness, the flakiness of a fragment is the mean ratio between the width and thickness of the fragments. TNC 73, 1979.
flame drill method, see '**jet piercing**'.
flameproof, ability to resist flames; a term descriptive of electrical machines, switches and fittings demanded legally for use in fiery mines in Great Britain. Enclosing boxes with accurately fitted wide flanges are used. Pryor, 1963.
flame sensitivity, see '**flammability**'.
flammability, the ease with which an explosive material may be ignited by flame and heat. Atlas Powder, 1987.
flange, projecting rim or collar on a pipe, beam or wheel used to facilitate attachment to another part. Atlas Copco, 2006.
flange half, one half of a pair of flanges. Atlas Copco, 2006.
flange joint, conventional reverse circulation drill pipe comes in two basic forms, with modifications. Pipe up to 200 mm is threaded and coupled pipe. In larger sizes reverse circulation drill pipe often has flanged joints held together by bolts and nuts. Stein, 1997.
flare, a pyrotechnic device designed to produce a single source of intense light. Atlas Powder, 1987.
flashing mixture (surrounding the fuse head), see '**fuse head**'.
flash over, sympathetic detonation of explosive charges put under separate delay or between charged blastholes. Olofsson, 1990.
flash over test, a test to determine the maximum distance for flash over between one charge and another. See '**gap test**'.
flash point, the lowest temperature at which vapours of a volatile combustible substance ignite in air when exposed to flame; determined in an apparatus specifically designed for such testing. Atlas Powder, 1987.
flat-bottom crown, see '**flat-face bit**'. Long, 1920.
flat drill, drill rods made by steel having a rectangular cross section instead of circular.
flat-face bit, a diamond core bit where the face in a cross section, is a square. Also called *flat-bottom crown*, *flat-nose bit*, *square-nose bit*. Long, 1920. Drill bit design in which all the cutting points or buttons are at the same height, forming a flat hole bottom drilling. Atlas Copco, 2006.
flatjack test, a method to measure the field stress over a larger area e g 300 × 300 mm. It includes drilling of two short holes and between them a slot is done where the flatjack

pressure cell can be inserted and grouted, see Fig. 31. The distance between the short holes are measured before drilling of the slot. The flat jack is pressurized until the distance between the holes is restored and thereby also the field stress. Brady, 1985.

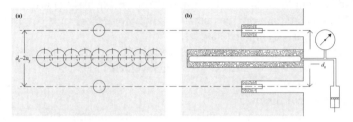

Figure 31 (a) Core drilling a slot for a flatjack test and (b) slot pressurization procedure. After Brady, 1985.

flat key, locking device made from a flat piece of metal. Atlas Copco, 2006.
flat-nose bit, see '**flat-face bit**'. Long, 1920.
flaw, a small defect in a material which when stressed may develop into a crack.
Flexirope (tradename), a special designed cable bolt to be grouted into drillholes. Yield strength for a 28 mm cable is dependent on quality and ranges from 50 or 500 tons. Stillborg, 1994.
flexural failure, in drifting, the rock excavation causes the roof normally to deflect downwards due to the surrounding stress and the gravity stress and this will cause a downwards bending of the roof especially when the roof is flat. In bench blasting the face in front of a blasthole is bent and vertical surface cracks are developed propagating from the face towards the blasthole.
flexural modulus of elasticity (E_F), **(MPa)**, the modulus of elasticity E_F of a material in Pa in the flexure test, a 3-point bending test with two support points and one load point in the middle of the span.

$$E_{FR} = 0,251 \frac{L^3 F}{lWH^3} \quad \text{for rectangular specimen} \quad \text{(F.3)}$$

$$E_{FC} = 0,425 \frac{L^3 F}{ld^4} \quad \text{for core specimen} \quad \text{(F.4)}$$

where L is width of the support span for the specimen in m, F is load in N, W is the width of specimen in m and H is the height of specimen in m and d is the diameter of the core in m. Hunt, 1965.
flexural rigidity (χ), **(N²/m)**, *second moment* of the section of a beam A in Nm multiplied by its *Young's modulus* E in MPa. Ham, 1965.

$$\chi = AE \quad \text{(F.5)}$$

flexural rupture theory, because of the fact that movement of the free surface in a bench is larger in the front of the blasthole than at the place where the side cracks will appear at the bench face, there will be a considerable bending of the bench face causing *bending cracks* to start at the free surface and propagate towards the blasthole. Many authors wrongly call these cracks tensile cracks but they are typical bending cracks. This flexural rupture was first discovered by Daw in 1898 and later on verified by Ash, 1985.

The flexure character of these cracks has been verified both in full scale and model single hole blasting tests (SHB). Rustan, 1987.

flexural strength or **stress (σ_{fb}), (MPa)**, nominal stress at fracture in a *bend test* (σ_{fb}) or *torsion* (σ_{ft}) *test* also called *modulus of rupture, bend strength* or *fracture strength*. For a rectangular sample under a load F in N in a 3 point the *bend test* the flexural strength is defined;

$$\sigma_{fb} = \frac{3L}{2WH^2} F \quad \text{for rectangular beams} \tag{F.6}$$

where σ_{fb} is the *flexural strength* or *modulus of rupture* in bending in MPa, L is the *width of support span* in m, W is the *width of the beam* in m, H is the *height of beam* in m. The flexural strength in bending is normally in excess of the tensile strength of rock and the shearing stress is slightly in excess of flexural strength in bending.

The maximum bending strength σ_{fb} in Pa, in a three point bending of a *drill core specimen* with the diameter d in m, is at load F in N

$$\sigma_{fb} = \frac{16l}{5\pi d^3} F \quad \text{for circular beams} \tag{F.7}$$

where l is the span of the support in m. The ratio between span and diameter should be larger than 3:1. Hudson, 1993.

flexure, bending or folding of strata under pressure. Standard, 1964.

floating sub, threaded connector on a drill rotation unit, usually splined, to allow limited axial movement when connecting and disconnecting drill rods. Atlas Copco, 2006.

floor, the bottom horizontal part of an excavation upon which haulage or walking is done. The lower surface of a bench, drift, tunnel or underground chamber.

floor holes (lifters), drill holes along the bottom (floor) contour line. Nitro Nobel, 1993.

floor rack, floor mounted frame along which something is moved. Atlas Copco, 2006.

flow ellipsoid or **draw ellipsoid,** the approximately ellipsoidal shape of the volume of caved or broken material which moves or "flows" when a bottom outlet is opened or draw occurs. The ellipsoid shape can only be observed in fine material. In mining the shape is in most cases *a drop turned upside down*.

flow meter, instrument for measuring the rate of flow of either a gas or fluid in a pipe. Atlas Copco, 2006.

flow stress or **deviator stress (σ_d), (MPa)**, $\sigma_f = \sigma_1 - \sigma_3$ where σ_1 is the major principal stress in MPa and σ_3 is the minor principal stress in MPa. The term is used in geological literature. Hudson, 1993. Also called '*deviator stress*' when determined on test samples in the laboratory.

flush hole, hole in the center of a drill steel making it possible to transport water or other drilling media from the drilling machine to the drill bit. LKAB, 1964.

flushing at drilling, the removal of drill cuttings with water, air mist, foam or other flushing agents. *Water* and *air* are the most common used methods. Air flushing is mainly used in surface drilling.

flushing device, a flushing tube (copper tube to avoid sparks) with a nozzle at one end and a hose for water at the other end. The purpose with the flushing device is to clean the drill hole with flushing water from drill cuttings or explosives remained after a failed blast.

flushing head, rotary joint that allows the flow of flushing fluid into the drill string. Atlas Copco, 2006.

flushing hole, hole in an object that allows the passage of a flushing medium. Drill bit have flushing holes that direct the flow of the flushing medium. Atlas Copco, 2006.

flush joint, drill rod connections where one tube is threaded into the other in such a way that there is no increase in the outer diameter of the rod at the coupling joint. Atlas Copco, 2006.

flush jointed rod, drill rod having the same diameter at the threaded joint as the rest of the tube. Atlas Copco, 2006.

flush pump, pump used to pump flushing fluid into a hole during drilling to remove the rock cuttings. Atlas Copco, 2006.

flyrock, any rock fragments thrown unpredictable from a blasting site. Flyrock may develop in the following situations: insufficient stemming, too small a burden because of overbreak from the preceding or proceeding blast, planes of weakness in rock which reduce the resistance to blasting, and finally existence of loose rock fragments on top of the bench. Total control of flyrock is only possible with the use of adequate blasting mats.

foam drilling, a method of dust suppression in which thick foam is forced through the drill rods by means of compressed air and the foam and dust mixture emerges from the mouth of the hole in the form of a thick sludge. With this method the amount of dust dispersed into the atmosphere is almost negligible. The amount of water used is about 3,8 l/h. Approximately 9–15 m of drilling can be done with one filling of the unit. Mason, 1951.

foamed explosive, an explosive to which foam is added to reduce the linear charge concentration. It may be used in controlled contour blasting or for the reduction of fines in production blasting.

foam stemming plug, a foam plug created in blastholes by mixing two organic components. The foam stemming plug seals to hold stemming and decks in place.

foliation, the *banding* or the *lamination* of metamorphic rocks as distinguished from the *stratification* of sedimentary rocks. Fay, 1920. a general term covering different kinds of structures produced in different ways. Slate cleavage, schistosity, gneissose, banding and sets of closely-spaced fractures are all examples of foliations. Ii is the result from segregation of different minerals into layers parallel to the schistosity. It is also a set of new planar surfaces produced in a rock as a result of deformation under geological times. Many rocks exhibit several generations of foliation which are distinguished chronologically (S1, S2, S3, etc.). During the evolution of the structural geological formation earlier foliations are deformed and cut by later foliations.

footage, number of feet drilled.

foot plate, plate mounted on the side of a machine or rig on which the operator can stand or climb into or on the machine. Atlas Copco, 2006.

footprint, a term used in block caving for the horizontal undercut area under a specific panel. MassMin, 2008.

foot valve, plastic (or aluminium) component mounted in a DTH-bit, or in some cases in the DTH-hammer. Atlas Copco, 2006.

footwall, the wall or rock under an inclined vein. In bedded deposits it is called the floor. Atlas Powder, 1987.

FOPS (acronym), see '**falling object protective structure**'. Atlas Copco, 2006.

Foraky boring method, a *percussive boring system* comprising a closed-in derrick over the crown pulley of which a steel rope is passed from its containing drum. The boring tools are suspended from the end of the rope and are moved in the hole as required by means of the drum. A walking beam, operated by driving mechanism, gives the boring tool a rapid vibration motion. Nelson, 1965.

Foraky freezing process, one of the original freezing methods of shaft sinking through heavily watered sands. Although the principle is the same today, the process has been improved in many respects. Nelson, 1965.

forbidden or **unacceptable explosives**, explosives that are forbidden or unacceptable for transportation by common, contract, or private carriers, by rail freight, rail express, on highways, by air or by sea, in accordance with the regulations of the U.S. Department of Transportation. Atlas Powder, 1987.

forble (old term), see '**fourble**'.

force (*F*), (**N**), basic physical quantity defined by the product of mass and acceleration. Force can be measured by a dynamometer. (1 N (Newton) = 1 kgm/s^2. Old units; 1 kilopond = 1 kp = 9,81 N, 1 pound force (lbf) = 4,45 N).

force of blow (F_b), (**N**), the force of the drill hammer acting on the rod.

forced-caving system, a stoping system in which the ore is broken down by large blasts into the stopes that are kept partly full of broken ore. The large blasts have the further effect of shattering additional ore, part of which then caves. USBM Bull. 419, 1939.

Forcit (finish tradename), low density perimeter explosive packed in stiff plastic pipes with outer diameter 17 mm and being used for controlled contour blasting.

foreset, temporary forward support, a middle prop under a bar. Mason, 1951.

fork, U-formed structure, used as an attachment and permits movement in a single plane. Atlas Copco, 2006.

forke the hole, to drill a second hole from some point within a complete borehole by deflection methods and equipment. Long, 1920.

fork half, one half of U shaped attachment that has been manufactured in two pieces. Atlas Copco, 2006.

form factor, see '**shape factor**'.

fourble, in rotary drilling four lengths of drill rod or drill pipe connected to form a section, which is handled and stacked in a drill tripod or derrick as a single unit. Also spelled forble (old term). Long, 1920.

four-cutter bit, see '**roller rock bit**'.

four-insert drill bit, drill bit with four linear and wedge shaped cutting surfaces.

four point bit, drill bit with four cemented carbide inserts. Sandvik, 2007.

four point integral drill steel, integral drill steel with four point bits. Sandvik, 1983. See also '**integral drill steel**'.

four section cut, a *parallel hole cut* of type *cylinder* used in drifting. Around an uncharged center hole 110 mm in diameter (first section) the first row has a square shape and includes six charged blastholes. The second row consists of four charged blastholes and a square shape that is rotated 45° compared to the first row. The third row (fourth section) includes four blastholes and the shape is a square and is rotated 45° to the second row. Langefors, 1963.

fractal, is a shape, made of parts similar to the whole in some way or an array of points which constitutes a subset of a particular space of interest. Examples of fractals of interest in geomechanics are the trace of a joint on a two-dimensional surface, the array of points that constitutes a joint surface embedded within three-dimensional Euclidean space, the array of points that comprises a complicated set of joints embedded within a three-dimensional Euclidean space or the array of points that constitutes the orientation diagram for poles to joints embedded in the Riemannian surface of a sphere. Hudson, 1993.

fractal dimension (D_f), (**dimensionless**), the fractal dimension is a quantification of the surface roughness (microstructure) of a particle or fragment. It is defined by

$D_f = 1 - m$ where m is the slope of the relation between the perimeter of the particle (fragment) and the selected *step length* for describing the surface roughness of the particle (fragment) by straight lines in a log-log diagram. For 15 selected minerals D_f varied from 1,07–1,11. The larger the D_f the more rugged surface. Some examples of measured D_f's for 0,13 mm particles are; iron ore 1,074 and 1.083, quartz 1,081, coal 1,083, limestone 1,087 and 1,094, granite 1,098 and diorite 1,100. Li Gonbo, 1993.

fracto-emission or **exoelectron emission (electric)**, the emission of electrons due to the fracturing of rock starting at micro seismic level of load. The emitted electrons will interact with the ambient air molecules and produce light that can be seen by the naked eye. Liinamaa and Reed, 1988.

fractography, the systematic examination, description and quantification of the topography of fracture surfaces. This knowledge is of important in fracture mechanics.

Descriptive treatment of fracture, especially in metals, with specific reference to morphology (photographs) of the fracture surface. Macro-fractography involves photographs of low magnification; micro-fractography involves photographs of high magnification, A.S.M. Gloss., 1961.

fracture, the character or appearance of a *freshly broken surface* of a rock or a mineral. The peculiarities of the fracture afford one of the means of distinguishing minerals and rocks from one another. Fay, 1920. There are several different natural mechanisms causing fracture in the rock mass,
1 Extension and shear fractures.
2 Joints due to erosional unloading in isotropic rocks.
3 Joints due to differential volume changes in heterogeneous bodies.
4 Joints due to regional deformation.

fracture (noun), the general term for any mechanical (displacement) discontinuity in the rock. It is therefore the collective term for *joints, faults* and *cracks,* etc. In most rock materials in conditions of interest in engineering fracturing occurs in a brittle or quasi-brittle fashion accompanied by little plastic flow. A fracture is defined by its length, width, depth, filling material and roughness.

fracture (verb) **(concerning rock)**, to break with only little movement of the broken pieces. The pieces have not been able to rotate and are still puzzled together.

fracture control blasting, a controlled contour blasting method where the necessary borehole pressure can be reduced considerably by creating two diametrical notches along the borehole before blasting. The notches can be created by mechanical tools. Fourney et al., 1984, linear shaped charges, Rustan et al., 1985, and water jets.

fracture energy or **fracture surface energy** (G), (J/m^2), see '**strain energy release rate**'.

fracture growth velocity c_c **(m/s)**, see '**crack propagation velocity**'.

fracture initiation, transitional phase in fracture characterized by the sudden extension of a pre-existing static crack in a material.

fracture mechanics, the continuum mechanics approach to the understanding and quantitative prediction of the phenomenon of fracture.

fracture propagation velocity c_c **(m/s)**, see '**crack propagation velocity**'.

fracture strength, depends on the loading conditions like compressive-, tensile-, shear-, bending-, torsion load etc. The fracture strength also depends on the rate of loading or strain rate. 5–13 times larger strength values have been measured at high strain rates than the corresponding static strengths. Rinehart, 1970. Also the confinement influences fracture strength. See also '**fracture toughness**'.

fracture stress, the maximum principal true stress at fracture. Usually refers to un-notched tensile specimens. ASM Gloss, 1961.

fracture surface energy or *fracture energy* (G), (J/m^2), see '**strain energy release rate**'

fracture system, group of fractures (faults, joints, or veins) consisting of one or more sets, usually intersecting or interconnected. McKinstry, 1948.

fracture test, breaking a specimen and examining the fractured surface with the unaided eye or with a low power microscope to determine such things as composition of minerals, grain size, case depth, soundness, or presence of defects. ASM Gloss, 1961.

fracture toughness (K_{Ic}), ($MN/m^{3/2}$), resistance of a material to fracture. Fracture toughness is determined in the laboratory by testing standardized fracture test specimen such as notched cylindrical cores or disc-shaped specimens. The chevron test makes use of V-type shape of the starter crack. International standard for testing fracture toughness, see ISRM, 1988 and 1995.

fracture toughness test, there are many test methods to test a material to fracture. Here are presented three test methods, each testing in a different plane in regard to the axis of the drill core. The y-z plane is defined as the plane perpendicular to the drill core and the x plane as parallel to the axis of the drill core.

1 The **disk-shaped compact test**. DC(T). Used in the metals standard E 399-81. The notch is cut along the side of the core and towards the center of the core in the X-Z plane, see Fig. 32a. Ouchterlony, 1983.

2 The **shot rod test** SR. The notch is oriented along the axis, but only at the edge of the core in the X-Y-plane, see Fig. 32 b. It was devised by Barker, 1977 a, and is the subject of standardization work within the ASTM. Ouchterlony, 1983.

3 The **straight edge crack round bars in bending** SECRBB test method, Fig. 32c, was introduced by Finn Ouchterlony and from a synthesis of it and the short rod test method the **chevron edge notch round bars in bending** CENRBB, see Fig. 33 was developed. In the SECRBB test the notch is oriented perpendicular to the axis of the core and a little deeper than to half of its diameter in the z-y plane. In CENRBB test the chevron notch cuts about 2/3 of the cross-sectional area of the core. Ouchterlony, 1983.

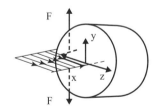

a) Disk shaped compact test (DC(T))

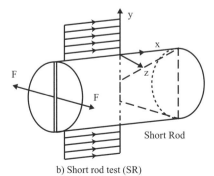

b) Short rod test (SR)

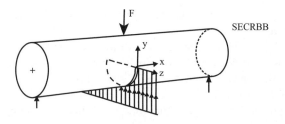

c) Straight edge crack round bar in bending (SECRBB)

Figure 32 A piece of rock core from which three fracture toughness specimens with mutually perpendicular crack orientations can be made. After Ouchterlony, 1983.

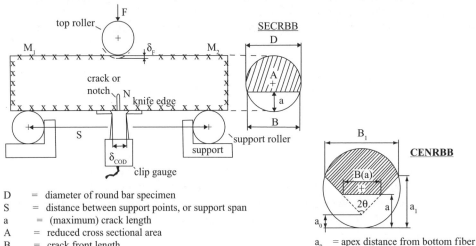

D = diameter of round bar specimen
S = distance between support points, or support span
a = (maximum) crack length
A = reduced cross sectional area
B = crack front length
F = load on specimen
δ_F = (ideal) load point displacement (LPD)
δ_{CMOD} = crack (mouth) opening displacement (CMOD)
N = notch tip, or front
M_1, M_2 = saddle points of measuring base
a_0 = initial notch depth

a) Single edge notch crack round bar in bending (SECRBB)

a_0 = apex distance from bottom fiber
a = (maximum) crack length
a_1 = chevron base distance
2θ = subtended chevron angle
B(a) = crack front length
B_1 = chevron base width

b) Chevron edge notch round bar in bending (CENRBB)

Figure 33 a) The Single edge crack round bar in bending (SECRBB) specimen configuration with basic notation. b) Chevron edge notch round bar in bending (CENRBB) variety. After Ouchterlony, 1983.

The work done per unit new crack surface or the fracture toughness G_{Ic} is related to K_{Ic} by the following formula in plane strain.

$$G_{Ic} = \frac{1-v^2}{E} K_{Ic}^2 \qquad (F.8)$$

where G_{Ic} (J/m²) is called *fracture toughness* or *critical energy release rate* and is the fundamental material constant, v is Poisson's ratio and E is Young's modulus. Some authors also refer to K_{Ic} (MPa/m$^{1/2}$) as the fracture toughness.

The fracture toughness refers to *mode I fracture* (the normal, opening mode or pure tensile load) but the fracture in the test is caused by bending and not pure tensile load. Fracture toughness differs from conventional bending test because there is made a notch where the fracture is expected to start. Fracture toughness is a necessary strength value, as input parameter in fracture modelling of a rock mass using the theory that fracturing starts at pores and flaws.

Some typical fracture toughness values K_{Ic} (MN/m$^{2/3}$) are for Siporex light weight concrete 0,08, magnetite concrete 0,23, Luossavaara magnetite 0,72, Öjeby granite 1,41, Kallax gabbro 1,90 and Henry quartzite 2,28 MN/m$^{2/3}$. A ratio of 29 between the highest and lowest value. Rustan, Vutukuri and Naarttijärvi, 1983.

fracture velocity (c_c), **(m/s)**, see '**crack propagation velocity**'.

fracturing, the process of rock breaking a solid material such as rock.

FRAGBLAST (shortened form), see '**International Symposium on Rock Fragmentation' by Blasting**.

fragment (noun), fractional piece of a larger block of rock generated by blasting or mechanical tools. The rock or *mineral particle* must be larger than a grain (0,06 kg). Bates and Jackson, 1987. A piece of rock or mineral. Von Bernewitz, 1931. Compare also with the definitions for 'block' and 'particle'.

fragment (verb), to disintegrate a solid material by blasting or mechanical tools.

fragmentation, the extent to which a rock mass or rock sample is broken into small pieces *primary* by mechanical tools (boring or full-face boring), blasting, caving and draw of initially *in situ* rock or ore, hydraulic fracturing and *secondary* by the breakage during loading and transportation of the material to the mill. It can be characterized by a histogram showing the percentage of sizes of particles or as a cumulative size distribution.

fragmentation assessment, see '**fragmentation measurement**'.

fragmentation gradient (*n*), **(dimensionless)**, *see* '**uniformity index**'.

fragmentation important explosive parameters, shown here are a summary of what researchers believe are the most important, see Table 23.

Table 23 Important explosive properties for rock fragmentation.

Reference	Year	Explosive density (kg/m³)	Detonation velocity (m/s)	Acoustic impedance (kg/m² s)	Detonation heat (MJ)	Gas volume (m³/kg)	Borehole pressure (MPa)	Energy at high pressure and high confinement
Langefors	1963	–	–	–	Yes	Yes. Gas volume given at 0° C instead of actual temperature.	–	–
Noren	1974	–	–	–	Not OK	Bubbel from pond test, not OK		Yes
Favreau	1979	–	–	–	–	–	Yes + γ	–
Paine/ Harries 1983	1983	Yes	–	–	–	–	–	–
Brady	1985				Yes		Detonation pressure	
Müller	1990	Yes	–	Yes	Yes	–	–	–
Qian Liu	1993	Yes	Yes	–	–	–	–	–
Rustan	2009	Yes	Yes	Yes	Yes	Yes	Yes	Yes

fragmentation important rock mass properties, acoustic impedance, number of joint sets. joint properties and joint directions are the most important properties known today and water content in the rock mass.

The influence of joint direction was carefully examined in laboratory test. Bhandari, 1990.

fragmentation measurement, measurement of the degree of fragmentation by means of qualitative, semi-quantitative or quantitative methods.

Qualitative methods

1. *Three point estimation* (visual estimation of k_{100}, k_{50}, and k_{10}).
2. *Rate of shovel loading* (This parameter depends on the type of loading machine and its condition, the skill of the operator, the shape and roughness of the fragments, the fragment size distribution, and the swell factor).
3. *Conventional and high speed photogrammetry*
4. *Bridging delays at the crusher*
5. *Microprocessor-based performance monitoring of loading equipment* as used to calculate a diggability index (DI) based on motor torques, velocities and digging trajectories, which combine to derive a measure of diggability which relates to fragmentation size, looseness and operator skill.
6. Production rate at the primary crusher.

Semi-quantitative methods

1. *Comparison of photographs* taken of the muckpile with photographs taken from standardized size distributions. (*Compaphoto method*). Cunningham, 1993.

Quantitative methods

1. *Screening.*
2. *Boulder frequency counting.* See '**boulder frequency**'.
3. *Boulder volume or mass* estimation (%).
4. *Quantity of explosives used for secondary blasting* (kg/t or kg/m3).
5. *Photographic method.* The size distribution is determined from photographs taken of the muckpile surface or a cross-section. A computer is used to digitize the photographs and calculate the size distribution. If a video camera is used the process can be carried out automatically. Note that the accuracy of this method depends very much on the quality of the photographs, the resolution of the video camera and the software for the following processing of the image. Systems available, see Franklin and Katsabanis, 1996.

fragment shape, see '**shape factor**'.

fragment size (W), (H), (L), (**m**), can be defined by the physical dimension of the fragment where W = width, H = height and L = length. In the field of rock blasting it is recommended to use square meshes and sieving to define the fragment size. When sieving can not easily be undertaken the rock fragments (boulders) can be defined by its maximum length (L), width (W), and height (H), when the fragment is placed on a horizontal surface and enclosed by a parallelepipedic box where $L > W > H$, see Fig. 34.

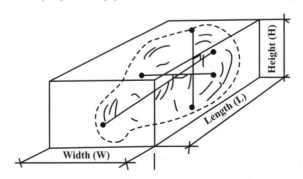

Figure 34 Definition of the size of a rock fragment on the basis of the length (L), width (W) and height (H).

fragment size (k_x), (m) and its dependence on strain rate (ε'), (s^{-1}), it is of great importance for a conceptual model of blasting to know the effect of strain rate on fragmentation at a defined energy impulse. Coarse fragmentation at high strain rates is expected because the strength of the material is increasing with the strain rate. Results from experiments on US oil shale using *block* (1 m³), *cylinder* and *gas gun* test, representing strain rates from ~10^1, 10^2 and 10^3 respectively corresponded to fragment sizes 13 mm, 10 mm and 1 mm respectively, see Fig. 35.

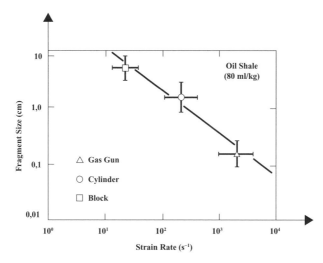

Figure 35 Influence of strain rate in US oil shale on fragmentation. Block = blasting in a '1 m³ block of oil shale with 80 g C-4 explosive'. Cylinder = a 'core of oil shale' and Gas gun is a 'gun projectile hitting the surface of oil shale'. Grady, 1980a.

fragment size distribution, relationship between the cumulative amount of fragments (%) passing different mesh sizes and the mesh size. A fragment size distribution can roughly be quantified by mesh sizes with 10, 50, and 100% of the mass passing (k_{10}, k_{50}, and k_{100}), or, if the distribution follows the Rosin-Rammler-Sperrling (RRS) distribution, by k_{50} and the slope of the distribution or uniformity index *n*. Upon calibration to the blast site a prediction of fragment size distributions in rock blasting can be based on formulas such as Kutsnetzov-Rosin-Rammler formula (Kuz-Ram) after Kuznetzov, 1973 and Cunningham, 1987 or the Swedish Detonic Research Foundation formula (SveDeFo-formula) after Ouchterlony et al., 1990.

fragment size distribution in situ, see '**block size in situ**'.

frame, fabricated structure normally made of metal used as mounting platform for a number of components. Atlas Copco, 2006.

frame grabber, in image analysis it is a term used for the two operations digitizing and storing the image in the memory.

frame work, skeleton like structure that can be fitted with a cover to form an enclosed space. Atlas Copco, 2006.

free face, an unconstrained surface almost free from stresses; e.g. a rock surface exposed to air or water or buffered rock that provides room for expansion upon fragmentation. Sometimes also called open face.

free-fall drill, synonym for '**churn drill**'.

free field stress, the stresses existing in rock before the excavation of any mine opening. Lewis, 1964. See also '**rock stress**'.

free surface velocity (v_{fs}), **(m/s)**, (bench blasting), the velocity of the free face in front of the blasthole or the top surface of the bench. Clay, 1965.

free swell (in sublevel caving), the volume of mainly ore but also in some blasts waste that flows into the drift after blasting of a sublevel caving round. MassMin, 2008.

freeze sinking, use of circulating brine in system of pipes to freeze waterlogged strata so that shafts can be sunk through them, established and lined. Pryor, 1963.

freezing and thawing test, cold-warm temperature test applicable only to nitroclycerin (NG) explosives of the gelatinous type. The explosive sample in the form of an unwrapped 100 gram cartridge is placed into a test-tube of suitable dimensions. The tube is closed with a velvet cork and kept at a temperature of –3° to –6°C for 16 hours, followed by 8 hours at room temperature. The test is repeated for 3 days and examined for any abnormality with respect to gel breaking/oozing out of the NG. Sandhu and Pradhan, 1991.

freezing of a blast (mechanically), the semifusing and non-ejection of the pulverized rock or ore in the cut portion of a blasting round, generally caused by providing insufficient void space (swelling) for the initial holes blasted in the cut. Reasons for freezing may be too little or no delay between charges, and/or excessive charge masses.

freezing of nitroglycerin, may occur already at +10°C. The frozen cartridges are unsafe in handling, while improvised thawing operations are risky. The freezing is prevented by adding nitroglycol to the nitroglycerin.

frequency (f), **(Hz)**, number of cycles per unit of time.

frequency of cracks (f_c), **(No./m)**, the number of cracks crossing a straight line per unit length.

frequency of discontinuities (f_d), **(No./m)**, the number of discontinuities crossing a straight line per unit length.

frequency of joints (f_j), **(No./m)**, the number of joints crossing a straight line per unit length.

fresh (in connection to erosion), no visible sign of weathering/alteration of the rock material.

friability or **fragility**, tendency for particles to break down in size (degrade) during storage and handling under the influence of light physical forces. Pryor, 1963 and Simha, 2001.

friability value (S_{20}), a measure of rock resistance to crushing by repeated mass-drop impacts. Procedure to determine the friabiltiy see '**brittleness (friability) test** (S_{20})'. Tamrock, 1989.

friable, easy to break or crumbling naturally. Descriptive of certain rocks and minerals. Fay, 1920.

frictional coefficient (μ), see '**coefficient of friction**'.

frictional ignition, the ignition of firedamp in coal mines by friction sparks, such as the rubbing of sandstone against sandstone or sandstone against rock bolts. Another cause to ignition is gas ignition. Nelson, 1965. Heat liberated from friction can also cause ignition of explosives.

friction bearing, the contact surfaces in a friction bearing consists of two cylinders which reduce the surface pressure. Friction bearings are used in roller bits. Sandvik, 2007. See also '**bushing bearing**'.

friction bolt anchored wire rope, a rock reinforcement method where a wire rope is friction anchored by a Swellex bolt at the collar of the hole. This allows the wire rope to slide at high loads and to absorb significantly more energy than either a Swellex bolt or a cement grouted able when used alone, see Fig. 36. Stillborg, 1994.

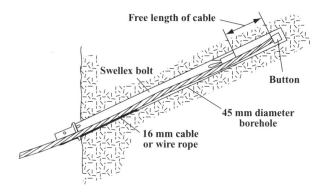

Figure 36 Wire rope anchored by Swellex rock bolt generates a frictional anchorage that allows the wire rope so slide at high loads. After Stillborg, 1994.

friction bushing, tubular bushing that is pressed into place and retained through frictional contact. Atlas Copco, 2006.

friction drum, a drum for the hoist rope with enough diameter and friction properties to have the hoisting rope laid over it only to 180°, with the cage at one end of the rope and the counter mass at the other end.

friction hoist, a mine hoist with an electrical engine coupled via a gear to the friction drum.

friction sensitivity or friction test of explosive materials, the sensitivity to friction can be identified by rubbing a small quantity of the explosive in an unglazed porcelain mortar. The sample being tested is compared against a standard specimen. Meyer, 1977. There are several test methods developed in the World but an international standard method is lacking. One method to test the sensitivity of an explosive with respect to stresses similar to those which occur when charging with a pneumatic cartridge loader is performed as follows; the test apparatus consists of a block of granite with a semi-cylindrical groove with a diameter of 30 mm, and a length of 50 cm. A cylindrical metallic slider with a diameter of 25 mm and a length of 80 mm is pressed by a pneumatic cylinder against the bottom of the groove where the explosive is being placed. The slider is pushed back and forth by another cylinder with an automatic valve until the explosive detonates. Johansson and Persson, 1970.

friction welding, a kind of welding of metal where heat is created and the two parts fused together through friction created by rotating at least one of the parts and forcing it against the other. Atlas Copco, 2006.

front caving (block caving), a variant of block caving in which the cave is retreated on one or more levels from an initiating slot in the centre or on the boundary of the block.

front caving (sublevel caving), a variant of sublevel caving where the draw on each level is done on a line perpendicular to the crosscuts, normally starting at the hanging wall and working towards the footwall. The strategy is to reduce waste rock dilution. The method was originally suggested by Prof. Ingvar Janelid, KTH, Sweden.

front end loader (LHD), loading machine with a bucket at the front used to load material such as broken rock or earth. The bucket can also be used for the transportation of the material and also to dump it.

front end loading machine (LHD), see '**front end loader**' or '**load haul dump unit**'.

front head, forward housing of a rock drill. Atlas Copco, 2006.

frost blasting, a process occurring mainly on the surface of earth but also underground in permafrost zones. The cracks in the rock mass on the surface of earth are successively widened by the forces that develops when the water in the cracks is frozen to ice, because ice needs physically a larger volume then water.

FSa (abbreviation), for '**fine sand**' defining particles with sizes from 0,063 to 0,2 mm which represents particles that can be sieved in dry conditions.

FSi (abbreviation), for '**fine silt**' defining particles with sizes from 0,002 to 0,0063 mm.

fuel, a chemical compound or an element in an explosive which reacts with an oxidizer to form gaseous detonation products and cause a release of heat. When aluminium is used as fuel, solid and liquid residual products are produced and energy is liberated.

fuel oil, the fuel, usually No. 2 diesel fuel, used in ANFO. USBM, 1983.

Fulcrum effect, a reamer behind the drill bit act will act as a hinge point for lever with the weight of drill collar above the reamer to provide a leverage force at down hole drilling. This causes the drill bit to bend upwards. Stein, 1997.

full face boring, the use of large boring machines using e.g. disk cutters or three cone bits rolling on the face crushing the rock beneath and thereby drilling the whole tunnel face in one step. The advance is mainly controlled by the hardness of the rock and the direction and number of joint sets in the rock mass, and varies up to a maximum of about 6 m/hour in soft rocks. Diameters of up to 20 m have been drilled by this method.

full face driving, method of blasting normally used for adits, tunnels, and drifts. The full section is drilled and then blasted in one operation. Fraenkel, 1952.

full-face machine, a machine for excavation of the whole tunnel or drift area by mechanical methods.

full-facer, a machine for full face boring (mechanical boring of the full face area).

full-facer tunnelling machine, see '**full-face tunnelling machine**'.

full-face tunnelling machine, a large boring machine that mechanically removes rock with the purpose to create an opening of a desired size in a single pass. There exists two types of machines, machines drilling *circular openings* and machines drilling *non-circular openings*.

full grip wrench, wrench that has contact around the full circumference of a pipe when used to connect and disconnect the threads. Atlas Copco, 2006.

full hardening, need a definition hear. Sandvik, 2007.

full hole bit, synonym for 'non coring bit'. Long, 1920.

full hole drilling, term that may be used to describe the boring of large diameter openings such as raises and tunnels in a single pass. Atlas Copco, 2006. Better to use the term '**full-face boring**'.

fully mechanized scaling, drilling and bolting, a special designed underground movable machine, equipped with boom mounted hydraulic breaker that performs the hazardous scaling job with the operator remotely located away from rock falls. Blast holes for installation of rock bolts are drilled with the help of a hydraulic boom, and all functions in the rock support process are also performed at a safe distance. The operator controls everything from a platform or cabin, equipped with a protective roof.

fulminate, an explosive compound of mercury, $HgC_2N_2O_2$, which is employed for the caps of exploders (blasting caps), by means of which charges of gunpowder, dynamite, etc, are fired. Fay, 1920.

fume classification, a quantification of the amount of fumes produced per kg of explosive or blasting agent detonated. Based on the result the explosives are divided into different fume classes. In the USA this classification is performed by the Institute of Makers of Explosives (IME). Se also 'IME fume classification'.

fume classification test, the fumes from explosives are normally examined by blasting a small sample of explosive in a tank or a sealed drift in a mine. Among the test are 1) *Bichel Gauge*, the 2) *Crawshaw-Jones Apparatus*, and 3) the *Ardeer Tank method*. Elitz and Zimmerman use a rock chamber with the volume 30–100 m^3 in which 7 kg of explosive is detonated. If the round generates less than 32 l/kg CO and less than 4 l/kg NO_x, the explosive is considered acceptable for underground use.

fume quality, a measure of the toxic fumes to be expected when a specific explosive is properly detonated. See also '**fume classification**'.

fumes, the gas and smoke (more specifically the noxious or poisonous gases) given off by the explosives upon detonation. The character of the fumes is largely influenced by the completeness of the detonation, the degree of confinement of the charge and the size of the detonator. The fumes depend also heavily on the composition of the explosive.

fundamental mode of vibration, that mode of a system having the lowest natural frequency. H&G, 1965.

fundamental strength, the maximum stress that a substance can withstand, regardless of time, under given physical conditions, without rupturing or plastically deforming continuously. Billings, 1954.

fuse, a long flexible small diameter tube containing a core of black powder used for igniting black powder or fuse type blasting caps. Aso called safety fuse, See '**safety fuse**'.

fuse blasting cap, see '**fuse cap**'.

fuse cap or **fuse detonator**, a detonator that is initiated by a safety fuse; also referred to as an ordinary blasting cap. Atlas Powder, 1987.

fuse cutter (tool), a mechanical device for properly cutting a safety fuse clean and at right angles to its long axis. IME, 1981.

fuse cutter (person), in metal mining, one who cuts blasting fuse to standard lengths, inserts fuse in open end of detonators or caps, and attaches it by squeezing the open ends with a pair of crimpers (special pilers).

fuse detonator, see '**fuse cap**'.

fuse gage, an instrument for cutting time fuses to length. Standard, 1964.

fuse head, the electric resistance wire in a blasting cap surrounded by a material (flashing mixture) which is sensitive to heat.

fuse igniter, a pyrotechnic device which burns with a very hot jetting flame which is used to ensure ready ignition of safety fuses. AS 2187.1, 1996.

fuse lighter, a pyrotechnic device for the rapid and reliable lighting of a safety fuse; also a stick of material which burns with an intense flame and will not be extinguished by the ignition flame from the fuse.

fusion piercing, drilling based on directing thermal energy onto the rock until it reaches *fusion point*, usually about *1 500° to 1 700°C*. Fusion piercing was the first thermal piercing method to be used. The necessary heat is usually generated by burning an oxygen gas flame at the end of a iron pipe. Sometimes the *oxygen* is mixed with *hydrogen gas*. In some case a *fluxing agent* is added. This method is used for piercing holes in reinforced concrete, etc, but has not been practically employed in rock blasting. Fraenkel, 1952.

gallery, several parallel drifts e.g. the crosscuts in sublevel caving.
gallery head, drift (tunnel) which will act as a main drift for several parallel drifts laying perpendicular or at an angle to the gallery head.
galvanic action, physical-chemical process characterized by the flow of a current caused when dissimilar metals contact each other or through a conductive medium. This action may create sufficient voltage to cause premature firing of an electric blasting circuit, particularly in the presence of salt water. Konya and Walter, 1990.
galvanometer *or blasters' galvanometer*, see '**blasting galvanometer**'.
gamma function, a probability distribution that can be used for the presentation of a rock fragment distribution developed from blasting.
gamma ray well logging, a method of logging boreholes by observing the natural, radioactivity of the rocks through which the hole passes. It was developed for logging holes which cannot be logged electrically because they are cased. A.G.I., 1957.
gangue, the mineral material in an ore, forming part of a vein or lode, that is not commercially useful. Gangue minerals are discarded as tailings as soon as they can be separated from the useful or valuable minerals, during the concentration process. Gregory, 1983.
gap, the gap is the greatest distance at which, under certain given conditions, a priming cartridge (*donor*) is capable of initiating a receiving cartridge (*receptor*).
gap sensitivity, see '**gap test**'.
gap test, a test to determine the greatest distance at which, under certain given conditions, a priming cartridge (*donor*) is capable of initiating a receiving cartridge (*receptor*). The same type of explosive is usually used as donor and receptor. The gap distance will be affected by any change in strength which may occur in the explosive. The test gives information about the sensitivity of explosives to moisture, temperature, etc. The gap test can be carried out with the cartridges unconfined in air or confined in water, in blastholes or in iron tubes. The gap may be filled with air or solid material. See also '**gap**'.
Gasbag (tradename), a nylon bag with an inner bag encasing an aerosol, e.g. propane or butane. The Gasbag is used for implementing air gaps in charge columns. It is for example used in airdeck blasting. Once inflated each bag is capable of withstanding pressures up to 10 MPa.
gas detonation system, a non-electric initiation of detonators by means of a plastic tube containing an explosive gas.
gas explosion, a major or minor explosion of firedamp in a coal mine, in which coal dust apparently did not play a significant part. See also '**coal dust explosion**'. Nelson, 1902.

gas ignition, the setting on fire of a small or large accumulation of firedamp in a coal mine. The ignition may be caused by a safety lamp, electrical machinery, explosives, frictional sparking, etc. Nelson, 1965.

gas ratio (R_g), (dimensionless), the ratio of the volume at atmospheric pressure of the gas developed by an explosive to the volume of the solid from which it was formed is called the gas ratio of the explosive. Many *commercial explosives* have a *gas ratios* ranging from *450 to 1 050*.

gas release pulse (GRP), (p_{og}), (Pa or dB), air blast overpressure produced from gas escaping from the detonating explosive through rock fractures, see also '**air blast**'. Atlas Powder, 1987.

gas venting, the uncontrolled escape of the confined borehole gases to a free face (through fissures, cracks or weak rock horizons), reducing the heave action of the charge, before adequate breakage and displacement can occur.

gateroad, a road through the goaf (caved hanging wall) used for haulage of coal from longwall working. Pryor, 1963.

gathering arm loader, a machine for loading loose rock or coal. It consists of a tractor mounted chassis carrying a chain conveyor, the front end of which is built into a wedge-shaped blade. Mounted on this blade are two arms, one on either side of the chain conveyor, which gather the material from the muckpile and feed it onto the loader conveyor. The tail or back end of the conveyor is designed to swivel and elevate hydraulically so that coal or stones can be loaded into a car or onto another conveyor. Nelson, 1965.

Gaudin-Meloy distribution function, is defined in the formula below. It has been widely used in mineral processing, and has an appropriate theoretical derivation. It has also been used to represent fragment size distributions after blasting.

$$y = 1 - \left(1 - \frac{x}{x_o}\right)^r \tag{G.1}$$

where cumulative mass fraction smaller than size x, x_o is particle size prior to fracture and r is an integer equal to the number of potential fracture surfaces along a characteristic dimension x_o. The most common fragment size distribution function is the Rosin-Rammler. Lovely 1973. The Schuhmann distribution function is identical to the Gaudin-Meloy function only with the difference that the exponent r is called the *Schuhmann distribution modulus*. It represents the slope at the fine end of the distribution. x_o is the theoretical (constant) intercept of the straight-line portion of the curve at 100% passing. Gaudin-Meloy, 1962 and Lizotte, 1990.

gauge button, cemented carbide button mounted in the outer peripheral row on a drill bit. Atlas Copco, 2006.

gauge grinding, adjustment of the diameter of a drill bit by grinding. Sandvik, 2007.

gauge row, the outer row of cutters on a full face boring head. Sandvik, 2007.

gauge row buttons, the periphery buttons on the drill bit. Sandvik, 2007.

gauge wear, decrease of the diameter of a drill bit with the length of drilling. Sandvik, 2007.

gauge (wire), a series of standard sizes of wires such as the American Wire Gauge (AWG), used to specify the diameter of a wire.

gear feed, see '**screw feed**'.

geology terms (of importance for drillers), the terms are shown in Table 24.

Table 24 Geology terms of importance for drillers.

Geological terms of importance to drillers		
Position of formation	Uniformity or structure of formation	Nature of formation
Outcrop	Faults	Massive
Basement	Folds	Cavernous
Bedrock	Bedding	Fractured/broken
Overburden	Foliation	Sheared
Hanging wall	Fissility	Consolidated
Footwall	Strike	Unconsolidated
	Dip	Fresh
		Weathered
		Decomposed
		Porosity
		Permeability

gel ampoule, a semi-rigid, blow moulded, polythene tube, 419 mm in length and 38 mm in diameter. A gel compound is filled under pressure and then sealed. The gel consists of 96% water and remaining chemical gelling and bactericides agents. Its length is 66 cm and diameter 38 mm. The explosive charge is placed inside the shot hole followed by the ampoule. The ampoule is slit longitudinally in 4 places, 100–125 mm from the in bye end of the ampoule. The area of slitting is marked on the ampoule sheath. Using a wooden tamping pole it is then pushed up to the explosive charge. By applying pressure from the tamping pole, the gel ejected out of the ampoule fills the shot hole completely. A solid plug of stemming is additionally used to keep the ampoule intact. Pradhan, 2001.

gelatin, see '**gelatin explosive**'.

gelatine dynamite, a *high explosive* consisting mainly of *nitroglycerin, sodium nitrate, meal, collodion cotton,* and *sodium carbonate*. Pryor, 1963. A dynamite produced by gelatinizing nitroglycerin with collodion cotton before adding an absorbent. Bennett, 1962. It is dense, plastic, and more water-resistant than straight or extra dynamite. Its relatively high velocity of detonation makes it ideal for hard rock and wet conditions, or for actual underwater blasting. Carson, 1961.

gelatin explosive, an explosive or blasting agent that has a gelatinous consistency. The term is usually applied to a gelatin dynamite but may also be used in combination with a water gel.

generator blasting machine, a blasting machine operated by vigorously pushing down a rack bar or twisting a handle. Now largely replaced by condenser discharge blasting machines. USBM, 1983.

genetic programming, a technique for automatic generation of computer program by means of natural selection. Poli, 2008.

geological discontinuity, see '**discontinuity**'.

geological strength index (GSI), a rock mass classification scheme developed by Dr E Hoek as an index for use in estimating the strength and deformability of jointed rock masses.

geomechanic, see '**geomechanics**'.

geomechanics, the mechanical properties of all geological materials, including soils. Brady et al., 1993.

geomechanics (in mining), the subject describing the occurrence, conditions and progress of the mechanical processes inside a geomaterial (soil, intact or broken rock) caused by mining activity.

geomechanics classification scheme for mining (modified), Laubscher, 1977 modified Bieniawski's gromechanics classification on the basis of experience gained in a number of chrysotile asbestos mines in Africa. The classification in Table 25 uses the same five classifications parameters as Bieniawski' scheme but involves differences in detail. Each of the five classes are divided into subclasses, A and B, new ranges and ratings for '*intact rock strength*' (**IRS**) in Table 25 are used, and the joint spacing and condition of joint parameters are evaluated differently.

The only discontinuities (joints) included in the assessment of RMR are those having trace lengths greater than one excavation diameter of 3 m, and those having trace lengths of less than 3 m that are intersected by other discontinuities to define blocks of rock. True spacings of the three most closely spaced discontinuities sets present in the rock mass are used in conjunction with Fig. 37.

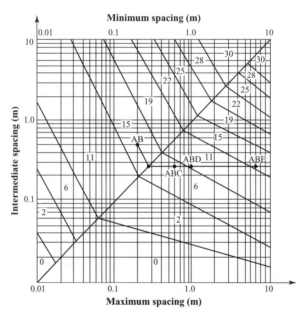

Figure 37 Joint spacing ratings for multi-joint systems. In the example, spacings of sets A, B, C, D and E are 0,2, 0,5, 0,6, 1,0 and 7,0 m, respectively; the combined ratings for AB, ABC, ABD and ABE are 15, 6, 11 and 15, respectively. After Laubscher, 1977.

The way in which the joint condition rating is influenced by a range of factors is set out in Table 26. A possible rating of 30 is assigned initially and, depending on the features exhibited by the joints, is then reduced by the percentages given in Table 25.

geometric damping (spreading), see '**attenuation**'.

geophone, a detector placed on the ground or mounted rigidly onto a rock surface or building (foundation) to measure the particle velocity of the structure. The measuring principle

Table 25 Modified geomechanics classification scheme. Laubscher, 1977.

Class	1		2		3		4		5	
Rating	100–81		80–61		60–41		40–21		20–0	
Descript.	Very good		Good		Fair		Poor		Very poor	
Subclass	A	B	A	B	A	B	A	B	A	B
"Item ratings"										
RQD (%)	100–91	90–76	75–66	65–56	55–46	45–36	36–26	25–16	15–6	6–0
IRS (MPa)	141–136	135–126	125–111	110–96	95–81	80–66	65–51	50–36	35–21	20–0
Rating	10	9	8	7	6	5	4	3	2	1
Joint spacing Rating	See Figure 37 in previous page. 30..0									
Joint condition Rating	45°.........................Static angle of friction. See Table 26 below..........................5° 30...0°									
Groundwater (Ratio joint water pressure/to major principal stress)	0		0		0,0–0,2		0,2–0,5		0,5	
Description	Completely dry		Completely dry		Moist only		Moderate pressure		Severe problem	
Rating	10		10		7		4		0	

Table 26 Assessment of joint condition—adjustments as combined percentages of a total possible rating of 30. Laubscher, 1977.

Parameter	Description	Percentage adjustment
Joint expression (large scale)	Wavy unidirectional	90–99
	Curved	80–89
	Straight	70–79
Joint expression (small scale)	Striated	85–99
	Smooth	60–84
	Polished	50–59
Alteration zone	Softer than wall rock	70–99
Joint filling	Coarse hard-sheared	90–99
	Fine hard-sheared	80–89
	Coarse soft-sheared	70–79
	Fine soft-sheared	50–69
	Gouge thickness < irregularities	35–49
	Gouge thickness < irregularities	12–23
	Flowing material > irregularities	0–11

most commonly used for measuring blast vibrations is based on electromagnetism such as created when a magnet is moving in a metallic coil and thereby inducing a current/voltage in the coil proportional to the particle velocity. Geophones are designed to measure vibrations within certain frequency ranges.

geophysical ground probing, the use of geophysical methods like *seismic*, *radar*, *electrical*, or *radioactive* methods to investigate rock conditions before tunnelling, shaft sinking metal or coal mining. Seismic methods can be used in wet rock conditions and radar is preferred to be used in dry rock conditions. Nord, 1991.

geophysical logging, see '**geophysical ground probing**'.

georadar, is a geophysical method which makes it possible to study, soil properties, soil depth, depth to rock, ground water surface, examination of road embankments and under ground piping, salt damages, concrete bridges and electricity cables, etc. Radio waves, with a frequency from 80 to 1 000 MHz are used to investigate properties of surface layers. The higher the frequency, the smaller is the penetration depth of the georadar. Good penetration is achieved in isolating materials like, gravel and sand. Salt water and clay which are good conductors, reduces the penetration depth. Example of maximum penetration depths are clay 2–4 m, silt 5–10 m, peat 12–20 m, moraine 15 m, fresh water 15–25 m, sand and gravel 60 m and rock 75–100 m.

geotechnical drill rig, drill rig designed for use in ground stabilization work etc. in the construction industry.

geotechnical properties, chemical, physical and mechanical properties of a geological material.

geotechnology, a term including *earth sciences*, *mineral engineering*, *mineral technology and mineral economics*. Bennett ed, 1962.

geotomography, investigation of soil or rock mass with non-destructive methods e. g. seismic velocity measurements or electrical resistance measurements between drill holes or the use of ground seismic or georadar. Kikuchi, 1991. The purpose is to scan a region and generate a map of the parameter of interest.

German cut, see '**pyramid cut**'.
glass micro-ballons, see '**micro-ballons**'.
Global positioning system, see '**GPS-based navigation**'.
glory hole, the conical surfaces created above mines with caving methods.
glory hole mining, a) a conical shaped excavation on surface with top of cone directed downwards and connected with a raise driven from an underground haulage level. The ore is broken by drilling and blasting in benches around the periphery of the cone. Lewis, 1964.
Glötzl pressure cell, see '**flatjack pressure cell**'.
goaf or **gob**, that part of a mine from which the coal has been worked away and the space more or less filled up. Fay, 1920. See also '**longwall mining**'.
gob, see '**goaf**'.
gopher hole blasting or *coyote blasting*, see '**chamber blasting**'.
gouge, a layer of *soft material along the wall of a vein*, favouring the miner, by enabling him after gouging it out with a pick, to attack the solid vein from the side. Fay, 1920.
gouge-filled discontinuity, crack, movement- or contact zone filled with a weaker rock type than the surrounding rock types. It could include clay, mica, chlorite, graphite or talc. TNC, 1979.
GPS (acronym), for 'global positioning system', see '**GPS-based navigation**'.
GPS-based navigation, the localization of an object in three dimensions by measuring the distance to known satellites positions instead of to fix points on earth. Peck, 1997. This technique can be used for determination of the position of mining equipment on surface like drill rigs, shovels, trucks or subsidence of the hanging wall etc. Positioning precision is down to ± 20 mm.
Gr (abbreviation), for '**gravel**' defining fragments with sizes from 2,0 to 63 mm.
gradation curve, a curve quantifying the size distribution of a material. Normally it shows the amount of material passing a certain mesh size as function of the quadratic mesh size.
grade (mineral content), **(weight-% or g/t)**, in mining this term refers to the relative content of valuable/cash ingredients contained in a material and is rated as high, low, etc. The classification of an ore according to the desired or worth less material in it or according to value.
grade, the elevation of a road, railway, foundation and so on in excavations. When given a value such as percent or degrees, the grade is the amount of fall or inclination as compared with a unit horizontal distance for a ditch, road, etc. 'To grade' means to level ground irregularities to a prescribed level. Konya and Walter, 1990.
grade control blasting, see '**selective mining**'.
grade level, the floor level in the open pit.
grade strength (%), makes it possible to relate the explosive performance for various explosives to ballistic mortar values corresponding to straight dynamites with a certain percentage of nitroglycerin. An explosive giving the same pendulum deflection as a 60% dynamite has a 60% grade strength. Persson, 1993.
grain, old measure for mass based on the mass of one grain (seed). 7 000 grains are equal to 0,454 kg or 1 lb. 'Grains per meter' is still used to quantify the charge concentration of detonating cord (grains/m). The recommended unit today is g/m.
grain (size), (**m** or **mm**), in petro physics a grain is defined as a piece of mineral defined by its boundary. The maximum size of grains is set to about 1 m.

grain (weight), (grain), unit of weight that equals 0,0648 gram. Fay, 1920.
grain boundary crack, a microcrack along a grain boundary. See also '**microcrack**'.
grain size distribution, distribution of the size of grains of minerals in rocks.
grain volume (V_g), (mm³), the volume of a grain including its open and closed pores.
Granby car, a type of automatically dumped underground mine car used for hand or power-shovel loading. In this type of car, a wheel attached to the side of the car body engages an inclined track at the dumping point. As the side wheel rides up and over the inclined track, the car body is automatically raised and lowered, activating a side door operating mechanism which raises the door. Permitting the car to shed its load.
granulometric analysis, determination of the size and shape of the essential rock forming minerals. The border of the size classes analysed are often doubled e.g. 1–2 mm, 2–4 mm, 4–8 mm etc.
gravel, rock grains or fragments with a diameter range from 2,0 mm to 63 mm. A subdivision of gravel is made in the classes: coarse gravel (CGr) >20 to 63 mm, medium gravel (MGr) 6,3 to 20 mm and fine gravel (FGr) >2,0 to 6,3 mm. See also '**particle size**'. SS-EN ISO 14688-1:2002, 2003.
gravitational acceleration (g), (m/s²), the acceleration caused by the mass of the earth.
gravitational stress, the stress state due to the weight of the super imcumbent rock mass. Hyett, 1986.
gravity draw, the movement of caved or broken material from a higher level to a lower level by using the gravity forces. The method is used in primary mining but also for transportation of waste or ore to lower levels through ore passes.
gravity flow, the downwards movement of broken material by gravity forces. See also '**drawbody**', '**loosening draw body**'.
GRC (acronym), for '**ground reaction curve**' or the relation between stress and strain of the rock mass. MassMin, 2008. (p. 793).
Green's function, Fourier analysis is particularly useful in analysing a blasting single-charge wave source in the rock mass and which gives the *effect of the travel path* (known as the Green's function) on the source impulse. By de-convolution (finding the origin from the tracks of a phenomenon) and using Fourier analysis, the seismic efficiency and firing times may be determined for a given shot. Anderson, 1993.
grey level, in image analysis, a term used to characterize the level or degree of darkness or brightness of a pixel.
Griffith crack theory, the necessary tensile stress σ_t in MPa to extend an existing pore with the length l_p and located perpendicular to the stress field is calculated by,

$$\sigma_t = \sqrt{\frac{4E\gamma_s}{\pi l_p}} \tag{G.2}$$

where E is the Young's modulus MPa, γ_s is the specific surface energy per unit area of the crack surfaces associated with the rupturing of the atomic bonds when the crack is formed in J/m² and l_p is initial pore (microcrack or flaw) length in m. Experimental studies show that, for rock, a pre-existing crack does not extend as a single pair of crack surfaces, but a fracture zone containing *large numbers of very small cracks develops ahead of the propagating crack*. In this case it is preferable to treat w_f as an *apparent surface energy* to distinguish it from the *true surface energy*, which may have a significantly smaller value.

For high stresses the Griffith crack theory is not valid. As the stresses increase into region of high compressive stresses, this criterion fails to take into account the crack closure and the consequent frictional effects between the crack surfaces. A *modified Griffith theory* was proposed to account for this effect. The modification resulted in a transition from Griffith's criterion at low stresses to the *Mohr-Coulomb criterion at high compressive stresses*. Both criteria predict that the intermediate principal stress has no effect on the strength. Hudson, 1993.

grill, grid arrangement made from metal or other hard material used as a protection cover. Atlas Copco, 2006.

grinding cup, cup shaped diamond grinding tool for dressing button bits. Atlas Copco, 2006.

grinding disc, disc shaped grinding tool. Atlas Copco, 2006.

grinding machine, machine specially designed for grinding drill bits. Sandvik, 2007.

grindstone, a large, circular, revolving stone used for sharpening tools and instruments. It is made from a tough sandstone of fine and even grain, composed almost entirely of quarts, mostly in angular grains which must have sufficient cementing material to hold the grains together but not enough to fill the pores and cause the surface to wear smooth. Sanford, 1914.

gripper, hydraulic jacks needed to fix the full-face boring machine to the walls to make it possible to create a thrust on the drill head.

gripper unit, in full-face drilling, hydraulic jacks on the machine being pushed perpendicular to the tunnel direction against the wall so a feed thrust can be applied from the machine on the face.

gripping tools, e.g. pipe wrenches (spanners) and pipe tongs are used in connection to drilling operations. Wrenches can be open-ended, tube or tubular box spanners, sockets (four, six or eight-point) and ring types. Stein, 1997.

grit, sand, especially coarse sand. AGI Suppl., 1960.

grit blasting, cleaning of drilling tools by 'grit blowing'. Sandvik, 2007. Grit stands for hard particles or sand. See also '**sand blasting**'.

grizzly, a structure made of parallel steel beams or quadratic or rectangular oriented bars through which broken rock of up to a maximum size may pass under the influence of gravity.

grizzly blasting, blasting of fragments which cannot pass the grizzly.

grommet, eyelet, ring or washer used to reinforce an opening in a material or protect something passing through the opening. Atlas Copco, 2006.

gross vehicle mass, (**GVM**), (**kg**), the total mass of the vehicle inclusive fuel and maximum permitted load. The term gross vehicle weight should not be used.

gross vehicle weight (GVW), (kg), see '**gross vehicle mass**'.

ground drill, drill used to drill holes in loose, unconsolidated material. Atlas Copco, 2006.

ground engineering equipment, equipment used for preparing earth or soils for a building or other structure. Atlas Copco, 2006.

ground fall, uncontrolled loosening of rock. It is often caused by geological structures and is generally associated with large deformations, which are almost instantaneous. As a consequence, insufficient time is available for their detection and for the implementation of remedial measures. Hudson, 1993.

ground fault, electric contact between a part of the blasting circuit and the ground (earth).

groundhog, see '**barney**'.

ground reaction curve (GRC), the relation between stress and strain of the rock mass. MassMin, 2008. (p. 793).

ground reinforcement, practice of increasing the bearing strength of soils and rock by various means. The rock surrounding mine openings and tunnels is generally reinforced by installing rock bolts along with various shaped plates, wire mesh or sprayed concrete. In soft rock mining external supports are common and in construction operations sometimes the ground is reinforced by injecting cement. Atlas Copco, 2006.

ground release pulse (GRP), (p_{og}), **(Pa or dB)**, air blast overpressure produced from gas escaping from the detonating explosive through rock fractures. See also '**air blast**'. Atlas Powder, 1987.

ground vibration (v), (mm/s), vibration of the ground induced by elastic waves emanating from a blast or pile driving, etc.; characterized by vibration *particle velocity*, and usually measured in millimetres per second (mm/s) regarding damages to rock and buildings. When measuring damage to electronic equipment such as computers vibration *acceleration* forms the basis of assessment and validation due to blasting. Most countries have developed their own rules for maximum allowed vibrations for different kind of buildings on surface, e.g. USA, USBM, Siskind, 1980; Germany DIN 4150 Teil 3, Anon., 1986; Sweden SS 460 48 66, SIS, 1991; Australia, Australian Standard, 2006; South Korea, The Ministry of Land Transports and Maritime Affairs (MTLM), Anon, 2007 and India, Singh, 2009.

ground vibration influence on reinforced concrete (RC), is dependent on the natural frequency f_{nrc} of the RC-structure in Hz, frequencies induced by blasting f_x in Hz, peak particle velocity from blast vibration v_{max} in mm/s, propagation velocity of wave motion through RC-structure c_p in m/s, peak particle displacement u_{max} in mm and acceleration a_{max} in m/s², density of the RC-structure ρ_{RC} in kg/m³, modulus of elasticity of the RC-structure E in MPa, Poisson's ratio or the structure v, distance of blast source from structure R in m, maximum charge weight per delay Q_{max} in kg and depth of RC-structure H in m. Isaac, 1981.

ground vibration particle velocity transformation to stress, anticipating a planar wave front in a 2-dimensional media the stress σ in Pa can be calculated according to the following formula.

$$\sigma = vc_p\rho_r \quad (G.3)$$

where v is the particle velocity in m/s, c_p is the longitudinal wave velocity in m/s, ρ_r density of rock in kg/m³.

ground vibration simulations from multiple holes, can be simulated if the wave form from the ground vibration from one single blasthole is known. By linear superposition of the waveforms from multiple holes activated at their detonation times the ground vibration can be calculated. One simulation e.g. shows peak amplitudes at 0 and 28 ms delay time and a minimum at 17 ms delay time. Hinzen, 1988.

ground vibration transmission calibration, determination of the ground vibration transmission characteristics of a region. Konya and Walter, 1990.

grout, special cement mixture used to affix something. Atlas Copco, 2006.

grouted bolt, a steel bar is inserted in a drill hole in the rock mass and grouted by cement or resin along its whole length in the hole with the purpose to reinforce the rock mass.
grouted cable bolt (steel cable), a steel rope is inserted and grouted along its whole length in a drill hole with the purpose to reinforce the rock mass.
grouted steel bar, see '**grouted bolt**'.
grouted wedge bolt, *friction type bolt* based on a reinforcing bar fastened at the bottom of the borehole with a wedge and grouted by cement into the drill hole.
grout hole drilling, drilling of holes for injection of cement to increase the stability of rock and to prevent water inflow to tunnels or drifts.
grouting, the act or process of injecting grout into crevices of the rock mass, usually through a borehole drilled into the rock to be grouted; also, the grout thus injected. Long, 1960. The purpose is to glue the crevices to make them water tight and to reinforce the rock mass. Some types of rock bolts are affixed in holes in rock by pumping grout around them.
grouting after blasting, when rock conditions are bad, the grout will be injected after the drift- or tunnel round has been blasted, mucked and reinforced by bolts or shotcrete.
grouting before blasting, when rock conditions are bad, it may be necessary to grout in advance before the drift or tunnel proceeds through the area. This is called pre-grouting. Especially when the water low is large pregrouting is preferable.
grout injection, the process of forcing, normally cement grout, with pressure pumps via a plastic hose into crevices, cracks and fractures in the rock formation usually through a borehole. The purpose with the action is to increase the strength of the rock by assisting self-supporting mechanisms such as arching and to prevent inflow of water.
grout injector, a machine that mixes the dry ingredients for a grout with water and injects it, under pressure, into a grout hole.
grout machine or **grouting machine**, see '**grout injector**'.
grout packer, tube fitted with an expandable sleeve that can be fixed in a drill hole so that grout can be pumped into the surrounding rock at high pressure. Atlas Copco, 2006.
grout pump, pump designed to pump cement grout. Atlas Copco, 2006.
GRP (*acronym*), (p_{og}), (**Pa or dB**), for '**gas release pulse**' and '*ground release pulse*'. See '**gas release pulse**' and '**air blast**'.
Grönlund cut, a parallel hole cut, type burn cut, using one charged center hole with ordinary diameter and a first row of holes in square shape and uncharged that are used as swelling volume. The second row is almost squared shaped and includes 8 charged holes and it is turned 45° to the first row. See Fig. 38. Langefors, 1963.

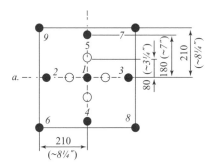

Figure 38 Grönlund cut, a parallel hole cut type burn cut used in drifting. After Langefors, 1963.

GSI (acronym), see '**geological strength index**'.
guard bellows, bellows like flexible guard. Atlas Copco, 2006.
guar gum, is made from crushed guar seeds and is used to thicken slurry and water gel products.
guhr, or silica guhr is a diatomaceous earth. See '**diatomaceous earth**'.
guhr dynamite, see '**dynamite**'.
guide, anything that is used to mechanically control the direction of a moving part. Atlas Copco, 2006.
guide bit, drill bit with a fluted skirt used with TDS tube drilling systems. Atlas Copco, 2006.
guide rail, long rail used to guide or steer something moving along it. Atlas Copco, 2006.
guide rod, a more stiff drill rod than the ordinary used coupled to and having the same diameter as a core barrel on which it is used. It gives additional rigidity to the core barrel and helps to prevent deflection of the borehole. Also called *core-barrel rod, oversize rod*. Compare 'drill collar'. Long, 1920.
guide rope, a gage guide. Standard, 1964.
guide shoe, a circular attachment to the cage made of metal and used as guide to follow the guide ropes for the cage and with the purpose to prevent too large lateral movement of the cage during hoisting.
guide sleeve, tubular device used to steer or determine the path of an object passing trough it. Atlas Copco, 2006. Barrels on the drill steel with the purpose to make the finished hole more straight.
guide strip, thin strip of material used as a guide. Atlas Copco, 2006.
guide tube, rigid tube used in the drill string to reduce deviation and high drill steel consumption when drilling in poor rock formations. The guide tube can replace the first rod after the bit. This provides extra support for the drill string, usually keeping *deviation within two percent of the hole depth*. Atlas Copco, 2006.
gull, a large fissure or chasm in strata, especially limestone, generally filled with earth or higher strata. Arkell, 1953.
gun, **butt, boot-leg,** or **socket**, that portion or remainder of a shot hole in a face after the explosive has been fired incompletely after the blast and that still may contain explosives and is thus considered hazardous. A situation in which the blast fails to cause total failure of the rock because of insufficient explosives for the amount of burden, or owing to incomplete detonation of the explosives.
gun drill, a drill, usually with one or more flutes and with coolant passages through the drill body, used for deep-hole drilling. ASM Gloss, 1961.
gunite, which pre-dates shotcrete in its use in underground construction, is *pneumatically applied mortar*. Because it lacks the larger aggregate sizes of up to 25 mm typically used in shotcrete, gunite is not able to develop the same resistance to deformation and load-carrying capacity as shotcrete. Brady, 1985. Cement sprayed onto mine timbers to make them fire-resistant. A mixture of sand and cement sprayed with a pressure gun onto roofs and ribs to act as a sealing agent to prevent erosion by air and moisture. To cement with a cement gun. See also '**shotcreting**'.
gunite gun, see '**cement gun**'.
guniting, see '**gunite**'. It is different from '**shotcreting**'.
gunpowder, see '**black powder**'.

Gurit (tradename), low density perimeter explosive delivered in stiff plastic pipes with diameters 11 and 17 mm and being used for *controlled contour blasting*.

GVM (acronym), **(kg)**, see '**gross vehicle mass**'.

GVW (acronym), **(kg)**, for '**gross vehicle weight**'. This term is recommended to be replaced by 'gross vehicle mass'. see '**gross vehicle mass**'.

gyratory breaker, see '**cone crusher**'.

gyratory crusher, see '**cone crusher**'.

gyroscope (survey instrument), are used in magnetically complex formations or when a survey must be conducted inside drill rod or casing, azimuth readings can be taken to orient the instrument. High speed gyroscopes (40.000 rpm) are available to run in holes as small as 40 mm, although 75 mm is standard.

H (abbreviation), see '**hardness**'.

hacked bolt, an anchor bolt with a barbed flaring shank which resists retraction when loaded into stone or set in concrete. Also called *rag bolt, Lewis bolt*. Webster 3rd ed., 1961.

hack hammer, a hammer resembling an adz, used in dressing stone. Webster 3rd, 1961.

hackle marks, in fractography, topographic feature used to denote conjugate crack surface markings formed by normal and/or shear forces. They are characterized by a plumose pattern diverging from the point of origin of fracturing in the direction of propagation. Carrasco and Saperstein, 1977.

half barrels, the percentage of identifiable blastholes barrels or half casts over the total number of meters of perimeter holes drilled in the walls and the roof of a tunnel- or drift round.

half cast factor (R_{hcf}), (%), the percentage of the visible length of identifiable blastholes 'barrels' or 'half casts' over the total numbers of meters of perimeter holes drilled.

$$R_{hcf} = 100 \frac{\sum_{i=1}^{n} L_i}{\sum_{j=1}^{n} L_j} \tag{H.1}$$

where R_{hcf} is the half cast factor in %, ΣL_i is the total length of visible half casts after blasting, and ΣL_j is the total drilled length of the wall and roof holes in the blast.

half pusher, a charging machine for large diameter cartridges (>100 mm) in up-holes developed by Nitro Nobel in the 1980's, see Fig. 39. It is an air powered unit, initially fastened to the charge, which pushes the charge a short distance up hole. The charge has wings so it cannot retract. Now the fastening part of the unit is deflated and the half pusher contracts. After that the fastening unit expands so the charge can be pushed a further distance in the hole. This operating cycle is repeated until the charge is in position. At that point the half pusher separates itself from the charge and is removed from the hole. The method is quick and reliable if the rock is not too jointed. Persson, 1993.

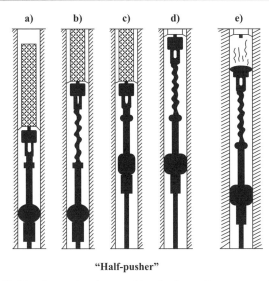

"Half-pusher"

Figure 39 Half-pusher a charging machine for large diameter cartridges in up-holes. After Persson, 1993.

half-second detonator or **timing (HS)**, initiation using delay detonators with a nominal half- second delay between sequenced numbers.

hammer, see '**hammer drill**'.

hammer and penknife test, a test to estimate the compressive strength of rock by scratching the rock with a penknife and blow it with a hammer. The result is listed according to Table 27.

Table 27 Field test estimation of compressive strength by penknife and geological hammer.

Properties of hand specimen, ~100 mm cube or sphere	Strength description (rock type)	Approximate compressive strength (MPa)	Class
– Easily scratched or scraped by penknife. – Shattered or indented by a single firm hammer blow.	**Weak** (chalk, mudstone, coal)	0–25	V
– Can be deeply scratched by penknife. – Breaks with a single moderate blow of hammer.	**Moderately strong** (siltstone, shale, sandstone)	25–50	IV
– Can be scratched by penknife. – Requires 3 or 4 moderate blows of hammer to break specimen.	**Strong** (sandstone, limestone, slate)	50–100	III
– Difficult to scratch by penknife. – Requires a number of heavy hammer blows to break specimen.	**Very strong** (limestone, granite, ironstone)	100–200	II
– Not scratched by penknife. – A number of heavy hammer blows only chips the specimen.	**Extremely strong** (quartzite, basalt, dolerite)	>200	I

hammer cut out switch, switch mounted in a system to shut down the hammer to prevent damage to it. The rock drill on a drill rig may be referred to as a hammer. Atlas Copco, 2006.

hammer drill (machine), a light, mobile, and fast cutting drill machine in which the bit does not reciprocate but remains against the rock in the bottom of the hole, rebounding slightly at each blow. There are *three types of hammer drills*, *drifter* (drilling horizontally), *sinker* (drilling downwards) and *stopper* (drilling upwards). Lewis, 1964. A drill machine where the impact energy is transmitted from the hammer via drill rods to the drill bit at a certain frequency. The driving mechanism may be of pneumatic or hydraulic type (oil or water).

hammer drilling, a drilling method where the drilling energy is transferred to the drill steel and drill bit by impact without a rebound of the drill bit. It is *normally combined with rotation of the drill steel* to increase the penetration rate and then it is called **percussive drilling**. Sandvik, 1983. This is the most common used drilling method.

hammer mills, an impact mill consisting of a rotor, fitted with movable hammers, that is revolved rapidly in a vertical plane within a closely fitting steel casing. Hammer mills are sometimes used for the size reduction of clay shale, glass cullet, and some of the minerals used in the ceramic industry.

hammer pick machine, a hydraulic operated hammer mounted on a digging crawler or rigidly mounted in connection to a crusher or chute and used to break the boulders by hammering a wedge into the boulder. The technique is used both on surface and underground.

hammer rock drill, machine used for hammer drilling. Sandvik, 2007.

hand drill, the operation to use a drill steel and a sledge where the sledge is used to strike the drill steel to achieve the necessary splitting force on the rock. The drill steel is slightly turned between each strike. It is a two person operation where one is holding the drill steel and the other is doing the striking.

hand feed, thrust of drill machine is caused by hand craft. Sandvik, 2007.

hand held drilling, normally drilling with an air pusher leg. Sandvik, 2007.

hand held hydraulic breaker, similar in construction to a pneumatic breaker but powered by oil hydraulics. Atlas Copco, 2006.

hand held motor breaker or **drilling machine**, similar use as a pneumatic breaker but the percussive mechanism is driven by the pressure from an internal combustion engine that is an integral part of the machine.

hand mucking, the loading of ore or waste by pickaxe and trough. LKAB, 1964.

handpicking or **sorting**, manual removal of selected fraction of coarse run-of-mine ore, usually performed on picking belts (belt conveyors) after screening away small material, perhaps washing off obscure dirt, and crushing pieces too large for the worker to handle. Pryor, 1963.

hand pneumatic breaker, hand held machine having a percussive mechanism but no rotation device and powered by compressed air. The pneumatic breaker is fitted with a chisel steel or moil point and used to demolish cement, asphalt and masonry structures. Atlas Copco, 2006.

hand steel, drill steel used for handheld drilling. Sandvik, 2007.

HANFO (acronym), see '**heavy ANFO**'.

hangfire or **misfire**, failure of a charge to explode at the time expected. After the unsuccessful initiation, a time period must elapse before the blasting area may safely be revisited, and precautions to deal with the misfire must be taken immediately. Misfires are always dangerous.

hanging wall caving, the usually progressive failure or caving of the hanging wall rock overlying a mined orebody. The caving is caused because the mined area is not stabilized by waste rock or hydraulic fill.

hanging wall, the wall or rock on the upper-side of an inclined vein. In bedded deposits it is called the roof. Atlas Powder, 1987.

hang-up or **blockage**, a situation in underground mining when several larger rock blocks form a bridge in a crater or chute used for the discharge of the ore or waste from a stope or raise. A *'golden rule of thumb'* is that the largest blocks should be less than 1/3 of the diameter of the raise or the crater outlet to avoid hang-ups.

HAR (acronym), for high aspect ratio, see '**aspect ratio**'.

hardening, metallurgical process in which iron or suitable alloy is quenched by abrupt cooling from or through a critical temperature range. Precipitation hardening. Pryor, 1963.

hard heading, a heading driven in rock. Nelson, 1965.

hardness or **abrasive hardness** (*A*), resistance of a material to indentation or scratching. Hardness is a concept of material behaviour rather than a fundamental material property. As such the quantitative measure of hardness depends on the type of test employed.

hardness scale, the scale by which the hardness of minerals is determined as compared with a standard. The unit of Mohs' scale are relative numbers with the 1 used for most soft mineral and 10 for the hardest 1) talc, 2) gypsum, 3) calcite, 4) fluorite 5) apatite, 6) orthoclase, 7) quartz, 8) topaz, 9) corundum, and 10) diamond. Fay, 1920.

hardness test, a test method for the resistance of a material to indentation or scratching. The hardness of rock is dependent on the type and quantity of the various mineral constituents of the rock and the bond strength that exists between the mineral grains. Three types of tests have been developed to measure the hardness of rocks and minerals: 1) *Indentation tests* (for example the National Coal Board has developed a cone indenter hardness index. Whittaker et al., 1992.), 2) *Dynamic or rebound tests,* the Shore scleroscope or Schmidt impact hammer test (see also 'point load strength'), 3) *Scratch tests* (Talmage and Bierbaum devices). See, 'Suggested Methods for Determining Hardness and Abrasiveness of Rocks'. ISRM, 1978.

hardpan, boulder, clay, or layers of gravel found usually a few feet below the surface and so cemented together that it must be blasted or ripped in order to excavate.

haulage, the drawing or conveying in cars or otherwise, or movement of men, supplies, ore and waste both underground and on the surface. Lewis and Clark, 1964.

haulage level, underground level either along and inside the orebody or closely parallel to it, usually in the footwall. On this level the mineral is gravitated or drawn (slushed) down from overhand stopes or raised from underhand stope is loaded into trams (rail cars or trucks) and sent out to the hoisting shaft. Pryor, 1963.

hazard, a potential occurrence or condition that could lead to injury or death of personnel, delay, economic loss or damage to the environment.

HCF (acronym), see '**half cast factor**'.

HCS (acronym), see '**hydraulic control system**'.

HDS (acronym), see '**high definition surveying**'.

HDT (acronym), see '**heat to detonation transition**'.

HE (acronym), see '**heave energy**' or '*bubble energy*'.

head, development opening in a coal seam. Pryor, 1963.

header, an entry-boring machine that bores the entire section of the entry in one operation. Fay, 1920.

headframe, the steel or timber frame at the top of the shaft or reinforced concrete building that carries the sheave or pulley for the hoisting rope, and also serves several other purposes. Tweney, 1958.

headgear, that portion of the winding machinery attached to the headframe, or the headframe and its auxiliary machinery. Fay, 1902.

heading, a passage leading from the gangway. The main entry in a coal mine is laid out with the precision of a main avenue in a city. From it, at right angles, headings run like cross streets, lined on each side with breasts, or chambers. Korson, 1938. A horizontal level, airway, or other excavation driven in an underground mine.

heading and bench, a method of tunnelling in hard rock. The heading is in the upper part of the section and is driven only a round or two in advance of the lower part or bench. Fraenkel, 1954.

heading and bench mining, a stoping method used in thick ore bodies where it is customary first to take out a slice or heading directly under the top of the ore and then to bench or stope down the ore between the bottom of the heading and the bottom of the orebody or floor of the level. The heading is kept a short distance in advance of the bench or stope. This system is termed *heading and bench* or *heading and stope mining*. USBM Bull. 390., 1936.

heading and stall, see '**room and pillar**'. Fay, 1920.

heading and stope mining, see '**heading and bench mining**'.

heading blast, see '**chamber** or **coyote blasting**'.

heading blasting, see '**chamber** or **coyote blasting**'. AS 2187.1, 1996.

heading overhand bench or **inverted heading and bench**, the heading is the lower part of the section and is driven at least a round or two in advance of the upper part, which is taken out by overhand excavation. Fraenkel, 1952.

heading wall, the footwall or lower wall of a lode along which the heading is run. Fay, 1920.

headroom, height between the floor and the roof in a mine opening. Long, 1920.

headrope, in any system of rope haulage, that rope which is used to pull the loaded transportation device toward the discharge point. In scraper loader work, the headrope pulls the loaded scoop from the face to the dumping point. Jones, 1949.

headway, primarily a road or gallery in a coal mine parallel to the principal cleat. Also called headway course. Arkell, 1953.

headwork or **headframe**, the headframe with the headgear. Webster, 1960.

healed joint, a joint which has been 'welded' together by a filling material such as quartz, calcite, epidote, etc.

healed microcrack, a microcrack which has healed in geological times by contact or deposition of a filler material.

heap leaching, a process used of the recovery of copper from weathered ore and material from mine dumps. The material is laid in beds alternately fine and coarse until the thickness is roughly 6 m. It is treated with water or the spent liquor from a previous operation. Intervals are allowed between watering to allow oxidation to occur, and the beds are provided with ventilating flues to assist the oxidation of the sulphides to sulphate.

The liquor seeping through the beds is led to tanks, where it is treated with scrap iron to precipitate the copper from solution. This process can also be applied to the sodium sulphide leaching of mercury ores. USBM Staff, 1968.

heat (Q), (J), is defined by 1 J = 1 Nm. Heat and work have the same unit and the difference between them is only the state of energy. Heat could be transferred to mechanical energy (work) and vice verse. (Energy in transit due to decrease in temperature between the source from which the energy is coming and the sink toward which the energy is going. Besançon, 1985).

heat and friction test, a test to examine the sensitivity of an explosive to heat and friction. Heat can affect every explosive and cause thermal decomposition which may lead to detonation. At room temperature, water based explosives do not burn stable under atmospheric pressure and therefore the tests must provide some degree of confinement to allow for pressure built up. Examples of heat and impact tests are Koenen test, Princess incendiary spark test, external fire test for UN Hazards Division 1.5, deflagration to detonation transition test, and Woods test. Persson et al., 1994.

heat of combustion (Q_{co}), (MJ/kg), represents the caloric equivalent of the total combustion energy of the given substance. It is determined in a calorimetric bomb under excess oxygen pressure. The heat of combustion is usually employed to find the heat of formation. The heat of combustion depends only on the composition of the material and not on any other factor such as loading density or other factors. Meyer, 1977.

heat of detonation (Q_e), (MJ/kg), see '**heat of explosion**'.

heat of explosion (Q_e), (MJ/kg), the heat liberated per unit of mass of explosive during the chemical reaction or decomposition of an explosive material, including explosive mixtures, gunpowder, or propellant, etc. The magnitude depends on the thermodynamic state of the decomposition products. Q_e can experimentally be determined by specific laboratory tests and can theoretically be predicted if the chemical products formed during the explosion process are known. Following the calculations recommended by different sources, varying results are obtained. Therefore figures from different sources cannot be compared reliably. The Q_e of ANFO (6% fuel oil) at a density of 800 kg/m^3 should be estimated to be about 3,8 MJ/kg. Some sources estimate 'useful energy' by assuming an arbitrary volume or pressure (100 MPa) below which the expansion of the gases fails to do extra work. This has the effect of increasing the relative strength of high density explosives. The explosive energy can be divided into 6 different parts (areas) see Fig. 40.

Zone	Energy
1	Transient energy associated with the detonation wave
2	Potential shock energy
3	Strain energy around the borehole behind the shock wave
2+3	'Brisance energy'
4a	Fragmentation and heave energy
4b	Strain energy in burden at escape
5	Lost energy
2+3+4+5	Absolute mass strength

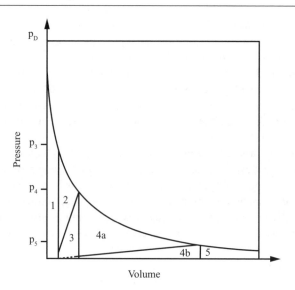

Figure 40 Pressure-volume diagram (p-V diagram) for a blasthole illustrating how the explosive energy can be divided into different areas. After Lownds, 1986.

where p_3 is the initial (explosion) state and p_4 is the equilibrium borehole state. The burden is compressed by gas in the cracks, more or less hydrostatically, to pressure p_5.

heat of formation or **heat of reaction (Q_f), (MJ/kg)**, the quantity of heat consumed or liberated in a chemical reaction, as heat of combustion, heat of neutralization or heat of formation. Hackh, 1944. For an explosive the difference in internal energy between the explosion state and standard temperature and pressure (STP). The energy which is bound during the formation of a given compound from its constituent elements at constant volume (energy of formation) or at constant pressure (enthalpy of formation, which includes the mechanical work performed at the stable state, 25°C and 0,1 MPa). The knowledge of the energies of formation of the presumed reaction products makes it possible to calculate the heat of explosion.

heat of reaction, see 'heat of formation'.

heat resisting steel, a steel with high resistance to oxidation and moderate strength at high temperature, that is, above 500°C. Alloy steels, of a wide variety of compositions, which usually contain large amounts of one or more of the element *chromium*, *nickel*, or *tungsten*, are used. C.T.D., 1958.

heat sensitivity test, a method to test the heat sensitivity of an explosive. There are several tests developed, see 'Koenen test' and 'Woods test'.

heat to detonation transition (HDT), a thermal quantity needed to start a detonation by heating the explosive rapidly.

heave, the extent to which the broken mass of rock is moved and displaced from its original location. AS 2187.1, 1996.

heave or **bubble energy (HE)**, the energy provided by the gas needed to lift and throw a mass of rock in blasting.

heave damage zone, see '**blast induced damage zone**'.

heavy ANFO (HANFO), an emulsion explosive which contains more than 50% (mass) of ANFO.

heavy duty, term used to describe an object that has been constructed to withstand unusually hard usage or unusual stress. Atlas Copco, 2006.
heavy hydraulic breaker, similar to hand held breaker in that they have a hydraulic oil powered percussive mechanism and no rotation device, but large. Generally mounted on some sort of boom for positioning. Atlas Copco, 2006.
heavy mud, a drilling mud having a high specific gravity, usually attained by adding minerals such as *grounded barite* or other heavy loaders to the usual ingredients making up a drill mud. Long, 1920.
heel, the mouth or collar of a borehole. See '**collar**'. Fay, 1920.
Heelan model of equations, advanced equations for calculation of the vibrations around a blasthole, see Blair, 2006.
heel row, the '**bottom row**' when blasting a bench with horizontal holes. Sandvik, 2007.
height (H), (m), the perpendicular distance between horizontal lines or planes passing through the top and bottom of an object. Lapedes, 1978.
height of abutment (H_{ab}), (m), the height from where the vertical wall ends in a drift, tunnel, incline, decline, room or stope and the curved roof starts to the highest point of the roof.
height of arch (H_a), (m), the vertical distance from the bottom of the arch to the top of the arch in a tunnel or drift.
height of bench (H_b), (m), the vertical distance between the floor- and top level of a bench.
height of draw (h), (m), the vertical distance from the draw point to the highest level from which a piece of rock has been drawn.
height of lip of crater (H_{lip}), (m), see '**crater blasting**'.
height of loosening zone (HLZ), (h_{loose}), (m), the vertical height at the base of the cave within which the draw produces a larger or small motion.
height of the interaction zone (HIZ) or **height of draw (H_{di}), (m)**, the vertical height at the base of the cave within which adjacent draw zones interact and produce motion of caved ore. When the draw body symmetry axis deviates from the vertical the height is defined is as the length of the movement of caved ore along the hanging wall in longitudinal sublevel caving.
height of tunnel, drive, crosscut, adit, incline or **decline (H_d), (m)**, the maximum vertical distance of the cross-section of an excavation opening.
height wear, frontal wear on cemented carbide inserts. Sandvik, 2007.
HEL (acronym), see '**Hugionot elastic limit**'.
helical screw compressor, uses two screws for transportation of the air.
helical screw pump, a pump using two screws for transportation of the fluid.
helper holes, the blastholes next to the contour holes in a round.
hemispherical ended borehole technique in hot rock, a method to determine rock stresses in hot rock, 50–80°C, by *over coring technique* of a drill hole where the strain gauges are glued to the hemispherical bottom of the hole. Obara, 1991.
hemispherical projection, a method to describe the dip of discontinuities in the rock mass. Imagine a sphere which is free to move in space so that it can be centred on an inclined plane. The intersection of the plane and the surface of the sphere is a great circle. A line perpendicular to the plane and passing through the centre of the sphere intersects the sphere at two diametrically opposite points called the poles of the plane. Because the great circle and the pole representing the plane appear on both the upper and lower parts of the sphere,

only one hemisphere need by used to plot and manipulate structural data. In *rock mechanics*, the *lower-hemisphere projection* is almost always used. The *upper-hemisphere projection* is often used in textbooks on *structural geology* and can be used for rock mechanics studies if required. Brady, 1985.

Hercudet (tradename), a non-electric initiation system. A special blasting machine is connected to the detonators by thin plastic tube which closes the circuit. The blasting machine injects a gaseous mixture of two components, oxygen plus a fuel gas, into the circuit, and the explosion starts when the whole line is filled with the mixture. The detonation is propagated at a speed of 2.400 m/s, initiating the detonators along its way, but not the explosive in contact with the tubes, which means that bottom priming is also feasible.

Herculare lining (tradename), a German method of lining road-ways subjected to heavy pressures. It consists of a closed circular arch of specially shaped pre-cast concrete blocks. The blocks, which are wedge-shaped, are made in two sizes for each lining and erected in such a way that alternate blocks offer their wedge action in opposite direction—the larger blocks towards the center of the roadway and the smaller outwards. This arrangement gives a double-wedge effect so that part of the lateral pressure exerted by the start on the lining is diverted axially along the roadway. Nelson, 1965.

Hertz (f), **(Hz)**, a unit used to quantify the frequency of vibrations in cycles per second. In blasting used in connection with ground vibrations and air blasts.

Hertz theory, is a soft contact formulation in which the assumption of elasticity is used to derive the normal stiffness of the contact between two deformable spheres. A further assumption is that the contact area is small compared to the radius of the spheres. The derived normal stiffness is nonlinear, and may be used directly in numerical simulations of sphere assemblies. Other type of interfaces, such as rock joints, may display similar behaviour because the asperities on a rough surface act as though they are an assembly of tiny spherical contact points. Hudson, 1993.

Hess cylinder compression test, a test method for explosive which defines the 'breaking power' of the explosive which is related to the fragmentation capacity of the rock. A cylindrical lead block (length 65 mm and diameter 40 mm) is placed with the end surface on a steel plate with a thickness of minimum 8 mm. A circular steel plate of 4 mm thickness is placed on the top of the lead cylinder and on that plate a 100 g of the explosive with a cylindrical shape and diameter 40 mm is placed and detonated by a blasting cap. The deformation of the lead cylinder is measured. Jimeno et al., 1995.

heterogeneous, micro structural property of materials showing different properties (mechanical, optical, electrical, etc.) at different points in the material. E.g. coarse-grained granite, breccia, etc.

HEX (abbreviation), see '**high energy explosive**' normally abbreviated to 'high explosive'.

hexagonal drill steel, drill steel with a hexagonal cross section. Sandvik, 2007.

hexogen, or cyclotrimethylenetrinitramine $C_3H_6N_6O_6$. It consists of colourless crystals with a heat of explosion of 6 025 kJ/kg, a density of 1 800 kg/m³, a melting point of +204°C and a detonation velocity (c_d) of 8 750 m/s. An explosive substance used in the manufacture of compositions B, C-3, and C-4. Composition B is useful as a cast primer.

HF (acronym), for '**high frequency**'.

HF surface-harden, high frequency surface hardening of drill steel. Sandvik, 2007.

high brisance explosive, see '**brisance**'.

high energy electric detonator, an electric detonator equipped with a fuse head which requires a higher current than normal for initiation. The initiation energy is normally quantified by the impulse, mJ/Ω. Standard detonators which require 0,5 A of current, fire at about 5 mJ/Ω. The most resistant detonators, type 3, require about 1 500 mJ/Ω, and 5 amps of minimum current.

high energy explosive, see '**high explosive**'.

high explosive, an explosive with a very high rate of reaction and pressure which creates a shock wave to propagate through the explosive column at a high velocity called the detonation velocity (c_d). High explosives require the use of a detonator for initiation. Upon detonation a shock propagates through the column of the explosive. Examples of high explosives are Nitroglycerin, TNT, slurries, and PETN. *Permitteds* (or *permitted explosives*) are a special type of high explosives for use in gassy or dusty coal mines. High explosives detonate as compared to low explosives which deflagrate and high explosives possess much greater concentrated strength than low explosives such as black powder.

high-order detonation, a detonation occurring at the maximum possible detonation velocity (c_d) in an explosive.

high pressure filter, filter used on the pressure side of a hydraulic system to remove impurities from the oil. Atlas Copco, 2006.

high pressure water-jet notching, see '**water-jet notching**'.

high speed camera, a camera that can take one or more pictures with very short delay times down to micro seconds. Some camera types are 1) Mirror camera with rotating prism 2) Crans-Shardin camera 3) Kerrcell camera.

high speed video camera, a video camera that can take several hundred of pictures per second. The result of the event can be immediately be reviewed.

high speed water burst, method to break boulders by a high speed water burst. The high velocity of the water is achieved by hydraulically compressing a nitrogen accumulator. Commercial applications of this principle has not been very successful. Tamrock, 1983.

high strength explosive, see '**high explosive**'.

high temperature blasting, blasting which is carried out in material at a temperature greater than 100°C. AS 2187.1, 1996.

high tension detonator, a detonator requiring an *electrical potential of about 50 volts* for firing. B.S. 3618, 1964.

high velocity detonation (HVD) explosives, a class of explosives characterized by a high velocity of detonation, e.g. military explosives. There is a graded transition from high to low velocity of detonation. Examples of explosives with high detonation velocities are military explosives. There is, however, no sharp limit between high and low velocity of detonation.

highwall (*bench, bluff or ledge*), a nearly vertical face at the edge of a bench on a surface excavation and most usually used in coal strip mining.

History of explosives (brief). Pradhan, 2001.
    1100th century Black powder was used in China.
    1300th century Knowledge of black powder came to West from China.
    1320 Black powder used as gun propellant.

1600th century Black powder used as a blasting agent.
1833 French chemist Henri Baronet prepared a flammable HNO_3 acid ester of starch.
1838 Trophies J. Delouse reported the explosive properties of nitrated paper.
1846 German Swiss chemist Christian F. Scobey nitrated cellulose with mixed HNO_3 and H_2SO_4. He demonstrated that Nitrocellulose was two to four times as effective as black powder in underground blasting.
1846 Italian chemist Ascension Sombrero discovered Nitro-Glycerine (NG). First publication in 1947, see Strandh, 1983.
1859 and 1866 NG as blasting agent was demonstrated by Alfred Nobel. Nobel added diatomaceous earth with NG to render NG less sensitive to shock. The explosive was marketed in sticks and called DYNAMITE.
1863 TNT or trinitrotoluene explosive was discovered by the German chemist J. Wilfred.
1875 lfred Nobel introduced blasting gelatine.
1895 Li**quid oxygen explosive (LOX).** Invented by German chemist Karl P. G. von Lined.
1899 Hexogen (RDX) explosive was known.
1920 PETN, pentaerythritoltetranitrate was known.
1924 Use of explosives for seismic examinations of oil fields in USA. Anon., 1941.
1940's AN. Two accidental explosions, one at Brest in France and the other at Texas City *1947* bring out the potential of ammonium nitrate AN as an explosive after being mixed with **diesel oil.**
1941 Use of seismic for the planning of work on a water power station in Sweden. Anon., 1941.
1955 ANFO (dry blasting agents). Akre and Lee patented the use of the fertilizer AN with solid carbonaceous fuel sensitizer – AKREMIT.
1957 Water gel explosives commercialised.
1961 Emulsion explosives discovered by Richard S. Egley.
1961–1964 Slurry explosives are patented by Cook and Farman**.**
1966–1967 Emulsion and ANFO. Bluhm patent on emulsions and Clay patent on methods of mixing emulsion and ANFO for higher explosive density and increased water resistance in wet boreholes.
1980 Heavy ANFO is made commercial.
History of initiation devices (brief). Pradhan, 2001.
1830 Moses Shaw patented a way of **electric firing black powder** (gunpowder) by an electric spark.
1831 William Bickford introduced **safety fuse** which consists of a core of black powder enclosed in textile sheaths and suitably waterproofed.
1830–1832 Robert Hare developed the **bridge wire method** of electric blasting.
1864–1867 Alfred Nobel developed a method of **initiating nitro-glycerine** by using safety fuse for initiating
1864–1868 Black powder detonators and later capsule or **mercury fulminate**. Those became the **first commercial detonators**.
1870's Julius Smith successfully introduced **bridge wire initiated electric blasting caps** and developed a **portable, generator type blasting machine**.
1895 Julius Smith introduced delay electric blasting caps utilizing **safety fuse as the delay element**.

1907 The **first detonating cord "Cordean"** invented by **Louis L'heure** in France using **TNT core**. Today normally PETN is used.

1930's Replacement of mercury fulminate in ignition and primer charges was begun with the use of a variety of more stable explosive compounds.

1937 Detonating cord with PETN explosive.

1946 Millisecond interval delay electric blasting caps introduced, having delay times millisecond rather second.

1948 Use of **capacitor discharge type blasting machines** began replacing a major share of the generator types with safe and more reliable power units.

1960 Low energy detonating cord was introduced which led to improved non-electric detonating systems.

1966/68 Shock tube NONEL system developed by Per-Anders Persson Nitro Nobel.

1973 NONEL delay detonator (chock tube system) was introduced. It improved timing and noise levels.

1981 Magnadet system is introduced.

1986 Electronic delay detonators introduced by ICI (UK).

History of rock mass classification (empirical design methods)

a   **Rock load classification method.** It was the first rational method of rock classification. It was developed for tunnels supported by steel sets driven during a period of 50 years. It is not suitable for modern tunnelling support methods like shotcrete and rock bolts. Developed by Karl von Terzaghi in 1946.

b   **Stand up time classification**. The main significance of this method is that an increase in tunnel span leads to a major shortening in the stand up time. This method has influenced the development of more recent rock mass classification systems. Developed by Lauffer in 1967.

c   **Rock quality designation (RQD)**, developed Deere in 1967.

d   **Rock structure rating** (RSR). Developed by Wickham in 1972.

e   **Rock mass rating (RMR)**, developed by Bieniawski South Africa, in 1973.

f   **Q-system**, developed by Nick Barton, Norway, in 1974.

g   **Size strength classification.**

HLZ (acronym), for '**height of loosening zone**'.

HMX (abbreviation), see '**octogen**'.

HNS (acronym), see '**hole navigation system**'.

Hoek and Franklin biaxial pressure chamber, an portable equipment for biaxial testing in the field. It consists essentially of a cylindrical steel jacket, an internal self-sealing loading membrane made of neoprene rubber and fittings for the attachment of a pressure gauge and a hydraulic pressure pump.

Hoek-Brown failure criterion, when the discontinuity spacing is small compared with the size of excavation under consideration, the rock mass will behave in a pseudo-homogenous manner and its strength can be defined by

$$\sigma_{1ef} = \sigma_{3ef} + (m\sigma_c \sigma_{3ef} + s\sigma_c^2)^{1/2} \tag{H.2}$$

where σ_{1ef} is the major principal effective stress at failure, σ_{3ef} is the minor principal effective stress or confining pressure, m and s are material constants, and σ_c is the uniaxial compressive strength of the intact rock (Specimen size 50 mm diameter and length 100 mm which is free from discontinuities such as joints or bedding planes). For disturbed rock masses m = exp[(RMR-100/14] and s = exp[(RMR-100/6). For undisturbed or interlocking rock masses m = exp[(RMR-100/28)] and s = exp[(RMR-100/9)]. Hudson, 1993. RMR is the strength value according to the Bieniawski rock mass classification system.

Hoek cell, a cell designed for triaxial testing of rock under stress. Hudson, 1993.

hole plug, conical plastic cup to seal the top of the blasthole in bench drilling with the purpose to avoid drill cuttings or other loose material and water to fall into the blasthole.

hog box, a concrete box in which water and dirt are mixed to be pumped to a fill. Nichols, 1965.

hoist, a drum on which the hoisting rope is wound in the engine house, as the cage or skip is raised in the hoisting shaft. Pryor, 1963.

hoisting, the vertical or near-vertical transport of broken ore up the shaft from an ore pocket to the ore bins on the surface or underground.

hole deviation, see '**drilling accuracy**'.

hole diameter (*d*), (mm or m), the mean cross-sectional width of the borehole.

hole director, see '**template**'.

hole-in, to start drilling a borehole. Also called '**collaring**', *spud*, *spud-in*. Long, 1920.

Hole liner, low density poly tube.

hole inclination, see '**blasthole inclination**'.

hole liner, thin plastic tubes (about 0, 25–1 mm) on roles lowered into blastholes before charging and filled with explosive. The purpose of the hole liner is to prevent water in the surrounding rock mass to penetrate into the explosive and reduce its strenth.

hole navigation system (HNS), a system where GPS-signals from the satellites are used to determine the collaring position of the blasthole with an accuracy of 10 cm in most situations. It is therefore no need to mark the holes on the bench top. Using information on the operators display, the operator can navigate the rig to the collaring position for the given hole, and the computer will provide the information to place and align the feed exactly over the collaring position. The system can be used only in surface drilling because the need for a direct contact with the satellites. Atlas Copco, 2006 b.

hole through, the successful meeting of two approaching tunnel heads, or of winze and raise. Pryor, 1920.

holing, cutting of coal. Mason, 1951.

holing pick, a pick used in holing coal. Standard, 1964.

holistic approach, a modelling of something where the whole problem is analysed trying not to forget any influencing parameters.

hollow, old abandoned working. Fay, 1920.

hollow core anchor, expander type rock bolt made from hollow rebar. The bolt can be installed to provide support and later permanently fixed by injecting cement grout through the center of the bar. Atlas Copco, 2006.

hollow drill, a drill rod or stem having an axial hole for the passage of flushing water or compressed air to remove cuttings from a drill hole. The chippings and the fluid and air return to surface on the outside. Also hollow rod and hollow stem. B.S. 3618, 1964.

hollow inclusion stress cell, a rock stress measurement cell based on strain gauges. The cell is inserted into a borehole while measuring. Hudson, 1993.

hollow reamer, a tool or bit used to correct the curvature in a crooked borehole. Compare reaming bit. Long, 1920.

hollow rod, see 'hollow drill'. B.S. 3618, 1964.
hollow stem, see 'hollow drill'. B.S. 3618, 1964.
Homalite 100 (tradename), a transparent plastic material made of polycarbonate (PMMA) and used for dynamic photo elasticity and high speed photography of fracture propagation for studies of fragmentation in model blasting and also the study of stresses around openings for the rock stability around underground openings. Rossmanith, 1987. The initiation toughness of cracks in Homalite 100 or the fracture toughness K_{Ic} is about 0,40 Nm$^{3/2}$ and is less than that found for most rock. It is therefore more brittle in behaviour than rock.
homogeneous, property of a material exhibiting the same properties at all material points. Opposite to heterogeneous.
hone, a block of fine abrasive, generally SiC. The block is appreciably longer than it is broad or wide. Hones are used for fine grinding, particularly of internal bores. Dodd, 1964.
Hooke's law, defines the relation between stress and strain in a body within the elastic limit of the body. The stress σ in MPa

$$\sigma = E\varepsilon \qquad (H.3)$$

is proportional to the strain $\varepsilon = du/u$ where u is the displacement. The proportional constant is called the Young's modulus E and is here given in MPa.
hook up, to connect the initiation starting points (tails) of all blastholes for the purpose of achieving a single source detonation.
hoop stress, tangential stress.
hopper, a vessel into which materials are fed, usually constructed in the form of an inverted pyramid or cone terminating in an opening through which the materials are discharged. B.S. 3552, 1962.
horizontal bench blasting, bench blasting by using horizontal drill holes instead of vertical or inclined drill holes.
horizontal pressure (p_h), **(MPa)**, pressure directed perpendicular to the gravity field
horizontal slicing (*ascending*), see '**overhand stoping**'. Fay, 1920.
horizontal slicing (*descending*), see '**top slicing and cover caving**'. Fay, 1920.
horse tails or **tail cracks**, the dilatant cracks formed in association with faulting are one of the most important features for fracture hydrology. These form oblique to faults in order to accommodate the deformation due to the displacement along the fault. Hudson, 1993.
hose, a strong flexible pipe made normally of polyvinyl chloride (PVC). For charging of ANFO special antistatic treated PVC is needed to prevent accidental explosions or detonations.
hose clamp, clamp made from metal or other hard material used to fasten a fitting to the end of a hose. Atlas Copco, 2006.
hose coupling, a joint between a hose and a steel pipe, or between two lengths of hose. Ham, 1965.
hose reel, powered drum like structure on which the water hose is wound for storage when tramming a drill rig under ground. Atlas Copco, 2006.
hose retainer, devise used to maintain a hose in a fixed position. The hoses used to supply hydraulic oil to a rock drill on a drill rig are often secured at a point on the fed using a hose retainer. Atlas Copco, 2006.

host rock, the rock surrounding an epigenetic orebody.

hose sock, woven wire mesh stocking that fits over a hose and grips it in such a way that it forms an anchor for the hose. Atlas Copco, 2006.

hot material, any material which at the time of charging is at a temperature of +70°C or over. FEEM. In Australia hot material has a temperature of 55° to 100°C. AS 2187.1, 1996.

hot spot, small air pocket in the explosive which collapses when the detonation front reaches the pocket. This collapse causes a large temperature and pressure increase in the explosive. A large amount of hot spots favours the detonation.

hot storage test, a test applied in order to accelerate the decomposition of an explosive material, which is usually very slow at normal temperatures, in order to be able to evaluate the stability and the expected service life of the explosive from the identity and the amount of the decomposition products. Various procedures, applicable at different temperatures, may be employed for this purpose. Meyer, 1977.

hour (*t*), (h), a unit of time equal to 3 600 s.

HRF (acronym), see '**hydraulic rock fill**'.

HS (acronym), for '**half second**' and **MS** for '**milli second**'. See '**half second detonator**'.

HSBM (acronym), see '**hybrid stress blasting model**'.

HS-detonator, see '**half second detonator**'.

HS-shank, special design of integral steel shank part where hexagonal shape turns round for the bumped up part where shank is exposed to heavy stresses. A design to avoid breakages. Sandvik, 2007.

HU-detonator, a detonator with very low sensitivity to electric current and high precision in timing.

Hugoniot curve, a pressure-volume curve which obeys the Hugoniot equation. I.C. 8137, 1963. A pressure-density curve could also be called a Hugoniot curve. Johansson and Persson, 1970.

Hugoniot elastic limit (HEL), (MPa), pressure above which the material behaves in a plastic rather than elastic manner. Experimental determined values for the HEL are of the order of 3 000 to 4 000 MPa. For granite a HEL-value of 6 000 MPa have been found. Johansson and Persson, 1970.

Hugoniot equation, a relationship between pressure (p) and density (ρ) of a material as determined by the shock wave equation for conservation of energy (W). The equation reads as follows;

$$W_2 - W_1 = \frac{1}{2}(p_2 + p_1)\left(\frac{1}{\rho_1} - \frac{1}{\rho_2}\right) \tag{H.4}$$

where subscript 1 and 2 denotes the state at time 1 (before detonation) and time 2 (after detonation).

human response to vibrations, the reaction of a person to different vibration levels. Konya and Walter, 1990.

hung shot, a shot which does not explode immediately upon detonation or ignition. See also '**hangfire**' and '**misfire**'.

hurdy-gurdy drill, a hand auger used to drill boreholes in soft rock or rock material, such as soil, clay, coal etc. Long, 1920.

hurricane air stemmer, a mechanical device for the rapid stemming of shotholes. It consists of a sand funnel connected by a T-piece to the charge tube, one end of which is provided with a valve and fittings to the compressed air column. The funnel is filled with sand, which is held uppermost, and the charge tube is inserted into the shothole. The sand is injected by the compressed air and the tube is gradually withdrawn as the hole is being filled. Nelson, 1965.

hutch, a car on low wheels in which coal is drawn and hoisted out of a mine pit. Webster, 1961.

HVD (acronym) for '**high velocity detonation**', see 'high velocity detonation explosives'.

H-wave, see '**hydrodynamic wave**'. A.G.I., 1957.

hybrid stress blasting model (HSBM), a computer code, internationally developed, to calculate the velocity of detonation in an explosive used in different blasthole diameters and confinements.

hydraulic and pneumatic diaphragm transducer, measures the quantity of fluid pressure which acts on one side of a flexible diaphragm made of a metal, rubber or plastic. Twin tubes connect the read-out instrument to the other side of the diaphragm. When the supply pressure is sufficient to balance the pressure to be measured, the diaphragm acts as a valve and allows flow along the return line to a detector in the read-out unit. The balance pressure is recorded, usually on a standard *Bourdon pressure gauge* or a digital display. The transducer is used for measuring water pressures, support loads, cable anchor loads, normal components of stress and settlements. Brady, 1985.

hydraulic backfill, see '**hydraulic fill**'.

hydraulic blasting, fracturing method using a *hydraulic cartridge*, a ram operated device used to split coal. Pryor, 1963.

hydraulic boom, drill rig boom that is supported and operated by hydraulic cylinders. Atlas Copco, 2006.

hydraulic breaker or **splitter**, a hydraulic percussion drill machine weighing up to about 500 kg and with a chisel mounted on a hydraulic arm and used to break boulders. It can be mounted on different kind of carriers, e g backhoe excavators larger than 6 t and all hydraulic 360° excavators. Commonly used on surface but also underground. Atlas Copco, 1982.

hydraulic cartridge, a device used in mining to split coal, rock etc., having 8 to 12 small hydraulic rams in the side of a steel cylinder. Fay, 1920.

hydraulic chock, a *steel face support structure* consisting of one up to four hydraulic legs or uprights. The four-leg chock is mounted in a strong fabricated steel frame with a large head and base plate. In one type, the chock can be set to a load of 11,2 t at 7 MPa to yield at 120 t. It is controlled by a central valve system that operates either on the four legs simultaneously or on the front and rear pairs separately. See also 'self-advancing support'. Nelson, 1965.

hydraulic conductivity or **permeability** (c_h), **(m/s)**, a statement in fluid dynamics. The velocity of flow of a liquid through a porous medium due to difference in pressure is proportional to the pressure gradient in the direction of flow. Webster, 1961. The law has also been used for cracked materials like rock. The hydraulic conductivity is the product of the permeability coefficient (k) and the hydraulic gradient (dp);

$$c_h = k \frac{dp}{dl} \quad \text{(Darcy's law)} \qquad (\text{H.5})$$

where c_h is the hydraulic conductivity or permeability in (m/s), k is the permeability coefficient in (m²/Pas), and dp/dl is the hydraulic gradient or pressure fall measured in (Pa/m). For rock a distinction must be made regarding the conductivity of the rock type (taking into account only pores and microcracks) and the conductivity of the total rock mass (including faults and joints). The hydraulic conductivity of crystalline rock varies normally, from the surface and down to 150–200 m depth, between 10^{-5} and 10^{-8} m/s. The effective porosity in intact crystalline rock is very small, only about 0,01%, and therefore the water flow mainly depends on the faults and joints and not on the pores and microcracks. The corresponding porosity for Cambrian sandstone is about 1% and for gravel deposits about 15%. A general rock structural scheme for igneous rock has been established with the following typical hydraulic conductivities, see Table 28. Pusch and Börgesson, 1992. The hydraulic conductivity is size dependent with smaller rock volumes yielding smaller values for the hydraulic conductivity because no large cracks will be included in small samples.

Table 28 Bulk hydraulic conductivities dependent on the rock mass volume considered and spacing between joints and faults. After Pusch and Börgesson, (1992).

Feature	Spacing between joint and faults (m)	Bulk hydraulic conductivity (m/s)	
		Range	Mean value
Low order (discontinuity)			
1st	3000–5000	10^{-7}–10^{-5}	10^{-6}
2nd	300–500	10^{-8}–10^{-6}	10^{-7}
3rd	30–150	10^{-9}–10^{-7}	10^{-8}
High order (bulk rock volumes with no breaks of lower order)			
4th	2–7	10^{-11}–10^{-9}	10^{-10}
5th	0,2–0,7	Hydraulically inactive in undisturbed state	
6th	0,02–0,1	10^{-12}–10^{-10}	10^{-11}
7th	<0,02	10^{-14}–10^{-12}	10^{-13}

The permeability in weathered rock is often less than in not weathered rock. Weathered zones close to surface therefore often has a lower permeability then deeper not weathered zones with not filled cracks. Pusch, 1974.

hydraulic control system (HCS), a system to control the drilling machines and arms by hydraulic. It was introduced in the 1970's in Sweden. It was replaced in the 1990's by Programmable Logic Control PLC and in 2002 the first SmartRigs (tradename) started to take over. Atlas Copco, 2006 b.

hydraulic cylinder, machine component to create a large linear force. A cylinder is fitted with a piston and piston rod that can be extended or retracted by pumping hydraulic oil under pressure into the cylinder. Atlas Copco, 2006.

hydraulic drilling, using a drilling machine where the power to the machine is created by high-pressure oil or water.

hydraulic drilling machine, a drilling machine where the power to the drilling is created by high-pressure oil or water. There are four main functions; *percussion*, *rotation*, *feed* and *flushing*. Percussion and rotation are independent of each other.

hydraulic drill rig, a drill rig where the drill machines are powered by hydraulic oil instead of compressed air or an electric motor. Advantages are a higher rate of drilling, less oil mist in the air, and lower noise.

hydraulic drill steel support, drill steel support used at drilling. It is opened and closed by means of a hydraulic cylinder. Atlas Copco, 2006.

hydraulic excavation, excavation by means of a high-pressure jet of water, the resulting waterborne excavated material being conducted through flumes (open channels) to the desired dumping point. Ham, 1965.

hydraulic feed (in drilling), a method of imparting longitudinal movement to the drill rods on a diamond, rotary type drill by a hydraulic mechanism instead of mechanically by gearing. See also 'feed cylinder'. Long, 1920.

hydraulic fill, waste material transported underground and flushed into place by use of water. Pryor, 1963.

hydraulic fracturing (*oil wells*), method in which sand-water mixtures are forced into underground wells under pressure. This pressure splits the petroleum-bearing sandstone, thereby allowing the oil to move towards the wells more freely. USBM, 1990. The length of the radial cracks will depend on the hydraulic pressure used.

hydraulic fracturing (*determination of the principal stress direction*), in rock mechanics a method to determine the principle stress direction. A part of normally a vertical borehole is pressurized and afterwards the borehole wall is studied with a camera or video camera. The two radial cracks developed marks the principle stress direction.

hydraulic hammer drill, a hammer drill machine powered by oil or water instead of air.

hydraulic hoisting, see '**hydraulic transport**'.

hydraulic impulse fracturing, see '**directed hydraulic impulse fracturing**'.

hydraulic measurement, see '**water head measurement**'.

hydraulic mine, a *placer mine* worked by means of a stream of water directed against a bank of sand, gravel, or talus. Soft rock is similarly worked. Hess, 1968.

hydraulic pressure cell, see '**flatjack test**'.

hydraulic prop, a prop consisting of two telescoping steel cylinders which are extended by hydraulic pressure which may be provided by a hand-operated pump built into the prop. The prop holds about 2 litre of oil and is fitted with a yield valve that relieves the pressure when the load exceeds that of which the prop is set. A hydraulic prop enables quicker setting, uniform initial loading, and it can be withdrawn from a remote safe position. The first hydraulic prop was used in a British coalmine in 1947. Nelson, 1965.

hydraulic radius (r_h)**, (m)**, the ratio of the surface area in m^2 and the perimeter in m of an excavation like a tunnel, drift, stope or the roof of an undercut.

hydraulic rock fill (HRF), a mixture of sand and cement pumped in pipelines or boreholes, sometimes from surface, down into the mined out stope to fill it and thereby avoid subsidence on surface.

hydraulic rotary drill, a rotary drill machine powered by oil or water instead of air.

hydraulic rotary drilling, method of drilling that uses rotating bits lubricated by a stream of mud. See 'rotary drilling'.

hydraulics, the branch of science and technology concerned with the mechanics of fluids, especially liquids. Lapedes, 1978.

hydraulic splitter, see '**hydraulic breaker**'.

hydraulic splitting unit, device consisting of a hydraulic cylinder attached to a set of feathers and wedge and a control mechanism. The feather and wedge arrangement is inserted into a drilled hole and as the hydraulic cylinder is activated it is forcing the wedge between the feathers to generate a lateral force that breaks the rock. Atlas Copco, 2006.

hydraulic transport, movement of fragmented ore, tailings, sand etc by water, flowing through pipelines. It may can also be used as a hoisting method in a mine.

hydrocode, class of computer codes used to model the deformation and flow of solids and fluids. The dynamic behaviour of a continuous medium is governed by conservation laws (specifically conservation of mass, momentum, angular momentum and energy) and by constitutive equations which give relations between the state variables which characterize the medium. Hydrocodes as a rule contain several fluid-related equations of state and elastic-plastic material models, and are designed in such a way that new constitutive equations can be added relatively easily. The term also denotes computer codes which use elemental connections between particles to trace the development of stresses and reactions in a body or fluid following initial stimuli.

hydrodynamic code, see '**hydrocode**'.

hydrodynamic fluid, a class of materials which behave similarly to an ideal incompressible fluid with little or no viscosity influences.

hydrodynamics, the study of motion of a fluid and of the interactions of the fluid with its boundaries, especially in the incompressible in viscid case. Lapedes, 1978.

hydrodynamic theory, in general, hydrodynamic theory refers to behaviour of solids above their HEL (Hugoniot elastic limit) and the solids can therefore be analysed in a similar way as fluids. The "*hydrodynamic theory of detonation*" is one typical example. Metals driven by detonation e. g. the jet formation in shaped charges is another example. Rock materials can turn into a hydrodynamic state at pressures on the order of 10,000 MPa. There is no evidence so far, that the rock in contact with the explosive during detonation is in a hydrodynamic condition.

hydrodynamic theory of jet formation, a theory which explains the formation of a metal jet when an explosive is detonated in close contact with a metal cone. A thin metal jet tip and a thick metal slug are formed that with high velocity can penetrate rock. In 1948 the first theory was developed by Birkhoff and McDoughall. Birkhoff, 1948.

hydrodynamic wave (H-wave), a *surface seismic wave* recognized by Leet. It is similar to a Rayleigh wave but moving in an opposite or counter-clockwise sense, so that the wave is moving forward at its maximum up position. A.G.I., 1957.

hydrometer, instrument to measure specific gravity of a fluid. Stein, 1997.

hydrophone, an underwater microphone. Hy, 1965.

hydrostatic pressure, the pressure is the same in all directions in a liquid. The same may occur also in rock.

hydrostatic stress field, a stress condition in the rock mass where the vertical principal stress equals the two principal horizontal stresses. In a hydrostatic stress field, the tangential stress to a horizontal cylindrical opening (drift or tunnel) or in a two-dimensional plate will be $2\sigma_v$ where σ_v is the vertical stress.

hydrostatic tension, is defined as three equal and mutually perpendicular tensile stresses acting on a solid body. ASM Gloss, 1961.

Hydrox (tradename), for a permitted device, used in some English coalmines, that resembles *Cardox (*tradename) in that a steel cylinder with a thin shearing disk is used. However, the charge is not liquid carbon dioxide but a powder composed chiefly of *ammonium chloride* and *sodium nitrate*. It is proportioned to give water, nitrogen, and salt as the products of combustion. On being ignited this powder is gasified and shears the steel disk, the gas escaping into the hole and breaking the coal. Lewis, 1964.

hypoelastic behaviour, refers to incremental forms of non linear elastic constitutive laws usually adopted when the stress increment is a function, both of the current stress state and of the strain increment. A basic difference between this and other non linear elastic-plastic stress-strain relationships is that even though both of them are non linear, the hypo elastic laws are reversible within the increment whilst the elastic-plastic ones are not. Hudson, 1993.

hypogenetic rock, a rock that was formed deep within the earth under the influence of heat and pressure. Compare with 'epigenetic'. Hess, 1968.

IBSD (acronym), for 'in-situ block size distribution' the block sizes defined by the weakness planes in the rock mass before blasting.

ideal detonation code, is the set of chemical and thermodynamic rules used to describe the transition of a chemical or chemical mixture from its stable pre-detonation state into its stable post-detonation state, calculating the physical conditions behind the reaction zone and, during the expansion phase, the temperature, products of explosion and energy status at any given pressure. Cunningham, 1991.

ideal detonation velocity or velocity of detonation (c_{di}), (m/s), see 'ideal velocity of detonation'.

ideal explosive, an explosive where the chemical reaction is completed at the end of the Chapman-Jouget zone, and which tends to detonate near its maximum detonation velocity (c_{dmax}) regardless of confinement and charge diameter. Military explosives are often ideal explosives compared with commercial explosives which are non ideal.

ideal gas law, the relation between pressure, volume and temperature when a gas is being compressed is called the general gas law. It is a combination of *Boyle-Mariottes law* and *Gay-Lussacs law*.

$$pV = RT \qquad (I.1)$$

where p is the pressure in Pa, V is the volume in m³, R is the gas constant and T is the absolute temperature in degree K (Kelvin).

ideal velocity of detonation (c_{di}), (m/s), the velocity at which the detonation front in a high explosive will progressively advance under ideal physical conditions such as large explosive diameter and large confinement. It is the detonation velocity calculated by means of a 'good' ideal detonation code, which should tally with the measured detonation velocity for a large diameter well-confined explosive. For ANFO of density 800 kg/m³, and with a 6% diesel fuel content for example, the ideal detonation velocity is about 4 800 m/s (IDEX code). Actual measurements based on energy release behind the Chapman-Jouget zone may be less than this.

identification sleeve, sleeve having alpha numerical marking that identify the part or machine. Atlas Copco, 2006.

IDEX (abbreviation), for "**Ideal detonation explosive**" an '*ideal detonation code*' developed by ICI for the calculation of *energy, temperature* and *detonation velocity* (c_d) of an explosive during its expansion phase in a certain confinement and blasthole diameter. Cunningham, 1991. See also '**ideal detonation code**'.

idler, roller that is not powered or driven. Atlas Copco, 2006.

IDZ (acronym), for '**isolated draw zone**'.

IExpE (abbreviation), see '**Institute of Explosive Engineers**'.

igneous rock, rock type created from a high-temperature, mobile mass of plastic solids, liquids, and gases that was generated within the depths of the earth and from which the igneous rock are derived by crystallization. C.T.D., 1958.

ignitacord, see '**igniter cord**'.

ignite or **initiate (verb)**, to bring the explosive to detonation by a smaller charge.

igniter cord, a small-diameter pyrotechnic cord which burns with an external flame at the zone of burning and at uniform rate, and is used to ignite a series of safety fuses. Depending on the manufacturer, the fast igniter cord burns at 0,1667–1,0 m/s (1 to 6 s/m), whereas slow cord burns at 0,0222 to 0,0333 m/s (30 to 45 m/s). Slow cord is usually more water-resistant than fast cord. All kinds of cords are susceptible to ignition by accidental blows from rock, steel, etc.

igniter cord connector, a slotted and recessed metal tube containing a pyrotechnic material which is used to connect igniter cord to safety fuse. AS 2187.1, 1996.

igniter fuse, see '**igniter cord**'. AS 2187.1, 1996.

igniting pill, see '**fuse head**'.

ignition, the process of starting a thermal reaction in an explosive material or fuel.

ignition, the act of igniting, or the state of being ignited. The term is specifically used in mechanics for the act of exploding the charge of gases in the cylinder of an internal-combustion engine. The firing of an explosive mixture of gases, vapours, or other substances by means of an *electric spark*. For detonation of explosives the correct term is '**initiation**'.

ignition cable, see '**shot firing cable**'.

ignition pattern, a drawing showing the delay times and initiation order for the blastholes in a blast. See also 'initiation sequence'.

ignition point, see '**ignition temperature**'.

ignition temperature, the temperature at which a substance ignites. Crispin, 1964.

ignition test, see '**standard ignition test**'.

ignition threshold, the minimum energy needed to ignite the explosive by impact or friction.

image analysis (technique for quantification of blast fragmentation), the technique to determine the fragment size distributions by identifying the contours of the fragments and calculating a possible size distribution; manually, half-manually or fully automatic. Carlsson, 1983. See also "fragmentation measurement". State of the art can be found in Franklin, 1996.

imbricate structure, parallel to the bedding there may be preferred orientation of platy minerals such as mica or clay minerals and times there may also be a preferred orientation of ellipsoidal sand grains. In some instances such a *preferred orientation may be oblique to the bedding* forming what is called an *imbricate structure* where ellipsoidal sand grains are stacked one against the other, inclined to the bedding but pointing in what was the downstream direction at the time of deposition. Such preferred orientations may produce a strong anisotropy in a rock mass that can at times control roof failure in coalmines. Hudson, 1993.

IMCO (abbreviation), for **Transportation Regulations for Shipment of Dangerous Goods.**

IME (acronym), see '**Institute of Makers of Explosives**'.

IME fume classification, a classification of fumes based on the amount of poisonous or toxic gases produced by an explosive or blasting agent. The IME fume classification is expressed by Table 29. IME, 1981.

Table 29 Classification of fumes from explosives. After IME, 1981.

Fume class	Poisonous gases per (32 × 203 mm) cartridge of explosive (Litre)
1	<4,7
2	4,7 to 9,6
3	9,6 to 19,6

Note: The U.S. Bureau of Mines limits poisonous or toxic gases to 107 litres per kilogram of permissible explosive.

IMM (acronym), see '**Institution of Mining and Metallurgy**'.

immiscible, two liquids which do not mix with each other e.g. oxidizers dissolved in water droplets and surrounded by an fuel in emulsion explosives.

impact breaker, the impact breaker or *double impeller breaker* uses the energy in falling stone, plus the power imparted by the massive impellers for complete stone reduction. Rock fed into the breaker falls directly onto the impellers which weigh up to 6,5 t, and rotate away from each other, turning up and outward, at speeds from 250 to 1 000 rpm, depending on the desired size of finished product. Pit and Quarry, 53rd, Sec. B, p. 25.

impact crusher, a machine for the reduction of the size of the rock. Hammers with cemented carbide inserts are mounted on a cylindrical drum with the aim to crush the feed by impact forces. This method of crushing is only used for softer rocks.

impact hammer, a hydraulic hammer with a chisel impacting boulder by percussion. The hammer is mounted on a hydraulic arm, which in turn is mounted to a mobile machine. It is used also for scaling underground and is then mounted on separate mobile scaling machine on rubber tyres.

impact sensitivity, see '**sensitivity**' (to the impact of an explosive).

impact sensitivity test, test method to assess the impact sensitivity of an explosive. A body of known mass is dropped from a known height onto a quantity of explosive kept in-between the surface of two stainless steel Hoffman rollers of 6 mm diameter, the test being repeated 10 consecutive times, with the detonation or decomposition being observed. This test characterizes the safety of an explosive in handling, transportation and use. The larger the resistance to initiation or decomposition by impact the higher is the safety.

impact shoulder, area of a drill bit designed to take the impact of the hammer. Atlas Copco, 2006.

impact strength, the impact strength of a material *increases with the impact velocity*. See '**dynamic strength**'.

impact test, a test to determine the behaviour of materials when subjected to high rates of loading, usually in bending, tension, or torsion. The quantity measured is the energy absorbed in breaking the specimen by a single blow, as in the Charpy or Izod tests. ASM Gloss, 1961. See also '**fracture toughness test**'.

impact toughness test, this method involves determining the impact toughness of a rock by dropping a weight from successively greater heights until such a height is reached that the specimen is fractured. Lewis, 1964.

impact wave, a wave characterized by the longitudinal wave velocity (P-wave) and this velocity is equal to the sound velocity for the material.

impedance (acoustic) (Z), (kg/m²s), see 'acoustic impedance'.

impedance for an explosive Z_e **(kg/m²s)**, is defined by the product of explosive density ρ_e in kg/m³ and velocity of detonation, c_d in m/s.

impedance mismatch (Z_R), see '**impedance ratio (acoustic)**'.

impedance ratio (acoustic), (Z_R), (dimensionless), the ratio of the acoustic impedances of two adjacent contacting materials. For the contact of rock and explosive one obtains;

$$Z_R = \frac{\rho_e c_d}{\rho_r c_p} \quad (I.2)$$

where Z_R is the impedance ratio, ρ_e is the density of the explosive in (kg/m³), c_d is the detonation velocity of the explosive in (m/s), ρ_r is the density of the rock in (kg/m³), and finally c_p is the longitudinal wave velocity in the rock in (m/s). In blasting, the ratio of impedance of explosive to rock is important in the assessment of transfer of explosives energy to the rock material blasted. Atchison et al., 1964a, Atchison et al., 1964b, Nicholls et al., 1965 and Thum and Leinz, 1971. The impedance ratio does not appropriately characterize the energy transmission from the explosive to the rock, because the angle of incident is seldom 90° (it is more likely about 45°), and depends on the ratio of the detonation velocity of the explosive to the longitudinal wave velocity in the rock.

impeller, the rotating member of a centrifugal pump.

implosion, a bursting inward, sudden collapse. The opposite of explosion. Standard, 1964.

implosion drill, if a hollow sphere is activated by high pressure fluid pressure and breaks or implodes, hydraulic stresses are caused in the liquid and these stresses were planned to fragment the rock. Using glass balls and 10 MPa pressure was not successful.

impregnated bit, a sintered, powder-metal matrix bit with fragmented bort or whole diamonds of selected screen sizes uniformly distributed throughout the entire crown section. As the matrix wears down, new sharp diamonds points are exposed, hence, the bit is used until the crown is consumed entirely. Long, 1920.

impregnated diamond bit, diamond bit where very small diamonds are mixed with various metal powders and sintered to a bit body. As the matrix wears away new diamonds are exposed. Atlas Copco, 2006.

impression block, incorporates a block of a plastic material, which can be indented by objects in the borehole. Some blocks, like the paraffin wax block, allow for marking by objects protruding from the walls. Others will only strike against objects sticking up from the bottom of the hole. Any material that will hold the impression of the down–hole-object can be used to make the impression block. Stein, 1997.

impulse or **momentum, (I), (Ns)**, a physical quantity defined by the product of mass times velocity. It can also be calculated by the integration of force over an interval of time.

The impulse from the explosive on the blasthole wall is important for the magnitude of the blast vibration and a formula for calculation of the impulse per unit of volume was presented by Müller, 1996.

impulse fracturing, see '**directed impulse fracturing**'.

in-bank factor, see '**in-place factor**'.

incendive spark, is an electric spark of sufficient intensity to ignite flammable material. ASA M2.1, 1963.

incendivity, the ability of an igniting agent (e.g. spark, flame, or hot solid) to induce an ignition.

incendivity test, a test made for all instantaneous electric detonators and delay electric detonators to be used in underground coal mines with the permitted type of explosives. Instantaneous detonators are fired singly in a steel chamber containing methane air mixture or natural gas air mixture. Sandhu and Pradhan, 1991.

incidence pressure (p_i), **(MPa)**, the pressure in a ground vibration wave before reflection or refraction at a discontinuity or at a free face.

incipient spall threshold, product of stress amplitude and pulse duration below which no damage occurs.

incline, in mines, a drift driven upwards at an angle from the horizontal.

inclined caving, block caving where the draw is undertaken on different levels instead of as normally only one level. In inclined caving, the draw points are normally located on the footwall. MassMin, 2008.

inclined drawpoint caving, see '**inclined caving**'.

inclined undercut, a form of undercut in which the floor and roof are inclined at significant angles to the horizontal and which takes a chevron or zig-zag shape in vertical section.

inclinometer or **clinometer**, an instrument for measuring angular deviation from horizontal or vertical in drilling.

incompressibility or **bulk modulus** (*K*), **(MPa)**, see '**bulk modulus**'.

indentation hardness, the resistance of a material to indentation. This is the usual type of hardness test, in which a pointed or rounded indenter is pressed into a surface under substantially static load. ASM Gloss, 1961. See also 'hardness tests'.

index, to divide into equal marked parts, such as quadrants or degrees of a circle. Long, 1920.

indication lamp, signal light used to indicate a planned change in movement direction of a vehicle. Atlas Copco, 2006.

indirect priming, see '**priming**'.

indirect stress, see '**rock stress**'.

induced block caving (IBC), in blocking caving the drilling and blasting of those parts of the orebody that does not break by gravity forces. Hydraulic fracturing may also be used in some cases. Synonym to 'cave inducement'.

induced bursts, rock bursts caused by stoping operation to distinguish these bursts from development bursts which are called *inherent bursts*. Spalding, 1949.

induced stress, the natural stress state as perturbed by engineering. Hyett, 1986.

induction drill, an *exotic drilling method* where a high frequency magnetic field is used, 240 kHz and 8 kA/m. The magnetic field is causing heating of the rock mass through eddy currents and hysteresis losses. The effect is increased with higher field strength and frequency. KTH, 1969.

indurated, means hardened. Applied to rocks, hardened by heat, pressure, or by the addition of a cementing ingredient, for example, a marl indurated by the addition of calcite as a cement. Fay, 1920.
industrial diamonds, crystalline and/or crypto crystalline diamonds having colour, shape, size, crystal from, imperfections, or other physical characteristics that make them unfit for use as gems. Long, 1920.
inert dust, any dust that contains no or only a small amount of combustible material. Rice, 1960.
inert gas, a gas (such as nitrogen or carbon dioxide) that is normally chemically inactive especially in not burning or in not supporting combustion. Webster, 1961.
inert primer, a cartridge of inert material or inert material containing an electric blasting cap. It may be used deliberately in shaft sinking to leave a butt. AS 2187.1, 1996.
inertia (I), (J), the reluctance of a body to change its state of rest or of uniform velocity in a straight line. Inertia is measured by *mass* when linear velocities and accelerations are considered and by total joint area per unit of volume for angular motions (that is rotations about an axis). C.T.D., 1958.
inferred ore, ore for which quantitative estimates are based largely on broad knowledge of the geologic character of the deposit and for which there are few, if any samples or measurements. The estimates are based on an assumed continuity or repetition for which there is geologic evidence, this evidence may include comparison with deposits of similar type. Forrester, 1946.
infilling, material used for filling in, filling. Standard, 1964. Discontinuities in the rock mass can have infillings.
inflatable rubber plug, a air decking plug developed in Australia and used for the purpose to create air decks.
infrared thermography, the use electromagnetic radiation waves, in the infrared band with the aim to measure the temperature e.g. of a rock surface.
infusion gun, see '**water infusion gun**'.
infusion shot firing, a technique of shot firing in which an explosive charge is fired in a shothole which is filled with water under pressure and in which the strata around the shothole have been infused with water. B.S. 3618, 1964.
ingate, the point of entrance from a shaft to a level in a coal mine. Standard, 1964. See also '**inset**'.
inherent burst, rock burst that occur in development. They may be divided into two classes, those due to *violent arching* and those occasioned by the *influence of fissures or adjacent excavations* on the stress distribution. Spalding, 1949.
inherent rock fragmentation, see '**block size in situ**'.
initiate or *ignite* (verb), the process of heating up the fuse head in the blasting cap by electric current or starting the detonation in the blasting cap by a shock tube or detonating fuse or cord. Also to bring the explosive to detonation by a smaller charge.
initiating explosives **or primary explosives**, are defined as explosives that can detonate by the action of a relatively weak mechanical shock or by a spark. When used in blasting caps, the primary explosive initiates the delay element which initiates the main explosive.
initiating machine, see '**blasting machine**'.
initiation (noun), the process of causing a high explosive to detonate. It is a rapid chemical reaction which releases energy from an explosive either by applying heat (ignition) or

by imparting a shock (detonation) which heats up the material rapidly. AS 2187.1, 1996. The initiation of an explosive charge requires an initiation point which is usually a detonator, a primer and detonator or a primer and detonating cord. Nelson, 1965. It is the first reaction in the portion of an explosive charge at the position of the initiating agent.

initiation double, initiation at two points along a linear charge to be sure that the charge is initiated if there is some failure of any of the two initiation points. E.g. the initiation can be done at the bottom of the blasthole and at the collar. Double initiation reduces the ground vibrations. Müller, 1991.

initiation pattern or **initiation plan**, scheme of time and space which defines the sequential initiation of charges. There are basically two different drilling patterns used in blasting, rectangular and staggered. The blasted burden and spacing can be manipulated by assigning certain delays to certain blastholes. When the initiation is made in the shape of a *V*, the initiation pattern is called 'closed chevron', and when the blast is initiated row by row sidewards the pattern is called 'open chevron'. The angle of the closed and open chevron to the bench face can be changed by initiating different holes with the same delay. The initiation patterns are called V0, V1, V2, etc., see Fig. 41.

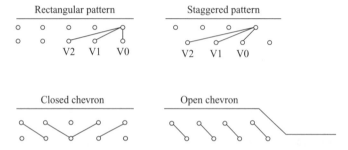

Figure 41 Initiation patterns. After Cunningham, 1993.

initiation sequence, sequential initiation of blastholes. In initiation plans the delay number contains information about the planned time of detonation of the detonators.

initiation system, incorporates the type of system and devices to initiate the blast, but does not include the blasting machine. The choice of initiation system is dictated by the explosive selected. Table 30 gives an overview of the most common initiation systems.

Table 30 Initiation systems.

	Caps		Starting systems for caps
Non electric fuse cap	**Electric cap**	**Special electric cap**	**Non-electric**
Conventional	Conventional delay	Exploding bridgewire (EBW) detonator	Safety fuse
	Ms delay	Magnadet detonator	Detonating cord
	Hs delay	High resistance safety detonators	Shock tube
			Gas tube
			Electromagnetic waves (wireless)

The interrelationship of the different initiation systems is shown in Fig. 42.

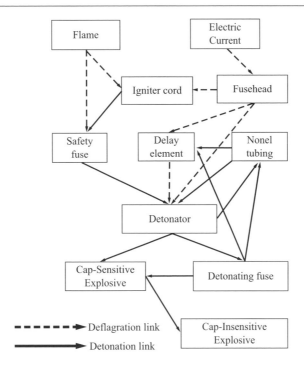

Figure 42 Initiation systems. After AECI, 1984.

initiator, a device or product used to transmit and/or supply heat and/or shock to start an explosion or detonation. Examples of initiators are detonators and detonating cord.
injected hole, a borehole into which a cement slurry or grout has been forced by a high-pressure pump and allowed to harden in adjacent cracks.
injection pump, is used to force foaming or misting compounds into air circulation at drilling. Stein, 1997.
injection test, see '**water head measurement**'.
inner tube, tube that fits inside another tube of larger diameter. Atlas Copco, 2006.
inner vane, blade like part used inside an air motor of a certain type. Atlas Copco, 2006.
in-place factor ($1/u$), (dimensionless), the inverse of the *swell factor u* which is defined as the ratio of volume of broken material to the volume of the material in its solid condition. It is also called the '*In-bank factor*'.
insert, a part formed from a second material, usually a metal that is placed in the mold or mould and appears as an integral structural part of the final casting. ASM Gloss, 1961. Cutting edge or tool as the carbides in a cruciform drill bit. Atlas Copco, 2006. A manufactured piece of *hard metal* inserted into a drill bit.
insert angle, front part of cemented carbide insert is normally V-shaped. The insert angle is the angle of the V-shape. Sandvik, 2007.
insert bit, a bit into which inset cutting points of various pre-shaped pieces of hard metal (usually a sintered, tungsten carbide-cobalt powder alloy) are brazed or hand peened into slots or holes cut or drilled into a blank bit. Hard-metal inserts may or may not contain diamonds. Also called slag bit. See also 'insert'. Long, 1920. Device of the solid center design, normally used for straighter holes or drilling parallel cut holes. Insert bits

(blade bits) are sometimes preferred in extremely soft or highly abrasive rock formations. Atlas Copco, 2006.

insert set, bits or reaming shells set with inserts. See also 'insert'. Long, 1920.

inserted blade cutters, cutters having replaceable blades that are either solid or tipped and usually adjustable. ASM Gloss, 1961.

inshot, device used for retrieving a core barrel in a drill hole containing no fluid. Atlas Copco, 2006.

inshot dry hole lowering tool, tool attached to a wire line, used in diamond drilling to lower the core barrel into a dry hole. Atlas Copco, 2006

inside gauge or *inside gage*, the inside diameter of a core bit as measured between the cutting points, such as between inset diamonds on the inside wall surface. Long, 1920.

inside thread, synonym for '*box thread*'. Long, 1920.

in situ (latin), meaning the natural or original position.

in situ block size, see '**block size in situ**'.

in situ fragmentation, the fragmentation of the rock mass before it has been disturbed by drilling and blasting or undercutting and caving.

in situ stress, the stress in the undisturbed (no excavation) rock mass. It can be quantified by the three orthogonal principal main stresses.

inspector of mines, one employed to make examinations of and to report upon mines and surface plants relative to compliance with mining laws, rules and regulations, safety methods, etc. State inspectors have authority to enforce State laws regulating the working or the mines. Fay, 1920.

instantaneous blasting cap, see '**instantaneous detonator**'.

instantaneous charge, total charge initiated in one delay. Charges fired on different delays can only be regarded separate if the delay time is roughly larger than 8 ms (a rule of thumb). In practice this value could be changed and is dependent on the *scatter in delay time* and also the physical *distance between the blastholes*.

instantaneous cut, all the holes in the opening of a drift, tunnel, raise or shaft sinking round that are initiated on the same delay time (instantaneous). These holes are drilled to the full depth of the round. There are four examples of these type of cuts, V-cut, Bålsjö cut, Pyramid cut and VP-cut. Langefors, 1963.

instantaneous detonator, a detonator without any inbuilt delay period between the passage of an electric current through the detonator and its explosion. B.S. 3618, 1964. A detonator that has an almost zero-delay firing time as compared with delay detonators with firing times from several milliseconds to several seconds. Atlas Powder, 1987.

instantaneous fuse, a rapid burning, ~10 m/s fuse, giving almost instantaneous initiation. Pryor, 1963.

Institute of Explosives Engineers (UK), a UK organisation for qualified explosive engineers.

Institute of Makers of Explosives (IME), a US trade organization which deals with the commercial use of explosives, and is concerned about safety in the manufacturing, transportation, storage, handling, and use of explosives. The IME publishes a series of blasting safety pamphlets. USBM, 1983.

Institution of Mining and Metallurgy (IMM), the London Institution of Mining and Metallurgy is the central British organization for regulating the professional affairs of suitably qualified mining engineers engaged in production and treatment of nonferrous metals and rare earths. Related bodies are those of Can. Inst. Min. Metallurgy in Canada,

Australia Inst. of Min. and Metallurgy in Australia, and finally South Africa. Inst. of Min. and Metallurgy in the Republic of South Africa. Pryor, 1963.

instrumented rock bolt or **measuring bolt**, a rock bolt equipped with strain gages to measure the load on the bolt for research purposes or detailed monitoring requirements.

intact rock, part of a rock mass not affected by the gross structural discontinuities.

intact rock strength (IRS), compressive strength in situ without any disturbances from excavations.

integral (drill) steel, a bar to the forged end of which is brazed a hard metal tip for drilling. B.S. 3618, 1964. Normally having a hexagon shape, with a blade like tungsten carbide cutting tool soldered into one end and the other designed to fit into the rock drill chuck. Atlas Copco, 2006.

integral (drill) steel set, a series of integral (drill) steels with increasing length for example 0,6, 1,2, 1,8 and 2,4 m. The length increment is usually determined by the wear of the bit and the feed length of the feeding device. Fraenkel, 1952.

inter-, prefix from the Latin word "inter" meaning '*between*'. Used for example as inter-row delay or delay between rows of blastholes as opposed to intra-row delay, expressing delays within a row of blastholes.

interactive flow, occurs in block- and sublevel caving when the draw bodies overlap. The advantage with interactive flow is that the ore rests are reduced compared to isolated flow. The broken ore or rock in adjacent draw columns may migrate from one draw column to the other. MassMin, 2008.

Interactive Spare Parts Application (ISPA), computer program that allows the user to select parts by pointing and clicking. Atlas Copco, 2006.

interferometer, an instrument used to determine the wave length of light by the production of interferences with waves of known lengths. The interferometer has been used to determine coefficient of expansion of enamels and glasses. Similar technique is used to measure small linear displacements in the field with very high precision. See also 'laser scanning'.

intergranular crack, a microcrack extending along a grain boundary between two adjacent grains. See also 'microcrack'.

intermediate cut, a type of cut used in coal mining. See 'middle cut'. Nelson, 1965.

intermediate level, a system of horizontal workings started from a raise and not connected directly to the main working shaft. Stout, 1980.

intermediate zone, the zone between the crushing zone close to the blasthole and the reflected wave fragmentation zone located close to the free surface. Clay, 1965.

intermittent cutters, coal cutting machines of the pick- and breast machine types. These cutters must frequently be reset compared to continuous cutters where the whole face can be cut without stopping the machine. Kiser, 1929.

internal damping, see 'attenuation'.

internal friction (ϕ), (°), see '**angle of internal friction**'.

International Conferences on Ground Penetrating Radar, the 8th Conference was hold in Australia in 2000. Noon, 2000.

International Congress on Rock Mechanics, conferences held every fourth year. See 'International Society for Rock Mechanics'.

International Drilling Conference & Exhibition, no known regular conferences of international character.

International Journal for Rock Mechanics and Mining Sciences, see '**International Society for Rock Mechanics**'.

International Rock Excavation Equipment Data Exchange Standard (RIDES), in order to facilitate use of different equipment from different producers in the same organization, Atlas Copco, together with other major machine manufacturers, mining and construction companies and third party suppliers, has established a standard for data exchange between rock excavation equipment and users' computer systems. It is the common language in data exchange for mining and tunnelling. Atlas Copco, 2007.

International Society for Rock Mechanics (ISRM), was founded in Salzburg in 1962 under the leadership of Leopold Muller, who acted as President of the Society until September 1966. Since 1966 the ISRM Secretariat has its headquarters in Lisbon Portugal at the Lab oratorio Nacional de Engenharia Civil. The number of members reached almost 5 000 in 48 national groups in 2008. The Society is a non-profit organization supported by the fees of its members and grants that do not impair its freedom of action. Its main objectives and purposes are stated in the statures an By-Laws. Congresses are held every *fourth year* since 1979. The 1st Congress was held in Lisbon Portugal in 1966, 2nd Yugoslavia in 1974, 3rd Switzerland in 1979, 4th Australia in 1983, 5th Canada in 1987, 6th Germany in 1991, 7th Japan in 1995, 8th France in 1999, 9th South Africa in 2003, and the 10th Portugal in 2007.

ISRM has developed *recommendations* for the following procedures

1. Petrographic description.
2. Monitoring rock movements and testing borehole extensometers.
3. Quantitative description of discontinuities.
4. Water content, porosity, density, absorption.
5. Pressure monitoring using hydraulic cells.
6. Strength in triaxial compression.
7. Rock anchorage testing.
8. Point load strength.
9. Deformabilitiy using a flexible dilatometer.
10. Fracture toughness.
11. Laboratory testing of argillaceous swelling rocks.
12. Hardness and abrasivity.
13. Determining sound velocity.
14. Tensile strength.
15. Uniaxial compressive strength and deformability.
16. In situ deformability of rock.
17. Geophysical logging of boreholes.
18. Surface monitoring of movements across discontinuities.
19. Deformability using a large flat jack.
20. Rock stress determination.
21. Seismic testing within and between boreholes.
22. Large-scale sampling and triaxial testing of jointed rocks.

International Society of Explosives Engineers (ISEE), is a non-profit association founded in 1974 by a group of blasters who recognized the need for an organization to promote the science and art of Explosives Engineering. Its name was originally the Society of Explosives Engineers. In 1992 it was internationalised. The founding members of SEE. established broad goals related to all aspects of blasting. Research, educational forums, governmental regulations, technical information, standardization, certification and international co-operation were initial concerns and continue to be the foundations of the Society's activities. The number of members, in 2009, was more than 4 900 from 90 countries. SEE and ISEE has been arranging annual conferences since 1974.

International Symposia on Rock Fragmentation by Blasting (FRAGBLAST), international rock blasting symposia arranged normally every *third year*. The 1th symposia was held in Luleå 1983, 2nd USA 1987, 3rd Australia 1990, 4th Austria 1993, 5th Canada 1996, 6th South Africa 1999, 7th China 2002, 8th Chile 2006 and the 9th in Spain 2009 and the 10th is planned to India. Fragblast was a Commission of ISRM during the years 1991–2003 and is since 2003 a section of ISEE, Int. Soc. of Explosives Engineers in USA.

International Symposia on Tunnel Safety and Security (ISTSS), the first symposia was held in Washington DC in 2004, the 2nd in Madrid in 2006 and the 3rd in Stockholm, Sweden in 2008. www.sp.se/fire/ISTSS2008.

International System for Units (SI), a system of units where m stands for meter, kg for kilogram, s for second, and A for Ampere.

International Tunnelling Association (ITA), founded in 1974 as a result of an OECD conference held in Washington 1970 regarding the use of underground space and its future potential. It has 2007 about 100 countries as members. It is organized by a General Assembly with one vote per member country. The board is selected for three years and includes 16 members. A meeting is held every year. A journal is produced with the title "Tunnelling and Underground Space Technology Journal with 5–6 issues per year. ITA have also several working groups. In 2007 there were 13 groups.

interval, delay time between detonators with different numbers.

in-the-hole drilling (ITH), drilling with a machine which operates inside the drill hole. When drilling downwards it is called down-the-hole drilling (DTH). See 'down-the-hole hammer'.

intra-, prefix from the Latin "intra" meaning '*within*'. Used for example as intra-row delay, expressing delays within a row of blastholes as opposed to inter-row delays, or delay between blasthole rows.

intrinsic, the same as 'real'.

intrinsic permeability, see '**Darcy's law**'.

intrinsic safety, in a circuit, safety such that any sparking that may occur in that circuit in normal working or in reasonable fault conditions, is incapable of causing an explosion of the prescribed inflammable gas. NCB, 1964.

intrinsic shear strength (C), (MPa), synonymous to '**cohesion**'.

inventory of explosives, a listing of all explosive materials stored in a magazine. Atlas Powder, 1987.

inversion, an exceptional atmospheric condition where the air temperature increases with altitude.

inverted heading and bench, the heading is the lower part of the section and is driven at least a round or two in advance of the upper part, which is taken out by overhand excavation. Synonym for '*heading overhand bench*'. Fraenkel, 1952.

IREDES (acronym), see '**International Rock Excavation Equipment Data Exchange Standard**'.

IRS (acronym), see '**intact rock strength**'.

ISANOL (Norway), a designation for a mixture of ANFO and polystyrene prills of varying proportions. ISANOL is used in controlled contour blasting to reduce the charge concentration in the borehole. Use 'ANFOPS'.

isentrope, in blasting the curve defining the relation between the pressure and volume at constant entropy, or pressure and density of the gas in a blasthole, assuming that no heat exchange occurs with the surroundings and no inner friction in the gas.

isentropic, thermodynamic property of a gas with a constant entropy process. A curve (such as on a pressure-volume diagram) which represents an isentropic process is called an isentrope. In both ideal and non-ideal detonation the post-CJ process is adiabatic but not isentropic, since chemical reaction continues beyond the CJ curve. In non-ideal detonation the material does not reach chemical equilibrium in the CJ zone, and reaction continues beyond the CJ-curve. In an ideal detonation, chemical equilibrium is attained at the end of the CJ zone. However, as the material cools under adiabatic expansion its equilibrium configuration changes, requiring continued reaction in order to maintain chemical equilibrium. Because it is easier to calculate an isentrope rather than an adiabatic non-isentrope, isentropic expansion is commonly assumed. The error introduced by this assumption is minimal. See also 'entropy' and 'adiabatic'. Sarracino, 1993.

iso chromatic lines, are lines of equal difference of principal stress in stress analysis performed by the photo elastic method. The iso chromatic lines appear as coloured streaks. Ham, 1965.

isolated draw zone (IDZ), a draw zone isolated from other draw zones as a result of drawing from an isolated drawpoint. MassMin, 2008.

isolated flow, flow of caved material through several draw points where there is now interaction between the flow in each draw point. Therefore there will be isolated volumes between the draw points where it will be no movement at all. MassMin, 2008.

isolated interactive flow, the case when there is a zone between the draw points where the flow is laminar or also called mass flow. It is laminar down to a certain level where the velocity will start to differ. MassMin, 2008.

isostasy, the equilibrium location of the earth crust. Where the earth has been covered with ice the crust has been pressed downwards and when the ice melts the crust starts to rise slowly to find its isostasy. Iso is Greek for 'equal' and stasy for 'location'. Skinner, 1989 and Fowler, 2005.

ISPA (acronym), see **'International Spare Parts Application'**.

ISRM (acronym), see **'International Society for Rock Mechanics'**.

ISRM (test methods), see **'International Society for Rock Mechanics'**.

ISEE (acronym), see **'International Society of Explosives Engineers'**.

issuing authority, a governmental agency, office, or official vested with the authority to issue permits or licenses. Atlas Powder, 1987.

ITA (acronym), see **'International Tunnelling Association'**.

ITH (acronym), for **'in the hole'** see **'in-the-hole drilling'**.

Izod test, a pendulum type, single-blow impact test in which the specimen usually notched, is fixed at one end and broken by a falling pendulum. The energy absorbed, as measured by the subsequent rise of the pendulum, is a measure of *impact strength* or *notch toughness*. ASM Gloss, 1961. Compare with the Charpy test and fracture toughness test.

jackhammer, a percussive type of automatically rotary rock drill that is operated by compressed air. It is light enough to be hand held.
jack hole, in coal mining a bolthole. Standard, 1964.
jack latch, a fishing tool for stuck drill tools. Long, 1920.
jackleg, a pneumatically actuated single rotary-percussion machine with a hinged air-assisted feed leg. Primarily used in small development headings and production stopes for drilling holes up to 45 mm in diameter. Atlas Powder, 1987. Also light supporting bar for use with jackhammer, Pryor, 1963.
jack prop, synonymous for *derrick, warwick, anchor prop* and *foreset*. Mason, 1951.
jadder, a cutting in a natural rock face, made fort the purpose of detaching shaped stone blocks. Pryor, 1963.
jag bolt, an *anchor bolt* with a barbed flaring shank that resists retraction when leaded into stone or set in concrete. Also called *hacked bolt* or *rag bolt*. Webster, 1961.
jar, an appliance to permit relative movement between the rope and rods in a cable drill. It reduces shocks and the risk of rod or chisel breakages. Nelson, 1965.
jar collar, a *swell coupling* attached to the upper exposed end of a drill rod or casing string to act as an anvil against which the impact blows of a drive hammer are delivered and transmitted to the rod or casing string; also, sometimes used as a synonym for drive hammer. Also called *bell jar, drive collar* and *jar head*. Long, 1920.
jar hammer, see '**casing jar hammer**'. B.S. 3618, 1963.
jar head, synonym for *bell jar, jar collar* or *drive head*. Long, 1920.
jar length, synonym for '**jar rod**'. Long, 1920.
jar rod, an extra heavy wall drill rod to which drive collars may be attached when using a drive hammer to jar or loosen drill string equipment stuck in a borehole. Also called '*drive rod*', '*jar length*', '*jar piece*'. Long, 1920.
jar piece or **jarring piece**, a piece of pipe used for the same purpose and in lieu of either a '*drive hammer*', '*drive rod*' or a '*jar rod*'. Long, 1920.
jar sleeve, see '**drive hammer**' and '**jar piece**'. Long, 1920.
jar staaf, a heavy bolt that forms a sliding connection between the jar and overshot heads of a wire-line core barrel. Long, 1920.
jar weight, see '**casing drive hammer**'. Long, 1920.
jaw breaker, see '**jaw crusher**'.
jaw crusher, a machine for reducing the size of a material by impact to crushing between a fixed plate and an oscillating plate, or between two oscillating plates, forming a tapered jaw. B.S. 3552, 1962. The jaw crushers are normally *primary crushers* and could take

blocks up to 1,5 m in diameter. The output size is normally less than 250–300 mm. A common used crusher type.

jaw-type gyratory crusher, a gyratory crusher with the throat extended on one side. See also '**cone crusher**'. Constable, 2009.

JCZ (acronym), for 'Jacobs-**Cowperthwaite-Zwisler**'.

JCZ3 (abbreviation), for '**equation of states for detonation**'.

Jeffrey crusher, whizzer mill, see '**swing hammer crusher**'.

jet corer, consists of a length of pipe which is lowered from a vessel in order to obtain samples. High, velocity water is pumped through the pipe and the jetting action of this water issuing from the lower end of the pipe very effectively cuts a hole in the unconsolidated overburden sediments. Once at bedrock, the *pipe is rammed into the rock* with sufficient force to obtain a plug of about 50–100 mm in length. Mero, 1965.

jet hole, a borehole drilled by use of a directed, forceful strain of fluid or air. Long, 1920.

jet hydraulic, stream of water used in alluvial mining. A 254 mm in diameter jet can deliver a thrust of over 27 t. Pryor, 1963.

jet loader, a system for loading ANFO into small blastholes where the ANFO is drawn from a container by the Venturi principle and blown into the hole at high velocity through a semi-conductive loading hose. USBM, 1983. The charging machine could also consist of a pressure vessel which is filled with a powdered explosive for example ANFO (ammonium nitrate and oil) and by adding a pressure to the vessel the ANFO is pushed through an antistatic treated PVC-hose to the blasthole for safety reasons.

jet nozzle, small hole through which high pressure fluid is pressed.

jet piercing, the use of high velocity jet flames to drill holes in hard rock, such as taconite (iron ore), and to cut channels in granite quarries. It involves the combustion of oxygen and a fuel oil fed under pressure through a nozzle to produce a jet flame generating a temperature of over 5 000° C. A stream of water joins the flame, and the combined effect rests on a thermodynamic spalling and disintegration of the rock into fragments which are then blown from the hole or cut. USBM, 1968. The burning device is essentially a long blowpipe consisting of three tubes equipped with jets at the bottom end. Two of the tubes carry *kerosene* and *oxygen*, which when jetted together and ignited, generate a flame having a temperature of about 2 760° C. This flame is directed downward against the rock, superheating a circular area. A following water jet cools the heated rock causing it to contract and spall, or, if partially molten, to granulate. The resulting steam evacuates the spall from the hole and also keeps the burner form melting. High velocity jet flames are used to drill holes in hard rocks, as taconite (hematite iron ore in the US Mesabi Range), and to cut channels in granite dimensional stone quarries. Fraenkel, 1952. Rustan, 2006.

jet propulsion pump, ejector pumps and carburettors are examples of jet pumps. They use a Venturi system to draw in water, build pressure and provide lift. Water must be moving through the pump before it can work. An air or water jet can be used to propel mud along a pipe or up the drill rod. Air has the advantage of both reducing pressure near the jet, and forming expanding bubbles to lift the mud above. Stein, 1997.

jetting removal of drill cuttings, is achieved by water circulation down through the rods washing cuttings from the front of the bit. The cuttings flow up the annular space and into a settling pit so that the water can be recirculated. Stein, 1997.

J-factor, see '**joint frequency**'.

joint, a fracture of geological origin generating a discontinuity in a body of rock occurring either singly or more frequently in a set or system, but not attended by a visible movement parallel to the surface of discontinuity. There is no resistance to the separation of open joints. Normal joints are perpendicular to the bedding, foliation, or layering. Joints, like stratification, are often called partings. Number of joints per cubic meter (J_V), (No. of joints/m^3) see '**volumetric joint count**'.

joint alternation number (J_a), **(dimensionless)**, see '**Q-factor**'.

joint block, a body of rock that is bounded by joints. Bates and Jackson, 1987.

joint compressive strength (σ_{cj}), **(JCS), (MPa)**, the 'joint wall compressive strength'.

joint diagram, a diagram constructed by accurately plotting the strike and dip of joints to illustrate the geometrical relationship of the joints within a specified area of geological investigation. ISRM, 1975.

joint frequency (f_j), **(1/m)**, the number of joints per meter.

joint-initiated fracturing, cracks initiated by blast waves at the planes of weakness in the rock. These cracks are oriented normal to the planes of weakness. Investigated on a laboratory scale in Homalite 100 by Fourney et al., 1979 and in bedded limestone on a full scale by Winzer and Ritter, 1980.

joint intensity (I_{jA}), **(No./m^2)**, number of joints per unit of area. Dershowitz and Einstein, 1988.

joint intensity (I_{jV}), **(No./m^3)**, number of joints per unit of volume. Dershowitz and Einstein, 1988.

joint intensity (I_{jtlA}), **(1/m)**, total joint trace length per unit of area. Dershowitz and Einstein, 1988.

joint intensity (I_{jAV}), **(1/m)**, total joint area per unit of volume. Dershowitz and Einstein, 1988.

joint or **fault set**, a group of parallel or quasi-parallel joints or faults, respectively.

joint or **fault system**, a system of two or more joint/fault sets or any group of joints/faults, respectively, with a characteristic pattern, e.g. radiating, concentric, etc. Both the mechanical behaviour and the fragmentation of a rock mass are dominated by the number of discontinuity sets which intersect one another, although the orientation of these discontinuities in relation to the face is considered of primary importance. The class numbers of the discontinuity sets occurring in a rock mass are given in Table 18.

joint persistence (k), **(m)**, see '**persistence**'.

joint plane orientation (JPO), (f_{JPO}), **(dimensionless)**, the joint plane direction is very important for the fragmentation and especially how the joints are oriented in regard to the blast direction. A very rough way is to classify the joint plane directions are *horizontal*, *dip out of face* or *dip into face*. See also '**blastability (index)**'.

joint plane spacing (JPS), (f_{JPS}), **(dimensionless)**, see 'discontinuity spacing' and '**blastability (index)**'.

joint roughness coefficient (JRC), (k_{JRC}), **(dimensionless)**, is quantified by comparing to a scale 1 to 20 where 1 is used for the smoothest and 20 for the roughest surfaces. Brady, 1985.

joint roughness number (J_r), **(dimensionless)**, see 'Q-factor'.

joint set, a group of more or less parallel joints. The term is used in geology and rock mechanics. Billings, 1954.

joint set number (J_n), **(dimensionless)**, see '**Q-factor**'.

joint shear strength (τ_{sj}), (MPa), a prediction formula for the determination of joint strength τ_{sj} in MPa has been developed by Barton and Choubey, 1977.

$$\tau_{sj} = \sigma_n \tan\left[k_{JRC} \log_{10}\left(\frac{\sigma_{cwj}}{\sigma_n}\right)\right] \qquad (J.1)$$

where σ_n is the normal stress acting on the joint plane in MPa, k_{JRC} is the joint roughness coefficient JRC and σ_{cwj} is joint wall compressive strength JCS in MPa.

joint spacing, (S_j), (m), the perpendicular distance between adjacent joints within a joint set. It is quantified in Table 31.

Table 31 Classification of joint spacing. After Deere, 1968.

Description	Spacing of joints (m)	Rock mass grading
Very wide	>3	Solid
Wide	1–3	Massive
Moderately close	0,3–1	Blocky/seamy
Close	0,05–0,30	Fractured
Very close	<0,05	Crushed and shattered

joint wall compressive strength (JCS), (f_{JCS}), (MPa), the compressive strength measured normal to the joint wall.

joint water reduction factor (J_w), (dimensionless), see '**Q-factor**'.

Jora method, a raise driving method where access is needed both at the top and the bottom of the planned raise. A larger hole about 100 mm in diameter is drilled either from the top or the bottom. A cable for a hoist cage standing at the bottom level is lowered through the whole and fastened to the top of the cage. The cage can be pulled by the cable to the top of the raise for drilling the blastholes. The cage will be sent down to the bottom of the raise and withdrawn to protect it from rock fragments falling down at blasting. The cable needs also to be withdrawn some meters to have it protected from blast damage. See also '**raise driving method**' for other methods.

journal, part of an axel or shaft that rests or turns on bearings. Atlas Copco, 2006.

journal bearings, sliding surface bearings capable of carrying high loads. At high speeds, they must be fed with oil under pressure. At lower speeds, a drip-feed oiler will suffice. Journal bearings carry radial loads only. The bearing metal could be made of bronze or white metal (slipper). Stein, 1997.

journal book, synonym for 'logbook' abbreviated to '**log**'. Long, 1920.

Joy loader, a loading machine that uses mechanical arms to gather mineral on to an apron that is pressed into the severed material. A built in conveyor then lifts the material into tubs or on to a conveyor. Use first in coal mining but later on also in metal mining. Pryor, 1963.

joy stick, electrical control device, used for control of several different functions on a drill rig, having a handle that can be gripped in the hand and positioned within 360°. Atlas Copco, 2006.

JPO (acronym), see '**joint plane orientation**'.

JPS (acronym), see '**joint plane spacing**'.
JRC (acronym), (k_{JRC}), **(dimensionless)**, see '**joint roughness coefficient**'.
jug, see '**seismometer**'.
jumbo or drill carriage, a machine on wheels containing two or more mounted boom-fed rotary-percussion drills, actuated either by compressed air or hydraulics. It is used in underground excavation of rock and generally capable of drilling holes of 45–90 mm in diameter. See also '**portal jumbo**'.
jump drilling, see '**cable churn drilling**'.
jumping mortar test, a method of testing the strength of permitted explosives. Two halves with finely ground exactly matching surfaces form a mortar with a borehole. One of the halves is embedded in the ground at a 45° angle, while the other half is projected in the manner of a shell, when the explosive charge is detonated in the hole. The distance to which it has been thrown is then determined. When high brisance explosives are tested, the disadvantage of the method is, that the surface must be ground anew after each shot. The method gives excellent results with weaker or permitted explosives. Meyer, 1977.
junking, the process of cutting a passage through a pillar of coal. Tweneys, 1958.
jute bag collection device, can be used to collect the coarse material from drilling for sampling purpose. The drill cuttings have to pass the jute bag which is placed on timbers. 1997.

k, prefix for kilo = 1 000.

K, a unit of temperature used for the *Kelvin degree scale*. Also a symbol for 1) *Bulk modulus* (the inverse of compressibility), 2) *Number of cycles at breakage of a structure* and 3) *Fisher constant*.

k_{50} **(m)**, mean fragment size measured with squared meshes.

Kaiser effect, a method based on acoustic emission, AE, to estimate the maximum stress a rock mass has been subjected to during its geological history (crustal stress). The rock sample is stressed in compression. The acoustic emission will start when the maximum history stress is reached. A complementary method in the cases where it is difficult to find the Kaiser effect is to load the rock twice and determine the difference between the first and second loading. Yoshikawa, 1981.

karst, are formed in carbonate rocks like dolomites and limestone's. Acid solutions dissolve large quantities of material so underground caves can be created causing a slow subsidence of the surface. Brady, 1985. It may be a problem to use conventional rock bolts e.g. mechanically anchored- or grouted rock bolts in karst rock. Swellex rock bolts are, however, unaffected by open cracks and smaller voids.

Kast cylinder compression test, a test to quantify the brisance of an explosive. An unconfined cartridge (enveloped in paper or thin metal sheet) acts upon a copper or lead cylinder (crusher). The loss of height of the crusher is a measure of the brisance of the test explosive. The brisance value in this test is the product of charge density, specific energy and detonation rate. Meyer, 1977.

Kelly bar, a two-piece drill steel. The inner steel may be withdrawn, allowing the loading of a cardboard casing and/or explosives, while the outer steel prevents rocks or cuttings from blocking the hole. Konya and Walter, 1990.

kerf, a slot cut into a face of coal or soft rock by means of a mechanical cutter to provide a free face for blasting.

kerf cutting, can be done by high-pressure water jet if the jet traverses the rock surface. Then the jet erodes the rock grains and cuts a shallow kerf in the rock face. The effectiveness of this erosional process depends both on the rock type and on the jet characteristics: pressure, flow rate, etc. Harries, 1974.

Kerr-cell camera, a high-speed camera that can take photos with very short time delays, down to microseconds. There are cameras in the market where 4 to 16 photos can be taken in one run. Persson, 1993.

keyhole notch, see '**Charpy test**'. Ham, 1965.

keystone, the stone in a natural arch underground that's give its bearing capacity. If this stone is moved the arch will collapse.

key twist blasting machine, A small handheld machine operated by a quick twist of the handle with one hand while the machine is held firmly on other hand. Normally up to 25 detonators each having a resistance of 4 ohms can be initiated.

kg (abbreviation), unit for mass and abbreviation for kilogram.

Kick's law, the amount of energy required to crush a given quantity of material to a specified fraction of its original size is the same no matter what the original size. CCD 6th ed., 1961. Some researchers combines the Bond and Kick's laws und use them for rock fragmentation by blasting calculations.

kinetic energy (W_k), (J), a physical quantity or state variable defined as the energy which a body possesses because of its motion; in classical mechanics, equal to one half of the body's mass times the square of its speed.

kieselguhr, see '**diatomaceous earth**'.

kilo (shortend form), for kilogram or 1 000.

kilogram, 1 000 g (2,204 pounds).

kindling point, the temperature at which a substance ignites. Crispin, 1964.

kindling temperature, the temperature at which a substance ignites. Crispin, 1964.

kink bands, megascopic deformation zones that form with the accompaniment of translation gliding within a mineral crystal such as is visible in mica, kyanite, and calcite. USBM Staff, 1968.

kinking failure, see '**kink bands**'.

Kjartansson transfer function, this transfer function is used to propagate observed pulse shapes from blast sources and thus match observed rise time slopes with distance. Kjartansson, 1979.

knapper, a stonebreaker, specifically one who breaks up flint flakes into sizes used for gunflints. Fay, 1920.

knapping hammer, a special hammer or machine to break rock and produce minimum of fine material. Tweney, 1958.

knapping machine, an instantaneous stone crushing machine, a stonebreaker. Standard, 1964.

knee hole, a blasthole in tunnelling situated in the blast row next to and above the floor holes (lifters) and with blasting action upwards.

knob, to remove knobs in rough dressing of stone in the quarry. Standard, 1964.

knobbing, the act of rough dressing stones in the quarry by knocking off the projections and points. Fay, 1920.

knock, chap or **scale**, to remove loose blocks from a mine roof with a scaling bar or hydraulic hammer to make safe working conditions. See '**scaling**'.

Knoop hardness, micro hardness determined from the resistance of metal to indentation by a pyramidal diamond indenter, having edge angles of 172° 30′ and 130°, making a rhombohedral impression with one long and one short diagonal. ASM Gloss, 1961.

knox hole, a circular drill hole with two opposite vertical grooves which direct the explosive power of the blast. Fay, 1920. Also an axial slotted blasthole for control of crack directions in controlled contour blasting.

Koehler lamp, a naphtha burning flame safety lamp for the use in gaseous mines. Fay, 1920.

Koenen test, a method to test the heat sensitivity of an explosive where the explosive is placed into a vented steel cylinder and heated by gas until it explodes. Koenen and Ide, 1956. A test for determining the combustion or thermal heat of explosives. Persson, 1993.

kt, unit for kiloton.

Kuznetsov-Ramler formulas, see 'Kuzntesov fragmentation formula' and the 'Rosin Ramler distribution function'.

Kuznetsov fragmentation formula, an empirical formula to calculate the mean fragment diameter k_{50} after blasting in a rock. The empirical values were gathered from blasting in blocks and full scale blasting with specific charges varying from 0,30 to 0,74 kg/m³.

$$k_{50} = A\left(\frac{1}{q}\right)^{4/5} Q^{1/6} \qquad (K.1)$$

A is the rock mass property factor with $A = 7$ for medium hard rocks and $f = 8 - 10$, $A = 10$ for hard but highly fissured rocks corresponding to $f = 10 - 14$ and $A = 13$ for very hard fissured rocks with $f = 12 - 16$. f is the numerical value for the *Protodyakonov hardness scale* ranging from 1 to 10, q is the specific charge in kg/m³ and Q is the weight of TNT in kg equivalent in energy to the explosive charge in one borehole. A is the a rock mass factor; Kuznetsov, 1973.

Kuz-Ram (Kuznetsov-Ramler) model, an empirical formula to calculate the mean fragment size after rock blasting and using the Rammler mathematical distribution function (simplified Weibull function) for calculation of the whole fragment size distribution. Kuznetsov, 1973 and Cunningham, 1987.

kW, the unit for **kilowatt**.

ladderway or *manway,* a vertical or inclined compartment, for personal transport to the stopes by ladders. The ladderways can also include ventilation ducts and water and compressed air pipes, etc. The drive may be a winze or a raise and its purpose is to give convenient access to a stope.

lagging, to secure the roof and sides behind the main timber or steel supports with short lengths of timber, sheet steel, or concrete slabs. Lagging wedges and secures the support against the rock and provides early resistance to pressure. Also called 'lacing'. Nelson, 1965. It is an old method and not so effective.

lag time, the total time between the initial application of current and the rupture of the circuit within the detonator. B.S 3618, 1964.

Lame's constants, in the case of isotropic material, all coefficients of the *elasticity matrix* may be expressed in terms of the *Lame's constants* λ and μ

$$\lambda = \frac{E\upsilon}{(1+v)(1-2v)} \quad \text{and} \quad \mu = \frac{E}{2(1+v)} \quad (L.1)$$

where E is the Young's modulus in MPa, v is the Poisson's ratio.

lampblack, carbon in form of a black or grey pigment made by burning low-grade heavy oils or similar carbonaceous materials with insufficient air, and in a closed system so that the soot can be collected in settling chambers. Properties are different from carbon black. Used as a black pigment for cements and ceramic ware, an ingredient in liquid-air explosives, in lubricating compositions, and as a reagent in the cementation of steel. CCD, 1961.

landslide failure surfaces, the surface developed when rock or soil starts to slide due to overriding the failure strength of the material. There are three types of failure surfaces; 1) *Buckling*, see 'buckling' 2) *Toppling* see 'toppling' or 'toppling of hanging wall' and 3) *Kinking,* see 'kink bands' are found in foliated and laminated rock. Hudson, 1993.

Langefors weight strength, see '**weight strength (defined by Langefors)**'.

LAR (acronym), for '**low aspect ratio**', see '**aspect ratio**'.

large diameter blasthole, a blasthole with a diameter larger than 100 mm underground, and 200 mm on the surface.

large diameter hole (on surface), in large open pit mines and quarries the most common method of primary blasting is by means of blastholes of >200 mm in diameter.

large hole, see '**large diameter blasthole**'.

large hole blasting, blasting methods using blastholes >100 mm in diameter underground and >200 mm on surface.

large hole cut, a parallel hole cut with the aim of creating an opening for the remaining holes in a tunnel, drift, raise or shaft round. The cut includes at least one uncharged hole with a diameter larger than the charged holes.

large hole drilling, the act of drilling large diameter holes; >100 mm under ground and >200 mm on surface.

large hole stoping or **big hole stoping**, a mining method using blastholes with a diameter >100 mm underground.

large scale mining, the use of production blastholes >100 mm underground and >200 mm on surface or a production volume >10 Mt/year.

larry, see '**barney**'.

laser drill, an exotic drilling method using a laser beam to heat the rock and vaporise it. Plans have been to combine this method with mechanical drilling methods. KTH, 1969.

laser plane drilling, a laser beam is used to define a horizontal plane on surface and from that plane each hole is drilled to a certain depth below this plane within an accuracy of 5 cm. A *laser beacon, mounted on a tripod*, generates a horizontal reference plane across the worksite by a rotating laser beam. A sensor fitted to the rock drill cradle, reacts to the signal generated by the rotating laser beam. Atlas Copco, 2006b.

laser scanning, a method to determine distances by measuring the time for a laser beam to travel from the source to the reflection point and back. Knowing the speed of light and time the distance can be calculated. This technique can be used both on surface and underground to determine the surface roughness of the contours after blasting.

Laubscher system, see '**geomechanics classification scheme for mining**' or '**cavability**'.

layflat tubing, thin walled plastic tubing used to contain and protect ANFO from water in wet holes. AS 2187.1, 1996.

LCC (acronym), for '**linear charge concentration**'. (kg/m).

lead, a small narrow uniformly trending passage in a cave. A.G.I., 1957.

lead azide, a primary explosive with the chemical composition $PbN6$ in the form of white needles which explode at +350°C. It is a very sensitive detonating agent used in blasting caps. CCD, 1961.

lead block test, or *Trauzl lead block test*, is a lead cylinder with a length of 200 mm and height of 200 mm. 10 g of the explosive being tested is inserted into the 25 mm blasthole, sand stemmed and detonated. The volume increase of the cavity formed is determined and the result is given in cm^3. This volume is a measure of the expansion work performed by the well-coupled explosive. Persson, 1993. The method was developed by Isodor Trauzl in 1885. This small-scale method can only be used for relative comparison of explosives because it is not reliable for test of the explosive strength in full scale rock blasting because of change in explosive properties with hole diameter and different reaction in different rock types, joint directions and joint frequency.

leading lines or *wires,* see '**lead wires**'.

lead wires or **connecting wire**, the wires connecting the electrical power source with the leg wires of the blasting cap or connecting wires of a blasting circuit. Also called firing line. USBM, 1983. New lead wires are used for each blast.

leakage resistance, the resistance between the blasting circuit (including lead wires) and the ground. Atlas Powder, 1987.

LEDC (acronym), see '**low energy detonating cord**'.

ledge height, see '**bench height**'.
Leet seismograph, a portable three-component seismograph designed primarily for registration of vibrations from blast, traffic, machinery, and general industrial sources. Leet, 1960.
LEFM (acronym), see '**linear elastic fracture mechanics**'.
leg wires, see '**cap wires**'.
length (*l*), (**m**), physical one-dimensional extension in space. In physics more precisely defined by a certain number of the wave lengths of a Krypton isotope. Length in m can be related to non-metric units as follows. (Old units; 1 inch = 25,4 mm, 1 foot = 0,305 m, 1 yard = 0,914 m, 1 mile = 1,61 km).
length density (J_d), ($1/m^2$), the length of the phase contained in the unit volume of the structure . Weibel, 1989.
length from collar to **column charge** (l_{co}), (**m**), the distance from the top of the column charge to the collar.
length of borehole or **blasthole** (l_{bh}), (**m**), the length of borehole or blasthole from the collar to the bottom of the hole.
length of bottom charge (l_b), (**m**), the axial length of the bottom charge in the blasthole.
length of column charge (l_c), (**m**), the axial length of the column charge in the blasthole.
length of charge (l_{ch}), (**m**), the sum of the lengths of the bottom and the column charge.
length of stemming (l_s), (**m**), the length of the non explosive material put at the collar of the blasthole to prevent early venting of the blast gases.
length of stemming between deck charges (l_{sd}), (**m**), the distance between neighbouring charges.
length of throw (*R*), (**m**), see '**distance of throw**'.
level, horizontal passages or drifts commonly spaced at regular intervals in depth in mines. They are customarily worked from shafts and are either numbered from the surface in regular order or designated by their actual elevation below the top of a shaft or the top of the mineralisation.
levelling, blasting in benches of a height less than twice the maximum burden ($2B_{max}$). Nitro Nobel, 1993.
Lewis bolt, a wedge-shaped bolt fastened in a socket by pouring in melted lead, and used in raising a heavy block, as of stone Standard, 1964. Steel bolt with roughened conical base, used for anchorage in concrete. Also called 'rag bolt'. Pryor, 1963.
LEX (abbreviation), see '**low energy explosive**'.
Leyner-Ingersoll drill, see '**water Leyner**'.
Leyner shank, a shank adaptor with only two lugs for rotation. Sandvik, 2007.
LHD (acronym), see '**load-haul-dump** unit'.
lift dog, simple tool used for handling rods in rod percussion drilling. Stein, 1977.
lifters, the lowest located drill holes in a drift or tunnel round, drilled slightly downwards in horizontal drifting and with breakage upwards.
lifting dog, any of various devices used for gripping something for the purpose of lifting it. Atlas Copco, 2006.
lifting eye, loop formed on the end of a lifting strap or wire rope used to attach to a lifting hook etc. Atlas Copco, 2006.
lift shot, the *initial cut* made in the floor of an open pit or quarry for the purpose of developing a bench at a level below the floor. Also called '**sinking cut**' or *drop cut*.
light emission (*at rock fracturing*), see '**fracto-emission**'.

light hydraulic breaker, hydraulic breaker weighing more than 30 but less than 670 kg. Atlas Copco, 2006.
lightning detector, see '**lightning warning device**'.
lightning explosion, an explosion of firedamp caused by electric current, during a thunderstorm entering a mine and igniting the gas. Fay, 1920.
lightning gap, a *break in the blasting cable* about *2 m in length* to prevent lightning discharges from following the cable into the mine. ASA M2.1, 1963.
lightning warning device, a device which is sensitive to electromagnetic disturbances in the atmosphere (thunder) and can give audible and/or visible warning of the approach of a storm.
light roof drilling rig, drill rig of light construction used for drilling holes above horizontal in an underground opening. Atlas Copco, 2006.
light sectioning method, a method to quantify the perimeter of a tunnel or drift by enlighten the perimeter with a light slot and photographing the slot casted on the wall. The drift area can be calculated by image processing. It requires less than 1 min per profile and the accuracy is within 2 cm. Franklin, 1989.
limit angle (α_{lim}), (°), the limit angle is defined as the angle measured from the horizontal and a line defined by one point on the deepest excavation level and the other point, the nearest point on surface where the vertical subsidence is zero. The soil on the bedrock, in regions with frost, normally have small vertical movements dependent on frost and to be sure that the subsidence is dependent on the mining activity, 2 cm e.g. was selected as minimum movement at the Kirunavaara mine. *Villegas, 2008*.
limit blast, see '**controlled contour blasting**'.
limit blast design (in surface mining), see '**buffer blasting**'.
limit border angle or **angle of limit border** (α_{lb}), (°), is the vertical tangent to the loosening drawbody at the outlet of the draw. It is important to know this angle especially in *sublevel caving* doing calculation of the drawbody shape by the Bergmark-Roos formula. Kuchta, 2004. Definition see Fig. 3 at the entry 'angle of caving'. This angle measured in the field may be in the order of ~70°.
limit ellipsoid, see '**loosening ellipsoid**'.
limit equilibrium method, see '**discrete element method**'.
lineament, a significant line of landscape that reveals the hidden architecture of the rock basement. Lineaments are character lines of the earth physiognomy. A.G.I., 1957.
linear charge, is normally a cylindrical charge in a borehole. Also detonating cord being used for controlled contour blasting is a linear charge when loaded into a contour hole. Linear charges are different to spherical charges or shaped charges.
linear charge concentration (LCC), (q_l), (kg/m), mass of charge per unit of length of blasthole.
linear charge concentration in the bottom of the blasthole (q_{lb}), (kg/m), mass of charge per unit of length at the bottom of the blasthole.
linear charge concentration in the column of the blasthole (q_{lc}), (kg/m), mass of charge per unit of length at the column of the blasthole.
linear elastic fracture mechanics (LEFM), the continuum mechanics approach to fracture based on the theory of linear elasticity. It is the classical branch of solid mechanics which addresses problems of crack propagation under certain restrictive assumptions. Ingraffea, 1993.
linear elastic-linear viscous rock mass behaviour, a model used to simulate time-dependent response of rock around supported drifts or tunnels. There is an analytical solution developed for this case. Hudson, 1993.

linear elastic non-linear viscous rock mass behaviour, a model used to simulate time-dependent response of rock around supported drifts or tunnels. There is an analytical solution developed for this case. Hudson, 1993.

linear variable differential transformers (LVDT), a displacement measuring instrument based on the linear displacement of an iron core along the axis of the instrument. Brady, 1985.

linear viscoelastic rock mass behaviour (Kelvin model), see '**viscoelastic model**'.

linear scale, a scale for the measurement of sound level that is non-weighted, so that there is little or no discrimination at low frequencies. Konya and Walter, 1990.

linear shaped charge, a shaped charge with linear extension used, e.g. for controlled contour blasting or demolition. Bjarnholt, 1981, Isakov, 1983, Rustan et al., 1985 and Holloway, 1986. Except the V-shape the main crosscut contour shape of the charge could be circular or elliptical. When two diametrically placed V-notches are used on the longest elliptical axel the charge is called elliptical bipolar shaped charge. Qin J.F., 2009.

lineation, the parallel orientation of structural features that are lines rather than planes. Lineation may be expressed by the parallel orientation of the long dimensions of minerals, long axis of pebbles, striae on slickensides, streaks of minerals, cleavages, and fold axis. Synonym for '*linear parallelism*' and '*linear structure*'. A.G.I., 1957.

line clinometer, a borehole survey clinometer designed to be inserted between rods at any point in a string of drill rods. An instrument for measuring angular deviation from horizontal or vertical in drilling. Compare 'end clinometer' or 'plain clinometer'. Long, 1920.

line drilling (for primary cut), in quarrying of dimensional stone: the method of drilling and *broaching* for the primary cut. In this method, holes are drilled close together in a straight line by means of a reciprocating drill mounted on a bar. The webs between the holes are removed with a drill or flat broaching tool; thus, a narrow continuous channel is cut. AIME, 1960.

line drilling (in controlled contour blasting), a method of controlling blast damage (or backbreak) by drilling a line of small diameter closely spaced holes at reduced burden behind the last row of main production blastholes in a blast. The shock energy from the main blast will, in theory, be sufficient to cause inter-blasthole splitting between the individual holes which have been drilled in line. Contour holes are normally drilled with a c-c distance of 0, 1–0, 2 m or 3–4 blasthole diameters apart.

line jumping, a premature initiation of a single hole leading to out-of-sequence initiation of a complete line of holes without any relief from the previous line as planned. Ford and Bonneau, 1991.

line lubricator, see '**line oiler**'.

line oiler, an apparatus inserted in a line conducting air or steam to an air- or steam actuated machine that feeds small controllable amounts of lubricating oil into the air or steam. Also called *airline lubricator, atomizer, line lubricator, lubricator, oiler, oil pot, (pineapple, pot, potato)*. Long, 1960.

liner pipe adaptor, is a casing adaptor. See '**adaptor**'. Atlas Copco, 2006.

liner plate, a steel plate segment, generally preformed to support a tunnel excavation. Bickel, 1982.

lining, a temporary or permanent concrete structure to secure and finish the tunnel interior. Bickel, 1982.

liquid explosives, numerous explosives materials are liquid. This applies, in the first place, to several nitric acid esters such as nitroglycrin, nitroglycol, diglycol dinitrate etc. Most of them are so highly sensitive to impact that they are converted to the less sensitive

246   liquid fuels – load distribution angle

solid state by absorption in nitrocellulose or gelatinization by it. As is well known, such processes form the subject of the pioneering patents of *Alfred Nobel*. It was shown by Roth that the impact sensitivity of explosive liquids is considerably enhanced if they contain air bubbles. Meyer, 1977.

liquid fuels, fuels in a liquid state used in combination with oxidizers to form blasting agents and/or explosives. IME, 1981.

liquidity index (I_L), (dimensionless), numerical difference between the natural water content and the plastic limit expressed as a percentage ratio of the plasticity index.

$$I_L = (w_m - w_P)/I_P \tag{L.2}$$

where I_L is the liquidity index (dimensionless), w_m is the water content in mass-%, w_P is the plastic limit in mass-% and I_P is the plasticity index in mass-%. SS-EN ISO 14688-2:2004, 2003.

liquid limit (w_L), (mass-%), *water content* (definition see '**water content**') at which a fine soil passes from the liquid to the plastic condition, as determined by the '*liquid limit test*'. SS-EN ISO 14688-2:2004, 2003.

liquid oxygen explosive (LOX), an explosive fabricated by packing sawdust, or other suitable carbonaceous materials, into a cartridge and dipping this into liquid oxygen before using it in blasting. This seldom used explosive becomes innocuous upon misfire. In the USA lamp black or carbon black was often the fuel of choice. The fuel together with oxygen forms the velocity of detonation (*cd*) of about 5 500 m/s and LOX is therefore regarded as a high explosive. The presence of liquid oxygen makes it one of the most hazardous explosives to handle. This explosive has been used in collieries as well as in quarries in India.

Livingstone crater formula, see '**crater blasting**'.

LM (acronym), see '**longwall mining**'.

load distribution angle, the half top conical angle of rock (60°) influenced by the stress from a pretensioned rock bolt. See Fig. 43.

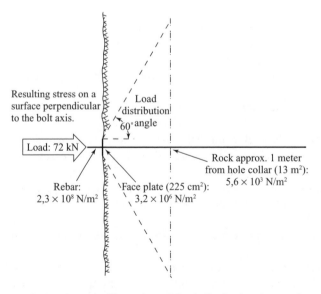

Figure 43 A schematic illustration of load distribution into rock as a result of rock bolt pre-tensioning. After Stillborg, 1994.

load or **charge (*Q*), (kg)**, mass of explosive used in a blast.
loader, see '**charging equipment**'.
loader, a mechanical shovel or other machine for loading coal, ore mineral, or waste rock and dumping the material on trucks or on a conveyor. See also 'power loader, scraper loader, shaker-shovel loader, shovel loader, cutter loader, gathering-arm loader'. Nelson, 1965. A machine for loading materials such as ore or coal into cars, trucks, conveyors or other means of convenience for transportation to the surface of the mine. ASA C42.85, 1956.
load haul dump unit (LHD), a machine to load, haul and dump broken rock. Normally a *front end loading machine* that can load a bucket, lift the bucket and transport the material to the dumping place.
loading, synonym to charging, see '**charging**'.
loading, the operation of a loader penetrating into a pile of rock (muckpile), filling the bucket with rock, and lifting the bucket with its content.
loading density (ρ_{ec}), **(kg/m³)**, see '**charged explosive density**'.
loading equipment, see '**charging equipment**'.
loading factor (q), **(kg/m³)**, see '**specific charge**'.
loading hose, see '**charging hose**'.
loading machine, see '**loader** or **charging equipment**'.
loading pan, a box or scoop into which broken rock is shoveled in a sinking shaft while the hoppit is travelling in the shaft. A small hoist is used to lift and discharge the pans into the hoppit at the shaft bottom. Nelson, 165.
loading pole or pole, see '**tamping bar**'.
loading tube, see '**charging hose**'.
load sensing hydraulic system, a regulation system for the hydraulic cylinders on the load haul dumper (LHD) that has two variable pumps working together. This provides exactly the right amount of flow and pressure at any time, distributing the power wherever and whenever it is needed. This system consumes little energy. A rather energy consuming system is the **open center hydraulic system** because the hydraulic pump constantly delivers a flow that is proportional to the speed of the engine. Atlas Copco, 2007.
local stress, the stress state in a small geological domain. Hyett, 1986.
lock-out assembly, assembly of part making up a device used to lock-out a function on a drill rig or other device. Atlas Copco, 2006.
lode, strictly a *fissure in the country rock filled with mineral*, usually applied to metalliferous lodes. In general miners'usage, a lode, vein, or ledge is a tabular deposit of valuable mineral between definite boundaries. Fay, 1920. A mineral deposit consisting of a zone of veins, veinlets, dissemination's; or any deposit which occurs in rock as opposed to a placer deposit. Bates and Jackson, 1987.
log, the registration of the changes of physical parameters along a specified distance.
log book, a book used for the notation of the changes of physical properties along a specified distance on a diagram sheet of paper.
long delay detonator, a detonator with a delay time larger than 500 ms used in half-second timing.
long hole blasting, an underground blasting technique in which holes with the length >20 m are used. The definition of minimum length required may change from time to time and from one mine to another.
long hole drilling, drilling of holes longer than 20 m underground. The definition of minimum length for the definition of a long hole may change from time to time and from one mine to another.

long hole drilling method (raises), see '**raise driving methods**'.
long hole drilling rig, any drill rig used for drilling long production blast holes in a mine. Normally from one level to the next. Atlas Copco, 2006.
long hole drill wagon, relatively simple and unsophisticated drill rig designed for drilling production blast holes in a mine. Atlas Copco, 2006.
long hole production drilling, process of drilling holes of extended length to excavate ore in certain mining methods. Atlas Copco, 2006.
longitudinal component, that component of vibration which produces motion in the direction of a line joining the vibration source and the seismograph. Konya and Walter, 1990.
longitudinal permeameter, see '**permeameter**'.
longitudinal pulse permeameter, see '**permeameter**'.
longitudinal trace, the line on the vibration record that records the longitudinal component of motion. Konya and Walter, 1990.
longitudinal wave, dilatational wave, compressional or **primary wave (P-wave)**, a mechanical wave in which the displacements are in the direction of wave propagation. Because of the fact that this wave has the highest velocity of the mechanical waves, it is called the primary wave (P-wave). See also '**seismic waves**'.
longitudinal wave velocity, (dilatational-, compressional- or **primary wave (P-wave) velocity)**, (c_p), **(m/s)**, the velocity of the longitudinal-, dilatational-, compressional- or primary wave from a disturbance of matter. See also '**seismic waves**'.
long round, a tunnel or drift round with a depth larger than 5 m in 2007. In the AECL Research Lab in Canada rounds up to 8,6 m have been blasted with success. Kuzyk, 1995. Long rounds are used with success at LKAB.
longwall advancing (LA), a system of longwall mining in which the faces advance from the shafts towards the boundary or other limit lines. In this method, all the roadways are in the worked-out-areas. Nelson, 1965.
longwall mining (LM), a coal mining method where the coal is cut along a horizontal line continuously and transported to surface and this causes the overburden to cave at a certain distance from the front. This will cause a small depression on the surface. Longwall mining is the most common coal mining method. The mining (working) of coal seams is believed to have originated in Shropshire, England, towards the end of the *seventeenth century*. The seam is removed in one operation by means of long working face or wall, thus the name. The workings advance (or retreat) in continuous line, which may be several hundreds of meter in length. The space from which the coal has been removed (called the *gob, goaf* or *waste*) is either allowed to collapse (caving) or is completely or partially filled or stowed with stone and debris. The stowing material is obtained from any dirt in the seam and from the ripping operations on the roadways to gain height. Stowing material is sometimes brought down from the surface and packed by hand or by mechanical means. See also '*longwall advancing*' and '*longwall retreating*'. Nelson, 1965. The longwall mining normally causes a subsidence of ground on the surface if not the mined areas are more or less completely filled with rock or other debris. Rustan, 2007. Longwall mining extracts the coal or ore along a straight front, with large longitudinal extension. The mining area close to the face is kept open, to provide space for personnel and mining equipment. The roof may be allowed to subside at some distance behind the working face. Development involves excavation of a network of haulage drifts, for access to production areas and transport of ore to shaft stations. As mineralization extends over a large area, haulage drifts are paralleled by return airways,

for ventilation of the workings. Haulage drifts are usually arranged in regular patterns, and excavated inside the coal or ore. Coal and gold longwall production techniques are similar in principle, but quite different in term of mechanization. In the coal mine shearers shuttle back and forth along the face, cutting coal and depositing it on chain conveyors. The gold reef conglomerate is much harder, and difficult to tackle. South African god mines have developed their own techniques, using handheld pneumatic rock drills in reefs as thin as 1,0 m, which constitutes a great challenge for equipment manufacturers to mechanize. Pillar of timber or concrete are installed to support the roof in the very deep mines. Longwall mining applies to thin bedded deposits, with uniform thickness and large horizontal extension. Typical deposits are *coal seams*, *potash layers* or *conglomerates*, and *gold reefs*. Atlas Copco, 2007.

longwall retreating (LR), a system of longwall mining in which the developing headings are driven narrow to the boundary or limit line and then the coal seam is extracted by longwall faces retreating towards the shaft. In this method, all the roadways are in the solid coal seam and the waste areas are left behind. Nelson, 1965.

longwall shearer, an excavation device, working at the face of a coal mine using the longwall mining method, equipped with narrow, rotating, pick equipped cutter heads, whose axes are perpendicular to the coal face and which are sometimes mounted on moveable hydraulic booms. This device is traversed across the coal face, while the cutter heads are rotating, to cut the coal.

lookout, in drifting and tunnelling, it is the practice of drilling the peripheral blastholes with a clearance angle (outward looking) for the drilling machinery, with respect to the axis of the tunnel, so that the subsequent rounds can be drilled to provide a constant minimum transverse dimension.

look-out angle, outward angle of the contour holes to the tunnel direction with the purpose of providing space for the rock drill when drilling the next round.

look-out assembly, hydraulic function on a drill boom that allows the feed to be tilted in such a way that holes can be drilled at an angle to the tunnel axis. Atlas Copco, 2006.

loose block, is created by open cracks, joints and other discontinuities that enable the block to dislodge.

loose block detector, an instrument to scan if a rock block is loose or not in the roof and walls in drifts and tunnels. One approach is to transmit an *artificial* vibration to the rock surface by for example a scaling bar combined with an electronic ear to analyse the sound when the scaling bar hits the rock surface. If the block is loose, the frequency in the reflecting signal will be lower.

loose block indicator, see '**loose block detector**'.

loosening drawbody, term used in *gravity flow of course material* to define the volume of rock mass that has been affected by a smaller or larger displacements. Its shape is like a drop turned upside down and can be calculated by the Bergmark-Roos formula derived from natural laws. See also '**angle of draw**'. Rustan, 2000.

loosening ellipsoid, term used in *gravity flow of fine material* in bunkers to define the volume of material that has been affected by a smaller or larger displacement when a certain volume has been drawn. Its shape is close to a rotational ellipsoid. Kvapil, 2004. The ellipsoid shape is only valid for fine materials like sand. For mines, where the fragmentation range is large, the shape is more like a '**drop turned upside down**'.

loose rock, rock blocks detached from the rock mass because of blasting cracks or discontinuities.

Los Angeles testing machine (abrasion), is in principle a *small ball mill* where the material is ground for a certain time and gives an idea of the *abrasion resistance*. It consists of a closed hollow steel cylinder 700 mm in diameter and 508 mm long mounted for rotation with its axis horizontal. The sample being tested and a charge of steel spheres are tumbled during rotation by an internal shelf. The abrasion values achieved are normally used in the construction industry road-, and railway bank material.

Love wave, a seismic surface shear wave which can only exist near a surface in a layer which covers a halfspace with dissimilar materials. The Love wave induces a horizontal particle motion transverse to the direction of wave propagation.

Love wave velocity (c_L), (m/s), the propagation velocity of Love waves. See '**Love wave**'.

Love wave (Q-waves), Love waves, also called Q-waves, are surface seismic waves that cause horizontal shifting of the earth surface during an earthquake or blast. The particle motion of the Love wave forms a horizontal circle or ellipse moving in the direction of propagation. The maximum particle motion decreases rapidly as one examines deeper layers. The amplitude on surface decays by $1/\sqrt{r}$ where r is the distance from the earthquake or blast. Lowe wave travels with a lower velocity than P-waves and S-waves but faster than Rayleigh waves. The existence of this wave form was predicted by A.E.H. Love mathematically in 1911.

low bench blasting, see '**levelling**'.

low brisance explosives, see '**brisance explosives**'.

low-density dynamites, dynamites containing up to 80% ammonium nitrate as the principal explosive ingredient. They have a minimum shattering effect and are designed for soft ores, soft limestone, and gypsum, and provide a maximum length of explosive column per unit weight of explosive. Lewis, 1964.

low density explosive, the density of an ordinary permitted explosive can be decreased in three different ways by 1) Loose packing, 2) An alteration in the granular state of the ammonium nitrate 3) The mixing of the explosive with bulking agents like sawdust, wood meal, bagasse, perlite, rice hulls, vermiculite and expanded polystyrene beads (EPS). These explosives are designed for the use in the mining of coal of soft and medium hardness, where it is required to blast with the least amount of shattering or in controlled contour blasting. Silvia and Scherpenisse, 2009. Example of low energy explosives are ammonium nitrate and fuel oil(ANFO) diluted by styropor beads and called Isanol when used in Norway. Heltzen and Kure, 1980. Another low density explosive is ammonium nitrate prills mixed with rubber and called ANRUB and used in Australia. Harries, 1993.

low energy detonation cord (LEDC), detonating cord with low linear charge concentration (less than 2 g/m) and minimal environmental effect. It is used to initiate non-electric blasting caps, usually at the bottom of a borehole. The use of LEDC guarantees that the explosive will not be affected by the detonation cord.

lower hemisphere equal angle projection, a method to represent joints on an imaginary hemisphere positioned below the plane of projection so that its circular face forms the projection circle. Hudson, 1993.

lower hemisphere equal area projection, a method to represent joints on an imaginary hemisphere positioned below the plane of projection so that its circular face forms the projection circle. Hudson, 1993.

lower yield point, the point where a metal under tensile strength begins to yield prior to breaking. Atlas Copco, 2006.

low (energy) explosive, an explosive which is characterized by deflagration, i.e. a low rate of reaction and the development of low pressure. The thermochemical transition of the explosive into a gaseous state occurs by burning and not by detonating as it is the case for high explosives. Blasting powder (synonymous to black powder and gun powder) is the only low explosive in common use. It does not require a blasting cap, but is ignited by means of a safety fuse. It is also called a propellant.

low freezing explosive, a particular explosive made from a low freezing point mixture and for use under cold conditions.

low incendivity explosive, an explosive that will not create an open flame outside a blast-hole. These explosives are used when blasting in dust generating non-coal mines, e.g. in sulphide or oil shale mines, where mineral dust can be initiated by an open flame and cause an explosion, causing damage to ventilation ducts and stoppings or ventilation doors. One critical factor affecting sulphide dust explosion is the sulphur content of the ore; ores of higher grade are more prone to explosion in that it requires less amount of particles suspended in the air to propagate an explosion. Weiss et al., 1992.

low-order detonation, see '**low order reaction**'.

low order reaction, occurs when the chemical reaction of an explosive and detonation velocity degrades to near deflagration without halting. The energy may be fully released, but the effect will be low-shock. Cunningham, 2007.

low profile mining, mining with mechanized equipment at roof heights varying from 1 m to 2,5 m.

low velocity detonation (LVD), a detonation which occurs at a low speed of propagation. Experimental measurements of the sonic wave speed and of the detonation wave velocity of a variety of materials have demonstrated that a necessary condition for stability in a low velocity detonation is that the detonation wave should be subsonic in relation to the wave velocity of the container material. If this condition is not met, a pulsating detonation will occur. Cook, 1974.

LOX (acronym), see '**liquid oxygen explosive**'.

lubricator, see '**line oiler**'.

lump, an aggregation, collection or clump of (large) fragments.

lump ore, an iron ore quality sold consisting of iron ore up to a size of about 100 mm.

lug, a replaceable cutting member on a 'expansion reamer'. Long, 1920.

lug (drilling), an ear on an adaptor. Sandvik, 2007.

lug (lifting), projecting part of an object often used as fastening point for lifting it. Atlas Copco, 2006.

LVD (acronym), see '**low velocity detonation**'.

LVDT (acronym), see '**linear variable displacement transformer**'.

LWM (abbreviation), for '**longwall mining**'.

macadam, crushed stone of regular sizes below 76 mm (3 inch) often used as road base or bound in bitumen to form a paved roadway.

machine gun drill, steel balls are used to impact the rock surface at high velocity ~30 m/s. The balls are recirculated with the flushing media until they are totally weared down. The energy efficiency is, however, only 4%. A 250 mm diameter hole in marble was drilled with a velocity of 2,5 m/hour. KTH, 1969.

macrocrack, a crack with a length comparable to the (smallest) characteristic structural dimension (e.g. thickness). A crack which can be seen by the naked eye.

magazine, a building, structure, or container especially constructed for storing explosives, blasting agents, detonators, or other explosive materials. USBM, 1983.

magazine shoes, shoes specially made without iron or steel in the soles and heels for wearing in magazines. AS 2187.1, 1996.

magazine, surface, a specially designed and constructed structure for the storage of explosive materials. Special rules apply for the construction and the location of a magazine.

magazine, underground, a specially designed and constructed space underground for the storage of explosive materials. Special rules apply for the construction and location of magazines.

Magnadet (tradename), a device to initiate electric detonators by a transformer coupling outside the blasthole. See '**detonator transformer coupled**'.

magneto or **dynamo exploder**, a blasting machine that produces an electric current to initiate electric detonators. Principle 1. By punching down a rack bar that through set of gears, spins the rotor in a DC generator. The electric energy from the generator is connected to the output terminals when the rack bar reaches the bottom where it closes a switch in its downward travel. Principle 2. A spring is first wound and with the help of the key, it is released suddenly. Easy to operate units, in which a set of gears, spins the rotor in a DC generator.

magneto restrictive cell, a cell developed by Hast in Sweden in 1950 and used for force displacement measurements in combination with the overcoring rock stress measurement method. Stephansson, 1993.

magnification factor (ground vibrations), used to assess the likelihood of resonance and is given by the following equation.

$$MF = \frac{1}{1-\left(f_i/f_n\right)^2} \quad \text{(M.1)}$$

where MF is the magnification factor, f_i is the frequency of the incoming blast wave in Hz and f_n is the natural frequency of the structure observed in Hz. Beer, 2004.

main explosive charge, the explosive charge which is expected to perform the work of blasting. The main charge usually consists of ANFO, dynamite, slurry/water gel, emulsion or a blasting agent which fills the borehole as compared to the primer or booster.

main gallery, usually a horizontal drift located along the strike of an orebody and connected to crosscuts.

main level, a level where ore is collected from shafts and transported by rail or trucks to a central crusher station.

mains firing, the use of electric power from a mains power supply for firing charges. AS 2187.1, 1996.

mains firing box, see '**mains firing**'. AS 2187.1, 1996.

major apex, the shaped structure or pillar formed between two adjacent drawpoints.

makeup gun, see '**breakout gun**'.

makeup tongs, a *heavy wrench*, usually mechanically actuated, used to couple or uncouple drill rods, drill pipe, casing, or drive pipe. Also called 'breakout tongs'. Long, 1920.

male thread, a thread outside of a drill rod. Synonym to 'pin thread'.

man cage, a special cage for raising and lowering men in a mineshaft. Fay, 1920.

manufacturing codes, code markings stamped on explosive materials packages indicating, among other information, the date of manufacture. Atlas Powder, 1987.

manway or **ladderway**, a raise or winze, vertical or inclined, for the accommodation of ladders, water and compressed air pipes, ventilation ducts etc for the purpose to give access to a stope.

marsh funnel test, a method to test the flow velocity of the drill cutting transportation media. The material being tested is filled through a screen into cone. The purpose with the screen is to prevent larger particles to flow into the cone. The cone has an outlet at the bottom that is sealed by a finger. The cone is hold vertical and the finger is removed and the time for filling 946 ml (1 US quarter) in a cup is taken. A second reading is taken who long time it takes to empty the cone. Stein, 1977.

mass (*m*), **(kg)**, a quantitative measure of a body's resistance to being accelerated. Measured in units of kilogram (kg) which is defined as the mass of the kilogram prototype in Sèvres, Paris. (Old unit: 1 pound (lb) = 0,454 kg).

mass blast, see '**mass blasting**'.

mass blasting, a term used for large-scale underground mining methods like sublevel stoping with large diameter holes underground >100 mm in diameter. In a mass blast several 100 kt is blasted in one blast but divided into different delays to keep the ground vibration problem under control. Especially the blasting of rib pillar and crown pillar causes the need of large amount of explosives in one single blast. Mass blasting has been used mainly in Australia, Canada, Chile and Sweden.

mass conservation, see '**conservation of energy laws**'.

mass detonate (mass explode), the explosive materials mass detonate (mass explode) when a unit or any part of a larger quantity of explosive material explodes and causes all or a substantial part of the remaining material to detonate or explode simultaneously. With respect to detonators, 'a substantial part' means 90% or more. IME, 1981.

mass detonation (mass explosion), act of mass detonation or mass explosion.

mass flow, the mechanism by which a volume of broken ore or rock due to gravity forces moves downwards uniformly, with the same vertical velocity and over the whole area, during draw.

massive (structure), homogeneous structure without stratification, flow-banding, foliation, schistosity, etc. Fay, 1920.

massive deposit, ore deposits are roughly divided into *massive* and *tabular deposits*. Massive deposits are developed in three dimensions with variable shapes. Tabular deposits are principally developed in two dimensions although some may have considerable thickness Higham, 1951.

massive rock, a rock mass having 1–3 joints per metre. See '**joint spacing**'.

MassMin (shortened form), for the International '**Mass Mining Conference**'. The First MassMin Conference was hold in Denver USA in 1982, 2nd Johannesburg South Africa in 1992, 3rd Brisbane Australia 2000, 4th Santiago Chile in 2003 and the 5th in Luleå Sweden in 2008.

mass mining, underground mining methods using large scale mining technique like in block caving, longwall mining, shrinkage stoping, sublevel caving and sublevel stoping.

mass of bottom charge (Q_b), (kg), the charge in the bottom of a blasthole which normally has a higher strength than the column charge.

mass of column charge (Q_c), (kg), the charge in the column of a blasthole which normally has a lower strength than the bottom charge.

mass of explosive charge (Q), (kg), the mass of an explosive charge. Don't use weight which has the unit (N).

mass rate of flow (q_m), (kg/s), mass flow per unit of time.

mass shooting, simultaneous detonation of charges in a large number of holes in contrast to firing in sequence with delay caps.

mass strength (MS), (s_{ma}), (MJ/kg), the strength of a unit mass of explosive used to compare to other explosives. See also '**absolute mass strength**'.

mast, tall open metal structure mounted on a drill rig to allow the adding and removal of drill rods. Atlas Copco, 2006.

mat, see '**blasting mat**'.

match fuse, see '**quick-match**'. AS 2187.1, 1996.

match head detonator, see '**electric igniter**'. AS 2187.1, 1996.

material damping, see '**attenuation**'.

Mathews stability graph, is a logarithmic diagram showing the relation between the *Q-factor* (Norwegian Geotechnical Institute (NGI) stability number 0,001–1000) and shape factor or *hydraulic radius* (1–100), see Fig. 44.

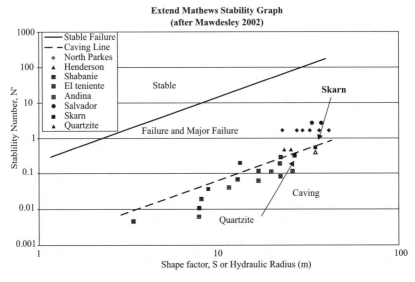

Figure 44 Mathew stability graph. MassMin, 2008.

Two, almost straight parallel lines in this diagram divide the area into three zones, a stable zone, a failure and major failure zone and a caving zone. The diagram is used to determine the minimum undercut area to start block caving. MassMin, 2008. See also '**Laubscher stability diagram**' that uses MRMR system for rock mass classification.

max (shortened form), for '**maximum**'.

maximum burden (B_{max}), (m), the largest burden distance which can be used for good fragmentation if all boreholes are situated in their correct positions (no borehole deviation). Because 'good fragmentation' is not properly defined in quantitative terms the use of the term maximum burden distance is not recommended.

maximum charge per delay interval (Q_{max}), (kg), largest amount of explosive per delay interval (kg/delay) in a round. The minimum delay time for charges considered to detonate separately is 8 ms (according to the recommendations of USBM).

maximum cooperating charge, see '**maximum charge per delay (interval)**'.

maximum (recommended) firing current (I_{max}), (A), the maximum value of the current (amperage) recommended for the safe and effective performance of an electric blasting cap. Too high current may damage the detonator and this may result in a misfire.

maximum fragment size (k_{100}), (m), the largest squared dimension of a fragment in a fragment size distribution. It is difficult to give an appropriate definition. The following procedure is suggested. Put the largest fragment on a horizontal surface and project it on the surface. Find the smallest square mesh where the projected surface will pass. This mesh size is defined as k_{100}.

maximum instantaneous charge (MIC), see '**maximum charge per delay**'.

maximum (peak) particle velocity (v_{max}), (mm/s), the maximum resultant peak particle velocity when measured in three orthogonal directions. It is also acronymized to PPV.

maximum radial strain (of the longitudinal wave) (ε_{rmax}), the maximum radial strain induced by the longitudinal wave emanated from an explosive charge placed in a blasthole in rock and completely filling the area of the blasthole. The maximum strain in the vicinity of a blasthole can be calculated by the following formula, Atchison, 1964 (a and b) and Nicholls, 1965.

$$\varepsilon_{rmax} = K_c \left(\frac{r}{R}\right)^{k_r} \tag{M.2}$$

where ε_{rmax} is the maximum radial strain of the longitudinal wave, and K_c is a constant that depends on the detonation pressure of the explosive and is called the *scaled strain intercept*. R is the radial distance from the blasthole (m), r is the radius of the blasthole (m), k_r is a constant depending on the physical properties of the rock, e.g. for Lithonia granite, $k_r = 1,88$, for Bucyrus limestone, $k_r = 2,10$, and for Marion limestone, $k_r = 2,64$.

maximum resultant ground vibration velocity (v_{Rmax}), (mm/s), the maximum resultant ground vibration velocity. See '**resultant ground vibration velocity**'.

maximum S-wave component of particle velocity (v_{Smax}), maximum of the S-wave component in a ground vibration.

maximum vertical component of particle velocity (v_{vmax}), (mm/s), the maximum value of the vertical particle velocity in any structure.

Maxwell liquid, the rock mass can be simulated as a Maxwell liquid. This model shows irreversible creep and a complete relaxation of stresses. This feature is not typical for the majority of rocks, which are mostly characterized by only a partial relaxation of stresses. Hudson, 1993.

MBI (acronym), see '**mechanised bolting unit**'.

mean fragment size (k_{50}), (**m**), that square mesh where 50% of the mass or volume of the fragmented material will pass.

measurement while drilling (MWD), a technique where the properties of the rock monitored and measured intermittently or at intervals are correlated to drill parameters like thrust, torque, penetration rate, etc.

measuring bolt or **instrumented rock bolt**, a rock bolt equipped with strain gages to measure the load on it for research purposes. TNC, 1979.

mechanical equipment for rock bolt installation, the equipment can be divided into three groups,
- A **Manual equipment.** Very light hand held drilling and installation equipment.
- B **Semi-mechanised equipment.** Mechanized drilling and manual equipment for rock bolt installation.
- C **Fully mechanised equipment.** Mechanized drilling and rock bolt installation controlled from an operator panel placed under a safe roof or in a cabin.

mechanical feed, see '**screw feed**'.

mechanical impedance (Z_m), (**Ns/m**), is defined as the ratio of the force F given in N divided by the velocity v in m/s and is a measure how much a structure resists motion.

$$Z_m(f) = \frac{F(f)}{v(f)} \qquad (M.3)$$

Its inverse is known as the *mobility* or *mechanical admittance*. The mechanical impedance is a function of the frequency f of the applied force and can vary greatly over frequency. At resonance frequencies, the mechanical impedance will be lower, meaning less force is needed to cause a structure to move at a given velocity. *Observe that mechanical impedance is differently defined compared to acoustic impedance!* There are three different kinds of impedances, *acoustic, electrical* and *mechanical*.

mechanical mucking at shaft sinking, the loading of rock or dirt by machines. Two main methods of mechanical mucking are in use in shaft sinking, 1) cactus grab used in shaft sinking 2) crawler mounted rocker shovel loaders. The cactus grab has a capacity up to about 0,8 m³. Some engineers prefer a circular ring on which the grab moves round the shaft bottom, others adopt a boom from which the grab is suspended and placed immediately above the broken rock. However, there is general agreement that if fast sinking rate is required, a cactus grab is essential. In tunnels, a fairly wide rang of machine are available of mechanical mucking, the most common type being the shovel loader. Nelson, 1965.

mechanical notching tool, a tool to create two diametrically located notches along a borehole before blasting. This technique has been tested in controlled contour blasting and its advantage is that borehole pressure and spacing between contour holes can be reduced.

mechanical scaling, the scaling with machines using impact or ripping tools instead of doing it by a scaling bar manually. See also '**mechanised scaling**'.

mechanical set bit, a diamond bit produced by mechanical methods as opposed to hand setting methods. Long, 1920.

mechanised bolting, can be divided into three mechanisation levels 1) *Manual drilling and bolting* using light hand held rock drills, scaling bars and bolt installation equipment in small drift and tunnels, 2) *Semi-mechanised drilling and bolting* using a hydraulic drill jumbo, followed by manual installation of the bolts by operators working from a platform mounted on the drill rig or a separate vehicle and 3) *Fully mechanised drilling and bolting* with a special designed rock bolter machine where drilling and installation of bolts can be made from the cabin of the machine.

mechanised bolting unit (MBU), a special drill rig built just for the purpose of reinforcing rock by bolts. The drilling of the hole, the instalment of cartridges (cement or resin) and bolts are performed from the cabin. Atlas Copco, 2007.

mechanised charging equipment, charging equipment where cartridges, liquid or pulverized explosive is transported with pumps or compressed air into the blasthole.

mechanised drilling, drilling performed with machines instead of manually. Sandvik, 1983.

mechanised explosive charging, is based on different charging principles, 1) compressed air for the transportation of pulverized explosives like ANFO or plastic cartridges like dynamite, 2) pumps to pump slurry and emulsion explosives 3) a mechanical climber to lift cartridges in up-holes larger than 100 mm in diameter. Principles 1 and 2 are used both underground and on surface whereas principle 3 is used only underground and in up-holes.

mechanised rock bolting, the drill rods and the insertion of rock bolts and cartridges are handled automatically. Also a special arm for the rock reinforcements screens can be added to the equipment.

mechanised rod handling, the drill rods are stored in a magazine and could automatically be added or reduced from the drill string without handling by hand. Tamrock, 1983.

mechanised scaling, a hydraulic boom to which an impact hammer is mounted is operated from the cabin mounted on a special designed vehicle. Three principles can be used for scaling, 1) *breaking*, 2) *scratching* and 3) *impact*. The hydraulic arm is operated from a sound isolated cabin. Stillborg, 1994.

mechanised stemming equipment, in open cast mines for example the use of a front end loader equipped with hydraulic cylinders to move the vessel for stemming to the top of the boreholes. A 315 mm blasthole in diameter can be stemmed in 15 min. Pradhan, 2001.

megalith, a huge undressed stone used in various prehistoric monuments such as the Menhir, Dolmen etc. (Lith from the greek lithos = stone and mega = large).

melting drill, an *exotic drilling method* using a cutting tool provided with a tungsten element for heating of the tool to 1 200–1 600°C. Under pressure the cutting tool is pressed into the rock that is melting away and is pressed through a short tube. When the melted rock is leaving the tool, it is cooled by helium or nitrogen gas and is solified to small balls which are blown out of the hole. KTH, 1969.

merge split, in image analysis a term used to denote the process of moving fragments closer to or away from each other.

meshing, the process of fastening steel mesh to rock bolts or short rock anchors.

mesh screen, square meshes made of reinforced steel.

mesh washer, specially designed perforated plate used to fasten wire mesh to rock bolts in the support of underground openings. Atlas Copco, 2006.

meta, a prefix which, when added to the name of a rock, signifies that the rock has undergone a degree of change in mineral or in chemical composition. One example of the prefix is the use in metamorphism. Fay, 1920.

metal jet, liquid metal with a drop-like shape and high velocity formed by conical or linear shaped explosive charges.

metallic slitter, a device containing a sharp edge, such as a razor blade, used for slitting open fibreboard cases. Atlas Powder, 1987.

metallized explosive, explosive sensitized or energized with finely divided metal flakes, powders, or granules, usually of aluminium. USBM, 1983.

meters drilled (L), (m), number of meters drilled during a certain time. Sandvik, 2007.

metric system, see 'Notes for user in beginning of this dictionary'.

MGr (abbreviation), for '**medium gravel**' defining fragments with sizes from 6,3 to 2,0 mm.

MIC (acronym), for '**maximum instantaneous charge**', see '**maximum charge per delay**'.

Michelson line or **Rayleigh line**, the relation between pressure and volume inside and after the reaction zone of an explosive based on the general conservation relations. The relation is linear, the slope of which is proportional to the product of (density of explosive and detonation velocity)^2. This can also be expressed by (explosive impedance)^2.

Michigan cut or **burn cut**, a cut with parallel boreholes. In the USA a cut which consists of drilling a hole with a large diameter, or a number of holes of smaller diameter at the centre of the heading and parallel to the direction of the tunnel. These holes are not charged. The remaining holes are then broken out towards these holes. Fraenkel, 1954.

micro-balloons, hollow spheres (e.g. made of glass) which are added to explosive's to increase their sensitivity and blasting strength by the creation of hot spots. This technique is used for example in emulsion explosives.

microcrack, a crack of small size when compared with a characteristic dimension of the structure of the material; often not visible to the naked eye. A crack-like opening in rock with an extension in one dimension much smaller than the planar extension and with a ratio of width to length, crack aspect ratio of less than 10^{-2} and typically 10^{-3} to 10^{-5} in rock mechanics. The length of a microcrack typically is in the order of the grain size or in the order of a couple of mm or less. More often it is found on a micro structural scale. Microcracks can be subdivided into *grain boundary cracks* (occur along grain boundaries), *transgranular cracks* (lying on the cleavage plane in a single grain), *intergranular cracks* (extending along a grain boundary into the two adjacent grains), and *multi-grain cracks* (cracks crossing several grains). Simmons and Richter, 1976.

microcrack density (ς_{mic}) (mm^{-1}), on micro level is defined for each grain as the sum of the intragranular and transgranular crack length in mm divided by the area of the grain. Montoto, 1982.

micro discontinuity, microcracks with variable length and usually with a width less than 1 μm located in intra, inter or transgranular positions. Miguel, 1983e.

micro failure, failures in solid bodies so small so a microscope is needed to see them.

microfissure, use the term '**microcrack**' instead because fissures cant be small due to its definition, see 'microcrack'.

micro fracture, fractures so small so a microscope is needed to see them.

micro fracture model, see '**NAG-FRAG micro fracture model**'.

micro joints, joints so small so a microscope is needed to see them.

micro pile, rock bolt used to stabilize soil.

micro pile drilling, drilling holes to insert micro piles. Several types of rock bolts can be used as micro piles. Atlas Copco, 2006.

micropile, see '**micropiling**'.
micropiling, a method of stabilizing soil by installing numerous micropiles vertically. The drill rod (length 15 m) and bit is one unit. The micropile system system is able to drill through a wide range of material from sand and clay to rock. The piles work as both tension and compression piles and can be installed on a vertical or inclined angle to provide group pile solutions. During installation, cement is simultaneously injected into the rod, out through the bit and distributed evenly into the surrounding ground. The rod and bit are then left in place to act as the pile's reinforcement element. Once the cement sets, the ground consolidates and forms a friction pile. Atlas Copco, 2009.
micro seismic activity also **called acoustic emission (AE)**, see 'acoustic emission'.
micro-sequential contour blasting, (*cut blasting initiation*), a controlled contour blasting method where the delay time is very short between the contour holes (~1,5 ms in tunnel blasting). The method is effective in reducing ground vibrations. For calculation of the delay time, see Rustan et al., 1985.
micro wave drill, an *exotic drilling method* using radar waves 2 500 MHz fed through a copper conductor to the rock surface where the rock is fragmented by heating. With a 10 kW source, boulder can be fragmented. KTH, 1969.
middle cut or **intermediate cut**, a machine cut in the midsection of a coal seam; sometimes adopted in thick seams (over 1,2 m) with a layer of dirt or inferior coal in the middle. A middle cut would be made with a turret coal cutter. Nelson, 1965.
mil, a thousandth of an inch.
mild detonating cord, see '**low energy detonating cord**'.
mild detonation fuse, see '**low energy detonating cord**'.
mill bit, see '**rose bit**'.
milling bit, see '**rose bit**'.
millisecond delay blasting, see '**short delay blasting**'.
millisecond delay detonator (MS), short delay detonator with a delay time between 0 and 500 ms. The detonator has an interval of less than 150 ms to the previous or the following delay number.
min (shortened form), for '**minimum**','**minute**' or '**minutes**'.
mineral, a naturally occurring inorganic compound with a definite chemical composition, characteristic crystalline form, and molecular structure.
mineral engineering, term covers a wide field in which many resources of modern science and engineering are used in discovery, development, exploitation, and use of natural mineral deposits. Pryor, 1963.
mine truck, a vehicle for transport of ore or waste. Mine trucks are often specially designed for underground use and the heights varies down to 1,5 m. Also surface trucks can be used as mine trucks underground.
miniaturized detonating cord or **low energy detonating cord**, detonating cord with a linear charge concentration of 1,05 g/m (5 grains/foot). Atlas Powder, 1987.
miniblast, small amount of explosives charged into short drillholes in boulders and initiated to break the boulder into smaller pieces. It is a *secondary blasting* method.
mini-dose, small charges (5–20 g) prepared on site by filling stiff plastic pipe, 20 mm in diameter with a dynamite explosive. The pipes can be connected to each other and used in borehole diameters 20–22 mm and borehole depths 0,35–0,5 m. The method is convenient for blasting in low benches and of boulders when the risk of flyrock is large. Recommended specific charge 0,02–0,04 kg/m^3. See also '**mini-hole blasting**'.

mini-hole blasting, the charges about 80 g, are prepared in small plastic wrappings of 17 × 275 mm with an prepared pocket for the detonator. The blasthole diameter recommended is 22 mm. These charges are used in certain environmental conditions such as in trench blasting for cables or pipelines, holes in rock for posts and beams etc. See also '**mini-dose**'.

mini pile drilling, drilling holes to receive supports that are larger than *micro-piles* but smaller than *standard piles*. Atlas Copco, 2006.

minimum booster test, a shock test for explosives where the minimum initiating charge (booster) for stable initiation is determined. This test is combined with measurement of the detonation velocity in a standard charge size. It is a part of production control. Persson et al., 1994.

minimum (recommended) firing current (I_{min}), **(A)**, the lowest level of firing current (amperage) necessary to initiate an electric blasting cap within a specified short interval of time. USBM, 1983.

minimum initiator, the smallest amount of the initiating explosive in a detonator that will reliably initiate the main charge. E.g. a standard No. 6 strength detonator which has a base charge of 0,35 g PETN is not sufficient for pneumatically charged ANFO, but is more than twice as strong as the one required for nitroglycerin and some water-based explosives. ICI Downline.

mining, the science, technique, and business of mineral discovery and exploitation. Strictly, the word connotes to underground work directed to severance and treatment of ore or associated rock. Practically it includes opencast work, quarrying, alluvial dredging, and combined operations, including surface and underground attack and ore treatment. Pryor, 1963.

mining geomechanics, the occurrence, conditions and progress of the mechanical processes inside the rock mass caused by mining activity. Ryncarz, 1989.

mining rate (v_m), **(m/day)**, how many, normally vertical, meters are mined of a block in block caving in a day.

mining rock mass rating (classification) (MRMR), a rock mass classification system developed by Laubscher, 1976. see '**rock mass rating**'.

minor apex, the shaped structure or pillar above and between drawpoint drifts, usually oriented lateral to, and smaller than, the major apex.

minor damage zone, see '**blast induced damage zone**'.

minor diameter or **box diameter** (d_i), **(mm)**, inside diameter of a bit. Sandvik, 2007.

misfire, the failure of an explosive charge to fire or explode properly when action has been taken to initiate it. Causes for misfire include unskilled charging, defective explosive, detonator or fuse, broken electric circuit and 'most dangerous' – the cutting-off of part or all of the charge due to lateral rock movement induced by the firing of other holes in the vicinity. Pryor, 1963.

misfired dynamite detector, see '**mite hunter**'.

misfire hole, see '**misfire**'.

misfired round, see '**missed round**'. Fraenkel, 1953.

misfire protection, e.g. special shatter proof glass installed in the windscreen of a drill rig to protect the operator from the accidental detonation of explosives remaining after the blast. Atlas Copco, 2006.

missed hole, see '**misfire**'.

missed round or **misfire**, a round in which all or part of the explosive has failed to detonate.

MIT (acronym), for '**Michigan Technology University**'.

mite hunter (*misfired dynamite detector*), a technique to indicate not detonated explosive after blasting developed in Japan (Taisei Corp). Permanent magnets, so called *misfired dynamite tracers,* are placed at the bottom of the blastholes and if the holes are not blasted properly the magnets will be intact and can be localized by a special instrument. Max detecting distance is about 40 cm.

MIT-model, a model showing how the strength of rock mass is varying with depth below surface and the fault zones. The model was developed at Michigan Technological University in the late 1970's. Hudson, 1993.

mixed explosive, a mixture of different explosives. Fraenkel, 1952.

mixed face, the portion of a tunnel where both rock and soft ground occur in the same cross-section. Bickel, 1982.

mixing and **charging truck**, a truck for both mixing and charging of the explosive on site.

mixing house building, a permanent, dedicated building used for the manufacture of explosives. AS 2187.1, 1996.

mobile boxhole rig, see '**raise driving methods**'.

mobile miner, a machine for excavating normally soft ores without drilling and blasting.

mobile mixing unit, a mobile unit (usually a vehicle) used for the manufacture of explosives. AS 2187.1, 1996.

modal analysis, the analysis of the percentage of minerals in rock samples.

mode of fracture, a geometric description of the basic deformation mechanism when a crack is subjected to loading, see Fig. 45.

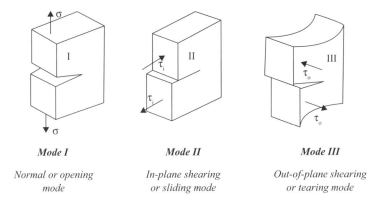

Figure 45 Modes of fracture.

In mode I, the opening mode, the crack surface displacements are perpendicular to the plane of the crack. This is the most commonly encountered mode of crack deformation in rock blasting applications. The mode I fracture toughness of a material is experimentally determined by means of standard laboratory fracture test specimens. In mode II, the sliding mode, the crack surface displacement occurs in the plane of the crack and perpendicular to the leading edge of the crack. In mode III, the tearing mode, the crack surface displacement occurs also in the plane of the crack but parallel to the leading edge of the crack.

Mode I fracture loading, is the normal or opening mode or tensile mode. See '**mode of fracture**'.

Mode II fracture loading, is the shearing or sliding mode of fracture. See '**mode of fracture**'.

Mode III fracture loading, is the out-of-plane shearing mode of fracture. See '**mode of fracture**'.

modified Griffith crack theory, see '**Griffith crack theory**'.

modulus of deformation (*D*), (MPa), when a piece of rock including joints is loaded uniaxially the axial strain ε will depend on the intact rock properties exclusive the joints but also on the joints. Mathematically this is expressed as

$$\varepsilon = \frac{\sigma_c}{E} + \frac{\sigma_c}{D} \qquad (M.4)$$

where σ_c is the compressive stress in MPa, E is the elastic modulus or Young's modulus in MPa and finally D is the modulus of deformation in MPa. The modulus D is similar to the modulus of permanent deformation. This module depends on joint orientation, spacing and deformability properties and can be calculated analytically. Hudson, 1993.

modulus of deformability, see '**modulus of deformation**'.

modulus of elasticity, elasticity modulus, or **elastic modulus, (MPa)**, the ratio of the increment of some specified form of stress to the increment of some specified form of strain, such as *Young's modulus (E)*, the *bulk modulus (K)*, the *shear modulus (G_{sh})* or the *torsion modulus (G_{tor})*.

modulus of elasticity in shear or *shear modulus (G_{sh})*, (MPa), see '**modulus of rigidity**'.

modulus of permanent deformation, see '**modulus of deformation**'.

modulus of rigidity (*G*), (MPa), is the ratio between shear stress and shear strain. Wikipedia, 2010. A measure of stiffness of a material subjected to *shear loading*. Usually the tangent (*G*) or secant modulus of elasticity (G_s) of a material in the torsion test. Also called *shear modulus of elasticity*; *modulus of elasticity in shear*; *torsional modulus of elasticity*; *modulus of elasticity in torsion*. H&G, 1965. See also '**shear modulus of elasticity**'

modulus of rupture in bending (σ_{fb}), (MPa), see '**flexural strength**'.

modulus of rupture in torsion (σ_{ft}), (MPa), maximum shear stress at the surface of a circular shaft computed by the torsion formula at the failure load.

Mohr-Coulomb failure criterion, under biaxial stress is defined as

$$\sigma_1 = k\sigma_3 + \sigma_c \qquad (M.5)$$

where σ_1 and σ_3 are respectively the major and minor principal stresses in the *x, y* plane in MPa, σ_c is the unconfined compressive strength of the rock in MPa and *k* (triaxial strength) is related to the angel of the internal friction (Φ) of the rock

$$k = (1 + \sin \Phi)(1 - \sin \Phi) \qquad (M.6)$$

and the cohesion *c* in MPa of the rock is related to σ_c and Φ by

$$c = \sigma_c \tan\Phi/(k-1) \qquad (M.7)$$

It is assumed that the stress in the z direction is a principal stress and is intermediate in value between σ_1 and σ_3. It is further assumed that σ_3 is limited by a maximum value for tension. Hudson, 1993.

Mohr-Coulomb-Navier law, a criterion governing shear failure used in block modelling. No tensile forces can act on joints and therefore negative normal forces are set to zero. Hudson, 1993.

Mohs' hardness scale, arbitrary quantitative units by means of which the *scratch hardness* of a mineral is determined. The units of hardness are expressed in numbers ranging from 1 through 10, each of which is represented by a mineral that can be made to scratch any other mineral having a lower-ranking number, hence the mineral are ranked from the softest, as follows; Talk 1, Gypsum 2, Calcite 3, Fluorite 4, Apatite 5, Orthoclase 9 Quartz 7, Topas 8, Corundum 9 and diamond 10. Long, 1920. Compare also with three other hardness scales *Vickers*, *Shore* and *Knoop*.

moil point, cylindrical pointed tool used in a pneumatic or hydraulic breaker and used to demolish hard material such as cement and asphalt. Atlas Copco, 2006.

mole, see '**full-face boring machine**'.

moment of force or **torque** (M), (Nm), the torque effect of a force on a body with respect to a point called the pivot point or fulcrum. In practice, the turning effect is commonly called leverage. For a single force, the cross product of a vector from some reference point to the point of application of the force with the force itself.

moment of inertia of mass (J), (kg/m^2), resistance (inertia) by a body against angular acceleration about a specific axis of rotation. Pryor, 1973. A simple definition of the **moment of inertia** (with respect to a given axis of rotation) of any object, be it a point mass or a 3D-structure, is given by:

$$I = \int r^2 dm$$

where I is the moment of inertia in kgm^2, m is the mass in kg and r is the perpendicular distance in m to the axis of rotation.

momentum (I), (Ns), see '**impulse**'.

monitoring systems in rock mechanics, monitoring systems for rock mechanic purposes are based on the following principles, 1) *Mechanical systems* 2) *Optical systems* 3) *Hydraulic and pneumatic diaphragm*, 4) *Electrical devices* 5) *Vibrating wire* 6) *Linear variable differential transformer* that can be used for different purposes like, a) *convergence measurements* with telescopic rod, invar bar or tape under constant tension, placed between two measuring point firmly fixed to the rock surface, b) *multiple point borehole extensometers*, c) *hydraulic pressure cells*, d) *micro seismic activity*.

monkey, see '**casing drive hammer**'.

monoblock drill, see '**integral drill steel**'.

monolith, a single stone or block of stone, especially one of large size, shaped into a pillar, statue, or monument. Webster, 1960.

monomethylaminenitrate, a chemical compound used to sensitize some water gels. USBM, 1983.

Monte-Carlo simulation, is defined as random sampling procedure of input data used to calculate a rock mass classification quantity like RMR. For example the RMR depends on several parameters and if these are varied randomly a frequency distribution of the RMR can be determined.

monumental stone blasting, special blasting technique employed when over shattering of rock is to be avoided for the production of rock blocks with minor damage and only for use in ornamental structure construction.

moraine, an accumulation of earth and stones carried and finally deposited by a glacier.

morphology, the science of classifying the form of external shapes of crystal, cracks, land and structure of organisms etc.

mortar, ballistic, see '**ballistic pendulum**'.

motor breaker or **tie tamper**, petrol powered breaker which can alternately be used for tamping work. Atlas Copco, 2006.

motor drill or **breaker**, gas (petrol) powered hand-held drilling and breaking equipment. It has a percussion mechanism with, and option of rotation, that can be used both to break or split a work object, or to drill holes, and is always used with an accessory such as a chisel, moil point or drill bit. It has a similar use as a pneumatic breaker but the percussive mechanism is driven by pressure from an internal combustion engine that is an integral part of the machine. Atlas Copco, 2006. Flushing is being provided by the exhaust gases or by compressed air produced in the machine. The total weight varies between 25 and 55 kg. Fraenkel, 1952.

motor hammer drill, see '**motor drill**'.

MOX (acronym), for '**metal oxidized explosive**' (USA). It is an explosive to which metallic powder has been added to increase the energy output of the explosive.

MPBX (abbreviation), see '**multiple point borehole extensometer**'.

MR (acronym), for 'mass rating', see '**Rocha classification**'.

MRMR (acronym), for '**Mining Rock Mass Rating**' a rock mass classification scheme developed by Dr D. H. Laubscher for use in mining, including block caving, applications. See '**rock mass rating**'. Laubscher, 1976.

MS (acronym), for '**mass strength**', see '**mass strength**'.

ms (abbreviation), for '**millisecond**'.

MSa (abbreviation), for '**medium sand**' defining particles sizes from 0,2 to 0,63 mm.

ms connector, a non-electric millisecond delay device used in combination with detonating cord to delay shots from the surface.

ms detonator, a detonator with minimum 25 ms delay time between intervals.

MSi (abbreviation), for 'medium silt' defining particles with sizes from 0,0063 to 0,02 mm.

muck, body or volume of rock or ore broken and displaced from its original position in the embedding rock mass by blasting or caving.

mucking, the operation of loading broken rock by hand or machine usually into shafts, drifts, stopes and tunnels.

muckpile, the pile of broken material or dirt in excavating that is to be loaded for removal.

mud blasting, see '**mud cap**'.

mud cap, adobe, dobie, plaster shot or **sandblast**, a charge of dynamite, or other high explosive fired in contact with the surface of rock after being covered with a quantity of wet mud, wet earth, or a similar substance. The slight confinement given to the dynamite by the mud or other material is responsible for that part of the energy of the dynamite is transmitted to the rock in the form of a shock. A mudcap may be placed on top or onto one side of, or even under a rock, if supported, with equal effect. Fay, 1920.

mud capping, the action of placing an explosive and covering it by mud on the surface of a boulder. See also '**mud cap**'.

mud drilling, usually a casing is driven when it is necessary to prevent a borehole from caving. Sometimes, particularly when drilling large diameter holes, it is preferable to stabilise the formation by filling the hole with mud. The idea is that the mud should fill cracks and caves and stabilize the borehole. Stein, 1997.

mud flush, see '**drilling fluid**'. Nelson, 1965.

mud rush, a sudden inflow of mud from a drawpoint or other underground openings. It may be followed by a large amount of water and is therefore very dangerous to men and equipment.

mud scow drilling, see '**scow drilling**'.

mule, see '**barney**'.

multigrain crack, a microcrack crossing several grains. See also '**microcrack**'.

multilevel monitoring of blast vibrations, advanced statistical technique that seeks to identify the hierarchical nature of various "error" components, which together cumulate to form the overall effectual error. This technique has been used on blast vibration data to reduce the scatter in data. Pegden, 2006.

multimeter, see '**blasting galvanometer**'.

multiple drill jumbo, a drilling vehicle with the possibility to use several drilling machines. Fraenkel, 1952.

multiple fuse igniter, a cardboard cartridge for igniting up to eight lengths of safety fuse simultaneously by means of a master fuse.

multiple grouting, a rock reinforcement method where the gluing substance, normally cement, for the cracks, and joints in the rock mass is injected through several holes instead of only one. TNC, 1979.

multiple-point bit, a drill bit with more than one insert. Sandvik, 2007.

multiple point borehole extensometer (MPBX), see '**extensometer**'.

multiple row blasting, blasting of several rows in a round.

multiple scabbing, when the outgoing compression wave from a blasthole in bench blasting is reflected in the free surface, the compression wave will change sign according to the reflection rules of waves and turn into a tension wave. Because rock is much weaker in tension than compression this may cause detachment of rock fragments from the free surface in one or several layers. Mr Clay called this fragmentation mechanism for the **release wave fragmentation zone.** The special term developed for this fragmentation is *scabbing*. There are formulas developed for the calculation of the velocity of the scabbing fragments and the number of scabbing layers. Observed initial fragment velocities are in the range 6–16 m/s corresponding to 0,04–0,11 MPa as peak pressure in the tensile wave. Clay, 1965.

Munroe effect, the jetting effect of a shaped explosive charge with a cavity in it.

mushroom starter, a device for the initiation of shock tubes by using a shot shell primer that is initiated by stomping on it.

MWD (acronym), see '**measurement while drilling**'.

NAG-FRAG computational fracture model, a computational model describing nucleation, growth, and coalescence of brittle microcracks and the production of fragments representing the brittle fracture in rocks. Gas penetration effects on crack growth, nucleation by shear strain, and a gradual crack-opening process are included in the model, and each can strongly affect the fracture process. Simulation and comparisons have been made with experiments for cratering in novaculite, quartz and basalt, radial cracking around boreholes in cylinders of oil shale, and laboratory and field explosions in ashfall tuff. The *micro fracture model* can treat fracture processes under widely varying material and fracture properties, cavity pressures, and scales. McHugh, 1980.

narrow flat undercut, a flat undercut whose vertical height is limited to approximately that of the drill drifts, often in the order of ~4 m.

narrow inclined undercut, an undercut that is both inclined and narrow.

narrow vein mining, see 'vein mining'.

NATM (acronym), see '**New Austrian Tunnelling Method**'.

natural arch, if a manmade cavern is made to a certain size underground it will continue to increase in size due to natural rock fall from the roof due to the increased rock stresses around the cavern. After a while this process can cease due to the stabilization of the rock mass due to the building of a natural arch like the stones in a stone bridge. Stresses are taken up by this arch and rock fall ceases. TNC, 1979.

natural frequency of a building (f_{nb}), **(Hz)**, the frequency at which the building vibrates when unconstrained and not excited by external forces. This frequency depends on many construction factors of the building e.g. the height of the building. A common range for natural frequencies for buildings are in the range of 10–20 Hz. Ground vibrations from blasting will enforce the vibration of buildings if they include these frequencies.

natural frequency or **natural vibration frequency** (f_n), **(Hz or cycles/s)**, the frequency at which a body or system vibrates when unconstrained and not excited by external forces.

natural logarithm (*e*), **(dimensionless)**, of the number x, is written lnx, and is called the logarithm of x with the base $e = 2{,}71828$.

natural slope angle (δ_{dump} or δ_{load}), (°), see '**angle of repose in dumping** or – **loading**'.

natural stone blasting, blasting of rock with the purpose of creating a large amount of boulders for break waters, piers, etc.

natural stress, virgin stress or **paleostress** (σ_{vir}), **(MPa)**, the stress state which exists in the rock prior to any artificial disturbance. Hyett, 1986.

NC (acronym), see '**nitro cellulose**'.

NCN (acronym), see '**nitro-carbon-nitrate**'.

near field, the distance from a point of interest in relation to a geometric quantity e.g. borehole diameter or hydraulic radius of a drift or tunnel. The limit between far, intermediate and near field should be defined for each special use of the term. Some examples; for large diameter hole blasting on the surface, the region smaller ~25 m from the source of the blast may be regarded as near field. In blasting underground, the near field ends at about two hydraulic diameters from the excavation drift or tunnel.

near-field stress (σ_{nf}), **(MPa)**, the stress state perturbed by a heterogeneity. Hyett, 1986.

neoprene plug, a plug made of neoprene used to seal the blastholes when they are open at the bottom, or to plug the blasthole using air decking technique.

Neuman condition, a boundary condition for the *diffusion equation* developed for *fluid flow in porous media*. Hudson, 1993.

New Austrian Tunnelling Method (NATM), a support system for rock in tunnels including systematic rock bolting in combination with shotcrete or concrete lining for the rock mass. Prompt installation of rock bolts is of key importance in order to maintain the integrity of the rock mass. Hudson, 1993.

NG (abbreviation), see '**nitroglycerin**' and '**NG explosive**'.

NG explosive, an explosive containing any amount of nitroglycerin (NG) or ethyleneglycoldinitrate (EGDN). These are collectively referred to as 'NG'. If the NG content is less than 12%, a powder explosive results. For concentrations greater than 25% NG, gel explosives may be formed. Intermediate NG contents form a semi-gel.

NGI (acronym), for '**Norwegian Geotechnical Institute**'.

NGI index, NGI system, or **NGI classification scheme**, a *rock mass classification system* developed at the Norwegian Geotechnical Institute, see '**Q-factor**'. Hudson, 1993.

NGI classification formula for roof support pressure (p_{sup}), **(MPa)**, the roof support pressure p_{sup} is calculated by

$$p_{sup} = \left(\frac{0,2}{J_r}\right) Q^{-0,33} \qquad (N.1)$$

where p_{sup} is the roof support pressure in MPa, J_r is the joint roughness number and Q is the NGI Q-factor, see '**Q-factor**'. Hudson, 1993.

nip, see '**squeezing rock or ground**'.

Nishimatsu's theory, a theory for rock cutting including an equation for the cutting force in rock using a wedge-shaped tool. Hudson, 1993.

Nissan RCB system, an electromagnetic firing method used for underwater blasting of rock where normal initiation systems are difficult to use due to e.g. rapid tidal current and deep water. The system consists of three components: an oscillator, an exciting loop antenna placed at the bottom of the sea, and detonators equipped with receiver coils for the electromagnetic waves created by the loop antenna. The firing element consists above the receiver coil, a diode, a firing capacitor and an electronic switch. By applying and alternating magnetic field with a frequency of 550 Hz using an alternating current through the loop antenna from a remote base, an alternating electric current is induced in the firing element. The alternating current is rectified to a direct current via a diode and charged into the capacitor. A firing signal of a different frequency triggers a separate circuit to discharge the capacitor through the bridge wires. The system shows promise of being simple and reliable to operate. A disadvantage is that it is not possible to use different delays for the blastholes. Nakano and Ueada, 1983.

nitrates, salts formed by the action of nitric acid on metallic oxides, hydroxides, and carbonates. Readily soluble in water and decompose when heated. The nitrates of polyhydric alcohols and the alkyl radicals explode with violence. C.T.D., 1958. Nitrates are components of many explosives like ANFO and emulsions and are dissolvable in water and therefore nitrates in mine water may increase above acceptable levels and have an impact on the environment.

nitro-carbon-nitrate (NCN), an explosive consisting of ammonium nitrate, carbonaceous material and nitro-compounds other than nitroglycerin and other liquid nitro-compounds, which is packed in approved sealed containers. AS 2187.1, 1996.

nitrocellulose (NC), organic compound consisting of $C_{12}H_{14}N_6O_{22}$. It consists of white fibres with density 1 670 kg/m^3, volume of explosive gases 841 l/kg, heat of explosion 4 052 kJ/kg. Used together with nitroglycerin in dynamites.

NITRODYNE, a *computer code* to model constant volume performance of an explosive. Persson, 1993.

nitrogen oxides (NO$_x$), poisonous gases (NO, NO$_2$, N$_2$O) created by detonating explosive materials. Excessive nitrogen oxides may be caused by an excessive amount of oxygen in the explosive mixture (excessive oxidizer), or by an inefficient detonation. USBM, 1983.

nitroglycerin (NG), trinitrate, trinitrin or explosive oil, a chemical compound (explosive) containing $CH_2NO_3CHNO_3CH_2NO_3$; molecular weight 227,09 and crystal form triclinic or orthorhombic when solid. The specific gravity is 1,5918 g/cm^3 (at +25°C, referred to water at +4°C). A pale yellow flammable thick liquid explosive and soluble in alcohol or ether but only slightly soluble in water; melting point 13,1°C and explosion temperature 256°C. Used as an explosive, in the production of dynamite and other explosives, as an explosive plasticizer in solid rocket propellants, and as a possible liquid rocket propellant. CCD, 1961 and CRC, 1976. The *Italian Sobrero* was the first to produce nitroglycerin in 1846. The *Swede Alfred Nobel* was the first man to use nitroglycerin for blasting purposes. In 1862 he filled a test-tube with nitroglycerin and placed a stopper in the tube. Then he placed the test-tube in a zinc container which he filled with gunpowder. After he inserted a fuse and tossed the tube into a water-filled ditch. The charge exploded with such force that it shook the ground. Hellberg, 1983.

nitroglycol, ethylene nitrate or 1,2 ethanedioldinitrate, a chemical compound (explosive) containing $C_2H_4(NO_3)_2$ with molecular weight 152,06, colourless or yellow; specific gravity 1,4918 g/cm^3 (at +20°C, referred to water at +4°C); melting point +22,3°C; boiling point 197°C (at 760 mm Hg). Insoluble in water but soluble in alcohol; decomposes in alkalis. Bennett, 1962 and CRC, 1976.

nitromethane, a liquid compound, CH_3NO_2, used as a fuel in two-component (binary) explosives and as rocket fuel. USBM, 1983. See also '**binary explosive**'.

nitropropane, a liquid fuel that can be combined with pulverized ammonium nitrate prills to make a dense blasting mixture. USBM, 1983.

nitrostarch, a solid light yellow explosive similar to nitroglycerin in function, used as the base of 'non-headache' powders. It is only used in the United States. Meyer, 1977.

No. 6 blasting cap, see '**test blasting cap No. 6**'.

No. 8 Star blasting cap, see '**test blasting cap No. 8**'.

No. 8 test blasting cap, see '**test blasting cap No. 8**'.

Nobel blastometer, a *galvanometer-type instrument* for testing continuity in detonators and blasting circuits. It has a special dry cell insufficient to fire a detonator and further reduced within the instrument by resistance to 10 mA. Nelson, 1965.

noise, any undesired level of sound (mechanical waves) in air, liquid or solids. For the human being it is the audible and infrasonic part of the spectrum, from 20 HZ to 20,000 Hz.

non caving mining methods, in steep ore bodies are *cut and fill mining with or without pillars* ascending (overhand stoping) or descending (underhand stoping), *sublevel stoping* with pillars, *shrinkage stoping* with pillars and finally in flat ore bodies normally *room and pillar mining* is being used.

Nonel (abbreviation), for '**non electric**'.

NONEL (trademark), see '**shock tube**'. Also abbreviation for '**non electric**'.

NONEL blasting machine, a special designed initiator for NONEL shock tube when blasting a whole round. Nitro Nobel have developed two types of blasting machines, one manually and one pneumatic actuated. They are designed to withstand cold, hot and humid climates encountered in all normal working conditions.

non-electric detonating device, a hand-held unit that initiates the shock tube by releasing the spring-loaded trigger.

NONEL trunkline delay (TLD), a unit consists of a plastic 'bunch block' connector which houses a Nonel delay detonator attached to a signal tube. The delay periods vary, depending on the particular manufacturers product. The relay type of each TLD, which functions unidirectional, ensures true hole-by-hole initiation with correct sequencing.

Nonel tracer line, the use of Nonel shock tube to check if all decks in a blasthole, detonated on separate delays, have detonated properly in time. One separate shock tube is inserted into each deck and led to surface and coiled on a stake at the collar. The lightning of each shock tube is recorded by a high-speed video camera.

non-electric delay blasting cap, a detonator with a delay element, capable of being initiated non-electrically, see '**detonating cord**', '**safety fuse**', '**shock tube**'.

non-electric detonator, see '**non-electric delay blasting cap**'.

non-explosive expanding agent, see '**expanding agent**'.

non-ideal detonation, explosives that detonate at a velocity lower than the maximum detonation velocity (c_{dmax}), see '**ideal velocity of detonation**'.

non-ideal explosive, an explosive where the chemical reaction continues beyond the Chapman-Jouget point. The maximum borehole pressure is therefore reached at a distance behind the Chapman-Jouget plane and therefore the borehole has started to expand. This results in a sub-ideal detonation velocity, varying with the diameter and strength of confining rock. Most commercial explosives exhibit non-ideal performance in small diameter boreholes. Performance is often enhanced by late energy release, especially in weaker rock types. Low detonation velocity can also mean failure to deliver explosive energy.

non-primary explosive detonator (NPED), is a detonator which does not contain a primary explosive such as lead azide. The substance has been replaced by a specially treated secondary explosive. Nitro Nobel, 1991.

non-sparking metal, a metal that will not produce a spark when struck with other tools, rock, or a hard surface, e.g. aluminium.

non-venting caps, blasting caps (detonators) which contain the hot gases from the burning of detonator delay elements so that they cannot cause premature initiation of the explosive charge. Du Pont, 1966.

Nordtest, a *standardized test method for explosives* used extensively in Scandinavia. Test equipment and test procedure is precisely specified. Persson, 1993.

normal force (F_n), **(N or MN)**, a force acting perpendicularly to a surface.

normal stiffness, ratio between load and deformation calculated perpendicular to a given surface. TNC 1979. The normal stiffness of intact rock is almost linear but for discontinuities it is hyperbolic. Goodman, 1976.

normal strain, see '**strain**'.

normal stress (σ_n), **(MPa)**, stress calculated from the force component which is perpendicular to that surface on which the force is acting.

Norwegian cut, a cut where the first drill holes are formed with a sharper angle towards the working face, which facilitates breaking. A type of fan cut with drill holes angled in two directions. This type of cut has been employed successfully in headings of a small section, the cut hole being blasted first, followed by the bench holes. In order to obtain the maximum possible advance the cut may also be deepened after blasting, during the first pause in working, for example, the whole section then being broken out simultaneously. Fraenkel, 1954.

NO$_x$ (abbreviation), for NO, NO$_2$ and N$_2$O, see '**nitrogen oxides**'.

nose (in drilling), the lead face of the crown of a diamond bit. Long, 1920.

nose row, the top row of cutters on a rolling bit. Sandvik, 2007.

notched blasthole, a borehole that has been notched by a *broaching tool*, *water jet* or a *linear shaped charge*. With these techniques it is possible to create normally two diametric opposite standing notches along the blasthole. By rotating a water jet with a sidewards acting nozzle, a disc shaped crack can be created at the bottom of the blasthole. The creation of the 7 mm deep notches enhances the fracture propagation and improves the final contour and reduces the explosive consumption. Hoshino, 1980. By using notched holes, the spacing can be increased by ~40% with half explosive consumption. Holloway, 1986. It is necessary to add some hard particles like iron slag to get the water jet technique economical. Nozzle pressures are in the range 69–140 MPa and corresponding transverse speeds 0,38–1,67 cm/s and iron slag consumption ~0,2–1,8 kg/min. Vijay, 1990.

NPED (acronym), see '**non-primary explosive detonator**'.

NTD, (acronym), for '**noiseless trunk delay**' see '**trunkline**'.

nuclear blasting, the use of nuclear charges in very large-scale blasting. For environmental and health reasons this technique is prohibited to be used today.

nuclear reactor drill, an *exotic drilling method* using a nuclear reactor for heating the rock. A patented construction is described having a diameter of 1 m and a capacity to heat the rock to the melting point, 1 000–2 000°C. KTH, 1969.

OB (shortened form), for '**overbreak**'.
octogen (HMX), colourless crystals of $C_4H_8N_8O_8$, cyclotetramethylene tetramine. An organic compound insoluble in water, molecular weight 296,2, volume of detonation gasses 0,782 m³/kg, heat of explosion 6 092 kJ/kg, the density varies with modification from 1 780 to 1 960 kg/m³ and finally the detonation velocity (c_d) is 9 100 m/s at a density of 1 900 kg/m³. Octogen mixed with aluminium powder is the main ingredient of the reactive powder in NONEL shock tubes and in this application the detonation velocity is reduced to ~2 000 m/s.
OD (acronym), for '**overburden drilling**'.
OD detonator, detonators equipped with heavy duty insulated leg wires for blasting underwater, in wet conditions, or in conductive rock masses such as magnetite ore. Persson et al., 1994.
ODEX (abbreviation), for '**overburden drilling with eccentric drilling**', see '**overburden drilling, ODEX method**'.
oedometer, an instrument for performing a *consolidation test*. Nelson, 1965.
oedometric modulus (*M*), (MPa), is defined by Young's modulus *E* in MPa and the Poisson's ratio *v* (dimensionless) as follows.

$$M = E \frac{(1-v)}{(1+v)(2-2v)} \qquad (O.1)$$

ohmmeter, a type of galvanometer used to measure (in the unit of ohm) the electrical resistance of a wire.
oiler, see '**line oiler**'.
oil mist lubricator, lubricating device that emits a mixture of oil and air in a fine mist. Atlas Copco, 2006.
oil pot, see '**line oiler**'.
Omega tube, plastic tube opened alongside, in which detonating cord and cartridges separated from each other at a set distance are placed. A method used to space the charges along the length of a blasthole in controlled contour blasting. Jimeno, 1995.
one boom rig, is a drill rig equipped with one boom having one drill feed and rock drill. Atlas Copco, 2006.
open cast, see '**open pit**'.
open center hydraulic system, a regulation system for the hydraulic cylinders on the load haul dumper that is rather energy consuming because the hydraulic pump constantly

delivers a flow that is proportional to the speed of the engine. A less energy consuming system is the 'load sensing hydraulic system' that has two variable pumps working together. This provides exactly the right amount of flow and pressure at any time, distributing the power wherever and whenever it is needed. Atlas Copco, 2007.

open cut, an open trench through a hill or mountain, access to a tunnel portal or any other surface excavation.

open cut mining, open cast mining, surface mining or open pit mining, see 'open pit mine'.

open pit, a surface operation by downward benching for the mining of metallic ores, coal, clay, etc. from open faces.

open pit mine, open cast mine, strip mine or **surface mine**, a mine working or excavation open to the surface. USBM, 1968.

open pit mining or **open cut mining**, the extraction of minerals from the surface of the earth by drilling and blasting, mechanical ripping or solving the minerals by water. Normally the mining of metalliferous ores as distinguished from the 'strip mining of coal' and the 'quarrying' of other non-metallic materials such as limestone, building stone etc.

open porosity (n_o), (%), the ratio of the *open pores* to the *bulk volume* expressed as a percentage. See also '**true porosity**'. Dodd, 1964.

open stope, an unfilled man-made cavity.

open stope and filling, see '**cut and fill**'.

operator's instructions, technical manual describing the operation and maintenance of a product. Technical documentation is compulsory under the EC Machinery Directive. It is considered a part of the product and has to be included in the delivery of the physical product. The operator's instructions gives warnings on possible dangers when operating a machine as well as practical advice on safe procedures for operation and maintenance of the machine. Atlas Copco, 2006.

optical detector method, a method that can be used to *measure the detonation velocity* of an explosive (c_d). Optical fibre is inserted in the explosive at certain distances. When the detonation front passes the fibre, optical light is produced which is recorded.

optimal blasthole diameter (d_{opt}), (m or mm), that diameter of a borehole which when loaded with an explosive will produce the maximum energy output per unit of volume of explosive used.

optimal delay time in controlled contour blasting (t_{opt}), (ms), the most economic delay time regarding for example low ground vibrations or high half cast factor. The highest half cast factor is achieved by instantaneous initiation and the lowest ground vibration with micro-sequential initiation. Rustan, 1996.

optimum breakage burden (B_{optb}), (m), in bench and crater blasting the burden distance which, for a certain charge size, gives the maximum amount of rock broken. In the interest of safe breakage, the burden distance in crater blasting should usually be slightly smaller than the optimum breakage burden.

optimum burden (B_{opt}), (m), the burden distance for which the combined cost of drilling, blasting, mucking, hauling, and crushing is a minimum.

optimum (breakage) burden (depth) ratio ($R_{opt/c}$), (dimensionless), the ratio between the optimum charge burden, defined by maximum broken volume, and the critical burden in bench blasting. The ratio between the optimum charge depth, defined by maximum broken volume, and the critical charge depth in crater blasting.

optimum fragmentation burden (B_{optf}), (m), the burden distance, which for a certain charge size, gives an acceptable surface area of fragmented rock.

ordinary blasting cap, see '**fuse cap** or **fuse detonator**'.
ordinary detonator, a blasting cap that is initiated by a fuse.
ore, a mineral aggregate of sufficient value as to quality and quantity to be mined at a profit.
orebody, a mineral deposit such as a vein containing ore. Gregory, 1980.
ore cluster, a group of ore bodies sometimes differing from each other in structure but interconnected or otherwise closely related genetically. Some ore clusters gather downward into a restricted root. A.G.I., 1957.
ore crusher, machine for disintegration of ore or and waste. The most common used crushers for primary crushing are *jaw* and *gyratory crushers* and for secondary crushing normally *conical crushers*.
ore deposit, a metalliferous mineral deposit sufficiently concentrated by nature to warrant extraction by mining. Gregory, 1980.
ore loss, ore which is fragmented but can not be economically mined and is lost in the mine. TNC, 1979.
ore pass, a vertical or inclined passage for the downward transportation of ore. It is equipped at the bottom with chutes, gates or other appliances for controlling the flow. An ore pass is driven in ore or country rock and connects one level with a lower level or a hoisting shaft.
ore rest, primary ore that is left in the mine but can be gained on a deeper level.
ore stamp (old), a machine for reducing ores by stamping. The most familiar form is the *stamp battery*, and a later form a powerful steam stamp. Standard, 1964.
orientation histogram, see '**direction histogram**'.
O-ring, ring shaped seal used in various applications to prevent the leakage of a fluid. Atlas Copco, 2006.
ornamental rock, see '**dimensional stone blasting**'.
ornamental stone blasting, see '**dimensional stone blasting**'.
orthogonal, a term meaning at right angles (90°).
orthogonal fractures, two or three fractures system each at a 90° angle to the two other systems. Hudson, 1993.
orthotropic, the description applied to the elastic properties of material, such as rock, that in many cases has considerable variations of strength in two or more directions at right angles to one another.
outburst, the violent evolution of firedamp usually together with large quantities of coal dust from a working face. Outbursts are known wherever coal is worked. Roberts, 1960. Stress-induced outburst also occurs in hard rock mines.
outcrop, the exposure of bedrock at the surface or the ground. ISRM, 1975.
outside gauge or *gage*, see '**set inside diameter**'.
overbreak or **backbreak**, the outfall of rock behind the lines of the intended section. The amount of overbreak depends on the blasting method, the strength and structure of the rock, and the size of the opening in underground excavations. Some correlation has been found between overbreak and rock mass quality (Barton Q-factor). Franklin and Ibarra, 1991.
overbreak area (A_o), (m²), the area outfall of rock behind the lines of the intended section of the tunnel, drift or room.
overbreak depth (l_o), (m), the perpendicular distance from the planned contour to the overbreak contour.

overbreak depth or **backbreak depth (average)** (l_{oa}), **(m)**, volume outside the projected area of the drift or tunnel divided by the projected area of the two walls and roof. Overbreak can be correlated to RQD and specific charge.

overbreak factor (f_o), **(%)**, the area of the overbreak in a section perpendicular to the tunnel, drift, room or stope divided by the planned area multiplied by 100.

overbreak volume (V_o), **(m³)**, the volume outfall of rock behind the lines of the intended section.

overburden, worthless surface material covering a deposit of useful material. Atlas Powder, 1987. The term is used by geologists and engineers in two different meanings: a) to designate material of any nature, consolidated or unconsolidated, that overlays a deposit of useful materials (ores or coal), especially those deposits that are mined from the surface by open cuts b) to designate only loose soil, sand, gravel, etc. that lies above the bedrock. The term should not be used without specific definition. Stokes and David, 1955.

overburden bit, a special diamond-set bit, similar to a set casing shoe, used to drill casing through overburden composed of sand, gravel, boulders, etc. Long, 1920.

overburden drilling, a method to penetrate overburden with a tube to make it possible to continue regular percussive drilling inside the tube when solid rock is reached. Sandvik, 2007.

overburden drilling equipment, equipment designed specifically for drilling in the earth and unconsolidated material overlaying bed rock. Atlas Copco, 2006.

overburden drilling (OD method), a method of drilling where charging and blasting are performed through the overburden which may consist of soil or water. The OD drilling equipment consists basically of an outer casing tube with a ring of cemented carbide at the lower end. The casing tubes encloses an inner drill string made up of standard drill steels with a cross bit. The casing tubes and the inner drill steels are of the same length and are jointed by coupling sleeves independently of the one another. The whole system is connected to the rock drill by a special shank adapter, which transfers both impact force and rotary force to the string of casing tubes and to the string of extension steels. Both the inner extension equipment and the outer tube equipment are connected to the adapter and drilled down by impact and rotation. The outer equipment can be disconnected from the adapter, and drilling can then be continued as ordinary extension drilling with the inner equipment. Atlas Copco, 1982.

overburden drilling (ODEX method), this method is based on the principle of under-reaming. The eccentric bit makes it possible to insert casing tubes into a hole at the same time as the hole is being drilled. During drilling the reamer on the ODEX bit swings out and drills a hole larger than the outer diameter of the casing tube. When the desired depth has been reached the equipment is rotated a couple of revolutions in the opposite direction, whereupon the reamer is folded in and enables the inner equipment to be retracted through the lining, which remains in the hole. If drilling is to be continued in solid rock, the ODEX bit is replace by an ordinary drill bit and drilling proceeds through the lining with extension equipment or with down-the-hole equipment. In prospecting operations, drilling is sometimes continued with diamond drilling. Atlas Copco, 1982.

overcoring of rock bolts, a rock bolt can be recovered to study the integrity of the grout or to check corrosion by diamond drilling technique. The rock bolt will be a part of the core.

overcut, a machine cut made along the top or near the top of a coal seam, sometimes used in thick seams or a seam with sticky coal. By releasing the coal along the roof it's mining becomes easier. Nelson, 1965.

over excavatuion, see '**overbreak**'.
overhand cut and fill, see '**ascending cut and fill**'.
overhand stope, one in which the ore above the point of entry to the stope is attacked, so that severed ore tends to gravitate toward discharge chutes so the stope is self-draining. Pryor, 1963.
overhand stoping, a mining method where the ore is being blasted from a series of ascending stepped benches. Both horizontal and vertical holes may be employed. Horizontal breast holes are usually more efficient and safer than vertical up-holes, although the latter are still used in narrow stopes in steeply inclined ore bodies. McAdam, 1958.
overhand stoping and milling system, see '**combined overhand and milling stoping**'.
overhand stoping on inclined floors, see '**rill mining**'.
overhand stoping on waste, see '**overhand stoping**'.
overhand stoping with shrinkage and delayed filling, see '**shrinkage stoping**'.
overhand stoping with shrinkage and no filling, see '**shrinkage stoping**'.
overhand stoping with shrinkage and simultaneous caving, see '**combined shrinkage stoping and block caving**'. Fay, 1920.
overhang, an undercut rock mass. In surface mining, e.g. a rock mass undercut unintentionally by a previous blast.
overhead loader or *overshot loader,* a machine which mucks the material into a bucket which is moved over the loader and dumps the material into its own container or into a neighbouring container. The bucket is following an overhead trajectory, hence the name. A loading machine used underground.
overhead shovel loader, see '**overhead loader**'. Atlas Copco, 1982.
overhead stoping, see also '*overhand stoping'*.
overlapping of consecutive delays, firing out-of-sequence which occurs when a later delay fires before an earlier delay.
overpressure (p_o), (Pa or dB), the pressure exceeding the atmospheric pressure and generated by sound waves from blasting.
overshot (blasting), condition resulting from the detonation of an excessive amount of explosives. Usually characterized by an excess of fragmentation, flyrock and noise.
overshot (drilling), a tool that fits over the end of and connects to a drill string that has become detached, for the purpose of recovering it. Atlas Copco, 2006.
*overshot loader, see '**overhead loader**'.
oversize, all fragments where secondary breakage is necessary before further handling. In underground mines this can be as little as 300 mm, while in open cast mines it is seldom greater than 1 000 mm.
oversize core, a diamond core cut by a thin-wall bit, as opposed to a standard-diameter of which is greater than a standard size. Long, 1920.
oxides of nitrogen, see '**nitrogen oxides**'.
oxidizer or **oxidizing material**, a chemical substance which serves as a source of oxygen to be consumed by a fuel. Common examples are ammonium nitrate (AN) and sodium nitrate (SN). Almost all explosive materials contain oxygen, which is needed for the explosive reaction to take place.
oxygen balance, a chemical state of equilibrium in an explosive. It is achieved when the amount of oxygen is just sufficient for complete combustion, keeping CO and NO_X at a minimum. Apart from minimizing toxic fumes, oxygen balance maximizes explosive performance (velocity of detonation and detonation pressure).

P

pack, a pillar, constructed from loose stones and dirt, built in the waste area or roadside to support the roof. Nelson, 1965.

packaged explosive, see '**cartridge**'.

packer, a device lowered into a borehole where it can be made to expand from surface with the purpose to produce a watertight joint against the sides of the borehole or the casing. Long, 1960. See also '**hydraulic fracturing**'. Pipe having a flexible outer sleeve that can be expanded to seal a drill hole while cement grout is introduced through the center pipe. Atlas Copco, 2006.

packing degree or **degree of packing** (*P*), (%), percentage of the volume of a blasthole occupied by explosives.

paddy, see '**expansion bit**'.

paddy bit, see '**expansion bit**'.

paleo, means '**old**'.

paleostress, natural stress or **virgin stress** (σ_{vir}), (**MPa**), a previously active in situ stress state no longer in existence. Hyett, 1986.

panel, a system of coal extraction in which the ground is laid off in separate districts or panels, pillar of extra size being left between. Fay, 1920.

panel barrier, the pillar of coal left between the adjacent panels. These pillars are often worked on the retreat after the coal in the panels has been extracted. Nelson, 1965.

panel caving, block cave mining where the area of the orebody is too large to be completely undercut because this will cause failure of pillars on the loading level. The orebody is therefore undercut progressively in a series of usually parallel panels often rectangular in shape and perpendicular to the strike.

panel retreat caving, see '**block caving**'.

paper cartridge, a cylindrical shaped piece of explosive wrapped in thin and flexible paper or in thick and hard cardboard.

paraffin impression blocks, see '**impression blocks**'.

parallel blasting circuit, see '**parallel (circuit) coupling**'.

parallel (circuit) coupling, blasting caps can be coupled in series or parallel. When parallel coupling is used the lead wire is extended by two bus wires. One leg wire from each cap is hooked to each of the bus wires. USBM, 1983.

parallel hole cut, method to blast a cut where a subgroup of parallel oriented holes is drilled perpendicular to the face in a tunnel. Some holes are charged with explosives and others are left uncharged to create the necessary swelling volume for the charged holes upon blasting. See Fig. 46.

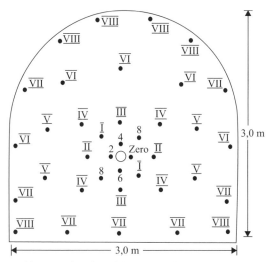

Arabic numerals = short delay cap number blasted first
Roman numerals = half-second delay cap number

Figure 46 Parallel hole cut. After Sen, 1995.

parallel-series (circuit) coupling, circuitry characterised by a parallel circuit coupling, involving two or more series of electric blasting caps. One end of each series of caps is connected to each of the bus wires.

partial mining, see '**strap mining**'.

particle, in mineral technology a single piece of solid material which can be defined as regards its size by triaxial measurement; by its mesh below some 2,5 mm in average width and above som 50 microns; and by micro measurement down to the resolving limits of a given magnifying system. Such a particle is not necessarily homogeneous. Also, in physics, the particle may be a molecule, atom or atomic component. Particle size distribution, assessed by screen analysis of sample through series of laboratory sieves, describes the percentage by weight of each size of particle in the range of screens used. Pryor, 1963.

particle acceleration (*a*), (mm/s^2 or m/s^2), the acceleration of a particle when subjected to a force.

particle emulsion explosive (PEE), see '**emulsion explosive**'.

particle flow code (PFC), a numerical analysis code based on Dr. P.A. Cundall's distinct element method in which the material is represented as an assembly of discrete particles and permitted to flow according to the laws of Newtonian mechanics. The code can be used to simulate throw of fragments at blasting and under the influence of gravity. The code can also be used to analyze gravity flow in caving. ITASCA in the USA have developed different codes.

particle shape, see '**shape factor**'.

particle size, is defined as the largest square mesh a certain particle can not pass. It is the fundamental basis for designating mineral soils or rock fragments using particle fraction to distinguish the soil or rock mechanical behaviour. Particles are divided into the following classes, see Table 32. SS-EN ISO 14688-1:2002, 2003.

Table 32 Terms to be used for each soil or particle fraction and its sub fractions together with the corresponding chosen range of particle sizes. SS-EN ISO 14688-1:2002, 2003.

Soil fractions	Sub fractions	Designation	Particle size (mm)
Very coarse soil	Large boulder	LBo	>630
	Boulder	Bo	>200 to 630
	Cobble	Co	>63 to 200
Coarse soil	Gravel	Gr	>2 to 63
	Coarse gravel	CGr	>20 to 63
	Medium gravel	MGr	>6,3 to 20
	Fine gravel	FGr	>2,0 to 6,3
	Sand	Sa	>0,063 to 2,0
	Coarse sand	CSa	>0,63 to 2,0
	Medium sand	MSa	>0,2 to 0,63
	Fine sand	FSa	>0,063 to 0,2
Fine soil	Silt	Si	>0,002 to 0,063
	Coarse silt	CSi	>0,02 to 0,063
	Medium silt	MSi	>0,0063 to 0,02
	Fine silt	FSi	>0,002 to 0,0063
	Clay	Cl	≤0,002

Guiding vocabulary for the division of mineral soils on a basis of the content of various fractions, see Table 33. SS-EN ISO 14688-2:2004, 2003.

Table 33 A possible soil classification, based on grading stone. SS-EN ISO 14688-2:2004, 2003.

Fraction	Content of fraction in weight-% of material ≤63 mm	Content of fraction weight-% of material ≤0,063 mm	Name of soil	
			Modified term	Main term
Gravel	20–40		gravely	
	>40			Gravel
Sand	20–40		sandy	
	>40			Sand
Silt + clay (fine soil)	5–15	<20	slightly silty	
		≥20	slightly clayley	
	15–40	<20	silty	
		≥20	clayley	
	>40	<10		Silt
		10–20	clayley	Silt
		20–40	silty	Clay
	>40			Clay

particle velocity (v), **(mm/s or m/s)**, the speed of movement of a particle when subjected to a force.

particle velocity measurement (**ground vibrations**), should be made in three mutually perpendicular directions. This allows the calculation of the resultant particle velocity, v_R, by the vector summation of the component velocities (v_x, v_y and v_z)

$$v_R = \sqrt{v_x^2 + v_y^2 + v_z^2} \qquad (P.1)$$

parting, rock mass located between two seams of coal.
parting plane, a joint or a crack in rock forming a plane of separation.
parts kit overhot assy, a parts kit for a drill rod recovery tool used in diamond drilling. Atlas Copco, 2006.
passage, a cavern opening having greater length than height or width, large enough for human entrance and larger by comparison than a lead. Also called drift, tunnel or roadway in metalliferous mines.
passive anchor system, rock reinforcement by not pretensioned rock bolts or cables.
paste fill, a fill consisting of non-cyclonized mill tailings mixed with cement and used for support in open stopes or as artificial pillars underground. Coarse tailings permit a very high solids content of up to 88% to be pumped at high pressure, and high setting strengths were achieved. Paste is currently being used as replacement for hydraulic fill, with the cement added at surface. It exhibits the physical properties of a semi-solid when compared to high density fill, which is a fluid. Atlas Copco, 2007. To some extent also slag and ash can be used as binding material.
pattern, a plan of holes laid out on a face or bench which are to be drilled for blasting. Burden and spacing should be given in metres. USBM, 1983.
pawl, part in a pneumatic rock drill mounted inside a toothed ring in such a way that it allows rotation in only one direction. Atlas Copco, 2006.
PCF (acronym), (q_p), (**kg/m³**), for '*perimeter charge factor*'. Recommended term to be use is '**perimeter specific charge**'.
peak particle velocity (v_{max}), (**mm/s** or **m/s**), the maximum particle velocity.
peak strain at rock blasting (ε_{max}) (**dimensionless**), the maximum strain measured at different distances from the blasthole. The peak strain ε_{max} around a confined cylindrical charge in a coupled hole in limestone varies as follows:

$$\varepsilon_{max} = k(r_c/r)^n \qquad (P.2)$$

where k is called *strain intercept* (dimensionless). k is, however, dependent on the detonation pressure of the explosive, r_c is the radius of the equivalent sphere from the cylindrical charge used in the blasthole in m, and r is the distance from the blast in m to the measurement point and n is the *distance exponent for peak strain. n is* dependent only on the rock properties. Atchison, 1964.
PEE (acronym), for 'powder emulsion explosive' see '**emulsion explosive**'.
peeled steel, drill steel manufactured by detaching one layer of a coating from another. Sandvik, 2007.
pellet powder, black powder pressed into pellets, 52 mm long and from 32 to 52 mm in diameter. USBM, 1983.

pendulum drift indicator, a pendulum is used to quantify the deviation from vertical in aimed vertical water well holes. Stein, 1977.

penetration rate (v_d), (m/min at hammer and rotary drilling), **(m/h** at full face boring), the linear velocity of the drill bit in drilling, rate of drilling or drilling rate.

penetration test, a test to determine the *relative values of density* of non-cohesive sand or silt at the bottom of boreholes. The standard penetration test is made by determining the number of blows required by a standard weight dropped through a standard height to produce a standard penetration of 304 mm (12 inches). The dynamic penetration test is used to determine the relative density of successive deposits by recording the penetration per blow or per specified number of blows. Nelson, 1965.

penstock, a sluice or gate for restraining, deviating, or otherwise regulating the low of water, sewage, etc. a floodgate. Webster, 1961. It is an important part of a hydro plant.

pentaerythritoltetranitrate (PETN), a chemical composition $C_5H_8N_4O_{12}$. It consists of colourless crystals; density 1 760 kg/m³, heat of explosion 5 895 kJ/kg, volume of detonation gases 780 l/kg and melting point +141,3°C. It is a high explosive used in some primers and boosters and in detonating cord.

pentaprism, a prism used in combination with a laser beam e.g. to level the drill rig in height and line the drill rig side wards and finally adjust the drill rig regarding the square, plumb (vertical) and plane. Irwin, 1979.

Pentolite, a 50%/50% mixture of PETN and TNT which, when cast, is used as a cast primer or in shaped charges. The density is 1 650 kg/m³ and the detonation velocity (c_d) 7 400 m/s. Meyer, 1977.

percussion drilling, a) a drilling method employing a string of drilling tools that are raised and lowered by a cable, '*cable drilling*' b) also a method in which hammer blows are transmitted by drill rods to a drill bit.

percussive drilling, a method of drilling where rock material is loosened by the impact and intermittent rotation of the drill bit onto the hole bottom.

perfobolt, two perforated cylinder half's with the length of the borehole are field with a cement mixture and closed before insertion into the borehole. A reinforcement iron is thereafter pushed into the cylinder and at this operation the cement is pushed through the holes of the cylinder and fills up the hole completely. Used mainly in civil engineering. Pusch, 1974.

performance of explosives, the relative performance of explosives in rock can be measured by *strain gauges* in a grout filled boreholes or *piezoelectric pressure gage* in a water filled boreholes. Rise times measured are in the order of 0,1 ms on distances 3–12 m. Bur, 1967.

perimeter, the borderline of an excavation.

perimeter blasting, see '**controlled contour blasting**'.

perimeter charge factor (q_p), **(kg/m³)**, the mass of explosive per intended volume rock broken by the perimeter holes. Use '**perimeter specific charge**' instead.

perimeter controlled blasting or -technique, see '**controlled contour blasting**'.

perimeter easer holes, the blastholes next to the perimeter holes.

perimeter quantification, see '**profile measurement**'.

perimeter specific charge (q_p), **(kg/m³)**, the mass of explosive per intended volume rock broken by the perimeter holes.

perimeter row, the last row of blastholes in a tunnel round.

period (*T*), **(s)**, the time interval passing during a full cycle of vibration or an oscillation of a wave.

period of vibration (T) **(s)**, see 'period'.
periscope, see 'borehole binocular'.
permafrost blasting, generally the strength of rock or soil will increase if the water content in the rock or soil freezes. Blasting in permafrost is therefore more difficult and the charge concentration must be more precisely calculated. The slightest deviation from the designed drilling and loading pattern in either direction could result in under blasting or putting the blast in the flyrock range. The physical properties, P-wave and S-wave velocity of rock and soil, will change a little if it is dry sample and temperature goes below zero but the velocities will increase a lot if it contains frozen water. In asbestos ore, the P-wave velocity increased from 2 470 m/s unfrozen to 3 325 m/s at −23°C that is about 35% increase. The excavation of frozen material therefore requires more energy. In mining, the cost of comminution (blasting and crushing) could double. Serious blasting problems began when the moisture content in the frozen asbestos ore is ~8% and the difficulties were as greatest when moisture content was 15–18%. Lang, 1976.
permanent bolting, see 'permanent rock reinforcement'.
permanent rock reinforcement, rock reinforcement (inside the rock mass) with the purpose to be active throughout the lifetime of the construction. TNC, 1979.
permanent support, rock support (outside the rock mass) with the purpose to be active throughout the lifetime of the construction.
permeability (hydraulic) (c_h)**, (m/s)**, see 'hydraulic conductivity'.
permeability coefficient (k)**, (m^2/Pas)**, the proportional constant in Darcy's law for gas or fluid flow in porous media. See 'hydraulic conductivity'.
permeability measurement, see 'water head measurement'.
permeameter, is a laboratory device to determine the *permeability of rock drill cores*. There are three kinds of permeameters the *longitudinal with* or *without pulses* and *radial*. In the longitudinal case the differential pressure acts on the two ends of the core sample. If pulses are used a transient longitudinal flow between upstream and downstream fluid reservoirs set at an initial pressure of head differential. In the case of radial permeability the core has to be prepared by an axial hole in the center of the core. Hudson, 1993.
permissible- or **permitted explosives** (adjective), a term used to describe a machine, material, apparatus, or device that has been investigated, tested, and approved by some national authority, and is maintained in permissible condition.
permissible blasting, blasting according to and satisfying the regulations of the national authority for underground coal mines or other gassy underground mines.
permissible diameter (smallest) (d_p)**, (mm)**, the smallest acceptable diameter of a permissible explosive, as approved by the Mine Safety and Health Administration (MSHA). Atlas Powder, 1987.
permissible explosive, an explosive characterized by a low flame temperature that does not ignite certain specified methane/air or coal dust/air mixtures. This kind of explosive is permitted for use underground in a methane or coal dust atmosphere.
permitted explosive, see 'permissible explosive'.
persistence (discontinuities)**,** (k)**, (m)**, a measure of the area extent or size of a discontinuity within a plane. It can roughly be quantified by observing the discontinuity trace length on the surface of exposure. It has significant importance for blasting, but it is one of the most difficult factors to quantify. The classification of persistence is shown in Table 34.

Table 34 Classification of persistence of discontinuities.

Description	Persistence (m)
Very low persistence	<1
Low persistence	1–3
Medium persistence	3–10
High persistence	10–20
Very high persistence	>20

Source: Brady and Brown, 1993.

PETN (acronym), see '**pentaerythritoltetranitrate**'.

petroleum, material occurring naturally in the earth composed predominantly of mixtures of chemical compounds of carbon and hydrogen with or without other non-metallic elements, such as sulphur, oxygen, nitrogen, etc. Petroleum may contain or be composed of such compounds in the gaseous, liquid, and/or solid state, depending on the nature of these compounds and the existent conditions of temperature and pressure. API Glossary.

petrography, the science of the properties, structure and composition of rock types. TNC, 1979.

petrology, the science of the creation and occurrence of rock types.

petrophysics, the science of the physical properties of rock and the petrographic interpretation of these properties. It is an interdisciplinary branch of science.

PFC (acronym), see '**particle flow code**'.

PFC2D (abbreviation), for '**particle flow code in two dimensions**'.

PFC3D (abbreviation), for '**particle flow code in three dimensions**'.

Philipp Holzmann measuring device, measures stresses in shotcrete lining of tunnels or drifts. The measurement cell is inbuilt in the lining. Hudson, 1993.

phlegmatization, deliberate reduction of explosive sensitivity, for example by adding dinitrotoluen (DNT) to a nitroglycerin explosive (NG explosive).

photoelasticity, a property of certain transparent substances which enables the presence of strain to be detected by examination in polarized light. If models of complicated engineering structures are made of such a substance, the stress distribution in the structure may be resolved. See also '**Homolite 100**' and 'iso chromatic lines'. Ham, 1965.

photogrammetric method of fragment size analysis, three dimensional study of the fragment size distribution in a muck pile. This method gives a higher precision than the photographic method.

photographic method of fragment size analysis, two dimensional photographs are taken from the surface and/or cross sections of the muckpile for further processing and interpretation manually, semi automatically or fully automatic.

physical properties, the properties, other than mechanical properties, that pertain to the physics of a material, for example density, electrical conductivity, heat conductivity, and thermal expansion. ASM, 1961.

physicochemical stress, the stress state set up as a result of chemical and/or physical changes in the rock. Hyett, 1986.

PIC (acronym), for '**plastic igniter cord**', see '**igniter cord**'.
pick cutting machine, uses a rotational head with picks for mechanical excavation of rock. The picks could also be mounted on a drum and is then used for shearing of coal.
pick hammer, a hammer with a point, used in cobbing. Fay, 1920.
picks, the steel cutting points used on a coal-cutter chain. See also '**coal-cutter picks**'. Nelson, 1965.
pickup, see '**seismometer**'.
piecework, work paid in accordance to the amount of work being done. Fraenkel, 1952.
Piezeodex system, a *water head measurement system* using a series of packers. The following advantages are achieved: 1) Great flexibility in selecting the number and length of measurement intervals, defined by a chain of packers, 2) Determination of the fluid pressure without any hydraulic connection between the sensor and the fluid in the measuring interval, 3) Sensitivity in reading of at least 0,02% and an accuracy of the measurements of 0,1% of the full selected measuring range. Hudson, 1993.
piezeometer, instrument used to obtain detailed fluid pressure distributions along a borehole. The *continuous piezeo meter*, with a single membrane along the entire length of a borehole, was developed in France, using a sliding probe to measure pressure at any location in the borehole. Patton developed the *modular piezeo meter* using a chain of packers and a sliding probe as a sensor. Hudson, 1993.
piezeo resistance dynamic pressure transducers, devices for measurement of stress or pressure in condensed media and blast waves using *manganin-*, *ytterbium* or *carbon piezeo resistance foils* in a variety of flatpack configurations. The change in resistance is proportional to the pressure. The manganin transducers work in a pressure range 3–100 GPa and pulses with rise times of tens of nanoseconds and total duration of microseconds and the ytterbium piezeo resistance foils in pressure range <3 GPa and with rise times of tens of microseconds and total duration of milliseconds. In the case of manganin there is a 15% difference between hydrostatic and uniaxial piezeo resistant response, with ytterbium, the difference is about 50%. E.g. Bridgeman used manganin coils in 1911 to measure hydrostatic pressure. Keough, 1979.
pile, long cylindrical pole, made of reinforced concrete, steel or wood, forced into earth or soil to support a building or other structure. Atlas Copco, 2006.
piling, process of installing piles. Atlas Copco, 2006.
pillar, a section or piece of ground or mass of ore or rock left in place to support the roof or hanging wall in an underground mine.
pillar and breast, a system of coal mining in which the working places are rectangular rooms usually five or ten times as long as they are broad, opened on the upper side of gangway. The breasts, usually from 5–12 m wide, vary with the character of the roof. The room or breasts are separated by pillars of solid coal (broken by small cross headings driven for ventilation) from 1,5 to 12 m wide. Fay, 1920.
pillar and chamber, a **room and pillar** mining method of working often adopted in extracting a proportion of thick deposits of salt or gypsum. The method may be adopted where the value of the mineral in the pillar is less than the cost of setting artificial supports. Nelson, 1965.
pillar and stall, a **room and pillar** mining method used in the coal mines in Australia, America and South Africa. Atlas Copco, 1982. See also '**bord-and-pillar**'.
pillar blasting, blasting of pillars located between the foot wall and hanging wall in an underground mine to increase the recovery of ore.

pillar caving, the ore is broken in a series of stopes or tall rooms, leaving pillars between. Eventually the pillars are forced or allowed to cave under the weight of the roof. McKinstry, 1948.

pillar extraction, mining of pillars in any mining method e.g. pillars left in room and pillar mining, sublevel stoping, cut and fill mining, sublevel caving and shrinkage stoping.

pilot adapter, threaded adapter to connect the pilot bit to threaded rod. Sandvik, 2007.

pilot bit, a cylindrical steel bar, fastened at the center and at the top of the drill bit and with a length of about 200 mm. It is put into the hole already drilled and planned to be reamed to a larger diameter. It acts as a guide for the reaming bit.

pilot drift, a drift that later on is enlarged to its full area.

pilot drill, a small drill used to start a hole in order to insure a larger drill running true to center. Crispin, 1964.

pilot heading, a drift or tunnel which later on is enlarged to its full area. Nitro Nobel, 1993.

pilot hole, a small hole drilled ahead of a full-sized, or larger borehole. Long, 1920.

pilot tunnel, a small tunnel excavated over the entire length or over part of a tunnel to explore ground condition and assist in final excavation. Bickel, 1982. The pilot tunnel is later on enlarged to its full area.

pinch, see '**squeezing rock or ground**'.

pinholing or *gas venting,* rupturing of a detonator by excessive blast energy delivered to the fuse head. This may disrupt the burning speed of the delay element or, in extreme cases, may cause a misfire. Sen, 1992.

pin in thread, is located on the outside surface of a cylindrical or tubular member. Long, 1920.

pinnate fractures, can form both in advance of the development of a through-going fault plane or during subsequent slip on a fault. They have been observed in experiments on a wide variety of rocks and other materials. Hudson, 1993.

pin to box, a coupling where one end is threaded on the outside (pin) and the opposite end threaded on the inside (box). Formerly designated as a male female coupling. Long, 1920.

pioneer bore, see '**pilot tunnel**'.

pipe, tubular object used for conveying gas, water oil, etc. Atlas Copco, 2006.

pipe charge, an explosive wrapped in a long cylindrical cover of stiff plastic or paper. The charge can be used for blasting of the bottom holes in a tunnel (lifters) but also in the roof and walls for controlled contour blasting. In the latter case, low (linear) charge concentrations (kg/m) must be used. The pipe charges used in controlled contour blasting should be coupled together and centred in the blasthole by plastic sleeves.

pipe half, one half of a pipe that has been divided along its length. Atlas Copco, 2006.

pipeline, a steel or PVC or glasfibre tube for the transportation of gas or oil. Special care should be taken when blasting close to pipelines. A particle velocity of 127 mm/s was set as a safe criteria from a nearby mining although the pipelines could stand 600 mm/s. Siskind, 1994. Buried pipelines can stand more vibrations than those on surface.

pipe stick, a support made of a steel pipe with a diameter typically 150 mm diameter and with a 4 mm thick wall. It can take an axial load of ~500 kN. Brady, 1985.

pit, a) the underground portion of a colliery (coal mine), including all workings. Used in many combinations, as pit car, pit clothes, etc. Fay, 1920. b) In surface mining it usually refers to an open pit operation in which the mine workings proceed to gradually lower elevations with respect to an initial elevation. Generally, pit means a local minimum elevation, the lowest point in a closed depression.

pitch, the angle to the horizontal of the movement of a steep orebody downwards. Synonymous with dip.

pitch diameter, the outer diameter of threads. Sandvik, 2007.

pit floor, the bottom level of a pit.

pit prop, a piece of timber used as a temporary support for the mine roof. Zern, 1928.

pit wall reinforcement, the reinforcement of pit walls with fully grouted anchors, cable bolts or swelling tubes (Swellex).

pixel, in image analysis a basic irreducible processing element. The smallest picture element which can be shown on a TV-screen. Each pixel can be given a certain grey level or, if a colour screen is used, a certain colour.

placard, sign placed on vehicles transporting hazardous materials, including explosives, indicating the nature of the cargo. USBM, 1983.

placer mining, see '**alluvial mining**'.

plain clinometer, an instrument for measuring angular deviation from horizontal or vertical in drilling. A clinometer having only its upper end threaded to fit drill rods. Also called 'end clinometer'. Long, 1920.

plain detonator, detonator designed to be fired by the flash from a safety fuse, and used only together with a safety fuse. It consists of a small cylindrical metal tube containing an explosive composition and with one end closed. The other end is open for the insertion of the safety fuse.

plain twin steel strand, two cables grouted into a drill hole in the rock mass for the purpose of rock reinforcement. Stillborg, 1994.

plane of weakness or **weakness plane**, a surface or planar narrow zone in the rock mass with a (shear or tensile) strength lower than that of the surrounding material.

plane strain, stresses and strains in two dimension.

plane wave, the wave front is planar. Many formulas developed assume a planar wave front. At larger distances this can be done without causing any large error in calculations.

plasma blasting, a non-chemical explosive means of rock fragmentation in which an electric probe, situated in a borehole and surrounded by a highly conductive and gelled electrolytic solution, is fed by a very high capacitance discharge. The very sudden transfer of energy to the electrolyte vaporizes the solution and creates a high energy plasma field which results in borehole pressures sufficient to cause rock fragmentation. The technique is still in a prototype phase, and hole lengths are limited to less than 1 m. Nantel and Kitzinger, 1990.

plasma drill, an exotic drilling method similar in design to 'electrical arc drill' only with the difference that the arc is created in high ionizised gas, that is streaming between the electrodes. A plasma is created at temperatures 8 000–16 000°C and it is heating up the rock. KTH, 1969.

plaster charge, see '**plaster shooting**'.

plaster of Paris, a plaster, made from gypsum by grinding and calcining. The name is due to its manufacturing near Paris, France. With water, it forms a paste which soon sets. Fay, 1920. This material has been used in model blast tests.

plaster shooting or **popping**, a method in which the explosive is detonated in contact with a boulder without the use of a borehole. It is used for secondary fragmentation or breakage of boulders produced during the pervious blast.

plaster shot, see '**mud cap**'.
plaster stone, synonym for '**gypsum**'. Fay, 1920.
plastic cartridge, a cylindrical charge enclosed by thin or hard plastic.
plastic deformation, a permanent change in shape of a solid that does not involve failure by rupture. In the narrowest sense, the change is accomplished largely by gliding within individual grains, but it also involves rotation of grains. In a large sense it includes deformation that is related to recrystallization. A.G.I., 1957. In the stress-strain behaviour of a material, it refers to the deformation that is permanent and cannot be recovered when unloading the material.
plastic drill hole plug, a hollow slightly conical device of plastic used at the collar of downholes to prevent drill cuttings to fall into the drillhole.
plastic explosive, an explosive with plastic consistency like dynamite. TNC, 1979.
plasticity index (I_p), (mass-%), numerical difference between the liquid limit and plastic limit of a fine soil.

$$I_p = w_L - w_P \tag{P.3}$$

where w_L is the liquid limit in mass-% and w_P is the plasticity limit in mass-%. SS-EN ISO 14688-2:2004, 2003.
plastic flow, when a material is strained beyond the elastic limit, but not to the extent of complete failure, it becomes plastic and behaves as viscous liquid. Extension under stress is then a function of time. Such movement is known as *plastic flow*. Spalding, 1949.
plastic igniter cord (PIC), see '**igniter cord**'.
plastic limit (w_p), (mass-%), *water content* (definition see '**water content**') at which a fine soil becomes to dry to be in a plastic condition, as determined by the '*plastic limit test*'. SS-EN ISO 14688-2:2004, 2003.
plasticity, the property of changing shape permanently without movement on any megascopically visible fractures. A.G.I., 1957. Property of a material to yield upon loading beyond the limit of elasticity and develop remnant shape deformation upon unloading. It is also a material that undergoes permanent deformation without appreciable volume change or elastic rebound, and without rupture. The plasticity depends on the mineral composition of the rocks and diminishes with and increase in quartz content, feldspar and other hard minerals. The humid clays and some homogenous rock have plastic properties. The plasticity of the hard rocks such as granite, schistoses, and sandstones becomes noticeable especially at high temperatures.
plasticity theory, the basic model is defined as isotropic hardening with the assumption of associativeness. Hudson, 1993.
plastic locking spring, a plastic plug with feathers to lock charges in upholes.
plastic pipe confinement, see '**plastic pipe test**'.
plastic pipe test, the velocity of detonation is sometimes determined in plastic pipes of different diameter resting in open air, to study the influence of diameter of the explosive on the velocity of detonation. These tests have, however, a lower confinement than blasting in rock and this will affect the velocity of detonation.
plastic strain, in rocks which are composed of many crystals which often belong to several mineral species, the term is conveniently applied to any permanent deformation throughout which the rock maintains its essential cohesion, and strength, regardless of

the extent to which local micro-fracturing and displacement of individual grains may have entered into the process. A.G.I., 1960.

plate dent method, a cylindrical charge is detonated upon a steel or aluminium plate. The dent formed in the plate gives a quantitative measurement of the energy of the detonation. The results of this test are subject to wide variations unless the geometry of the explosive charge, and system of initiation are maintained identical, and are favourably biased towards explosives with a high strain wave energy. Jimeno et al., 1995.

PLC (acronym), for '**programmable logic control**'. Atlas Copco, 2006b.

Plexiglas (tradename), see '**PMMA**'.

plough cut, see '**V-cut**'.

plow cut, see '**V-cut**'.

plunge of a line (β), (°), the acute angle, measured in a vertical plane, between the line and the horizontal. A line directed in a downwards direction has a positive plunge; a line directed upwards has a negative plunge. $-90° \leq \beta \leq 90°$. Basic term used in hemispherical projection. Brady, 1985.

PLS (acronym), (I_s), (**MPa**), see '**point load strength**'.

plug, plugs or stemmings are used in blastholes using separate charges that are initiated on separate delays with the purpose to avoid instantaneous initiation of all charges. In vertical crater retreat mining plugs are used to keep the explosive in place. Plugs can be a spacer made of two circular disks and a spacer between the disks, inflateable rubber balls, crushed rock, jute bags with sand, air bags chemically inflated etc.

plug hole role, type of integral drill rod, used for the drilling off plug holes. Atlas Copco, 2006.

plug subsidence, see '**subsidence**'.

plug subsidence, a form of discontinuous subsidence in which a plug of material overlying an underground opening subsides into the opening suddenly. See '**subsidence**'.

PM (acronym), for '**preventive maintenance**'. Atlas Copco, 2006.

PMMA, also called **Plexiglas** (tradename) is a polymeric transparent material consisting of *polymethylmethacrylate* with a density of 1 190 kg/m³, P-wave velocity 2 230 m/s, uniaxial compressive strength 137 MPa and elastic modulus 5,9 GPa. It is a commonly used material for scaled laboratory studies of shock waves, stress waves, cracking and fragmentation because of its photoelastic properties.

PN (acronym), for '**potassium nitrate**'.

pneumatic, a technique where compressed air is used as energy source and media distributed in pipes and hoses to a pneumatic motor, drill machine etc.

pneumatic blowpipe, a long metal pipe used for cleaning blastholes. Connected to the air supply to blow out dust and chippings from the vertical blast holes at quarries. Nelson, 1965.

pneumatic breaker, demolition machine having a percussive mechanism but no rotation device and powered by compressed air. A pneumatic breaker is fitted with a chisel steel or moil point and used to demolish cement, asphalt and masonry structures. Atlas Copco, 2006.

pneumatic cartridge loader, a machine using compressed air for the transportation of cartridges into the borehole. It is widely used for underwater blasting, for blasting without removing the overburden, and for long-hole blasting. It is also being used increasingly in tunnelling and other sorts of rock blasting especially in wet boreholes. Langefors, 1963.

pneumatic charger *or pneumatic loader*, a machine for charging of explosives by compressed air. Some of the machines are designed for bulk blasting agents like ANFO, emulsion and others for cartridged explosives.

pneumatic charging *or pneumatic loading (or placing)*, the loading of explosives or blasting agents into a borehole using compressed air as the loading force. IME, 1981.

pneumatic churn drill, the drilling effect is achieved by crushing the bottom of the drill hole. The drill steel and bit are moved by a compressed-air-operated piston, which drives them against the rock. The cuttings are removed by continuous flushing, usually with compressed air. Bits are generally of the 4-point type with tungsten carbide inserts. Rock drills of this type are naturally large and heavy. They are self-propelled and mounted on caterpillar tracks. Compressors are accommodated on the machine itself, although compressed air can also be supplied from a stationary plant. Fraenkel, 1952.

pneumatic diaphragm transducer, see '**hydraulic diaphragm transducer**'. Brady, 1985.

pneumatic drill, compressed air drill worked by reciprocating piston, hammer action, or turbo drive. Pryor, 1963.

pneumatic drilling, drilling holes using a pneumatic drilling machines. Atlas Copco, 2006.

pneumatic drilling machine, a drilling machine powered by compressed air. TNC, 1979.

pneumatic drill leg, see '**air-leg support**'. Nelson, 1965.

pneumatic hammer, a hammer in which compressed air is utilized for producing the impacting blow. USBM, 1990.

pneumatic hammer drill, a compressed air driven hammer drill machine.

pneumatic hoist, a device for hoisting operated by compressed air. Standard, 1964.

pneumatic injection, a method for fighting underground coal fires developed by the USBM. This air-blowing technique involves the injection of incombustible mineral like rock wool or dry sand, through 150 mm boreholes drilled from the surface to intersect underground passageways in the mines. USBM, 1990.

pneumatic lighting, underground lighting produced by a compressed-air turbo motor that drives a small dynamo. Pryor, 1963.

pneumatic machine, a machine operated by compressed air.

pneumatic packer, an instrument inserted into a borehole and by which a plastic film can be pressed against the borehole wall to detect new cracks. The impression of the wall is done before and after blasting to find out what new cracks have been developed during blasting.

pneumatic pusher, see '**air leg**'.

pneumatic stowing, a system of filling mined cavities in which the crushed rock is carried along a pipeline by compressed air and discharged at high velocity into the space to be packed. The intense projection ensuring a very high density of packed material. Nelson, 1965.

pocket charge, a small explosive charge placed centrally within the stemming column and with the purpose to break strong surface material in bench blasting. Sen, 1995.

pocket priming, initiation carried out by placing the primer cartridge at the bottom or on the top of the explosive column. Jimeno et al., 1995.

point charge, a spherical charge or as near spherical as possible. In crater blasting and when relating to the far field, a cylindrical charge with a maximum length of six times the diameter is also considered as a point charge.

point initiation, initiation of an explosive charge at one point in the explosive column. For safety reasons, and/or improving the performance of the explosive, several point charges can be installed in an explosive column; bottom, middle or top initiation. The initiation can be made by a cap and for non-cap sensitive explosives by a priming charge called a primer. Commonly used primers have a mass of 50 g to 5 kg. The opposite to point initiation is *side initiation* (*radial initiation*), which can be achieved by a detonating cord.

point load anisotropy (I_a), (dimensionless), the ratio between the maximum and minimum point load indices for a particular kind of rock. $I_a = I_{s(max)}/I_{s(min)}$. Broch, 1974. See also 'point load strength test'.

point load index, see 'point load strength test'.

point load strength or **index (I_s), (MPa)**, a method of determining an approximate compressive strength in the field or in the laboratory. A point load is applied diametrically to a rock core. The force (F) needed to break the rock and the length (l) between the loading points ($l = d$, where d is the diameter of the core) determines the point load index $I_s = F/d^2$. The test procedure is described by ISRM, 1985. The point load strength depends very much on the water content in the rock and a variation of the point load index of about 35% can be expected between 0 and maximum water content. The compressive strength can be calculated approximately by the formula $\sigma_c = 24 I_s$. Broch and Franklin, 1972.

point property, a rock property not depending on discontinuities. Examples of such properties are density, primary porosity, permeability of intact rock, point load strength, cutability etc. Hudson, 1993. Compare with volume property!

Poisson's ratio (ν), the absolute value of the ratio of lateral strain to corresponding axial strain for a material subjected to axial loading or tension below the proportional limit. Hunt and Groves, 1965. It can be calculated as follows;

$$\nu = \frac{c_p^2 - 2c_s^2}{2(c_p^2 - c_s^2)} = \frac{\lambda}{2(\lambda + G)} \quad (P.4)$$

where ν is Poisson's ratio, c_p is the longitudinal wave velocity in (m/s) and c_S is the transverse wave velocity in (m/s), λ is Lame's constant and G the shear modulus in (Pa). Hudson, J.A., 1992.

pole adaptor, a mechanical connector to extend a tamping pole. One example of such connector is the aluminium ball and socket connector.

poling boards, timber or steel planks driven into the soft ground at the tunnel face over supporting steel or timber sets to hold back soil during excavation. Bickel, 1982.

polyaxial compression, strength tests carried out on cubes, or rectangular prisms of rock with different normal stresses ($\sigma_1 > \sigma_2 > \sigma_3$). Brady, 1985. A special case is the loading of cylindrical specimens where two of the stresses are the same. Spathis, 2007.

polycorder, a pocket-sized microprocessor which is connected to a pressure transducer which in turn is coupled to a valve leading to the leg cylinders hydraulic fluid chamber. Each polycorder can serve up to 10 transducers. Hudson, 1993.

polycrystalline, composed of many crystals; an aggregate, as distinct from a single crystal. Rolfe, 1955.

polymethylmethacrylate or **PMMA**, see 'PMMA'.

polystyrene-diluted ANFO, (ANFOPS), a quality of ANFO where the strength of the explosive has been reduced by dilution with polystyrene prills. ANFOPS is normally used in controlled contour blasting, but can also be used in production blasting to reduce the amount of fines.

polytropic exponent (γ), (dimensionless), an exponent used to correct the pressure p versus the volume V relation at explosion in the relation pV^γ = constant.

pond test, see '**underwater explosive test**'.

pop, see '**popping**'.

popping or **pop shooting**, breaking boulders by using small explosive charges confined in short blastholes, see also '**secondary blasting**'. Sen, 1992.

pop shot, in mining, a shot fired for trimming purposes. B.S. 3618, 1964.

pore, elementary volume of interstitial space of a matrix skeleton through which the exchange of fluid matter with the exterior may occur. The pore volume in intact hard rock is normally only a few percent and therefore the fluid matter exchange with the exterior is limited. Pores are almost equant dimensioned cavities with rounded, polygonal or irregular shapes, ranging from micro to mega pores depending on their size. Pores are normally not closed when they are stressed compared to cracks which can be closed. Porosity has a high influence on the elastic wave velocities in most materials. The compression wave velocity is being more affected than the shear wave velocity. The presence of fluids in rock voids changes many properties and elastic constants of the rock. Water saturation is more favourable for the transmission of elastic waves. Miguel, 1983. Pores can be quantified by its *size distribution*, *shape* and their location in relation to the rock texture and mineralogy (intra-, or intergranular, associated or not to zones of mineral alteration, etc.). Montoto, 1983. *Micro pores* have size of 300 Å (Ångström) to 7,5 μm and *micro cavities* are >7,5 μm. Goni, 1970. The amount of porosity is very important for the uniaxial compressive strength of rock. For example one granite with 1% porosity could have an unconfined compressive strength (UCS) of 160 MPa when another granite with 5% porosity has a UCS of 50 MPa. Miguel, 1983.

pore pressure (p_p), (MPa), see '**pore-water pressure**'.

pore-water pressure (p_p), (MPa), the water enclosed by pores may have a pressure acting on rock. The *effective pressure* in the rock mass is the total or applied pressure plus the pore water pressure. Brady, 1985.

porosity (*n*), the ratio of the volume of the pore space (V_p) in the in situ rock mass to the total volume of rock and pores (V_T), $n = V_p/V_T$. For fragmented rock this term does not include the space between the fragments. Instead swell and swell factor are used to quantify the space between rock fragments.

portable magazine, a container used to store limited amounts of explosives in a building that is not a magazine, or to contain explosives during transport.

portailing, the process of starting a new tunnel from surface. The entrance to the tunnel is called '**adit**' or a '*portal*'.

portal, the entrance to a tunnel.

portal jumbo, portal jumbos are custom-built drilling units for drifting usually designed as track-bound but could also be delivered with rubber tyres. Portal jumbos have an opening, or a portal through which other traffic can move. Thereby the jumbo does not have to be moved when loading etc., but only pulled back a safe distance when blasting. Portal jumbos are mainly used in large or medium-size tunnels, but could be applied down to 10 m² tunnels. Tamrock, 1983.

portal point of attack, see '**portailing**'.

post, a mine timber, or any upright, timer, but more commonly used to refer to the uprights which support the roof crosspieces. Commonly used in metal mines instead of the leg which is the coal miners term, especially in the Far West regions of the United States. B.C.I., 1947.

post and cap or **prop and lid**, simplest support of the back consisting of a single upright with a plate above. Brady, 1985.

post-and-stall, see '**bord and pillar**'.

postblast assessment, quantification of the blast result by assessing, for example, fragmentation, diggability, overbreak, underbreak, throw of rock, etc.

postblast rock assessment, any factor or variable which serves the characterisation of rock or rock mass that can be measured and which is a direct result of blasting. Worsey, 1996.

post caving, mining methods where a stope is allowed to cave after mining e.g. sublevel stoping or shrinkage stoping. The two mentioned mining methods can also be used without post caving.

post- or **conventional undercut**, an undercutting strategy in which the undercut is mined after the development of the underlying extraction level, including the draw points, has been completed.

post-detonation, delayed detonation of an explosive charge. The delay may be due to a faulty ignition or to a deflagration process, taking place for a limited period of time in stead of the immediate detonation. Meyer, 1977.

post pillar mining (PPM) (acronym), a combination of *room and pillar mining* (*RP*) and *cut and fill* mining (CAF). Post pillar mining recovers the mineralization in horizontal slices, starting from a bottom slice, advancing upwards. Pillars are left inside the stope to support the roof. Mined out stopes are backfilled with hydraulic tailings, which might contain cement for added strength, to allow the next slice to be mined working on the fill surface. Pillars are extended through several layers of fill, so that the material contributes to the support permitting a high recovery rate. The fill allows the stope layout to be modified to suit variations in rock conditions and ore boundaries. Atlas Copco, 2007.

post-splitting, '**controlled contour blasting**' and especially **smooth wall blasting**'.

potential energy (W_p), **(J)**, the form of mechanical energy a body possesses by virtue of its position in gravity, magnetic or electromagnetic field. If for example a body is being dropped from a higher to a lower position in a gravity field the body is losing potential energy, but if the body is being raised, then it gains potential energy.

pothole, a circular or funnel-shaped depression in the surface caused by subsidence. Hudson Coal, 1932. A rounded cavity in the roof of a mine caused by a fall of rock, coal, ore, etc. Fay, 1920.

power or **effect** (*P*), **(J/s** or **W)**, physical quantity defined by the energy per unit of time.

power operated supports, see '**self-advancing supports**'.

power shovel, an excavating and loading machine consisting of a digging bucket at the end of an arm suspended from a boom, which extends cranelike from that part of the machine which houses the power plant. When digging the bucket moves forward and upward so that the machine does not usually excavate below the level at which it stands. Bureau of Mines Staff, 1968.

power source, the source of power for energizing electric blasting cap circuits. IME, 1981.

powder chest, a substantial, non-conductive portable container equipped with a lid and used at blasting sites for temporary storage of explosives. USBM, 1983.

powdered emulsified explosive (PEE), is made up through pulverization or "spin flash vaporization" or "post-cooling-solidifying comminution" of the water/oil emulsion which is composed of low water content or melted oxidizer solution as dispersion phase and carbonaceous combustible component and emulsifiers as continuous phase and formed by "forced shearing" under some special conditions. The content could be 87–93% ammonium nitrate, 0–8% water, 1,5–2,5% emulsifier, 3,6–5,5% compound oil-phase and 0,1–0,5% additives. Examples of physical properties of PEE are density of 0,9–1,05 t/m^3, detonation velocity (c_d) 4 200 m/s, sympathetic detonation distance 16 cm and lead block test 0,352 litre. Xuguang, 2002.

powder explosive, any solid (condensed or cast) explosive. Usually abbreviated to powder. USBM, 1983.

powder factor or explosive loading factor (q)**, (kg/m^3 or kg/t)**, see '**specific charge**'.

powder house, explosive store, **explosive magazine**, a surface building at a mine or construction site where explosives and detonators may be kept. It must be at a certain minimum distance from other buildings and the maximum quantity of explosives that may be kept is fixed.

powder magazine, a portable box or permanent structure used to store explosives and blasting accessories. Jimeno et al., 1995.

powderman, a man responsible for charging (loading) of the blastholes. USBM, 1983.

powder punch, is made like a screw driver but machined to a sharp point for easy penetration. The metal must be made of a non-sparkling material like brass. Iron or steel is forbidden.

power loader, a generic term for any power operated machine for loading coal or any other material into mine cars, conveyors, road vehicles, or bins. The term is also applied loosely to the man, in charge of such a machine. In Great Britain's coalmines, more than 50% of the deep mined output comes from some 1.400 power loaders mostly working on longwall faces with prop-free fronts. Nelson, 1965.

PPD (acronym), for '**peak particle displacement**'.

PPM (acronym), see '**post pillar mining**'. Also acronym for parts per million.

PPV (acronym), (v_{max})**, (mm/s)** for '**peak particle velocity**'.

practical burden (B_p)**, (m)**, the planned burden used in blasting, which is derived by subtracting the greatest blasthole deviation (caused by drilling) from the maximum burden. For vertical benches practical burden is equal to drilled burden.

practical spacing (S_p)**, (m)**, the planned spacing used in blasting which is determined after the practical burden has been calculated. One criteria for the calculation of the practical spacing may be the achievement of a certain specific charge. The calculated practical spacing will always be the same as the drilled spacing, $S_p = S_d$

preblasting, a blasting method where by using explosives the natural fractures of a rock mass is increased with very little displacement. This improves the production rates of the operation at a relatively low cost. Jimeno et al., 1995.

preblast rock assessment, any factor or variable in the rock or rock mass that could be measured, that has been determined to, or could potentially have an effect on the blast result.

preblast survey, a survey conducted to document the existing condition of a structure. It is used to determine whether subsequent blasting will cause damage to the structure.

precharging, charging of more rows of blastholes than that or those being blasted. The purpose of precharging of the production blastholes is that the vibration from the blasted holes should not destroy holes behind and make it impossible to charge them. The technique is used at LKAB.

precision drilling, drilling where the total hole deviation is less than 1%. Sandvik 1983.

preconditioning, to determine conditions and limitations before work starts.

preconditioning drilling, a term used at block caving in Codelco:s Andina mine in Chile for drilling of long vertical holes upwards with lengths up to 115 m and with a diameter of 146 mm in diameter. The ore is then fractured either by pressurising the drill hole with water, or blasting the hole with explosives. The cracking of the block will improve its caving properties. The method sorts under the category *induced block caving*. Anon., 2006.

precut, cutting a rock face to create a new free face prior to blasting. A term used in underground coal mining. Sen, 1992.

preignition, see '**premature blast or explosion**'.

premature blast,-detonation,-firing or **-explosion**, the accidental initiation of a blast prior to the expected time, due to either lightning, heat liberated by minerals or malpractice. In sulphide-bearing ore bodies exothermic oxidation of pyrite (FeS_2) forms ferrous sulphate ($FeSO_4$), which reacts with ANFO-based explosives exothermically, and the associated elevated temperatures can set off detonators and cause explosions in blastholes. Miron, 1992.

premium system, a mixed system with one part as payment by hour and the other part payment by piecework. Fraenkel, 1952.

prenotching, the act of making notches by mechanical tools, linear shaped charges or water jets. The notches are made along blastholes, on diametrically opposite sides, with the aim of lowering the necessary blasthole pressure for breakage. The borehole pressure needed to create radial cracks to the neighbouring holes will be reduced, and therefore less blast induced damage due to vibration in the surrounding rock is caused.

pre-reinforcement, installation of reinforcement in a rock mass before excavation commences.

preshearing, fracturing of rock by shearing (sliding).

preslotting, sawing a slot by mechanical tools around the excavation (tunnel) before blasting the main part of the round.

presplitting, a blasting method used to create a crack along the contour before blasting the actual round. The tensile cracks for the final contour are created by firing a single row of closely spaced holes (10- to 25-hole diameters apart) simultaneously prior to the initiation of the remaining holes in the round. The method is used in controlled contour blasting mainly on the surface but seldom underground, because of the strong air blast effect caused when many holes are initiated at the same delay time. *Degree of freedom in presplitting is* defined as the ease with which the split can be formed by wedging action. *Zero degree of freedom* means that the rock body cannot accommodate significant movement, so that the crack will be hairline. *First degree of freedom* means that the rock can give way due to intrinsic porosity, cracks, etc., forming a significant parting. The rock mass is, however, fixed on both sides. *Second degree of freedom* means that a free face is close enough, and the fixation is sufficiently loose, for the whole rock mass is able to slide away on one side, forming a major rift. *Third degree of freedom* means that there is a free face so close that there is danger of a presplit fragmenting the rock and disturbing it beyond further working. This is basically the situation of postsplitting. AECI, 1993.

press fitting of buttons, manufacturing technique to fit cemented carbide buttons to bit body by frictional force. Sandvik, 2007.

pressiometer, instrument, developed by Ménard, to measure the Young's modulus in situ by inserting a pressurizing tube in a borehole in the rock mass and apply a pressure on the borehole walls and measure the radial deformation. From the data measured the Young's modulus can be calculated by the following equation,

$$E = 2\pi r^2 (1+\upsilon) \frac{p_1 - p_2}{V_1 - V_2} \qquad (P.5)$$

where E is Young's modulus in MPa r is the borehole diameter in m, υ is Poisson's ratio and p_1 starting pressure and p_2 final pressure in Pa and V_1 starting volume in the cell and V_2 final volume in the cell in m^3. Pusch, 1974.

pressure (p), (Pa), force per unit of area. (1 Pa = 1 N/m^2, non-metric units 1 bar = 0,1 MPa, 1 lbf/in^2 (psi) = 0,00689 MPa, 1 atm = 0,101 MPa).

pressure cavity pump, type of single screw pump used for liquids with significant amounts of solids such as cement or sand slurry. Atlas Copco, 2006.

pressure desensitization of explosive, an explosive can become insensitive if it is compacted to a too high density before initiation. This may happen if the explosive is set under static and/or dynamic pressure before and during the blasting process.

pressure drilling, a process of rotary drilling in which the drilling fluid is kept under pressure in an enclosed system. Brantly, 1961.

pressure front, see '**shock front**'.

pressure measurement around a blasthole, a 0,1 W (watt) composition *carbon resistor*, 5 mm long and 1,7 mm in diameter can be used to measure pressure in the range of 1–70 MPa behind blastholes. Watson, 1967.

pressure measurement in a borehole, if the rock is saturated with water both the pressure from the initial blast wave and also the quasi static pressure can be measured in the neighbouring blasthole. Preston, 1984.

pressure meter, see '**dilatometer**'.

pressure of explosion (p_b), **(MPa)**, see '**borehole pressure**'.

pressure relief valve, valve that opens to relieve pressure in a system when it exceeds the desired level. Atlas Copco, 2006.

pressure transducer for blast waves, see '**pressure measurement around a blasthole**'.

pressure vessel, a part of a pneumatic charging machine for explosives, usually ANFO, where the explosive is contained in a sealed vessel made of stainless steel, to which air pressure is applied, forcing the ANFO through the semi-conductive hose into the blasthole. Also known as *pressure pot*.

pretensioned rock bolt or **cable**, after insertion of a rock bolt or cable they are pretensioned to 1/3 or 1/2 of the failure load. Pusch, 1974.

pre-undercut, an undercut strategy in which the undercut is mined before the development of the underlying extraction level and the draw points.

prill, in blasting a small bead of a chemical, formed to ensure free-flowing characteristics. Typically porous prills of AN are used for manufacturing ANFO.

prilled ammonium nitrate, ammonium nitrate in pelleted or prilled form. Atlas Powder, 1987.
Primacord (tradename), see '**detonating cord**'.
Primadet (tradename) **noiseless trunkline delays (NTD)**, an initiation system where detonating cord is used in the blastholes and non-electric initiation tube systems on surface. Persson, 1993.
primary blast, see '**primary blasting**'.
primary blasting, the main blast executed to sustain production. The purpose is to facilitate subsequent handling and crushing. Compare with secondary blasting.
primary charge, see '**primary explosive**'.
primary drilling, the process of drilling holes in a solid rock ledge in preparation for a blast by means of which the rock is thrown down. Fay, 1920.
primary explosive or **initiating explosive**, an explosive sensitive to spark, friction impact or flame, which is used, e.g. in a detonator to initiate an explosion.
primary fragmentation, the fragmentation defined by the blocks in the vicinity of the cave back as they separate from the cave back when the undercut is mined and caving is initiated and proceeding upwards.
primary state of stress or **virgin-, natural-** or **paleostress**, stress in undisturbed parts of underground, before any manmade excavations like drifts, shafts, tunnels, winzes, raises, stopes or rooms have been driven. TNC, 1979.
primary support, see '**support**'.
primed, the condition of an explosive when fitted with an initiating device. AS 2187.1, 1996.
primer or **primer charge**, unit package or cartridge of explosives (booster) used to initiate other explosives or blasting agents, and which contains, a detonator, or a detonating cord to which is attached a detonator designed to initiate the explosive. Atlas Powder, 1987.
primer cartridge, an explosive charge of high strength and sensitivity into which the detonator is placed. It is used to increase the energy of the explosive.
primer charge, see '**primer**'.
primer cord connector, plastic hook up device for connection of detonating cord downline (in the blasthole) to the trunk line (un surface).
primer location, the preferred primer location is at the bottom of the hole to ensure that the bottom of the charge detonates.
priming, the action of insertion of a detonator or primer in the explosive often placed in a blasthole. The priming can be done in two ways; *direct priming* where the detonator points towards the body of the charge and indirect priming where the detonator points away from the body of the charge. For greatest efficiency, direct priming is required, but for safety reasons *indirect priming* is often adopted. It is general experience that with non electric initiation systems, turning the detonator in the blasthole to face the column of explosive, makes the detonator difficult to extract again, and can cause misfires through stress at the fuse or tube entry point.
priming charge, see '**primer**' and '**ASA**'.
Princess incendiary spark test, a heat test to access the case of ignition of an explosive substance by incendiary sparks produced by a length of safety fuse. The test will be compulsory in the future for United Nation classification. Persson et al., 1994.
principal stresses (σ_x, σ_y, σ_z), **(MPa)**, the normal stresses on three mutually perpendicular planes on which there are no shear stresses. ASM Gloss., 1961. When defining principal

stresses in the earth crust, *y* corresponds to the *north direction*, x corresponds to the east direction and *z* to the radial direction from center of earth.

probability of damage, based on a large amount of ground vibration measurements in different structures and their correlation to the damage in these structures, it is possible to establish probability curves for damage especially in buildings associated and inflicted at certain vibration levels.

production blast, a blast whose purpose is directly associated with the production of a commodity and it is the primary activity at a mine, quarry or construction site. It is distinct from other forms of blasting, e.g. secondary blasting or seismic blasting, in that these operations are not uniquely concerned with the production of a commodity.

production drift, one of the major set of parallel excavations or drifts on the extraction or production level through which the drawpoint drifts are accessed and ore is transported away from the drawpoints.

production drilling, drilling of those blastholes needed for winning the main volume of ore. The development drifts placed in ore before the production drilling starts are therefore belonging to *development drilling*.

profile blasting, see '**controlled contour blasting**'.

profile gauge, a tool to measure the surface roughness of rock with the help of closely spaced pins which are pushed onto the joint surface. The resulting image can be traced for further analysis. Several shorter sets of pins can be joined together to form 1 m long profile gauge.

profile measurement, method to determine the contour after blasting. In drifts or tunnels normally perpendicular to the axis of the drift or tunnel. The most accurate method today is to use laser scanning. Hertelendy, 1983. A light slot can also be projected on the tunnel perimeter and photographed. Franklin, 1989.

profile recorder, the profile of drifts and tunnels can be measured by a laser beam. The measurement distance for one instrument is 1–10 m and the accuracy ±0,5 cm and it takes 15 s (seconds) to measure 1 profile with 400 point. Some points may give disturbing reflections. Celio, 1983.

profilometer, profile measurement instrument. Several techniques are available for measurement of the contour of the rock surface after blasting; a) Theodolite and distance measurement equipment (laser), manual or fully automatic, b) The light slot method where a thin light slot is projected on the perimeter of the tunnel and photographed. This method can be used underground to determine the real contour in drifts or tunnels. c) Stereo photography.

programmable logic control (PLC), a system started to be used in the 1990's to control drill rigs. Atlas Copco, 2006b.

projectile impact test or **armor plate test**, test developed in the USA to study the behaviour of a given explosive which is charged into a test projectile which is fired from a 'gun' against a steel plate. The impact velocity which causes the charge to detonate is determined. Meyer, 1977. Another variation of this test is to shoot brass bullets with a diameter of 15 mm onto the explosive. Persson et al., 1994.

prop, underground *supporting post set* across the lode, seam, bed, or other opening. Pryor, 1963. In mining, a roof support is usually temporary. B.C.I., 1947. In tunnel construction work a '*strut*' or '*post*' either vertical or inclined, usually of round timber, used as a support or *stay*. Fay, 1920.

prop and **lid**, see '**post and cap**'. Brady, 1985.

propagation blasting, closely spaced and sensitive charges placed in ground for ditch blasting in damp ground. The shock from the first charge propagates through the ground, setting off the adjacent charge, and so on. Only one blasting cap is required. USBM, 1983.

propagation velocity of a crack (c_c), (m/s), the velocity of the crack front of a propagating crack. Crack velocities found in the laboratory varies from a couple of 100 m/s to a maximum of 1 500 m/s and is in rock typically 1 000 m/s. The velocities are not related to any material property. Some methods to measure crack velocities are as follows,
   1   From high-speed photographs, if the frame frequency is known.
   2   Breaking of conducting strips glued to the surface of the rock specimen.

propagation velocity of seismic waves (c_P, c_S, c_L and c_R), (m/s), the different waves are mentioned in order of falling velocities; longitudinal wave c_P > transversal- or shear wave c_S > Love wave c_L and finally Rayleigh wave c_R.

propagation velocity of shock wave (c_{sh}), (m/s), the velocity of a mechanical longitudinal wave, when its velocity is greater than the sound velocity in the material (supersonic velocity).

propellant, see '**propellant explosive**'.

propellant actuated power device, any tool or special mechanized device or gas generator system which is actuated by a propellant or which releases and directs work through a propellant charge. IME, 1981.

propellant explosive, an explosive which under normal conditions deflagrates, and is used for propulsion. USBM, 1983. It may be a class A or B explosive depending upon its susceptibility to detonation. Atlas Powder, 1987.

prospect drilling, the exploratory drilling of boreholes in the search for minerals, oil and gas.

prospect exploration, see '**prospect drilling**'.

protective roof, a roof structure built onto a drill rig to protect the operator from falling objects. Atlas Copco, 2006.

Protodyakonov scale of strength, is based on the compressive strength of small cubes, with flat ground faces. The scale is defined as the compressive strength value in MPa divided by 10 and gives a convenient scale of 1 to 20 for the general range of rocks, with values less than 1 quoted for very weak rocks such as some coals. Hudson, 1993.

Protodyakonov's rock stress theory, the load on the underground drift or tunnel is described by,

$$\sigma_v = \frac{2}{3}\rho_r ah, \quad h = aH, \quad h = \frac{a}{2f} \tag{P.6}$$

where σ_v is the vertical stress on the drift in MPa, ρ_r is the density of rock in kg/m³ and h is the height of the parabolic pressure arch above the drift in m, a is the width of the drift in m, H is the vertical depth of the roof below the ground surface in m, f is the index of rock strength given by the following two expressions where ϕ is the angle of internal friction and σ_c is the immediate compressive strength of the rock in MPa.
   $f = \tan\phi$ for weak loose rocks
   $f \approx 0{,}1\ \sigma_c$ for strong competent rocks

PSC (acronym), (q_p), (kg/m³), for '**perimeter specific charge**' the mass of explosive per intended volume rock broken by the perimeter holes.

PSD (acronym), for '**particle size distribution**'.
PTZ (acronym), for '**lead-titanate-zirconate**'.
pull or **advance per round** (A_r), (**% of drilled depth** or **m/round**), achieved blasted depth of a drift, tunnel, raise or shaft round.
pulley, a wheel that carries a cable or belt on part of its surface. Nelson, 1965. Wheel used with a chain or wire rope to change circular motion into straight motion. Atlas Copco, 2006. A sheave or wheel with a grooved rim, over which a winding rope passes at the top of the headframe. Fay, 1920.
pulling casing, removing pipes from a drilled well. Hess, 1968.
pullout tests for rock bolts, with the help of a hydraulic jack system a tensile force is applied to the rock bolt head until the rock bolt breaks. Pull out force and displacement are recorded. Stillborg, 1994.
pull tester (rock bolts), device consisting of a pump, hydraulic cylinder, an attachment rod and a spacer, used to exert a tensile load on installed rock bolts to test the quality of the bolt and installation. Atlas Copco, 2006.
pulsating detonation, the property of certain explosive to alter the detonation velocity (c_d) (decrease and increase) between certain limits.
pulse, a disturbance propagated in a medium in a similar manner as a wave, but not having the periodic nature of a wave, it may represent the envelope of a small wave group. ASM Gloss, 1961.
pulse rise time (t_r), (**ms**), the interval of time required for the leading edge of a pulse to rise from ~5% to 95% of its peak value. Hunt, 1965. See also '**rise time**'. The rise time at the blasthole wall was never measured but an extrapolation would give a rise time of about 10 μs and the strain rate will be $10^5 s^{-1}$. Spathis, 1988. The pulse rise time measured in model blast tests was 35 μs corresponding to a strain rate of $2,9 \cdot 10^4 s^{-1}$. Stecher, 1981.
pulsed infusion shotfiring, this blasting technique combines the effect of an explosive charge in coal mine blasting with the effect of water pressure. The borehole is loaded with the explosive charge, after which water is pressed into the borehole with the aid of the so-called water infusion pipe, and the charge is ignited while maintaining the water pressure. The pressure shock in the water causes the coal to disintegrate into large lumps. In addition, the water fog, which is produced at the same time, causes the dust to settle on the ground. Meyer, 1977.
P-wave (primary wave), compressional or *longitudinal wave*, see '**seismic waves**'.
pulsed water injection by explosives, explosives can be used for pulsed water injection, see '**water injection**'. Fraenkel, 1952.
pulverizer rod, large rod used in a rig mounted hydraulic breaker used for demolition work. Atlas Copco, 2006.
pulverizing rod, pad used in demolition work. Atlas Copco, 2006.
pulverizor rod, see '**pulverizer rod**'.
push leg feed, see '**pusher leg**'.
pushback, the volume of rock removed from the hanging wall in an open pit mine to make it possible to deepen the pit. Atlas Copco, 2006b.
pusher leg, a pneumatic cylinder used for creating thrust to the drill bit. The leg is connected to the light or handheld drill machine by a joint. Sandvik, 2007.
pusher leg drill machine, a drill machine attached by a joint to a pneumatic pusher leg used for creating the necessary thrust for drilling. Sandvik, 2007.

PVC (acronym), a plastic material consisting of polyvinyl chloride, a non-flammable substance, used widely for mine belt conveyors, ducting, etc. Nelson, 1965.

PVC-ampoules, stemming material wrapped with PVC for use in underground coalmines of degree II and III gassiness. Depending on the content of the ampoules their use significantly contribute to the safety by reducing the risk of methane ignition or coal dust explosion, reduction of dust and fume, reduces risk of deflagration, reduces the fragment size and the energy transformation to rock is improved by 60% when water stemming is used.

pyramid, *diamond or German cut,* a method of making a cut in a drift or tunnel. Three or four holes (placed at the corner of a triangle or square) in the cut are obliquely drilled towards a common point in the rock thereby forming a pyramidal shape, see Fig. 47.

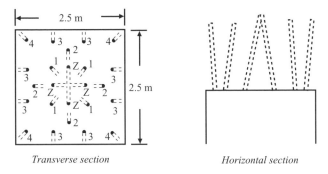

Figure 47 Pyramid cut. After Sen, 1995.

The pyramid cut is mainly employed in raises and for shaft sinking, but is not recommended for horizontal tunnels where a machine set-up for a definite direction of the four holes cannot easily be obtained.

P-wave velocity (c_p), (m/s), primary wave-, compressional wave- or longitudinal wave velocity, see 'seismic waves'.

pyrotechnic delay, delay interval in a detonator created by pyrotechnics. The delays cannot be manufactured with high precision and the scatter in timing increases with the delay number. Overlapping of intervals, however, is prohibited. When high precision is required, such as in controlled contour blasting, it is recommended to use electronic detonators.

pyrotechnics, a class of chemicals (combustible or explosive composition) employed for the purpose of producing heat, light, smoke, gas or noise. Pyrotechnics is commonly referred to as fireworks. It is also the science of low explosives, especially timing devices with slow burning powders.

Q-factor (Q), a rock mass classification factor initially designed for the purpose of rock support, see Table 35. Barton et al., 1974. Overbreak in tunnelling can be correlated to the Q-factor, see '**overbreak**'.

Table 35. Rock mass classification with respect to rock support. After Barton et al., 1974.

Description	Q-factor
Extremely bad	<0,01
Very bad	0,01–0,10
Bad	0,10–1,00
Acceptable	1,00–4,00
Good	4,00–10,00
Very good	10,00–100
Extremely good	100–400
Exceptionally good	>400

Q can be calculated with the following formula

$$Q = \left(\frac{f_{RQD}}{J_n}\right)\left(\frac{J_r}{J_a}\right)\left(\frac{J_w}{f_{SRF}}\right) \qquad (Q.1)$$

where f_{RQD} is the RQD-value (Rock quality designation), J_n is the joint set number (0,5 to 20), J_r is the joint roughness number (1 to 4), J_a is joint alternation number (0,75 to 20,0), J_w is the joint water reduction factor (0,05 to 1,0), f_{SRF} is the stress reduction factor SRF 0,5–20. Barton, 1974.

The P-wave velocity c_p and Young's modulus E can be approximately calculated by the following formulas

$$c_p \sim 1000 \log Q + 3500 \text{ (m/s)} \qquad (Q.2)$$

$$E \sim 25 \log Q$$

quality in tunnelling, is measured in form of the *advance rate, deviation in height and side* of the tunnel to the intended, *overbreak, half cast factor* (sensitive indicator of the quality of blasting). Innaurato, 1998. Also *underbreak*, *misfires*, *bootlegs*, *fragmentation*, *throw*, *clean floor* are important indicators of quality. Rustan, 2008.

quantity-distance table, a table listing minimum recommended safe distances from stores of explosive materials of various mass to a specific location.

quantity of heat or **heat** (Q), **(J)**, see '**heat**'.

quarry, a surface operation for the extraction of stone for construction and ornamental purposes, rather than for extracting minerals.

Quarryman (tradename) **profiling instrument**, an instrument for measuring profiles in quarrying with the help of a laser system which reflects a pulsed laser beam from the rock face. An internal electronic clock measures the 'time of flight' of the pulse and thus the distance is calculated. Simultaneous electronic vertical and horizontal angles are measured indicating the direction of the observation. For other methods see '**profilometer**'.

quasi-static pressure (p_{qs}), **(MPa)**, time-varying pressure conditions which do not require fully dynamic analysis.

quick-match, a rapid burning pyrotechnic fuse consisting of one or more strands of string covered with gunpowder and the whole being encased in a loose fitting paper sleeve. AS 2187.1, 1996.

quicksand, sand saturated with water, 1) underground quicksand tends to run into any opening and must be well supported at all points. 2) at the surface, quicksand has no bearing capacity, but the bearing can be improved through draining.

Q-wave, see '**Love wave**'.

radar (abbreviation), for '**radio detection and ranging**'.
radar wave velocity (c_{ra}) **(m/s)**, an electromagnetic wave with the same speed as light 3×10^8 m/s.
radar waves, electromagnetic waves with a wave length from about 1 cm to a 1 m and the corresponding frequencies are 3×10^{10} and 3×10^8 p/s.
radial coupling ratio (R_{cr}), ratio of area (or volume) of the explosive to the total area (or volume) of the blasthole in a section of the charged length of the blasthole.
radial cracking method, a method for estimation of the fragmentation in blasting due to radial cracks extending from the blasthole wall and terminating at planes of weakness in the rock mass. The blocks formed by the radial cracks and the planes of weakness constitute much of the fragmentation, but not the very fine material created in the crush zone near the blasthole. Harries and Hengst, 1977.
radial decoupling ratio (R_{dcr}), ratio of the total cross sectional area (or volume) of the blasthole to the cross sectional area (or volume) of the explosive in the section of the charged length of the blasthole.
radial initiation, see '**side initiation**'.
radio-frequency energy, the energy in the radio frequency spectrum ($3 \times 10^5 - 3 \times 10^{12}$ Hz) transferred by electromagnetic waves through air, liquid or solid material. Under ideal conditions, normally in air, this energy can fire an electric blasting cap. The national authorities therefore recommend safe distances from transmitters to electric blasting caps (depending on energy and wave length).
radio frequency transmitter, an electronic device for transmitting a radio frequency wave, i.e. a radio transmitting station, mobile or stationary. IME, 1981.
radio probe system, a radio transmitter probe is lowered into the borehole and the receiver is held at the front of the burden. The distance between the transmitter and receiver can thus be determined in meters or feet. The system is used on the surface to measure the true position of the front row boreholes to see if the burden is too small to be fully charged. Cunningham, 1988.
radius (r), **(m)**, the length of the line segment joining the centre and a point on a circle or sphere. Lapedes, 1978.
radius of crushing zone around a blasthole or spherical charge (r_{cr}), **(m)**, the radius of the crush zone is defined as the maximum radius of the zone of complete disintegration of rock into powder and/or fine fragments around e.g. a blasthole or spherical charge.
radius or macrocracks zone (r_{mac}), **(m)**, the maximum radius of the new area of macro cracks created around e.g. blasthole or spherical charge. This radius is determined by

the explosive strength, blasthole diameter, decoupling, rock properties and the rock pressure in the rock mass.

radius of microcracks zone (r_{mic}), **(m)**, the maximum radius of the area where new microcracks have been created in the surrounding rock by blasting of e.g. a blasthole or spherical charge. This radius is determined by the explosive strength, blasthole diameter, decoupling, rock properties and the rock pressure in the rock mass.

radius of plastic deformation zone (r_p), **(m)**, the maximum radius of the non elastic zone around a blasthole. This is an important part of the damage zone around a blasthole. Damage will also occur outside this zone due to the dilatation of rock along existing joints or cracks. The size of rp is dependent on the explosive strength, blasthole diameter, decoupling of the explosive in the blasthole, rock properties and the stresses in the rock mass.

radius of radial cracks zone (r_r), **(m)** the largest radius for a radial crack tip from the centre of the blasthole. In hard rock (e.g. granite and gneiss) this radius is approximately 10 times the diameter of the blasthole for a fully confined and coupled blasthole. Siskind, 1974.

rail, specially shaped steel bar which when laid parallel on crossties and fastened thereto form a track for vehicles with flanged wheels. Bureau of mines Staff, 1968.

raise, a vertical or inclined mine opening, drift, tunnel or shaft, driven upward from a level to connect with the level above, or to explore the ground for a limited distance above one level. After two levels have been connected, the connection may be a winze or a raise, depending upon which level is taken as the point of reference. The inclination of a raise is equal to or larger than 45° from the horizontal. See '**raise driving methods**'.

raise boring, method to excavate raises using the full face boring technique. BorPak (tradename) is one such machinery developed.

raise drilling, **raise driving methods**.

raise driving methods, there have been several methods developed during the years trying to mentioning them in chronological order and the methods are described more in detail under its separate entry.
1 Cribbing method 1940s
2 Alimak method 1950s
3 Longhole drilling 1960s
4 Boxhole drilling or 'mobile boxhole rig' drills a pilot hole which is later on reamed up. Developed during the 1990s
5 BorPak is a full-face boring method developed in the 2000s.

ramp (incline or decline), on surface a short incline to reach a higher surface by foot or by vehicle. Underground it might be a short non horizontal access drift to a stope.

rarefaction, the process or act of making rare or less dense; increase of volume, the mass remaining the same; now usually of gases; also, the state of being rarefied, as the rarefaction of the atmosphere on a high mountain. Standard, 1964. Rarefaction is the opposite to compression. A natural example of this is a phase in a sound wave or phonom. Half of a sound wave is made up of the compression of the medium and the other half is the decompression or rarefaction of the medium.

rarefaction waves, when a media suddenly changes its density, mechanical waves will be created in the surrounding media. When e.g. the stress suddenly changes in the rock mass, rarefaction waves will be created. These waves are also called '*release waves*' (release of load waves).

RAS (acronym), for 'rod adding system' a system designed for automatically change of drill rods at drilling.

Rayleigh wave or **R-wave**, a wave propagating along a free boundary of a solid. The wave is characterised by exponentially decaying amplitude of stress and displacement as a function of depth. A R-wave is an in homogeneous combination of a P-wave and a S-wave. Plane R-waves travel without dispersion along free straight boundaries. Rayleigh waves carry 90% of their energy within a shallow layer below the surface. The Rayleigh wave is important for the damage to constructions from blast vibrations on surface.

Rayleigh wave velocity (c_R), **(m/s)**, the velocity of one of the surface waves. The velocity strongly depends on the Poisson's ratio and is about 10% lower than the S-wave velocity.

RBS (acronym), (s_{br}), **(MJ/m³)**, for *'relative bulk strength'*. This term should be replaced by **'relative volume strength'**, see **'relative volume strength'**.

RCB-system, see **'Nissan RCB-system'**.

RDX (abbreviation), for **'hexogen'**.

reblast, make a new blast after a misfire.

receptor (acceptor), an explosive charge receiving an impulse from an exploding donor charge.

recess, a groove or depression in a surface. ASM Gloss, 1961.

reciprocating compressor, an air compressor which has a piston moving under pressure within a cylinder.

recoil system, in deep holes drilled by cable churn drilling the drill cable provides sufficient elasticity to make the recoil system work. In shallow holes, additional springiness is provided by mast head recoil rubbers.

recovery of down hole items, equipment used for this purpose are *impression blocks* to obtain an imprint of an object in the hole, a *rope spear* to recover broken rope or cable, a *whipstock* for wedge to deflect the hole, a *casing spear* suitable for placing a liner in the hole, a *screen* suitable for setting, using a latch on the bottom bail and finally a *jetting tool* used for developing through the screen, Stein, 1977.

reduced ANFO, ANFO diluted by a material to lower its strength. (e.g. polystyrene beads, sawdust, etc.). See **'ANFOPS'**.

reduced burden (B_r), **(m)**, the reduced burden is defined as $B_r = \sqrt{B_p S_p}$ where B_p is the practical burden and S_p is the practical spacing. Formulas for determination of reduced burden are presented under the term *'burden, reduced burden'*.

reel, cylindrical drum for storing shot firing cable or wires.

reflected pressure (p_r), **(MPa)**, pressure from the ground vibration wave or air blast wave after reflection at a discontinuity (ground vibration) and after reflection at discontinuities representing different densities of air (inversion) or at a free faces (ground vibration) and different topography (air blast).

reflection coefficient, (R), the amplitude ratio between reflected and incident wave for a discontinuity e.g. crack, joint, fault etc.

regulatory authority, an authority which specifies rules for blasting work.

reinforcement, insertion of a material to strengthen a structure. In a rock mass it means the improvement of overall properties of the rock mass by drilling of holes for the insertion of reinforced steel bars or cable bolts.

relative bulk strength RBS, (s_{br}), *(dimensionless)*, see **'relative volume strength'**.

relative mass strength (RMS), (s_{mr}), **(dimensionless)**, is the mass strength of an explosive related to the mass strength of a standard explosive. Nowadays ANFO is often used as the standard reference explosive.

relative vibration velocity (R_v), **(dimensionless)**, the ratio between the amplitude of the vertical particle vibration velocity (A_v) and the P-wave velocity of the ground (c_p) on which a building is founded. Damage associated with buildings and due to ground vibration can be characterized by this quantity.

relative violence of rupture, a measure of the impulsive rebound of the bed or a testing machine, caused during the failure of a rock specimen in a uniaxial compression test. The test is based on the assumption that the rebound is proportional to the degree of violence during the failure of the rock specimen. Singh, 1989.

relative volume strength (RVS), (s_{vr}), **(dimensionless)**, the volume strength of an explosive related to the volume strength of a standard explosive. Nowadays ANFO is often used as the standard explosive.

relative weight strength (*RWS*), (s_{wr}), **(dimensionless)**, use '**relative mass strength**'.

relaxation, the decrease of load support and of internal stress because of viscous strain at constant deformation.

relaxation time (t_{relax}), **(s)**, time for a wave to decay from its maximum to zero.

relay, see '**relay connector**' or '**delay relay**'.

relay connector, a connector to obtain delay periods between successive charges in shock tube or detonating cord initiation systems.

release of energy, sudden release of strain energy stored in the rock mass due to stresses from rock pressure or blasthole pressure.

release of load fragmentation, see '**stress release wave fragmentation**'.

release wave, see '**stress release wave**'.

release wave fragmentation, see '**stress release wave fragmentation**'.

relief, the effective (shortest) distance from a blasthole to the nearest free face. Atlas Powder, 1987.

relief drilling, secondary drilling in order to expose a misfired charged hole. Sen, 1992.

relief hole, this term has many meanings, for example a borehole that is charged and fired for the purpose of relieving or removing part of the burden for the charge to be fired in the main blast. Fay, 1991. A blasthole drilled behind a misfired hole to enable this to be blasted out safely without the undetonated explosive being subjected to handling hazards. Not to be confused with easer holes, which are positioned in a round to assist the breaking of perimeter holes. A hole drilled in close proximity to a hole containing a misfire, which when fired will either sympathetically initiate or dislodge the explosives in the misfired hole. (Editors note: Remote controlled drilling necessary because of the risk of initiation of the explosive during drilling). AS 2187.1, 1996. Uncharged hole used for the swelling of rock in drift rounds. Hagan, 1979.

relief hole diameter (d_r), **(m** or **mm)**, the diameter of a relief hole in a cut.

relief wave, see '**reflected wave**'.

relievers, in drifting, the holes closest to the cut used to enlarge the opening formed by the cut.

relieving hole, see '**relief hole**'. AS 2187.1, 1996.

REMIT (acronym), see '**rock engineering mechanisms information technology**'.

rescue elevator, an elevator that should only be used for bringing people to a safe place when in danger.

residual pressure in the borehole (p_{res}), **(MPa)**, a pressure lower than the borehole pressure. Harries, 1973.

residual strain (ε_{res}), (residual strength), (dimensionless), the strain in a solid associated with a state of residual stress. ISRM, 1975.

residual stress (σ_{res}), (MPa), the stress which exists in an elastic solid body in the absence of, or in addition to, stresses caused by an external load. Ham, 1965. The stress state remaining in the rock mass even after the originating mechanism has ceased to operate. The stresses can be considered as within an isolated body that is free of external tractions. Hyett, 1986.

residual subsidence, see '**subsidence**'.

resin cartridge, a chemical substance that hardens very fast when it is mixed with another chemical substance. It is enveloped by thin plastic to a cylindrical shape. It is used to fasten rock bolts when the support is needed immediately. When the bolt is rotated into the hole with the cartridges in front of the rock bolt the two chemical substances in the cartridges are mixed rapidly and hardens in a short time.

resin grouted anchor, see '**resin grouted rock bolt**'.

resin grouted rock bolt, a friction bolt based on a reinforcement bar glued into the borehole by a resin. TNC, 1979.

resin grouting injection, reinforcement method where other chemicals are used than cement e.g. plastic- or silica compounds. Synonymous with chemical grouting. TNC, 1979.

resistance (current), (R), (Ω), 1 Ohm is defined as that resistance that will let 1 A through it when the potential over the resistance is 1 V.

resistance of rock, see '**blastability**'.

resistance to blasting, see '**blastability**'.

resistivity (electrical), (Φ), (Ωm), is defined by

$$R = \Phi \frac{l}{A} \quad\quad (R.1)$$

where R is the resistance in Ω, Φ is the resistivity for the conductor, a proportional constant with the unit Ωm, l is the length of the conductor in m and A is the area of the conductor in m^2. The resistivity of a metal is increasing with temperature.

resistivity tomography, measurement of the electric resistivity of the rock mass between electrodes placed into two drill holes or on the surface "*crosshole resistivity tomography*". Noguchi, 1991.

resistor method, a method to measure the *detonation velocity* of an explosive (c_d). Resistors are inserted in the explosive at certain known distances and a current is passed through the resistors. When the detonation front passes the resistor string the resistance is changed by the destruction of the resistors in the path of the detonation front and the ionisation at the front ensures a continuous circuit whereby the remaining resistance may be recorded with respect to distance and time. The velocity of detonation is inferred from this information.

response spectrum, a spectrum showing the frequency response of a structure to different frequencies forced upon the structure.

resultant ground vibration velocity (v_R), (mm/s), if the ground vibration velocity has been determined in three mutually perpendicular directions, the resultant vibration can be determined to size and direction.

RETC (acronym), see '**Rapid Excavation and Tunnelling Conference**'.

retrac bit, a drill bit which has grooves and teeth on the back part which permit backward drilling in poor ground which tends to cave in.

retraction bit, drill bit that has been designed with cutting edges on the shoulders of the bit to ease extraction from holes drilled in fractured rock. Atlas Copco, 2006.

retraction of stuck drill steel, special fishing tools have been developed for extraction of stuck drill steel. TNC, 1979.

return filter kit, filter mounted in the return line of a hydraulic system. Atlas Copco, 2006.

reverse circulation drilling (RC), a complement to diamond drilling in prospecting an orebody. Hammer drilling with reversed circulation of drill cuttings is used to get samples of the rock cut in crushed form instead of getting a diamond core. In 2006 it is estimated that prospecting used 50% RC-drilling and 50% diamond drilling. Atlas Copco Craelius, 2006. This equipment is also used for exploration drilling, when accuracy of the sample taken is of utmost importance. A RC-hammer is a DTH-hammer that brings the flushing air from the outside of the bit, and forces the cuttings to be removed through an inner tube inside the drill string. Conventional DTH-hammers can be used in combination with dual wall drill pipes, and a collector-sub. Atlas Copco, 2006. 2 400 m long boreholes can be drilled with RC-drilling. The technique was transformed from the US oil industry to the mining industry in the 1970s.

revolutions per second (n), **(1/s** or **rps)**, the number of turns around an axes per time unit.

rheological classification of rocks, *analytical models* proposed to *simulate the properties of a rock mass*. See '**rheological models**'.

rheological models (constitutive equations simulating the rock mass), in geomechanics the rock mass behaviour can be simulated by different models. The simplest models are; 1) *Hooke* (elastic model), 2) *Newton*, 3) *St. Venant, Kelvin* (visco-elastic model), 4) *Maxwell and Bingham* (visco-plastic model) see Fig. 48. The most common used models in rock mechanics are the Hooke and Kelvin models. Hudson, 1993.

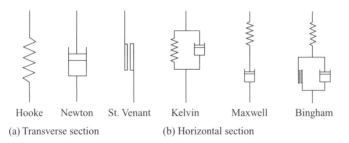

Figure 48 The simplest rheological models for simulating a rock mass. After Hudson, 1993.

There are even more models like; Poynting-Thomson-, Zener- and finally Burgers models. Pusch, 1974

rheology, the engineering discipline covering flow of materials, particularly the visco-elasto-plastic flow of solids and the flow of non-Newtonian liquids.

rib holes or **side holes**, the holes at the sides of a tunnel or drift round, which determine the width of the opening.

rib marks, see '**surface marks**'.

rib pillar, may be left in coal mining between adjacent longwall faces to protect an existing road way from the excessive displacement that can be associated with the mining of an adjacent face, to isolate a particular panel where unfavourable geological structures or fire, water or gas hazards exist, or to control surface subsidence by practising partial extraction. Brady, 1985. Rib pillars are also used between stopes in sublevel stoping.

rib ratio (R_{rib}), (dimensionless), the ratio of actual support required to that developed from Terzaghi's formula for determining roof loads for loose sand below the water table, expressed in percent. Bickel, 1982.

RID (abbreviation), for '**International Regulations Concerning the Transport of Dangerous Goods**'. Meyer, 1977.

ridge tile, a special fired clay or concrete shape for use along the ridge of a pitched roof. There are a number of varieties, for example, *segmental, hog's back, wind ridge* etc. Dodd, 1964. A *rock mass reinforcement method* for adits, drifts and tunnels.

rifle bar, device with spiral grooves used to generate rotation of the drill steel in a pneumatic hand held rock drill. Atlas Copco, 2006.

rifling, ejection of stemming from a blasthole.

rift, a quarryman's term for the natural joint in the quarry rock. The rift is the easiest direction for splitting for excavating the rock. Nelson, 1965.

rig, equipment for some purpose such as a surface-, underground- or diamond drill rigs. Atlas Copco, 2006.

rig control system (RCS), computerized control of drill rigs. It offers simplified fault detection, operator interactivity, and the basis for logging, storing and transferring of bolt installation production and quality data. The PC-based RCS platform and its associated modules, considerably reduce the wear and tear to which mining equipment and personnel are normally subjected. They make operations faster and lighter, and make the equipment easier to handle and maintain. Atlas Copco, 2007.

rig mounted pulverisers, rig mounted heavy demolition equipment. Atlas Copco, 2006.

rig mounted shears, rig mounted cutting tool used in demolition work such as cutting rebar when removing reinforced concrete structures. Atlas Copco, 2006.

rig remote access (RRA), an electronic interface that allows the user to connect the drill rig to an existing network system via LAN or WLAN. RRA is used for remote supervision when drilling unmanned in full fan drilling (sublevel caving) automation, as well as for transferring drill plans and log files and handling messages from the rigs' control systems. Atlas Copco, 2007.

rigidity, the properties by solid bodies whereby they offer an elastic resistance to deformation. Holmes, 1928.

rigidity modulus (G), (MPa), see '**shear modulus**'.

rill mining (RM), a *cut and fill mining method* normally used in thin ore bodies. The method was already used in the 1960s in the thick siderite iron ore mine Eisenerz in Austria. A slice of ore is blasted by inclined up-holes and the ore falling down by gravity is loaded and transported to an ore pass. The space created by mining is filled by waste rock or fillings from the surface so when the blasting is undertaken the ore is blasted against the fill very similar to the situation in sublevel caving. A disadvantage with the method is of course the mixing of waste with ore. The method has been used in many countries and different names were developed for the method like "*Stope mining with fill*" in Austria, "*AVOCA mining*" from the name of Avoca mine on Ireland and finally in Sweden "*Cut and fill with slice method*". Rustan, 1990.

rill stoping, see '**rill mining**'.

rim, the top edge of the vertical face of a bench (ledge).

ring, the set of blastholes drilled in a radial direction from a central location, usually a drift excavated specifically this purpose (i.e. sublevel drift).

ring bit, drill bit designed to drill only a thin band of material at the peripheries of a hole while the center portion is removed by another bit. Atlas Copco, 2006.

ring blasting, holes drilled transverse and radially from a horizontal drift, charged and blasted towards an open or rock filled room. It is a production blasting method where the ring pattern normally is vertical (sublevel stoping).

ring drilling, method of mining that requires that holes be drilled at interval in a 360° pattern. Atlas Copco, 2006.

rip (ripper, ripping), mechanical means for fragmenting rocks which are not sufficiently massive to require explosive fragmentation. The technique is unique to surface operations, and is usually undertaken by a large bulldozer to the rear of which are attached one or more large steel blades. The blade(s) are forced into the exposed rock and then dragged behind the bulldozer, thus ripping apart the intact rock mass.

rippability, the ease by which the material can be ripped is determined by many rock properties of which the two most essential are uniaxial compressive strength and fracture- and joint frequency, see Table 36 below and see Fig. 51 under the entry 'rock mass classification', for determination if the material can be ripped. Other important properties are deterioration, seismic speed, abradability, opening of the discontinuities, fracture toughness and joint strength frequency.

Table 36 Chart of suggested rippability rates. After Singh et al., 1983.

	Classes of rocks				
	1	2	3	4	5
Parameter					
Tensile strength (MPa)	<2	2–6	6–10	10–15	>15
Rate	0–3	3–7	7–11	11–13	13–15
Deterioration	Completely	Highly	Moderately	Slightly	Unaltered
Rate	0–2	2–6	6–10	10–13	13–15
Seismic speed (m/s)	400–1100	1100–1600	1600–1900	1900–2500	>2500
Rate	0–6	6–10	10–14	14–18	18–20
Abradability of the rock	Very weak	Weak	Moderated	High	Very high
Rate	0–7	7–9	9–13	13–18	18–20
Opening of discontinuities	<0,06	0,06–0,03	0,3–1	1–2	>2
Rate	0–7	7–15	15–22	22–28	28–30
Total rate	>25	25–50	50–70	70–90	>90
Appreciation of rippability	Easy	Moderately	Difficult	Marginal	Blast
Tractor recommended	Without or D7	D7–D8	D8–D9	D10 or blast	–
Power in C_V	200	200–300	300–400	700	–
Weight (t)	23	23–38	38–50	97	–

ripping blast, blast with very little displacement, which helps increase natural fracturing or rock swell when preparing for posterior breakage and loading.

ripple marks, the wavy surface of some beds of sandstone and mudstone produced by gentle movements in shallow water when these rocks were in a soft condition. Fay, 1920.

rip rap, coarse-sized rocks used for river banks, dams, piers etc., to provide stabilization and to reduce erosion by water flow.

rise time (t_r), (s), the interval of time required for the leading edge of a pulse to rise some specified small fraction to some specified large fraction of the maximum value. ISRM, 1975. The rise time t_r changes with distance from explosive source and can be calculated according to the following function.

$$t_r = t_0 + \frac{k}{cQ} \qquad (R.2)$$

where t_0 is the rise time at the source in ms, k is a seismic parameter about 0,5 but a little dependent of the kind of explosive used, c is the wave velocity in m/s and Q is the rock quality factor. The variation of Q values in the granite rock dependent on frequency was found to be lognormal. Civets, 1989.

risk, formally the product of the probability of occurrence of a hazard and the magnitude of the consequences of that occurrence.

risk analysis, a structured process which identifies both the likelihood and the consequences of the hazards arising from a given activity or facility. For example a method to analyse the potential possibility of damage to any construction or building in the surroundings of a blast site. Usually a risk analysis will be the base for recommendations concerning the level of ground and air blast vibrations that can be tolerated, and concerning the necessity of using blasting mats to prevent flyrock.

risk assessment, the comparison of the results of a risk analysis with risk acceptance criteria or other decision parameters.

Rittingers law, the energy required for the reduction in particle size of a solid is directly proportional to the increase in surface area. See also '**Kick's law**'. CCD 6th ed., 1961.

rivet, a round bar of mild steel having a conical, cup or pan-shaped head, which is driven while red hot into a hole though two plates of steel which have to be joined together. Aluminium, copper, and other materials are also used for rivets. Ham, 1965.

RM (acronym), see '**rill mining**'.

RMD (acronym), for '**rock mass description**', see '**blastability index**'.

RMI (acronym), see '**rock mass index**'.

RMR (acronym), see '**rock mass rating**'.

RMS (acronym), see '**relative mass strength**' or 'rock mass strength'.

roadheader, a mechanical device, with a rotating, pick equipped cutter head on a hydraulically moveable boom, used for excavating rock. The cutter head axis can be parallel or perpendicular to the boom axis. It is used for mechanical breakage of rock in tunnelling. It is a full face tunnelling machine often used in civil engineering projects.

roadway, drift or tunnel accessible by a vehicle. In caving mines the term is sometimes used to describe the surface on which vehicles travel on the extraction or production level.

robbing, see '**selective mining**'.

robot charger (loader), a pneumatic charger (loader) for inserting explosive cartridges into the drill holes. USBM, 1990.

Rocha classification (MR), the rock mass classification system that quantifies information on the *joint spacing, joint sets, shear strength* and *water pressure* to determine an index rating (MR). This can be used in the determination of the rock load requiring support or can be used directly to assist in the choice of support strategy. The system is used in Portugal but has not received much attention elsewhere. Hudson, 1993.

rock, any consolidated or coherent and relatively hard naturally formed mass of mineral matter. Rock materials with uniaxial compressive strength less than 1 MPa are regarded as soils. Bieniawski, 1973.

rock anchor, see '**rock bolt**'.

rock anchorage, the process of inserting rock anchors. TNC, 1979.

rock blasting, the art of utilizing explosives for breaking of rock.

rock blasting journals,

Austria-USA (Academic journals)

1997–2006, The Int. J. For Blasting and Fragmentation Rock. Short name: Fragblast Journal. Four volumes per year. Editors-in-Chief Hans Peter Rossmanith Austria and William Fourney USA. Publishers: A. A. Balkema, Netherlands (1997–1999), Swetz & Zeitlinger, Netherlands (2002–2003) and Taylor & Francis (2004–2006).

2007 Blasting and fragmentation. Editors-in-Chief Hans Peter Rossmanith Austria and William Fourney USA. Publisher: Int. Soc. for Explosive Engineers.

USA (Not academic journal)

1983 The Journal of Explosive Engineering published by International Society for Explosives Engineers in USA. About 50 pages inclusive advertisements. Publisher: Int. Soc. for Explosive Engineers, USA.

rock bolt or **roof bolt**, a bar usually constructed of reinforcing bar or glass fibre rod or a steel tube, which is inserted into pre-drilled holes in rock and secured for the purpose of ground control. Rock bolts are classified according to the means by which they are secured or anchored in rock. In current usage there are mainly five types, namely, 1) see '*expansion shell bolt*' 2) see '*wedge bolt*' 3) see '*grouted bolt*' and 4) friction type bolts, see '*split-set bolt*', '*Swellex bolt*' and 5) '*rock bolts explosive anchored*' 6) sliding rock bolts. There is also possible to combine the methods e.g. expansion shell and grout and wedge and grout.

rock bolt design principles, there are three possible main methods 1) **Analysis of structural stability** that can be divided into three sub methods a) *limited rock block stability analysis*, b) *beam or slab concept for bedded rock* and c) *rock arch concept* 2) **Empirical assessment** and 3) **Numerical models**. Stillborg, 1994.

rock bolt, explosively anchored, a device developed by the USBM to give better support in underground mining operations. It can be anchored more firmly than conventional bolts because the principle of explosive forming enables the anchor to grip the sides of the borehole along its entire length, if necessary. The key to the design is a seamless steel anchoring tube, welded to the threaded end of the bolt. Exploding a small charge inside the tube makes it expand to get it tightly in the borehole. Water, wax or a similar buffer surround the charge to distribute the force of the explosion evenly and prevent it from rupturing the tube. USBM, 1990. Compare with 'Swellex bolts'.

rock bolt reinforcement, see '**rock bolt**'.

rock bolter, machine designed specifically for drilling holes and installing rock bolts. Atlas Copco, 2006.

rock bolting density (Ψ), (No./m^2), the number of rock bolts per square meter. Choquet and Charette, 1988 determined the rock bolting densities in ten Quebec hard rock mines, in more than 57 drift portions, assessing ground conditions by means of the Norwegian Nick Barton Q-factor with symbol Q and the mining RMR (MRMR) with symbol R_{MRMR}. The formulas of the minimum number of bolts per square meter of roof or wall Ψ_{min} required are,

$$\Psi_{min} = -0{,}227 \ln Q + 0{,}839$$
$$\Psi_{min} = -0{,}0214 R_{MRMR} + 1{,}68 \qquad (R.3)$$
$$S = 1/\Psi^{0,5}$$

where S is the spacing between the bolts in m.

rock bolting, the process of rock bolting consists of 1) anchoring, 2) applying tension to the bolt to place the rock under compression parallel to the bolt, and 3) placing the bolts in such a pattern that they will properly support the rock structure. Rock may be supported by bolts in five ways, 1) *suspension*, 2) *beam building*, 3) reinforcement of arched opening requiring support, 4) reinforcement of an opening otherwise self supporting, and 5) reinforcement of walls against shear and compressive action. Lewis, 1964.

rock bolting, the act of installing rock bolts. Atlas Copco, 2006.

rock bolting drill rig, drill rig designed specifically to drill holes and installing rock bolts in them. Atlas Copco, 2006.

rock bolting equipment, there are three different technology levels for installing rock bolts, 1) handheld, 2) semi-mechanized equipment for grouted rebars or Swellex bolts (inflatable tubes), and 3) fully mechanized equipment for cement grouted rebars or Swellex bolts.

rock boring equipment, equipment used to create an underground opening or tunnel by mechanically boring out the whole area of the tunnel or drift.

rock breaker or better *rock crusher*, see '*jaw crusher*', '*gyratory crusher*', '*cone crusher*' and '*swing hammer crusher*'. Nelson, 1965.

rock burst, a sudden explosive-like release of energy due to the failure of a brittle rock of high strength due to very high rock mass stress exceeding the rock strength. It creates a sound where part of it is in the human audible range. A micro rock burst may also be caused by minor slippage along rock contacts. Rock bursts most often are induced by mining activities.

rock burst-efficiency ratio (η_b), (dimensionless), the ratio η_b between the throw energy of rock fragments W_t in J (Joule) after failure of a specimen under uniaxial compression to the maximum elastic strain energy stored in the rock W_{smax} in J (Joule).

$$\eta_b = \frac{W_t}{W_{smax}} 100 = \frac{W_t}{\sigma_{fail}/2E} 100 \qquad (R.4)$$

σ_{fail} is the rupture stress in Pa and E is the Young's modulus in Pa.

rock bursting, see '**rock burst**'.

rock breaking capacity of explosives, can be defined in a pressure-volume diagram of the explosive gases by the grey area, see Fig. 49. Breakage is judged to have ceased when the explosive gases have expanded to 10–20 times their original size in the blasthole.

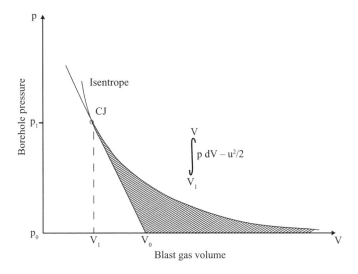

Figure 49 The rock breaking energy of an explosive is defined by the grey area. After Persson, 1993.

rock bump, see '**rock burst**'.
rock caving, the breakage and movement of rock mass due to gravity forces where the relative position of rock mass considerably is changed.
rock characterization, see '**rock mass classification**'.
rock chute, see '**chute**'.
rock chute mining, a **room and pillar mining method** used in coal mining, see also '**board and pillar mining**'.
rock classification, see '**rock mass classification**'.
rock cleavage, any structure by virtue of which a rock has the capacity to part along certain well-defined planes more easily than along others. Geologists usually employ the terms for secondary structures produced by metamorphism or deformation rather than for original structures such as bedding or flow structures. Stokes, 1955. Compare with 'rift' and 'joint'.
rock conceptual model (vibrations), the rock mass can best be simulated regarding blast vibrations by a Kelvin and Voigt-element (a viscous-elastic material) which means that the rock can be engineered as an combination of a elastic spring in parallel with a viscous element.

$$\sigma = E\varepsilon + E\eta \frac{d\varepsilon}{dt} \qquad (R.5)$$

where σ is the stress induced in the material in Pa, E is the Young's modulus in Pa and ε is the strain (dimensionless) and η the energy dissipation in s (seconds). By, 1980. In normal calculations the second part in the equation is neglected.
rock cone bi, (synonym), see '**roller rock bit**'.
rock constant (*c*), (**kg/m³**), a numerical value characterizing the blastability of rock. The rock constant expresses how much explosive is necessary just to detach homogeneous rock in a bench of 1 m height with 1 m of burden and blasthole diameter of 33 mm. Fraenkel et al., 1952. Extreme values of the rock constant are 0,2 kg/m³ (easy blasted)

and 1,0 kg/m³ (difficult to blast). Standard value 0,4. The rock constant is used for calculation of the maximum burden according to the following empirical formula given by Langefors and Kihlström, 1978.

$$B_{max} = 0,958d \sqrt{\frac{\rho_e s}{(S_b/B_b)c_0 f_{bi}}} \tag{R.6}$$

where B_{max} is the maximum burden in (m), d is the blasthole diameter in (m), ρ_e is the density of the explosive (kg/m³), s is the weight strength of the explosive (related LFB dynamite), S_b is the blasted spacing in (m), and B_b is the blasted burden in (m). When B_{max} is between 1,4 and 15 m, c_0 (the corrected blastability factor) is $(c + 0,05)$. When B_{max} is less than 1,4 m, $c_0 = c + 0,07/B$. c is the rock constant (kg/m³). f_{bi} is an inclination factor for the blastholes. $f_{bi} = 1$ for vertical blastholes when the bench height $H = 1,3\ B_{max}$. The Langefors and Kihlström formula can only be used for borehole diameters in the range ~30 to 75 mm in diameter, because for larger blasthole diameters the relationship between burden and blasthole diameter is non-linear. For other hole diameters see '*burden (reduced burden)*'.

rock crusher or *rock breaker*, a machine for reducing rock or ore to smaller sizes. Four principal types are available; see 1) '*jaw crusher*' 2) '*gyratory crusher*' 3) '*cone crusher*' 4) '*swing hammer crusher*'. Fay, 1920.

rock cutting, the process of disintegration of rock with mechanical tools like drill bits, tricone bits, wedges etc.

rock cuttings, see '**cuttings**'.

rock damage, the blast induced damage in the remaining rock mass behind the perimeter holes. Rock damage may mean any blast disturbance which changes the rock mass continuity, i.e. radial cracks, change of permeability, change of stability, overbreak, etc. Blast induced rock damage depends on the *intensity of the ground vibration*, which is dependent on the (linear) *charge concentration* in the near field, the *strength of the explosive*, the *decoupling ratio*, the *confinement of the charge* (burden, number of free surfaces, etc.), the *inter-row delay time* (delay time between neighbouring contour charges), *the rock mass properties* (including *rock stresses* and the *method of support*) and the *size of the opening*. The maximum charge per delay is important for the blast induced damage in the far field.

rock damage criterion (v_{max}), (mm/s), the measured standards for the likelihood of damage caused to a rock body. These are generally expressed in terms of *maximum (peak) particle velocity*, v_{max} or (PPV). For hard rock with strong joints v_{max} is ~1 000 mm/s, for medium hard rocks with no weak joints v_{max} is 700–800 mm/s, and finally for soft rock with weak joints v_{max} is ~ 400 mm/s. An alternative damage evaluation criteria is based on the change in *fracture frequency* before and after blasting. Holmberg, 1984. For more detailed damage classification see, 'blast damage index BDI' and 'blast induced damage zone'.

rock debris, after excavation of the fragmented rock in a round there will still be some fine material left that is called debris. TNC, 1979.

rock drift, a horizontal mine passage, a drift cut in solid rock. A.G.I. Supp., 1960.

rock drill, a machine for boring a hole in rock for blasting purposes. It may be a *sinker* (drilling downwards), jackhammer and drifter (drilling horizontally) and finally a *stopper* (drilling upwards). Rock drills are always used together with drill rods and drill bits, either detachable or integral, and can have either a rotary, percussion or percussive mechanism. They are either hand-held or rig-mounted.

rock drilling, use of different mechanical techniques to create holes in rock. The purpose is to use the holes for examination of the rock mass, placement of explosives or anchor- or rock bolts, injection of water, examination or permeability, measurement of vibration and pressure etc.

rock drivage, hard heading or **stone drift**, a drift excavated in rock, as from the surface down to a coal seam. Nelson, 1965.

rock dynamic index, the ratio of elastic energy recovered to the energy dissipated in shock loading in a Hopkinson bar test. Singh, 1989.

rock engineering, the field of engineering concerned with excavation and construction in the rock mass for mining-, civil- or military purposes.

rock engineering mechanisms information technology (REMIT), a method for the identification of relevant rock engineering parameters and their interactions—as applicable to different engineering objectives and circumstances. An interaction matrix is developed. Hudson, 1993.

rock factor, see '**blastability**'.

rock failure, up to a certain value of stress, which is known as the elastic limit, the deformation is purely elastic and the mass will return to its original shape if the stress is removed. Above this stress value, the material is permanently deformed, it acquires permanent set. Finally, if the stress is still further increased, the material ruptures or fails, and the value or stress required to cause rock failure is known as the *ultimate strength of the material*. Rice, 1960.

rock fall, the relatively free falling of a nearly detached segment of bedrock of any size from a cliff, steep slope, cave or arch. A.G.I., 1957. Rock fall also occurs in drifts and stopes.

rock fall restraining net, a *wire net* used as a fence to catch falling rocks from slopes close to highways. Duffy, 1991.

rock fill, see '**fill**'.

rock filling, the operation to use waste rock, to fill up worked-out stopes with the purpose to support the roof.

rock flow, a slope failure, when there is a general breakdown of the rock mass. When such a rock mass is subjected to shear stresses sufficient to break down along weakness planes and/or through or around grains or to cause crushing or the angularities and points of the rock block, the block will move as individuals and the mass will flow down the slope, or will slump into a more stable slope.

rock fracture models, the rock fracture model depend on how the rock is disintegrated by stress relief, mechanically or by explosives. It needs a conceptual model for each process.

rock fracturing, the process of disintegration of rock by rock mass stresses, blasting, drilling, other mechanical tools like impact hammer etc. The result due to blasting depends on many parameters which can be grouped into four main groups; 1) Rock properties, 2) In situ rock stress and confinement, 3) Explosive properties and initiation properties. When rock breaks, light is emitted but this could not be seen by the naked eye.

rock hardness, the resistance of the rock to the intrusion of a foreign body. Stoces, 1954. See '**hardness**'.

rock hardness test, see '**hardness test**'.

rock head, the boundary between the superficial loose deposits and the underlying solid rock.

rock impact hardness, see '**hardness**'. Hudson, 1993.

rock loader, any device or machine used specifically for loading slate in coalmines or any rock inside a metal mine. There are three types of loaders, shovel loaders, gathering arm loader (Joy loader or Hägglund loader) and front-end loaders (LHD).

rock mass, rock including its structural discontinuities in situ.

rock mass characterization, the systematic process of describing a rock mass both quantitatively and qualitatively for engineering purposes. See '**rock mass classification**'.

rock mass classification, a method of assigning numerical values to a range of characteristics considered likely to influence the engineering behaviour of a rock mass and of combining these values into one overall numerical rating. A rock material can be classified with regard to its physical properties heterogeneity, anisotropy and permeability, with regard to its mechanical properties or with respect to its resistance to a method of excavation, (e.g. blastability or rippability). The Q-system, developed by Barton, 1974, and the RMR-system (Rock Mass Rating System), developed by Bieniawski, 1973 are recommended to be used for blasting purposes. The following correlation exists between the Q and RMR-values, see Singh et al., 1986.

$$RMR = 19 \ln Q + 26 \qquad (R.7)$$

A linguistic assessment of the rock mass is given in Table 37.

Table 37 Rock mass classification. After ISRM, 1981.

Class	Description
Massive	Few joints or very wide spacing
Blocky	Approximately equal dimensions
Tabular	One dimension considerably smaller than the other two
Columnar	One dimension considerably larger than the other two
Irregular	Wide variations of block size and shape
Crushed	Heavily jointed

The two most important parameters for the quantification of the strength of a rock mass are the uniaxial compressive strength of the intact rock and the spacing of the joints, see Fig. 50.

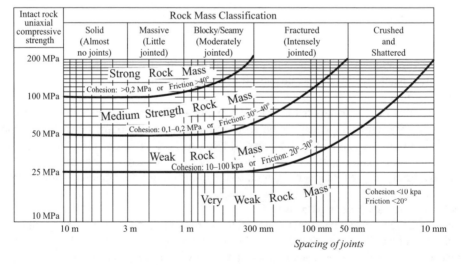

Figure 50 Strength diagram of jointed rock masses. Bieniawski, 1973 and modified after Müller et al., 1970.

The interaction of the rock mass properties, the surrounding environment, the geometry, blasting, support and excavation methods on slope stability or underground excavation stability in rock has been qualitatively and systematically described in the form of matrixes. Hudson, 1992. The quality of the rock determines the method suitable for rock breakage. An indication of what method should be used can be gained if the mean fracture spacing and uniaxial compressive strength are known, see Fig. 51.

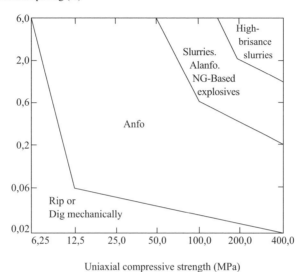

Figure 51   Methods for fragmenting rock. After Franklin et al., 1971.

rock mass classification (history), the following systems have been developed; 1946 Terzaghi (tunnelling with steel support), 1958 Lauffer (stand up time), 1967 Deere (RQD), 1972 Wickham et al., (RSR) 1973 Bieniawski (RMR) and 1974 Barton et al., Q-system.

rock mass description (RMD), (f_{RMD}), (dimensionless), see '**blastability index**'.

rock mass rating (RMR), (R_{RMR}), (dimensionless), a rock mass classification system developed by Bieniawski, 1973 who developed his scheme on a scale 0–100 using data obtained mainly from civil engineering excavations in sedimentary rock in South Africa. Bieniawski's scheme uses five classification parameters. Bieniawski, 1973.
1   Strength of the intact rock material.
2   Rock quality designation.
3   Conditions of joints.
4   Groundwater conditions.

rock mass strength, see '**rock mass classification**'.
rock mass structure, see '**rock mass classification**'.

rock material characterization, the determination of the mechanical properties of the rock mass, especially deformation, stability and strength.

rock matrix, the fragmented rock material.

rock mechanics (mining), the field of mechanics associated with the analysis of mechanical processes of rock and rock mass (due to mining activity) and the methods used to reinforce the rock mass.

Rock Mechanics Journals (Academic) (history),

Austria Journals

1929 Geologie und Bauwesen **(Geology and Construction)**. Vienna, Austria. Editor Josef Stini.

1962 The first mentioned journals name was changed to Felsmechanik und Ingenieurs-Geologie **(Rock Mechanics and Engineering Geology)**. Vienna, Austria. Editor Leopold Müller.

1969 Name change of the **Rock Mechanics and Engineering Geology Journal** to **Rock Mechanics Journal**.

1983 Further change of name from **Rock Mechanics** to **Rock Mechanics and Rock Eng**.

United Kingdom Journal

1964 International Journal of Rock Mechanics and Mining Sciences. Founded by Albert Roberts and published by Pergamon Press. Initially the journal was also Mining orientated but this has disappeared.

1974 A Geomechanics Abstract section is added to the Int. J. of Rock Mechanics and Mining Sciences.

rock noise, see '**acoustic emission**'.

rock oil (synonym), for '**petroleum**' see '**petroleum**'. Fay, 1920.

rock pass, vertical or inclined raise or shaft for gravitational transportation of waste or ore. TNC, 1979.

rock pressure pulse (RPP), (p_{or}), **(Pa or dB)**, (*in connection with air blasts*), air blast overpressure produced from the vibrating ground, see also '**air blast**'. Atlas Powder, 1987.

rock (pressure) stress, the litho static (virgin) rock (pressure) stress is dependent on depth below surface and the plate tectonic forces that determines the direction for the maximum horizontal stress. When underground excavation start, this stress field is disturbed locally.

rock properties, see '**rock mass classification**'.

rock protrusion, a smaller part of rock which is being left inside the theoretical contour of the blast. Normally this rock has to be removed and usually in a later blast.

rock quality, see '**rock mass classification**'.

rock quality designation (RQD), single parameter for classification of the discontinuity spacing in the rock mass. It is defined as follows;

$$RQD = \frac{100\sum x_i}{L} \qquad \text{(R.8)}$$

where x_i are the lengths of individual pieces of core in a drill run with a length of 0,1 m or greater and L is the total length of the drill run. The *RQD* varies between 0–100%, where 100% is used for the most competent rock, see Table 38.

Table 38 Classification of rock mass according to discontinuity spacing. After Deere and Miller, 1966.

RQD (%)	Rock quality
<25	Very poor
25–50	Poor
50–75	Fair
75–90	Good
90–100	Very good

There exist also other definitions for the characterisation of discontinuity spacings, but *RQD* is the international accepted system. *RQD* values smaller than 50 indicate difficult conditions for controlled contour blasting.

rock quality factor (Q), (attenuation of blast waves), attenuation factor for seismic waves in rock, determined by the fractional loss (W_L) of maximum stored energy (W_{max}) per cycle, when acoustic waves move through the rock.

$$Q = \frac{2\pi W_{max}}{W_L} \tag{R.9}$$

The formula is valid for an acoustic wave propagating in an imperfect elastic medium where energy is lost due to the presence of imperfections in the media. For dry rocks Q is independent of the frequency over a reasonably wide frequency range 10^{-2}–10^7 Hz. Adapted from McKenzie et al., 1982.

rock quality factor (Q), (rock mass support, Barton Q-factor), see '**Q-factor**'.

rock quality index in drilling (RQI), the rock quality index in drilling is defined as the ratio between the hydraulic pressure of the drill machine and the mean penetration rate in the borehole. Jimeno et al., 1995.

rock reinforcement, techniques to keep the strength of the rock mass by rock bolting, stamping, shotcreting and cable bolting.

rock rheology, see '**rheology**'.

rock slotting, the use of high-pressure water jet to create slots at the perimeter of a tunnel. El-Saie, 1980. Also to created two diametric longitudinal slots in blastholes for the purpose of activating radial cracks in certain directions.

rock stabilization, combined application of *rock reinforcement* (placed inside the rock mass) and *rock support* (placed outside the rock mass) to prevent failure of the rock mass.

rock strength, the rock strength was initially viewed as a strict material property but this is relaxed because rock strength is today regarded as a *system property*, the system itself being defined by a particular *geometry*, a set of *boundary constraints* and a set of *constitutive equations* describing the material. This less conventional viewpoint combines concepts from both '**bifurcation theory**' and from '**constitutive theory**'. In blasting e.g. the strain rates are very high close to the explosive and the tensile strength of the intact rock may be 5 times higher compared to static conditions. An example of how the fracture strength (tensile strength) varies with the load rate is given under the entry 'dynamic tensile strength'. The rock mass strength is also dependent on the static three-dimensional rock stress. The larger the stress the larger the strength, see Fig. 52.

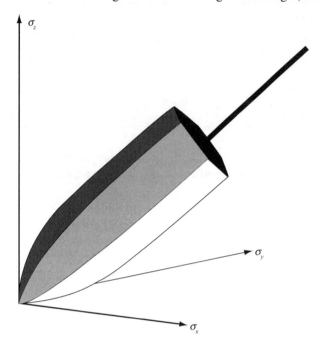

Figure 52 A model of the tri-axial strength of rock. σ_x, σ_y and σ_z are the three principal stresses. The space diagonal represents the hydrostatic pressure ($\sigma_x = \sigma_y = \sigma_z = p$). After Lundborg, 1972.

rock stress (σ), (MPa), the stresses in the rock crust due to depth and horizontal movements of the plates of the crust. Many mining problems are directly concerned with stresses which may cause mine openings to collapse. Two phases of occurrence of rock stresses are important, the stresses existing in the rock mass before excavation of the mine opening, that is, 1) *free field stress* and 2) *indirect stress* within the rock mass caused by the mine excavations.

The rock stress is increasing with depth due to the weight of the rock mass. The horizontal stress measured in Sweden on surface is about 18 MPa and at 1 000 m depth it is 100 MPa (seems a little too large) where the vertical stress is 27 MPa. Measurements in Australia and South Africa shows that the horizontal stress is smaller than the vertical stress and normally 0,5 to 1,0 of the vertical. Pusch, 1974.

rock structure rating (RSR) (acronym), **(R_{RSR}), (dimensionless)**, a *rock classification method* originally intended for *steel-supported tunnels*. It is a three step method that is relatively easy to apply. The three parameters assess the 1) *Geology* (rock type and folding or discontinuities), 2) *Joint pattern* and *joint orientation* and 3) *Water inflow and joint condition*. The summation of these parameters is the rock structure rating or RSR on a scale 1 to 100. This method is not used so much any longer when shotcrete and bolting have become more popular to use. Hudson, 1993.

rock support, is defined as the external support to the rock mass. Examples of rock support are the application of a *concrete lining*, *steel sets* or any other type of engineering structure which will restrict movements of the rock.

rock talk, see '**acoustic emission**'.

rock temperature (T_r), (C°) or (K), the rock temperature increases with depth, about 1°C for each 100 m in depth. The increase in temperature is due to the still hot core of the earth and also the decomposition of radioactive minerals in the crust which liberates heat.

rock testing, includes determination of density, compressive-, tensile-, shear- and bending strength, fracture toughness, porosity, joints, joint surface properties, water content in the rock mass etc.

rock throw, the movement of the material disintegrated and fragmented in blasting. Single stones thrown longer distances are called flyrocks.

rock type, the earth crust is divided into different rock types dependent on its mineral composition and its origin. TNC, 1979.

Rockwell, a unit of hardness as determined by a Rockwell hardness tester. See '**Rockwell hardness**'. Compare with 'Knoop hardness'. Long, 1920.

Rockwell hardness, a measurement of metal hardness which interprets resistance to penetration by indentation. Bureau of mines Staff., 1968.

Rockwell hardness test, a method to determine the relative hardness of metals and case-hardened materials. The depth of penetration of a steel ball (for softer metals) or of a conical diamond point (for harder metals) is measured. See '**scleroscope hardness test**'. Nelson, 1965.

Rockwell hardness tester, a machine for testing the indentation hardness o materials by means of a diamond point. See '**Brinell hardness tester**'. ACSG, 1963.

rod bitch, simple tool used for handling rods in rod percussion drilling. Stein, 1977.

rod percussion drilling, see '**cable drilling**'.

rod, see '**capped fuse**'. AS 2187.1, 1996.

rod room, see '**capping station**'. AS 2187.1, 1996.

roller rock bit, a rotary bit fitted with two or more hardened steel or tungsten carbide tipped rollers of cylindrical or conical form. Variously known as two-cone bit, three-cone bit, four-cone cutter bit, etc. B.S. 3618, 1963.

ROM (acronym), for '**run of mine**' or better, finished mining products leaving the mine.

ronometer, see (**convergence meter**).

roof, the upper surface of an underground excavation.

Roofex (tradename), a rock bolt made of an inner steel bar and surrounded partly by a plastic sleeve that is glued into the drill hole. The bolt has an inbuilt sliding property and can thereby absorb rock movements. The Roofex bolt provides support for forces up to 80 kN before the sliding effect and the frictional system comes into effect. At the load of 80 kN, the bolt stats to slide along with the rock deformation, slowly dissipating the energy of the rock. The sliding length could be 60 cm and when this length has been reached the bolt will work as a conventional rock bolt up to a maximum load capacity of 100 kN. A special designed version could show how much the bolt has slided which can be seen by the naked eye by different colour of the end of the rock bolt which after mounting can be seen outside the borehole. The sliding bolt and its different versions are shown in Figure 53a, b and c.

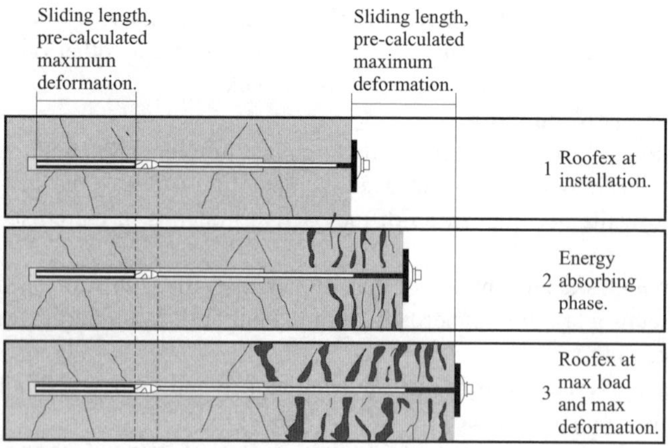

Roofex

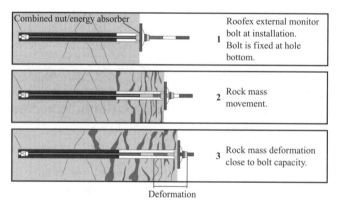

Figure 53a The heart of the Roofex rock bolt mechanism is a frictional steel-to-steel sliding system. This consists of a steel bar sliding inside a plastic sleeve, and an energy alsorber bonded to the wall of the hole. Once the rock starts to move, the steel bar is pulled through the energy absorber, creating friction and absorbing energy from the rock mass movement. After Neugebauer, 2008.

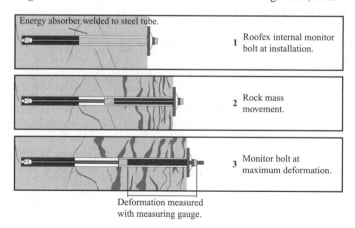

Figure 53b Roofex external monitor bolt. After Neugebauer, 2008.

Figure 53c Roofex internal monitor bolt. After Neugebauer, 2008.

roof hole, see '**back hole**'.
roof support pressure, see ' **NGI classification formula for roof support pressure**'.
room, a wide working place or a chamber in a flat underground mine, see '**stope**'.
room and pillar mining (RPM), a mining method for horizontal and somewhat inclined ore bodies. The ore is extracted by drifting and sometimes also by vertical or horizontal benching. Pillars are left for the support of the back.
rope spear, a tool to recover broken rope or cable from a drill hole. Stein, 1997.
ropeway, an aerial tram way. Fay, 1920.
rose bit, a hardened steel or alloy noncore bit, with a serrated face to cut or mill out bits, casing, or other metal objects lost in the hole. Also used to mill off the rose-bit dropper on a Hall-Rowe wedge. Also called *mill, milling bit*. Long, 1960.
Rosin-Rammler-Sperrling (RRS) distribution, a mathematical distribution function to describe the amount of rock of different sizes as obtained after blasting. The RRS-distribution is a special case of the more general Weibull distribution and it is the most commonly used in blasting and comminution. The RRS-distribution is defined as;

$$y = 1 - e^{-\left(\frac{x}{k_c}\right)^n} \qquad (R.10)$$

where y is the cumulative fragment fraction smaller than x, x being the fragment size in (m), k_c is the characteristic size of the fragment size distribution in (m), (could be taken as any point on the distribution function but preferably a k_c-value should be selected, where half of the mass is passing), and n is the uniformity index.

rotary bit, a drill bit that is rotated under pressure against the rock surface. The main type of rotary bits are: *blade and drag bits, roller bits, hammer bits* and *diamond bits*. Special purpose bits are: *coring bits, pilot bits* and *reaming bits*.
rotary compressor, a machine using a two rotors one female and one male rotor to compress air. It is also called screw compressor. Atlas Copco, 1982. The machine is compressing the air by lamellae principle.
rotary drill machine, various types of drill machines that rotate a rigid, tubular string of rods to which is attached a bit for cutting rock to produce boreholes. The bit may be a *roller cone bit*, a *toothed or fishtail drag bit*, an *auger bit*, or a *diamond bit*. USBM, 1990.
rotary drilling, a form of drilling in which the rock is penetrated by the continuous rotation of the bit under heavy thrust. Fraenkel, 1954. There are two principles developed the *crushing action* from three cone bits and the *abrasive action* where the rock is removed by friction and finally *cutting action*.
rotary drilling (*abrasive drilling*), drilling method based on the abrasive action of the drill steel or drilling medium which rotates while being pressed against the rock. The methods included under this heading are 'diamond' and 'steel shot drilling'. Fraenkel, 1952.
rotary drilling (*crushing*), also called roller bit drilling, a drilling method based on the crushing or grinding action of a roller bit which rotates while being pressed against the rock. The drill bit consists of three conical-shaped toothed rollers which rotate in reverse direction against the bottom of the drill hole when the drill pipe is turned. The hardness, length and shape of the roller teeth depend on the hardness of the rock. Cuttings are removed by water or air flushing. The borehole diameter range is from 100 to 600 mm and the feed force is about 1 t per each 25 mm diameter of the borehole. Fraenkel, 1952.

rotary drilling (*cutting*), is effected through the cutting action of the drill steel which rotates while being pressed against the rock. The drilling machines are powered by electricity or by compressed air. Fraenkel, 1952.

rotary drilling parameters, the measured rotary drilling parameters may be used for determination of the *specific charge in controlled contour blasting* (buffer blasting) on surface. From the drilling parameters a rock quality index I_{rc} can be calculated and it was found that the I_{rc} can be correlated to the specific charge q in kg/t.

$$I_{rc} = \frac{p}{v_d} \quad \ln q = \frac{I_{rc} - 25000}{7200} \tag{R.11}$$

where I_{rc} is the rock quality index in kPa · min/m, p is the pressure from the drill bit on the rock in kPa, v_d is the drilling rate (penetration rate) in m/min and q is the specific charge in kg/t. The second of the two above given empirical formula will be different for different drill rigs and drilling equipment. Leighton, 1982.

rotary drilling unit, drill rig specifically designed for drilling holes by applying a force and rotating a drill bit. Drill bits may be of the roller (crushing) or drag types (cutting).

rotary hardening, process to harden steel where steel is rotated during the process. Sandvik, 2007.

rotary percussion drilling, a drilling method using a roller type bit. This method is used for oil drilling in USA. Fraenkel, 1952.

rotary percussive drilling, a method of drilling in which repeated blows are applied to the bit which is continually rotated under power. B.S. 3618., 1964.

rotary table, part of a drill rig where where rotary energy is transformed to the drill steel. Stein, 1977.

rotary tachometer, instrument to measure the rotation of the drill steel in rpm.

rotasonic drilling, this is a form of rotary vibracoring that uses high-frequency mechanical oscillations to transmit resonant vibration (50–120 Hz) and rotary power through the drill pipe to the bit (carbide tipped). The action fluidizes and pushes soil particles away from the bit. Relatively rapid penetration rates ca be achieved in unconsolidated and lightly cemented granular materials without drilling fluids, air, or the large-diameter disturbance of augers, to over 120 m. It is preferred on a performance base over hollow stem augers in deep, fluid formations such as grave aquifers. Stein, 1977.

rotational firing, a delay blasting system in which each charge successively displaces its burden into a void created by an explosive detonated on an earlier delay period. USBM, 1983.

rotation mechanism, device consisting of several parts used to generate a rotary motion. Atlas Copco, 2006.

rotation per minute (*n*), (rpm), unit for rotation where rpm stands for rotation per minute and rps rotation per second.

rotation per second (*n*), (rps), unit for rotation where rps stands for rotation per second.

rotation unit, set of gears enclosed in a housing, with a protruding threaded spindle that attaches to the drill tubes, and is rotated by a hydraulic, electric or pneumatic motor. Atlas Copco, 2006.

round, a group or set of blastholes drilled, charged with explosives and fired instantaneously or with delay detonators in order to break a certain volume of rock. Also used as a term for the amount of rock broken.

round washer, flat circular steel disc having one perforation and fastened by a nut to the threaded end of a rock bolt with purpose to transform load bearing capacity from the bolt to the rock surface. Atlas Copco, 2006.

row, drill holes located in a plane parallel or at an angle to the free face. In blasting a row is defined by the plane consisting of all holes detonated at the same time.

RPCF (abbreviation), for '**rotation pressure control of feed pressure**'.

RPM (acronym), for '**room and pillar mining**'.

rpm (acronym), unit for '**rotation per minute**'.

RPP (acronym), see '**rock pressure pulse**'. See also '**air blast**'.

rps (acronym), unit for '**rotation per second**'.

RQD (acronym), for '**rock quality designation**'.

RQI (acronym), see '**rock quality index**' in drilling'.

RRA (acronym), see '**rig remote access**'.

RRC (acronym), see '**radio remote control**'.

RRS (acronym), see '**Rosin-Rammler-Sperrling distribution**'.

RSR (acronym), see '**rock structure rating**'. See also '**rock mass classification**'.

rubber boss, protuberance or raised circular projection such as the rubber boss on the front of a drill feed that is pressed against the rock. Atlas Copco, 2006.

rubber hose guard, rubber sleeve placed over a hose to protect it from abrasive wear. Atlas Copco, 2006.

rubber mat (*drilling*), flat sheet of rubber placed on the floor of a drill rig for the operator to stand on. Atlas Copco, 2006.

rubber mat (*blasting*), rubber mat is made of used cut tyres and tied together with cables. It is an effective protection against flyrock when blasting close to infrastructure. Persson, 1993.

running board, narrow platform or step just above the ground on the side of a vehicle, used as a step for entering. Atlas Copco, 2006.

run of mine (ROM) (acronym), fragmented ore leaving the mine and used as incoming raw material for mineral processing. Also called '*crude ore*'.

run-up, run-down zone, the length of blasthole along which the detonation velocity (c_d) gradient is changing (rising or falling). One reason for a run-up or run-down zone may be that the detonation velocity of the explosive is different to that of the primer. In large blastholes, the detonation velocity may fall owing to the distance from the primer until the shock wave encounters the blasthole wall, then stabilize and build up again.

rupture, discontinuous deformation characterized by loss of cohesion on a surface in the interior of a body culminating in complete separation.

rupture energy, see '**strain energy release**'.

rupture limit, the stress at which a material breaks into pieces.

rupture-, failure- or breakage waves, develops when the rock mass ruptures, fails or breaks, at blasting due to shock- and seismic waves. These waves can be measured by accelerometers and geophones.

RVS (acronym), see '**relative volume strength**' which replaces '*relative bulk strength*'.

R-wave (abbreviation), for '**Rayleigh wave**'. See '**Rayleigh wave**'.

RWS (acronym), for '*relative weight strength*' which should be replaced by '**relative mass strength (RMS)**'.

S

saddleback bit, see 'concave bit'.

safe distance (R_s), (m), minimum distance beyond which the explosive does not appreciably affect the surroundings.

safe limit, the amplitude of vibration that a structure can safely withstand. Vibrations below this limit have a very low probability of causing damage whereas vibrations above this limit have a reasonable probability of causing damage.

safe-T-Cut system, a system to quickly fell trees which have fallen on electrical wires during storms. The system comprises a length of PETN explosive mounted on a wooden clamp, which can be attached to a telescopic fibreglass pole to put the explosive at the place where the tree is wanted to be broken. The initiation of the PETN is made by a detonator and shock tube (NONEL). Nitro Nobel, 1991.

safety belt, a belt fastened to the human body with the purpose to fasten a one end of the rope to the safety belt and the other end is secured to something stable. The purpose is to prevent fall to lower level.

safety catch, a safety appliance which transfers the weight of the cage onto the guides if the winding rope breaks. Fay, 1920.

safety distance, the separation distance that will prevent the immediate direct propagation of explosives or fire from one magazine to another by missile, flame or blast and will not cause damage to protected works or injury to persons. AS 2187.1, 1996.

safety factor (R_{sf}) (dimensionless), the ratio of breakage resistance to load in rock mass reinforcement or support. In rock bolting it is strived for a safety factor of 3. Pusch, 1974.

safety fuse, string of black powder covered by textile and waterproof material. The safety fuse is used to initiate plain detonators. Its burn rate falls within defined limits and it does not explode. The burning will not be communicated laterally to itself or other similar fuses.

safety fuse tests, methods to test a safety fuse are: 1) measurement of diameter, 2) test for burning time, 3) accelerated ageing test, 4) gap test, and 5) lateral transmission test etc.

safety hat, a cap or hat with a hard crown worn by miners and which will resist blows against it. B.C.I., 1947.

safety hook, hook normally attached to an anchored line and fastened to an object to prevent it from falling. Atlas Copco, 2006.

safety instructions, safety instructions are either a separate document supplied with the machinery or included in the Operator's instructions. Safety instructions convey important information on safe operation of the machinery or on important safety issues (warnings). Atlas Copco, 2006.

safety shoe, a well-built shoe of leather or rubber provided with a sheet-steel or other strong, stiff toe. USBM Staff, 1968.

safety standard, suggested precautions related to the safety practices to be employed in the manufacture, transportation, storage, hauling, and use of explosive materials. Atlas Powder, 1987.

sales leaflet, small sheet of printed matter containing pictures and text to promote the features of a product. Atlas Copco, 2006.

saltpetre, sodium nitrate (SN) or potassium nitrate (PN). One of the principal ingredients of black powder.

sampling device (*fitted to the dust precleaner*), equipment used to take automatic samples of dust at drilling.

sand, detrial material of size 0,063–2,0 mm. A subdivision is made into coarse sand (CSa) >0,63 to 2,0 mm, medium sand (MSa) 0,2 to 0,63 and fine sand (FSa) 0,063 to 0,2 mm. SS-EN ISO 14688-1:2002, 2003.

sandblast, see '**mud cap**'.

sandblasting, a method of cleaning metal and stone surfaces with sand sprayed over them through a nozzle at high velocity. Ham, 1965.

sand fill, sand or plant tailings conveyed to underground by water or by gravity to support cavities left after extraction of ore. The purpose is to support, and avoid subsidence up to surface.

saver sub, see '**sub**'.

SB (acronym), see '**submarine blasting**'.

scabbing, the spalling of rock by the high amplitude tensile wave which is created upon the reflection of the compressive wave at a rock/air interface (free surface).

scabbing fragmentation, a reflected wave is developed when the body waves, P and S-waves, are hitting an interface between solid-gas or solid-fluid. The P-wave or compression wave will be reflected as both a tensile wave and compression wave, and if the amplitude is large enough the tensile wave can detach pieces from the solid material. This effect is called *scabbing*. It is recommended to call the fragmentation created by this mechanism for *scabbing fragmentation*.

scabbling, a quarrying term used to describe the process of *trimming blocks of stone*. It, may be done by hand with a scabbing pick, with circular saws, wire saws, heavy iron disks provided with cutting tools, diamond-toothed drag saws, or scabbing planers. AIME, 1960.

scabbling hammer, a hammer with two pointed ends for picking the stone after the spalling hammer. Fay, 1920.

scabbling planer, a planar consisting of massive blades, which scrape the surfaces being scabbled. Such blade will remove 6–12 mm of stone at each cut. AIME, 1960.

scaffold, the framework in the drill tripod on which the helper stands to couple and uncouple drill rods or casing. Also called safety board, safety platform. Long, 1960.

scaled distance SD, (D_{scaled}), **(dimension depends on what scaling factor being used)**, a factor relating similar blast effects e.g. ground vibrations or air blasts from various sizes of charges of the same explosive at various distances. Scaled distance is obtained by dividing the distance (R) in m from the blast to the point of observation by a root of the explosive mass per delay (Q) in kg. Usually the ratio of $R/Q^{1/2}$ or $R/Q^{1/3}$ is defined as the scaled distance. In crater blasting the scaled distance is defined as any linear measure (e.g. burden) divided by the cubic root of the explosive charge.

scaled strain intercept (K_c), (dimensionless), proportional constant for the calculation of the maximum strain of a longitudinal wave.

scaled volume (V_{scaled}), (m³/kg), a term used in crater blasting for the *ratio* between the amount of *material removed*, and the amount of *explosive used*. It is the same quantity as inverted specific charge (powder factor) (q).

scale invariance, the property that the rock structure looks the same independent of scale, and it is therefore necessary to include a scale (pen knife, hand lens or coin) when photographing rock structures. Hudson, 1993.

scaling, in blasting the removal of rock that is loose or in poor condition from the side walls, face and roof after blasting. A metal bar (e.g. iron or aluminium) with an angled and flattened iron at the end is used in manual scaling. Hydraulic hammer mounted on a boom is used in mechanised scaling.

scaling bar, a bar-like implement for removing loose rock in drift, tunnels, rooms and shafts. Also called *scaling rod*.

scaling law, a mathematical formula which permits the prediction of effects of large explosions on the basis of the effects of small explosions.

scaling rod, see '**scaling bar**'.

scalper, heavy screen shielding fine screen for separating differently sized particles. Bennett 2nd ed., 1962.

scalping, the removal, by screen or grizzly of undesirable fine material from broken ore, stone or gravel. Nelson, 1965.

scalping screen, a coarse primary screen, or grizzly, usually a vibrating grizzly. Nichols, 1965.

scanline mapping, see '**scanline survey**'.

scanline survey, a method to determine the joint frequency in the rock mass. On surface three orthogonal lines will be made and each crossing of joint along the lines will be noted. The data will be used to calculate the actual joint frequency in three to each other perpendicular directions. In drifts underground, scanlines can only be done in two directions. Maynard, 1990.

scanning electron microscope (SEM) (acronym), instrument using electronic beams instead of light waves to magnify objects. It can dissolve details down to 3×10^{-7} mm and larger atom nucleus can be reproduced.

scantling, the dimensions of a stone in length, breadth, and thickness. Standard 1964.

scatter in timing, the variation in actual firing times for detonators with the same nominal delay period. If pyrotechnical delays are used the scatter normally increases with an increase in nominal delay time. Half-second delayed detonators are therefore not recommended for high precision controlled contour blasting. For that purpose electronic delayed detonators which give a much lower scatter in timing are recommended.

SCB (acronym), see '**semi conductor bridge**' ignition.

schist, a crystalline rock that can be readily split or cleaved because of having a foliate or parallel structure, generally secondary and developed by shearing and recrystallization under pressure. Fay, 1920.

schistosity, the variety of foliation that occurs in coarser-grained metamorphic rocks. General it is the result of the parallel arrangement of platy and ellipsoidal mineral grains within the rock substance. ISRM, 1975.
Schmidt hammer test, a device for the non-destructive testing of concrete or rough rock surfaces. It is based on the principle that the rebound of a steel hammer, after impact against the concrete or rock, is proportional to the compressive strength of the concrete. Dodd, 1964.
Schmidt impact hammer test, see '**Schmidt hammer test**'.
Schmidt rebound hammer, see '**Schmidt hammer test**'.
Schuman distribution function, a distribution function for rock fragments after blasting. See '**Gaudin-Meloy distribution**'. Lizotte, 1990.
science, from *Latin "scientia"* meaning *having knowledge*. A branch of study concerned with observation and classification of facts, principles and methods and accumulated systematized knowledge. A science must pass through various stage of maturity, namely description of phenomena with implied reproducibility by independent observers and quantification. Webster, 1986. In short, science can be defined as the subject of *describing*, *understanding*, *explaining* and *predicting* phenomena.
sclerograph, see '**scleroscope**'.
sclerometer, an instrument for determining the degree of hardness of mineral using a sharp diamond point to scratch the specimen. The method, the mechanical mechanisms used, and the measured quantity used to express the hardness vary greatly. The Talmage and Bierbaum devices are among the better known. USBM has developed a pendulum sclerometer. Hudson, 1993.
scleroscope (hardness test), the instrument consists of a small diamond-tipped "hammer" which is dropped from a standard height of 251,2 mm on the prepared smooth surface of a specimen (min 40 cm^3 and 50 mm thick). The rebound height is a measure of the hardness and 165 mm is taken to be 100 scale units, this being the rebound height of quenched tool steel. Could also be used for rock. Hudson, 1993. This technique was first proposed by A.F. Shore in 1906.
scooptram (LHD), see '**load-haul-dump**' unit.
scout drilling, this is an investigation process involving the drilling of small diameter holes to locate the water table and/or the depth of sands and clays. The drilling tools may be hand operated. If they are to be engine driven, a highly mobile or portable rig is essential. Stein, 1977.
scow drilling, a drilling method where mud from the bottom of the drill hole is brought to surface by a *mud scow* (hollow cylinder) with a flap valve at its bottom. The lowest part of the mud scow is called the cutting shoe. Stein, 1977.
scraper, 1) a *steel tractor driven surface vehicle*, 6–12 m^3 capacity, mounted on a large rubber-tired wheels. The bottom is fitted with a cutting blade which, when lowered is dragged through the soil. When full, the scarper is full it transports the material to the dumping point where the material is discharged through the bottom of the vehicle in an even layer. Used for stripping and releveling topsoil and soft material at opencast pits. Nelson, 1965. 2) a *metal bucket* which is moved across the surface of a rock pile by wires which are operated from an electrically driven drum. The bucket is filled with rock or ore and the material is transported by sliding over the rock surface to an ore pass.

scraper bucket, in coalmines, the scraper bucket is a bottomless, three-sided box, with a hinged back. The hinge operates in a forward direction so that on the return journey on the coalface, the back opens allowing the box to remain empty. On the loading journey, the coal closes the hinge and the material is drawn or scraped forward to the point of discharge. Also called 'scoop'. Nelson, 1965.

scraper loader, a machine used for loading coal or rock by pulling an open-bottomed scoop back and forth between the face and the loading point by means of ropes sheaves, and a multiple drum hoist. The filled scoop is pulled on the bottom to an apron or ram where the load is discharged into a car or conveyor. Jones, 1949.

scraper ripper, a piece of strip-mine equipment that assists breaking coal, loading, and hauling. Features of the scraper ripper include ripping teeth on the lip for breaking the coal and a flight conveyor for carrying the broken coal away from the lip. As the ripper teeth bite into and loosen the coal, the conveyor seeps the loose coal upward and prevents build up ahead of the lip. Coal Age, 1966.

scratcher, a mechanical device to break up the mud cake and stir the mud, which has gelled agianst the wall. Stein, 1977.

scratch hardness test, see '**sclerometer**'.

scratch sclerometer, see '**sclerometer**'.

screen, a large sieve made of steel or rubber for grading or sizing coal, ore, rock or aggregate. It consists of a suitably mounted surface of quadratic woven wire or of a punched plate or rubber. It may be flat or cylindrical, horizontal or inclined, stationary, shaking or vibratory, and either wet or dry operation. The holes are sometimes made circular in steel plates or rubber.

screen or better **wire mesh handling arm**, a special hydraulic arm used in connection to mechanized rock bolting rig with the purpose to pick up a *wire mesh section* and fix it into the correct position along the roof and the walls. Atlas Copco, 2007.

screen analysis, grading rock into sizes by use of screens. Screen analysis is a laboratory or full scale procedure widely used in ore testing and plant control, in which a sample is screened for a prescribed period on a series of sieves and the mass on each sieve and the rest passing through is recorded. The size distribution of the rock fragmentation is determined by a series of square meshes normally where the size of the next larger sieve normally is twice the size of the proceeding screen e.g. 0,064, 0,128, 0,256, 0,512, 1,024 m.

screen cloth, a woven tissue suitable for use in a screen deck. B.S. 3552, 1962.

screening of blast waves, occurs if a physical slot is made between the blast source and the recorder, or if the strength of the rock mass is reduced, this will act as a filter and screen the blast waves.

screening or **wire meshing**, the process of reinforcing the rock surface by a steel net which can hold smaller stones in place or by using reinforcement steel mats made of 5 mm reinforcement bar and size normally 1,5 × 3,3 m which are fastened to the rock by rock bolts and shotcrete. It is a time consuming operation, but common practise in Canada and Australia. Atlas Copco, 2007.

screw compressor, a machine used to compress air by a screw.

screw feed, a system of gears, ratchets, and friction devices, or some combination of these parts, in the swivel head of a diamond drill, which controls the rate at which a bit is made to penetrate the rock formation being drilled. When controlled by a feed gear, the bit maintains the same penetration rate per revolution regardless of drill-stem revolutions per minute.

Also called '*gear feed*', '*mechanical feed*'. Long, 1920. Drill feed in which the drill carriage is moved along the feed beam by means of a long threaded rod passing through it and driven by an air or hydraulic motor. Atlas Copco, 2006.

screw feeder, an auger-type screw to transfer material from one piece of equipment to another. ACSG, 1963.

scrubber, device in which coarse and sticky ore, clay, etc., is washed free of adherent material or mildly disintegrated. The main forms are the wash-screen, wash trommel, log washer and hydraulic jet or monitor. Scrubbers or scrubbing towers are also used to separate soluble gases with extracting liquids, or to remove dust from air by washing. Pryor, 1963. Device used to separate unburned material from exhaust fumes or water from air. Atlas Copco, 2006.

sculpture blasting, artistic forming of rocks to a sculpture by blasting.

SD (acronym), see '**scaled distance**'.

SE (acronym), see '**shock energy**'.

sealed microcrack, a microcrack sealed with a material that is different from the surrounding material.

sealed porosity or *closed porosity* (n_s), **(dimensionless)**, the ratio of the volume of the sealed (closed) pores to the bulk volume expressed as a percentage. See also '**true porosity**'. Dodd, 1964.

sealing of swelling clay zones, when a clay zone crosses a drift it may swell into the drift and cause rock mass movement and thereby lowering of the rock mass strength. This is a very dangerous situation and must be corrected immediately. When the zones are narrow or <0,2 m, the two rock surfaces are cleaned and covered by shotcrete. When the zone is wider, reinforcement irons are mounted in 45° angle to the both sides and armed shotcrete is used to cover the clay zone together with the reinforcement irons. Pusch, 1974.

seal off, see '**blankoff**'.

seam, a stratum or bed of coal or other mineral, generally applied to large deposits of coal. Also a stratification plane in a sedimentary rock deposit.

seam blast, a blast made by placing explosives along and in a seam or crack between the solid wall and the stone or coal intended to be removed.

seam cut, a parallel hole cut used in drifting using 2 uncharged holes for swelling and 9 charged holes se Fig. 54.

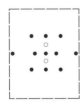

Figure 54 Seam cut, a parallel hole cut used in tunnelling. Brännfors, 1973.

secant modulus of elasticity (E_s), **(MPa)**, materials such as rock, concrete or wire have a variable Young's modulus E so that the particular value of E adopted must be either the slope of the tangent of the stress-strain curve or that of the secant. Ham, 1965. E.g. the secant modulus E_s, is the slope of a straight line joining origin of the axial stress-strain curve to a point on the curve at peak strength, see Fig. 55.

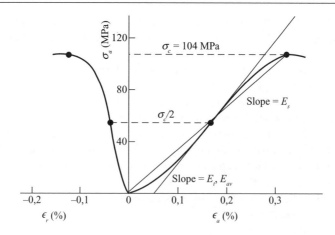

Figure 55 Definition of secant modulus of elasticity from results obtained in uniaxial compression test. After Brady, 1985.

secondary blasting or **boulder blasting**, use of explosives to break up larger blocks of rock resulting from the primary blasts, the rocks of which are generally too large for easy handling. Different methods for boulder blasting are 1) *Popping*. Drilling a hole into the boulder, charging, and blasting, 2) *Plaster shooting*. Putting a charge in close contact with the boulder and blasting, 3) *Mud capping*. This method is the same as method 2, but the explosive is covered with mud to increase its breaking efficiency 4) *Shaped charges*. They are located close to or several meters from the boulder to be blasted. The last method is used in underground mining for hang ups.

secondary breaking or **breakage**, the additional breaking of caved or broken ore or rock by mechanical means or by explosives to reduce large fragments to the sizes required for loading and transportation. Secondary breaking may also be required to treat hang-ups after blasting so the majority of fragments are of acceptable size for further processing, hauling and crushing.

secondary crusher, crushing and pulverizing machines next in line after the primary crushing to further reduce the particle size of rock. ACSG, 1963. This group of machines includes the finer types of *jaw crusher* and *gyratory crusher*, *single roll crusher* and also *crushing rolls, hammer mills*, and *edge runner mills*. Dodd, 1964.

secondary drilling, the process of drilling the so-called "pop holes" for the purpose of breaking by explosives the larger, unacceptable masses of rock resulting from the primary blast.

secondary explosive, explosives in which detonation is initiated by the detonation impact of an primary explosive. Insensitive materials such as ammonium nitrate or ammonium perchlorate are classified as tertiary explosives. Meyer, 1977. Examples of *tertiary explosives* are nitroglycerin (NG), ethyleneglycoldinitrate (EGDN) or pentaerythrotetranitrate (PETN), mixed with other explosives and blasting agents. Brady, 1985.

secondary fragmentation, the fragmentation produced in caved ore or rock during residence in the ore column and the following draw.

secondary state of stress or **induced stress**, disturbed stress field in the rock mass due to manmade excavations.

secondary support, see '**support**'.
second degree of freedom, see '**presplitting**'.
second moment of area (A), (Nm), the movement of inertia I of the plane area of section. Ham, 1965.
SECRBB (acronym), for '**straight edge crack round bars in bending**' see '**fracture toughness**'.
SEE (acronym), for '**Society of Explosives Engineers**'. The title has been changed to '**International Society of Explosives Engineers**'.
seed waveform, the waveform caused in the surrounding rock mass close to the blasthole. See also '**signature hole vibration monitoring**'.
segmental concrete, see '**concrete segmental lining**'.
segmentation, in image analysis a term used for the technical procedure of avoiding a fragment associated with two or more distinct grey levels being analysed as two or several fragments.
seismic detonators, these detonators are required to have a high degree of uniformity of firing and their reaction time is less than 1 ms. They are made of aluminium tubes and strong enough to withstand high hydrostatic pressures. They are protected against static electricity. The fuse heads are surrounded by a neoprene sleeve to afford additional security against static discharge between the fuse head and the detonator wall. They are fitted with special insulated extra-strong tinned-copper leading wires short-circuited sheathed at the ends.
seismic blasting, the use of explosives to initiate a shock wave which after some attenuation changes to a seismic wave which is used to investigate the geological formation of interest to be examined.
seismic cord, a detonating cord with high linear charge concentration used for special blasting in seismic prospecting.
seismic detonator, see '**blasting cap**'.
seismic evaluation (of rock mass quality), the relationship between seismic velocities measured in the field and corresponding values measured upon intact rock samples in the laboratory can be used as and index of the rock mass fracture state. Onodera, 1963 suggested the ratio of the *compression wave velocity* measured in situ, c_{Pi}, to that measured in the laboratory, c_{Pl}. Subsequently, Deere et al., showed that the square of this ratio (velocity index) is numerically *approximately* equal to Rock Quality Designation RQD, which is dependent upon fracture state. For boreholes, RQD is expressed as the percentage of solid core recovered greater than 0,1 m in length.

Table 39 Seismic evaluation of rock mass quality. Hudson 1993.

Rock quality classification	RQD (%)	Fracture frequency (m^{-1})	c_{Pi}/c_{Pl}	$(c_{Pi}/c_{Pl})^2$
Very poor	0–25	15	0–0,4	0–0,2
Poor	25–50	8–15	0,4–0,6	0,2–0,4
Fair	50–75	5–8	0,6–0,8	0,4–0,6
Good	75–90	1–5	0,8–0,9	0,6–0,8
Excellent	90–100	1	0,9–1,0	0,8–1,0

c_{Pi} = P-wave velocity measured in situ, c_{Pl} = P-wave velocity measured in lab.

seismic explosive, an explosive used to create seismic waves in rock for the purpose of prospecting for geological deposits, particularly oil horizons, or the preinvestigation of the rock mass before making constructions in it. In oil prospecting, the seismic explosive must detonate even under high hydrostatic pressure.

seismic imaging, the use of seismic waves to characterize the rock mass before and after blasting of excavations underground and on surface. See '**cross hole seismic**', 'seismic refraction technique' (surface seismic) etc.

seismicity (f_{sei}), **(No./year and 100 km²)**, a quantity of the measure of the frequency of earthquakes, for example, the average number of earthquakes per year and per 100 square kilometres. Schieferdecker, 1959.

seismic methods, can be used to quantify the hydraulic conductivity of the rock mass and rock stresses. Shear wave attenuation and polarization may be used to determine the location and predominant orientation of fluid-filled fractures since these features selectively filter the shear waves. This directional polarization may be used to gain an understanding of crack presence or anisotropy in geological media. In observing the frequency content of compression waves, modification by viscous damping across fractures or in porous media may enable mechanical characteristics of fractures or the porous medium to be conceivably determined. Good correlation has been found between the maximal radial amplitude and that of maximum horizontal stress. Hudson, 1993.

seismic ray transmission tomography, see '**seismic evaluation of rock mass quality**'.

seismic refraction method, in refraction shooting, the detecting instruments are laid down at a distance from the shot hole that is large compared to the depth of horizon to be mapped. The explosion waves travel large horizontal distances through the earth and the time required for travelling gives information on the velocity and depth of certain subsurface formation. Dobrin, 1960. The refraction method can be used to investigate the blast induced damage in the wall and roof of drifts and tunnels. The seismic source is a 3 kg hammer blow on the wall of the drift and a seismic sensor is placed some meters away and used to measure both the direct signal from the hammer blow, but also the refracted waves which appear if there are loose blocks near the surface.

seismic crosshole method, see '**crosshole seismic**'.

seismic strength test, a method to test the strength of an explosive. A certain amount o charge is detonated in an isotropous rock medium, and the seismic disturbance produced at a determined distance. The standard explosive is usually ANFO and it is supposed that the variation in vibrations is proportional to the energy of the explosive raised to 2/3. This test is not considered very adequate in measuring the available energy of an explosive. Jimeno et al., 1995.

seismic transmission tomography (crosshole seicmic), use of seismic waves to detect fracture in the rock mass. The compression wave also called the P-wave velocity can be used to detect changes in fracture frequency. S-waves are attenuated too much. Hudson, 1993.

seismic velocity, see '**seismic waves**'.

seismic waves, dynamic disturbance in the earth due to mechanical disturbance on the surface or underground. Seismic waves are utilized in geophysical exploration. Several types of waves are produced: two types of body waves 1) *P-wave* (Longitudinal or dilatational wave); 2) *S-wave* (Transverse, distortional or shear wave); 3) Several types of surface and interface waves 1) *Rayleigh waves,* 2) *Love waves,* Love waves occur only if a horizontal layer covers a semi-space, 3) *Stonley waves* etc. (exists at the interface of two dissimilar media). The speed of propagation is characteristic for each type of wave and rock, depending largely on its compactness (density) and water content. In sandy clay the speed of the P-wave is about 1 200 m/s, in sandstone 3 000 m/s, and in igneous rock up to 6 600 m/s. The water content and porosity of rock has a large influence on the seismic velocity. Ramana, ACSG, 1963. 1973. The P-wave velocity might increase up to 90% at increased water content. Broch, 1974, and Lama and Vutukuri, 1978. For a planar wave, the P- and S-wave velocities can be calculated from the elastic properties of the rock.

$$c_P = \sqrt{\frac{\lambda + 2G}{\rho_r}} = \sqrt{\frac{E}{\rho_r}\left(\frac{1-\nu}{(1+\nu)(1-2\nu)}\right)} \qquad (S.1)$$

where c_P is the longitudinal wave velocity (P-wave velocity) in (m/s), λ is the Lame's constant, $\lambda = \nu E/(1 + \nu)(1 - 2\nu)$ in (Pa), G is the shear modulus in (Pa) or $G = E/2(1 + \nu)$ where E is Young's modulus in (Pa), $\nu =$ Poisson's ratio, and finally ρ_r is the density of the rock in (kg/m³).

$$c_s = \sqrt{\frac{G}{\rho_r}} = \sqrt{\frac{E}{\rho_r}\frac{1}{2(1+\nu)}} \qquad (S.2)$$

The S-wave velocity is approximately half of the P-wave velocity. The Rayleigh wave velocity depends on Poisson's ratio and is about 10% lower than the S-wave velocity. The velocity of the Love wave is larger than the Rayleigh wave velocity. As the attenuation of the Rayleigh wave is of lower order than P- and S-waves, at larger distances from a blast, Rayleigh waves account for most of the damage done to buildings.

seismograph, an instrument that measures and supplies a permanent record of earthborne vibrations induced by earthquakes or blasting. For blasting other types of instruments are used today. Usually, the particle velocity or acceleration is measured. Previously, the displacement and vibration frequency were recorded and from these values the particle velocity was calculated. For more complete characterization of a seismic/blast it is necessary to measure and record the seismic/blast wave in three mutually perpendicular directions.

seismology, the science of earthquakes and attendant phenomena. Schieferdecker, 1959.

seismometer, an instrument buried in the ground which transforms the mechanical effects of earth shocks into electrical energy. This is transmitted by a circuit to seismograph placed above ground which records the impulses. Ham, 1965. Synonym for *geophone, detector pickup* and *jug.* A.G.I., 1957.

selective bolting, bolting only in these areas where it is judged necessary. TNC, 1979.

selective mining, a method of mining whereby ore of unwarranted high value is mined in such manner as to make the low-grade ore left in the mine incapable of future profitable extraction. In other word, the best ore is selected in order to make good mill returns, leaving the low-grade ore in the mine. Frequently called *robbing a mine.* Fay, 1920.

selective mining methods, methods with the aim to avoid mixing of high grade ore with low grade ore or waste. One example is the 'soufflé mining method'.

self-advancing support, an assembly of hydraulically operated steel hydraulic supports, on a long-wall face, which are moved forward as an integral unit by means of a hydraulic ram coupled to the heavy steel face conveyor. Nelson, 1965.

self-destruction detonator, a detonator which destructs itself after a certain time if ordinary detonation has failed.

self-hardening steel, see '**air-hardening steel**'.

self-ignition, see '**spontaneous combustion**'.

SEM (acronym), for '**scanning electron microscope**'.

semiconductive hose, a hose, used for pneumatic loading of ANFO, which has an electrical resistance chosen in order to ground static electric charges and to limit stray electric currents. In the USA, for example, the hose must have a minimum electrical resistance of 1 000 Ω/m, 10,000 Ω of total resistance, and a maximum total resistance of 2 MΩ.

semiconductor bridge (SCB) ignition, the fusehead in the detonator consists of a heavily doped polysilicon bridge which is much smaller than the conventional bridge wires. Passage of a current pulse with significant less energy than that required of a hot-wire ignition produces a plasma discharge in the SCB which ignites the explosive in contact with the bridge, producing an explosive output in a few microseconds. Persson et al., 1994.

semigelatin, dynamite containing ammonium nitrate as the main explosive ingredient and plasticized with a blasting gelatin. Semigelatin is more resistant to water than ammonia dynamites, but less than gelatin dynamites.

semi-mechanised drilling and bolting, the drilling is mechanised using an *hydraulic drill jumbo*, followed by *manual installation* of the *bolts* by operators working from a platform mounted on the drill rig, or on a separate vehicle. The man-basket, as a working platform, limits both the practical working space and the retreat capability in the event of falling rock. In larger tunnels the bolt holes are drilled with the face drilling jumbo. Atlas Copco, 2007.

sensitiveness, a measure of an explosive's cartridge-to-cartridge propagation ability under certain test conditions. It is defined as the distance through air at which a primed half-cartridge (donor) will detonate an unprimed half-cartridge (receptor).

sensitivity to propagation, sensitivity to propagation of an explosive can be ascertained by a method called the *Ardeer double cartridge*, or ADC test. The test consists of firing an explosive cartridge with a standard detonator and determining the maximum length of the gap across which the detonation wave will travel and detonate a second, receptor, cartridge. Both the primer and the receptor cartridges should be of the same composition, diameter, and weight. McAdam, 1958.

sensitivity to the impact of an explosive, the property of an explosive which determines its susceptibility to initiation by heat, light, shock or other applied energy. Also a measure of its ability to propagate the detonation.

sensitizer, ingredient used in explosive compounds to make it more easy to initiate or to improve the propagation of an explosive reaction.

sensor, device capable of receiving or responding to a stimulus such as the proximity switches on a drill feed. Atlas Copco, 2006.

separation distances, recommended minimum distances for accumulation of explosive materials to certain specific locations. IME, 1981.

separator, device used to separate two substances such as air and water. Atlas Copco, 2006.

sequence firing, see '**firing sequence**'. AS 2187.1, 1996.

sequential blasting or **delayed blasting**, the normal initiation technique is to initiate blastholes on separate delays and in some cases also the charge within a blasthole is divided into separate charges initiated under separate delays. The purpose with sequential blasting is to improve fragmentation and displacement while minimizing offsite blast effects such as air blast, flyrock, and ground vibration.

sequential blasting machine, a blasting machine, in which the delays between series of blasting caps are set. The series are coupled to different channels through which the initiation current is distributed with a prefixed delay of milliseconds. It is often used in combination with delay blasting caps. The main purpose of using sequential blasting machines is to set each blasthole and deck under separate delays and thereby reduce ground vibration and improve fragmentation.

sequential contour blasting, see '**micro-sequential contour blasting**'.

serial blast or **series blast**, see '**series blasting circuit**'.

series blasting circuit, a circuit of electric blasting caps in which each leg wire of a cap is connected to a leg wire from the adjacent caps, so that the electric current follows a single path through the entire circuit. USBM, 1983.

series connection, see '**series blasting circuit**'.

series electric blasting cap circuit, see '**series blasting circuit**'.

series in parallel circuit, see '**parallel series circuit coupling**'. USBM, 1983.

series in parallel electric blasting cap circuit, see '**parallel series circuit coupling**'. USBM, 1983.

set bit, a bit insert with diamonds or other cutting media. Long, 1920. Diamond drilling bit set with a large number of individual diamonds. Atlas Copco, 2006.

set casing, the cementing of casing in the hole. The cement is introduced between the casing and the wall of the hole and then allowed to harden, thus sealing off intermediate formations and preventing fluids from entering the hole. It is customary to set casing in the completion of a producing well. Williams and Meyers, 1964.

set diameter, the inside and/or outside diameter of a bit measured from the exposed tips of diamonds or other cutting media inset in the wall portions of the bit crown. Long, 1920.

set i.d. (abbreviation), (d_i), **(mm or m)**, for 'set inside diameter'. Long, 1920.

set inside diameter (d_i), **(mm or m)**, the minimum inside diameter of a set core bit. Usually written '**set i.d.**' in drilling industry literature. Also called *bore, center bore, inside gage, set i.d.* Long, 1920.

set o.d. (abbreviation), (d_o), **(mm or m)**, see '**set outside diameter**'. Long, 1920.

set outside diameter (d_o), **(mm or m)**, the maximum outside diameter of a set bit. Usually written '**set o.d.**' in drilling industry literature. Also called *outside gage*. Long, 1920.

sets, steel ribs or timber framing to support the tunnel excavation temporarily. Bickel, 1982.

setting-out error (D_s), **(m)**, deviation of the marked collaring point from the intended position of the hole.

setting-out of holes *or staking*, to mark the collar centre of the holes to be drilled.

setting rod, a special diamond-drill rod used to set a *deflecting wedge* in a borehole. Long, 1920.

SGI, (acronym), for '**specific gravity influence**'.

shaft, an excavation, generally vertical, in a mine extending downward from the surface or from some interior point (blind shaft) as a principal opening through which the mine is supplied and exploited. A shaft is provided with a hoisting engine at the top for handling men, rock, and supplies. It may also be used for rock transportation by gravity forces, for ventilation purposes or pipelines for water. Openings other than vertical used for such purposes are termed *inclined shafts*, *inclines* or *slopes*.

shaft plumbing, the operation of transferring one or more points at the surface of a vertical shaft to plumb line positions at the bottom of the shaft. A method to ensure that a shaft is sunk in the true vertical line. Nelson, 1965.

shaft set, supporting frame of timber, masonry, or steel which supports sides of shaft and the gear. Composed of two wall plates, two end plates, and dividers which form shaft into compartments. Pryor, 1963.

shaft sinking, driving a shaft from the top downwards.

shaker conveyor, a conveyor consisting of a length of metal troughs, with suitable supports, to which a reciprocating motion in imparted by drives. In the case of a down hill conveyor, a simple to-and from motion is sufficient to cause the coal to slide. With a level or a slight uphill gradient, a differential motion is necessary, that is, a quick backward and slower forward strokes. The quick backward stroke causes the trough to slide under the coal, while the slower forward stroke moves the coal along to a new position. Also called jigger. Nelson, 1965. A type of oscillating conveyor. ASA MH4.1, 1958.

shaker-shovel loader, a machine for loading coal, ore, or rock usually in headings or tunnels. It consists of a wide flat shovel, which is forced into the loose material along the floor by the forward motion of the conveyor. The shaking motion of the conveyor brings the material backwards and it is loaded into cars or a conveyor. It works at its maximum efficiency in upward inclined or in flat tunnels. The American version of the shaker-shovel is called the *duckbill loader*. See also '**loader**'. Nelson, 1965.

shaking-conveyor loader, the broad tapering shovel like end of a shaking conveyor that is thrust suddenly under the coal and slowly withdrawn, so as to carry the coal which has been lifted toward the dumping point. Zern, 1928.

shank, the end of the drill steel or rod fitted into the drilling machine. The shank is struck by the hammer in the drill machine. Lewis, 1964. Short steel rod that connects the drill string to a rock drilling machine. Atlas Copco, 2006.

shank adaptor, metal connector piece between the drill machine and the drill string or rods that transmits energy from the drill hammer to the drill bit.

shank end rod, drill rod equipped with a shank end, usually of hexagonal section. Atlas Copco, 2006.

shank rod, the drill steel including a shank which is inserted in the drill machine. Sandvik, 2007.

shaped charge, in general any special shape of explosive charge. More specifically, the term refers to cavity charges with metal liners, usually of conical or linear design, and used to produce a high velocity cutting or piercing jet of molten metal. '**Conically shaped charges**' can be used in boulder blasting and '**linear shaped charges**' in controlled contour blasting or demolition blasting. The penetration in granite by jets from shaped-charge liners of six materials was tested by Rollins and Clark, 1973. For maximum penetration the cone or the V must be covered by a metal. The first linear shaped charge for controlled contour blasting the '**circular bipolar shaped charge**' is shown in Figure 56a with two Al- metal V:s, for the forming of a metal slug with large penetration depth (5 mm) compared to the '**elliptical bipolar shaped charge**' in Figure 56b developed and used in China. Qin, 2009.

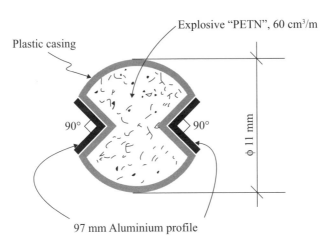

a) **Circular bipolar linear shaped charge**

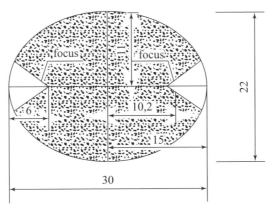

b) **Elliptical vbipolar linear shaped charge**

Figure 56 Comparison of section of a 'Circular bipolar linear shaped charge' (CBLSC). Rustan, 1983 compared to a 'Elliptical bipolar linear shaped charge' (EBLSC). Qin, 2009.

shape factor (*L*), (dimension varies with type of factor), a quantification of the shape (form) of a fragment. The general shape of particles (fragments), could be ellipsoidal, pyramidal, sphere, flat, rhombohedral etc.

Table 40 shows the European standard vocabulary for particle shape. SS-EN ISO 14688-1:2002, 2003.

Table 40 Terms for designation of particle shape. SS-EN ISO 14688-1:2002, 2003.

Parameter	Particle shape
Angularity/roundness	Very angular
	Angular
	Sub angular
	Surrounded
	Rounded
	Well rounded
Form	Cubic
	Flat
	Elongate
Surface texture	Rough
	Smooth

The shape can be defined in many ways and there are totally more than a hundred definitions. In the field of blasting, however, there is no international standard for shape classification. Some definitions used for classification in image analysis and for aggregates used for road and house building purposes are as follows:

$L_1 = 4A/P^2$
$L_2 = 1/L_1$
$L_3 = A/4d_{max}d_{min}$
$L_4 = L_{max}/(d_{eqv})^3$
$L_5 = L_{max}/W$ (Schistosity, used for road building purposes)
$L_6 = $ elongation ratio $R_{elo} = W/L$
$L_7 = $ flatness ratio $R_{fla} = H/W$
$L_8 = $ shape factor $F = R_{fla}/R_{elo}$

L_1 to L_8 are shape factors
$A = $ projected area of fragment (m^2)
$P = $ projected perimeter of fragment (m)
$d_{max} = $ maximum diameter of fragment (m)
$d_{min} = $ minimum diameter of fragment (m)
$H = $ height of fragment (m)
$L = $ length of fragment (m)
$L_{max} = $ maximum length of fragment (m)
$d_{eqv} = $ diameter of equivalent volume (m)
$W = $ width of fragment (m)

The shape can also be defined by the length, width and height of a parallelepiped which could circumscribe the fragment, see Fig. 34 at the entry 'fragment size'. Another way of classifying shape is by manual comparison of its sphericity with known shapes, see Fig. 57.

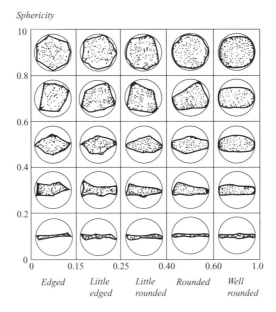

Figure 57 Shape of fragments. After Wahlström, 1955.

Fourier analysis can be used in defining the 2D-shape of fragments. A frequency spectrum can be determined by regularly measuring the distances from the centre of the point of gravity of the fragment to the edge of the fragment at regular angles in that figure created when the fragment is projected on a planar surface.

shape of fragment (L), the contour of a fragment when projected on a planar surface. See '**shape factor**'.

shatter cut, see '**burn cut**'.

shattering cut, see '**burn cut**'.

shatter zone, the zone adjacent to the blasthole and including the crushing zone. Clay, 1965.

SHB (acronym), see '**single hole blasting**'.

shear angle or **angle of internal friction**, (ϕ), (°), the angle between the axis of normal stress and the tangent to the Mohr envelope at a point representing a given failure-stress condition for a solid material. ISRM, 1975. See Fig. 58.

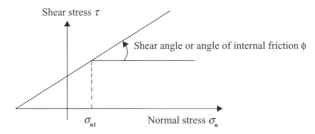

Figure 58 Shear stress τ versus normal stress σ_n-graph illustrating the shear angle ϕ at normal stress σ_{n1}.

shear box test, a standard method of measuring the shear strength of soil in a box, split in half, to which vertical pressure is applied combined with shearing force horizontally. The box can be either 152,4 mm (6 inch) or 304,8 (12 inch) mm in square. Although this is a very simple shear test, the distribution of stress on the shear plane is difficult to assess, and it is often superseded by the triaxial compression test. Ham, 1965. There are also available portable shear boxes that can test discontinuities contained in pieces of drill core or as an in-situ test on samples of larger size. Brady, 1985.

shear fractures, the result of differential movement of rock masses along a plane, are commonly observed in the field. Such structures are generally of tectonic origin and may range in size from faults with an extent of many tens of miles to small-scale structures observable in hand specimen or under the microscope.

shearing mixing of particles, improved cement injection mixture as the cement particles disintegrate in three steps through a shearing action. Atlas Copco, 2006.

shearing V-cut, see 'V-cut'.

shear modulus of elasticity (modulus of rigidity), **(G), (MPa)**, a quantity that defines the shear stress necessary to cause a specified change in the shape of a solid material. The shear modulus can be calculated by either a) $G = \rho c_s^3$ where ρ is the density of the material in kg/m³ and c_s is the S-wave velocity of the material in m/s or b) $G = E/2(1 + v)$, where E is Young's modulus, and v is Poisson's ratio.

shear stiffness (K_s), (MPa), is the rate of change of shear stress τ in MPa with respect to shear displacement u (dimensionless).

shear strain (γ), (dimensionless), angular displacement of a structure member due to a force acting across it measured in radius. Ham, 1965. In plane wave propagation, the shear strain γ caused by the ground waves can be approximated by

$$\gamma = v_{maxp}/c_P \tag{S.3}$$

v_{maxp} is the maximum particle velocity in the P-wave in m/s and c_P is P-wave velocity in m/s.

shear strength (τ_s), (MPa), the stress or load at which a material fails in shear. This property of a material is determined in a shear test where the shear strength is determined under different normal stress, see Fig. 59.

shear strength of soil, is determined in situ by the Vane shear test. The test apparatus consists of a vane in the form of crossed stainless steel plates. Torque rods connect the vane to the torque measuring instrument at the surface of the hole. Stein, 1977. See also **'confined shear strength of brittle materials'**.

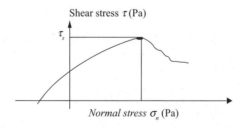

Figure 59 Definition of maximum shear strength.

shear stress, *shearing stress, tangential stress,* **(τ), (MPa)**, a stress in which the material on one side of a surface pushes on the material on the other side of the surface with a force which is parallel to the surface. Lapedes, 1978.

shear thinning, the fluid for transportation of chips is thin where shear action is occurring, that is, at the walls of the hole.

shear wave, see '**seismic waves**'.

shear wave velocity, transverse wave velocity, S-wave velocity (c_s), **(m/s)**, see '**seismic waves**'.

shear wave polarization, see '**seismic methods**'.

sheathed explosive, a permitted explosive surrounded by a sheath containing a non-combustible powder. The powder acts as a cooling agent and reduces the temperature of the resultant gases of the explosion, and therefore reduces the risk of these hot gases causing a firedamp ignition. Sheathed explosives were first introduced into British coal mining in 1934. Nelson, 1965.

sheave, a wheel grooved around its circumference, that guides and support a cable or rope. between the load and the hoisting engine. Long, 1920.

sheeting driver, an air *percussion hammer attachment* that fits on *plank ends* or *sheat piles* (sheat pile driving) so they can be driven without splintering into the earth.

sheeting rig, rig designed specifically for driving sheet piles. Atlas Copco, 2006.

shelf life (days), length of time an explosive can be stored without becoming unsafe or otherwise failing to meet specified performance requirements. The time period is normally stipulated by law and depends on the storage temperature. High storage temperature and humidity reduce the shelf life of an explosive. Some examples of shelf life of various explosives and initiation materials are listed in Table 41.

Table 41 Shelf life of various explosives and initiation material. After Pradhan, 1996.

Explosive or initiation material	Approximate age
Nitroglycerin based	>12 month
Slurry explosives	>12 month
Site mixed slurry	2–4 weeks
Cast booster	>5 year
Detonator	>2 years
Shock tube	<2 years

shell, a tool to remove soft strata already loosened by chisels from the bottom of the hole in rod percussion drilling. Stein, 1977.

shield, a steel cylinder with open or closed face equal to the tunnel diameter for tunnel excavation in soft ground. Artificial roof normally made of steel is used to prevent rock to fall down and injure or kill personal at drifting or tunnelling with drill- and blast method or full face boring.

shield-driven tunnel, artificial roof normally made of steel is used to prevent rock to fall down and injure or kill personal at drifting or tunnelling with drill- and blast method or full face boring.

shielding, use of protective materials to prevent throw of fragmented rock from a blast. Examples of protective materials are *blasting mats* consisting of rubber, rope, steel wire or timber.

shock, a pulse or transient motion or force lasting thousands to tenths of a second which is capable of exciting mechanical resonances; e.g. a blast produced by explosives. Lapedes, 1978.

shock absorber, device for deadening a shock or sudden impact, as the shock absorber on a automobile or truck. Atlas Copco, 2006.

shock absorbing sub, see '**sub**'. Atlas Copco, 2007.

shock and impact test, test of explosives regarding accidental shock or impact during manufacturing or transportation. There are many tests developed such as air *gap* test, *minimum booster test, cap sensitivity test, BAM fall hammer test, projectile impact test* and *BAM 50/60 steel tube test.* Persson et al., 1994.

shock desensitization, see '**pressure desensitization of explosives**'.

shock energy (SE), (W_{sh}), **(MJ)**, the shock energy of an explosive is defined as the local energy released by the explosive during the time interval between onset of detonation and instant of maximum expansion of the blasthole given by;

$$W_{sh} = \frac{p_b V_b - p_e V_e}{\gamma - 1} 10^{-6} \qquad (S.4)$$

where W_{sh} is the shock energy in (J), p_b is the borehole pressure in (Pa) and V_b the volume of the blasthole before detonation of the explosive in (m³), p_e is the equilibrium pressure in (Pa) and V_e is the volume of the blasthole measured at the time instant of maximum expansion of the blasthole in (m³), and γ is the adiabatic exponent. Lownds, 1975. Shock energy is the energy in the compression wave induced by the detonation pressure in the blasthole. The shock wave is attenuated very fast and about 0,2 to 0,5 m from the blasthole, dependent on the blasthole diameter, it will change from a shock wave to a conventional compressive wave or P-wave. Lownds, 1986.

shock front, the outer side of a shock wave whose pressure rises from zero to its peak value. Also known as *pressure front*. Lapedes, 1978.

shock tube, initiation accessory in blasting consisting of a hollow plastic tube (inner diameter ~1,5 mm and outer diameter ~3–4 mm) containing a low core load of reactive material, such as an explosive composition (in NONEL, HMX is used) dusted lightly on the inner walls (~20 mg/m) or filled with a gas. It is used to transfer an ignition signal to the detonator. It is sealed at one end and has a delay detonator crimped to the other end. The sealed end of the tube is initiated by a spark or detonation, and the detonation is transferred through the tube without destroying the tube. The flash at the end is used to initiate a delay detonator. It is a sort of dust explosion travelling at about 2 000 m/s inside the tube. The original shock tube developed was given the name NONEL. Today there are several shock tube systems available on the market. Advantages of the system are the elimination of electric circuits for delay detonators and the hazards associated with these, and the ability of relay units to achieve infinitely large delayed blasts.

shock wave, a transient compressive wave moving in a non-linear (at high strain rates) and elastic solid (at low strain rates) faster than the velocity of sound in the material. The non-linear relationship between stress and speed causes shock waves to 'shock up' and thereby the amplitude increases at the shock front. Density, pressure, temperature and particle velocity change drastically. Attenuation (geometrical and material damping) causes that shock waves exist only in the near vicinity around a blasthole. When its speed diminishes, the shock wave decays to become an elastic wave (seismic wave). A shock wave generally is considered to have a pressure-diagram characteristic in which the pressure falls exponentially with distance behind the shock front following an equation of the form.

$$p = p_{max} e^{-t/\tau} \qquad (S.5)$$

where p_{max} is the maximum pressure at the shock front in MPa, t is the time for a given characteristic in the wave to pass a fixed point in s (seconds) and τ is the relaxation time in s (seconds). Clay, 1965.

shootability, property of a shotcrete mixture. The goal is to find the property that it can be placed overhead with minimum rebound.

shooter, see '**shotfirer**'.

shooting, blasting in a mine or construction work.

shore, prop or **strut**, mine timber used to prop, brace or stay working places Pryor, 1963.

Shore hardness test, see '**scleroscope**'.

Shore scleroscope, see '**scleroscope**'.

short circuit, an abnormal connection that occurs when a current is not traversing the full, intended electrical circuit. AS 2187.1, 1996.

short delay blasting or **millisecond delay blasting**, a method of blasting where several charges are detonated in a short time sequence, usually 25 to 500 ms between the charges. The purpose of short delay blasting is to improve fragmentation, and reduce vibration and throw.

shorted location indication by frequency of electrical resonance (SLIFER), a method for continuous measurement of the velocity of detonation (c_d). A coaxial cable is placed along the charge column (in a blasthole) and connected with an instrument which registers the change in resonance frequency as the detonation front short circuits the cable.

short-flame explosive, see '**permissible explosives**'.

short rod test (SR), see '**fracture toughness**'.

short skirt, the part of a drill bit, below the bit head, that normally contains a female thread or a tapered section. Threaded and tapered bits may be available with shorter skirts than is standard. Atlas Copco, 2006.

shot, see '**blast**'.

shot anchor, a device that anchors explosive charges in the borehole, preventing the charges from being blown out by the detonation of other charges. Shot anchors can be made of wood, plastic, etc., but not iron.

shot bit, a short length of heavy-wall steel tubing, ranging from less than 76 mm to more than 1,8 m in diameter. With diagonal slots cut in the flat-faced bottom edge. The replaceable flat-faced shoe on a shot-drill core barrel. See also '**shot drill**'. Long, 1920.

shot blasting, a method similar to sand blasting for cleaning the surface of metals, using broken shot or steel grit instead of sand. It is less effective than sandblasting because the peening effect of the shot tends to drive unwanted deposit such as oxides into the surface. See also '**descaling**'. Ham, 1965.

shot boring, the act or process of producing a borehole with a shot drill. See also '**shot drill**'. Long, 1920.

shot boring drill (synonym), to shot drill, see '**shot drill**'. Long, 1920.

shotcrete, a pneumatically applied concrete used to provide passive support to the rock surface or structural coating. Shotcrete is prepared using either the *dry-mix* or the *wet-mix process*. In the *dry-mix process*, dry or slightly dampened cement, sand and aggregate are mixed at the batching plant, and then entrained in compressed air and transported to the discharge nozzle, where water is added through a ring of holes at the nozzle. In the *wet mix process*, the required amount of water is added at the batching plant, and the wet mix is pumped to the nozzle where the compressed air is introduced. The addition of 50 mm long and 0,4–0,8 mm diameter steel fibres has been found to improve toughness, shock resistance, durability, and shear and flexural strengths of shotcrete, and to reduce the formation of shrinkage cracks. *Fibre-reinforced shotcrete* will accept larger deformations before cracking occurs than will un-reinforced shotcrete.

shotcreting or **guniting**, the act of applying a concrete on the roof and wall of underground openings and drifts. It can also be used to reinforce shafts and bins underground. It is important to shotcrete new opened rock surfaces immediately to reduce the movement of the rock surface.

shot drill, a core drill generally employed in rotary-drilling boreholes of less than 76 mm to more than 1 800 mm in diameter in hard rock or concrete, using chilled-steel shot as a cutting medium. The bit is annular shaped, flat-face, steel cylinder with one or more diagonal slots cut in the bottom edge. As the bit and attached core barrel are rotated, small quantities of chilled-steel shot are fed, at intervals, into the drill stem with water. The shot works its way under the flat face of the bit and wears away the rock as the bit rotates. At intervals the core is removed from the borehole in somewhat the same manner as in diamond core-drilling operations. Also called *adamantine drill, calyx drill or chilled shot drill*. Long, 1920. *shot firing*, see '**blasting**'.

shot-drilled shaft, shafts up to 1,5 m in diameter drilled through rock to a maximum depth of 360 m by means of a shot drill. The latter makes use of shot for cutting a circular groove in the rock being penetrated, from which solid cores are extracted. Ham, 1965.

shotfirer (UK), '**blaster**' **(USA)** *or shooter*, the licensed person in charge of and responsible for the loading and firing of a blast. Atlas Powder, 1987.

shot firing cable, a pair of insulated copper or iron conductors which lead from the blasting machine (*exploder*) to the blasting circuit, including the detonators. Iron is used where copper cannot be tolerated in the muckpile, as it can be removed by magnets. It may consist of a twin core (both conductors contained in the one cable) or two separate conductors twisted around each other. A cable with such a quality that it can be used for several blasts.

shot firing lead, see '**shot firing cable**'.

shot hole, see '**blasthole**'.

shothole drill, see '**blasthole drill**.

Shot Shell Primer (Tradename), a device for the initiation of shock tubes by stomping on it.

shovel loader, a loading machine mounted on driven wheels by which it is forced into the loose rock at the tunnel face. A bucket hinged to the chassis, scoops up the material which is elevated over and discharged behind the machine. There are two types 1) the bucket is discharged directly into a mine car behind the machine, and 2) a short conveyor, built into the loader, receives the dirt from the bucket and conveys it back into a car or conveyor. See also '**mechanical mucking**'. Nelson, 1965.

shrinkage mining, see '**shrinkage stoping**'. Atlas Copco, 1982.

shrinkage stoping (SS), an underground mining method where the ore is mined out in successive flat or inclined slices, working upward from the level. After each slice is blasted down, enough broken ore is drawn off from below to provide a working space between the top of the pile of broken ore, and the back of the stope. Usually about 40% of the broken ore will have been drawn off when the stope has been mined to the top. BuMines Bull., 1936.

shrink fitting button, reducing the size of the button by cooling it before it is inserted into the bit. Sandvik, 2007.

shunt, the piece of metal or metal foil which short-cuts the ends of cap leg wires to prevent stray currents from causing accidental detonation of the blasting cap. The leg wires can also be short-cutted by twisting the leg wires.

shuttle car, a vehicle on rubber tires or caterpillar treads and usually propelled by electric motors, electric energy which is supplied by a diesel driven generator, by storage batteries, or by a power distribution system through a portable cable or a diesel motor. Its chief function is the transfer of raw materials, such as coal and ore, from loading machines in trackless areas of a mine to the main transportation system. ASA C42.85, 1956.

shuttle train, consists of a number of long rail cars, each fitted with a conveyor in the bottom. The cars overlap each other at the ends. It is thus possible to load the front car and by the conveyor move the rock to fill the complete train. The number of cars in a train can be chosen to accommodate the complete blasting round, half the round or less. Atlas Copco, 1982.

SI, (abbreviation), for French 'Système International d'Unités' (In English: International System for Units). See '**Systéme International d'Unités**'.

Si (abbreviation), for 'silt' defining particles from 0,002 to 0,063 mm.

SIC (acronym), for '**strain induced crack**'.

side dumper, an ore, rock, or coal car that can be tilted sidewise and thus emptied. Fay, 1920.

side bolt, specially designed bolts used to fasten together the sections of a rock drill. Atlas Copco, 2006.

side hole helper, row of holes next to the side holes and blasted before the side holes in smooth blasting.

side holes or **rib holes**, row of holes forming the wall contour in a tunnel or drift round. Singh and Wondrad, 1989.

side initiation, initiation of the explosive charge by a detonating cord (downline) of sufficient strength. This can result in reduced explosive performance, especially if the initiation process is marginal.

side rock, the rock mass on both sides of an excavation on surface or underground. Atlas Copco, 2006.

side wall, the lateral face of an excavation.

Siewers *J*-value or **SJ value**, a method to determine the sliding wear of a material. The SJ value is determined by drilling with a miniature drill, 8,5 mm in diameter and tipped with tungsten carbide, on a pre-cut rock sample. The force acting on the drill bit is caused by a mass of 20 kg. The SJ value is a measure of the depth drilled expressed in 1/10 of mm after 200 revolutions. The mean value for 4 to 8 drill holes is selected as the SJ value. The orientation of the pre-cut rock surface to the foliation is essential for the measurement. Drilling is usually performed parallel to the foliation for determination of the SJ value. Tamrock, 1989.

SIF (acronym), see '**stress intensity factor**'.

sign (explosive), see '**placard**'.

signal tube, see '**shock tube**'.

signal tube detonator, a detonator which has been crimped to a length of signal tube.

signal tube starter, a device for initiating the detonation in a shock tube (signal tube). AS 2187.1, 1996.

signature hole blasting, shooting a single representative borehole and analysing the waveform caused by the single hole blast. Anderson, 2008.

signature hole vibration monitoring, measurement of the ground vibration from one production blasthole at different distances and using the waveform to calculate the ground vibration from multiple blastholes with the same charge configuration.

silcrete, a superficial quartzite formed by the cementation of rock fragments (as soil, sand, or gravel) by silica. Webster 3d, 1961.

silent drilling, drilling methods where the noise from drilling has been reduced. *Active noise reduction* is made by changing the technique of mechanical drilling and *passive noise reduction* is defined as reducing existing noise by different technical shielding methods. An example of active noise reduction is to use in-the-hole drilling machines and to use different media for the powering of the drilling machine. An example of passive noise reduction is to use an isolated cabin for the operator. The noise can also be reduced closer to the source by shielding of the drill machine and drill rods by conveyor belts hanging on four sides of the drilling boom at surface drilling, see Atlas Copco, 2001 or by a special designed silent cover Atlas Copco, 2006.

sill, a minor intrusion into the crust of the earth concordant to the general structure of the rock compared with sills, which are discordant with the rock structure. They are both essentially parallel-sided, sheet-like bodies. Park, 1989.

silo, a tall tower, usually cylindrical and of reinforced concrete, in which grains, cement, rock or other similar bulk material is stored. Ham, 1965.

silt, fine grained sediments, mainly mud with sizes between 0,002 to 0,063 mm, namely **coarse silt** (CSi) >0,2 to 0,063 mm, **medium silt** (MSi) >0,0063 to 0,02 mm and **fine silt** (FSi) >0,002 to 0,0063 mm. Silt is closely linked with alluvium or marsh land. SS-EN ISO 14688-1:2002, 2003.

silver chloride cell, a low-current cell used in a blasting galvanometer and other devices to measure continuity in electric blasting caps and circuits. USBM, 1983.

single boom rig, drill rig equipped with one boom and having one drill feed and rock drill. Atlas Copco, 2006.

single hand drilling (old), rock drilling by hand, for example, in narrow reefs. A drill steel held in the left hand is struck by blows with a 4-pound hammer, the drill being turned between the blows. The drilling is very slow and laborious. Nelson, 1902.

single hole blasting (SHB), a method to examine the blastability of rock. In the beginning single holes were blasted in a bench with height 1 m, hole depth 1,33 m and 1 m burden and blasthole diameter 33 mm. The ratio of amount of explosive to the volume broken needed to create a throw of 1 m, called specific charge q (kg/m^3), was searched for. Fraenkel et al., 1952. Because of the fact that the detonation properties is dependent on the diameter of the charge, these small scale blasting tests are not representative for large diameter holes and therefore large diameter single hole blast tests (SHBT) were introduced, see '**single hole blasting test**'. Single hole blasting is also used to determine the *signature of blast vibrations* from only one hole. With this knowledge synthesized ground vibrations from many holes can be simulated by the computer.

single hole blasting test (SHBT), a method to test the blastability of rock. Single hole tests are done in the full scale with that diameter and bench height which is to be used in production blasting. Different burdens should be tested, and the vibration, throw, fragmentation, angle of breakage and overbreak measured. Rustan, 1987, Bilgin, 1991 and Bilgin and Paþamehmetoðlu, 1993. It is an extension of the single hole blasting test procedure in the half scale developed by Langefors. Fraenkel, 1954.

single hole vibration monitoring, see '**signature vibration monitoring**'.

single layer bit (synonym), see '**surface set bit**'.

single packer, see '**packer**'.

single pass drilling, term used in open pit mining and quarrying to denote that the drilling of the blasthole is done with a drill steel or rod that has the same length as the blasthole. Atlas Copco, 2006b.

single pulse shock tube, see '**shock tube**'.

single roll crusher, a crushing machine consisting of a rotating cylinder with a corrugated or toothed outer surface which crushes material by pinching it between the teeth and stationary breaking bars. ACSG, 1963.

single shot, a charge in one drillholes only fired at one time as contrasted with a multiple shot where charges in a number of holes are fired at one time. Fay, 1920.

single shot exploder, a blasting machine of the magneto type. It is operated by the twist action given by a half-turn of the firing key. A magneto blasting machine consists essentially of a small armature, which can be rotated between the poles of a set of permanent magnets. The armature is rotated by means of toothed gear wheels, which are actuated by the movement of the firing key. The electric circuit between the blasting machine and the detonator is completed by means of an automatic internal switch operating at the end of the stroke, or contract may be made by means of push button. McAdam, 1958.

single-toggle jaw crusher, a jaw crusher with one jaw fixed, the other jaw oscillating through an eccentric mounted near its top. This type of jaw crusher has a relatively high output and the product is of fairly uniform size. Dodd, 1964.

sinker, one who sinks mine shafts and puts in supporting timber or concrete.

sink holes, subsidence features associated with pre-existing solution cavities in dolomites and limestones. Brady, 1985.

sinking, making a vertical or inclined shaft through soil or rock. TNC, 1979.

sinking bucket, a large steel bucket for hoisting broken rock in shaft sinking.

sinking cut, the *initial cut* made in the floor of an open pit or quarry for the purpose of developing a new bench at a level below the floor. A round drilled, charged, and timed to be lifted vertically, due to the fact that no vertical face is available. The ground surface is the open face. Also called '*drop cut*' or '*lift shot*'.

site characterization, see '**rock mass classification**'.

site mixed emulsion (SME), emulsion explosive mixed at the site where it is going to be used.

site mixed explosive or **site mixing system for explosive (SMS)**, the components of an explosive are transported separated from each other in a truck to the place where it is going to be used. At the blasthole site the ingredients are mixed to a free flowing explosive which is pumped into the borehole.

site mixed slurry (SMS), slurry explosive mixed at the site where it is going to be used.

site sensitized emulsion (SSE), components needed to manufacture an explosive transported by a truck to the blast site where the components are mixed transforming it to an explosive. The advantage is that the transport can be made safe and the strength of the explosive can be varied at the blast site. There are trucks available that can charge tunnel blast holes with five different linear charge concentrations. The charging technology is called '*string charging*'. Hedin, 2003.

Six Sigma, a methodology that seeks to indentify and remove the causes of defects and errors in manufacturing and business processes. It uses a set of quality management tools, including statistical methods, which creates a special infrastructure of experts within the organization. The method was originally developed by Motorola in the USA in 1986. Keller, 2001.

size effect, influence of a specimen's size on its strength or other mechanical parameters. ISRM, 1975.

size-strength classification system, is a simplified classification system for rocks, developed by John A. Franklin and Louis, whereby the *block size* and *point-load strength* can be used to arrive at a single value of support. The method allows for a support-excavation strategy and specific support determination. The system can be used to assist in the selection of support at the early stages of design and its correct application is restricted to shallow tunnels, less than 300 m deep. Hudson, 1993.

SJ-value, see '**Siewers J-value**'.

skid, one of a pair of metal rails placed under and used to move a heavy object by dragging it. Heavy pump units and cement mixing equipment is often mounted on a skid frame. Atlas Copco, 2006.

skid mounted, group of components forming a unit, mounted on a skid frame. Atlas Copco, 2006.

skin grouting, filling the annular gap between a steel lining and the concrete encasement with grout. While the allowable grout pressure is dictated by the steel lining slenderness, it is well known that skin grouting has occasionally led to buckling. Pressures used are from 0,15 to 0,5 MPa. Hudson, 1993.

skip, a guided steel hoppit usually rectangular with a capacity from 4 to 10 t and used in vertical or inclined shafts for hoisting coal or mineral. It can also be adapted for man riding. The skip is mounted within a carrying framework, having an aperture at the upper end to permit loading and a hinged or sliding door at the lower end to permit discharge of the load. The cars at the pit bottom deliver their load either direct into two measuring chutes located at the side of the shaft or into a storage bunker from which the material is fed to the measuring chutes. Nelson, 1965.

skirt, the lower part of the drill bit that has a smaller diameter than the face of the bit. Sandvik, 2007. The skirt normally contains a female thread or a tapered section. Threaded and tapered bits may be available with shorter skirts than is standard. Atlas Copco, 2006.

slabbing, the loosening and breaking away of relatively large flat pieces of rock from the excavated surface, either immediately after or sometime after excavation. Often occurring as tensile breaks which can be recognized by the sub-conchoidal surfaces left on remaining rock surfaces. ISRM, 1975.

slab blasting, blasting of a slice of rock or ore where the blasthole is open at both ends.

slag bit, see '**insert bit**'.

slaking, disintegration of tunnel walls in swelling clay zones due to inward movement and circumferential compression. Stokes, 1955. Also, loosely, the crumbling and disintegration of earth materials when exposed to air or moisture. Stokes, 1955.
slapper, see '**exploding foil initiator**'.
slashing, or *chopping*, refers to the widening of a drift or stope by blasting parallel holes along the surface.
SLBC (abbreviation), see '**sublevel block caving**'.
SLC (abbreviation), see '**sublevel caving**'.
sledge, a long-handled heavy hammer used with both hands. Crispin, 1964.
sleeper, the pressure-creosoted wood, steel, or precast concrete beams laid crosswise under the rails of a rail track and holding them at the correct rail gage. Also called 'sole plate'. Nelson, 1965.
sleeping detonator, detonator which has been initiated but is waiting for its time to detonate due to its inbuilt delay.
sleep time, the time between explosives being loaded into a blasthole and their initiation.
slicing method, a descending cut and fill mining method. Removal of a horizontal layer from a massive orebody. In top slicing extraction retreats along the top of the orebody, leaving a horizontal floor, which becomes the top of the next slice. A timber mat separates this from the overburden, which caves downward as the slices are made. Other methods attack from the bottom (sublevel caving) or side. Pryor, 1963.
slickenslide (smooth joint), surface in the rock mass where movements has created a blank or low friction surface consisting of flaky minerals like chlorite and mica. Also surface in rock that has been grinded blank without any creation of new minerals. TNC, 1979.
slick line, pipe or hose inserted between the rock surface and casting forms to place concrete lining. Bickel, 1982.
slide, the descent of a mass (as of earth, rock or snow) down a hill or mountain side. Webster, 1961.
sliding, relative displacement of two bodies along a common surface, without loss of contact between the bodies. ISRM, 1975.
sliding angle (δ_{load}), (°), angle at or above which rock in movement will continue to slide, but less than the angle needed to initiate movement from rest. Some angles measured on clean steel are given in Table 42.

Table 42 Sliding angles for different minerals.

Material	Starting	Continuing
Coal	16°–25°	14°–22°
Limestone	20°	18°
Sandstone	23°	20°
Hematite	23°	21°

sliding micrometer, a rock mechanic instrument to *measure deformation* in boreholes with high precision. Hudson, 1993.
sliding wedge bolt, a bolt used for rock reinforcement using a double wedge at the bottom to fasten it. Fraenkel, 1952.

SLIFER (acronym), for '**shorted location indication of frequency of electrical resonance**'. Method to measure the velocity of detonation of an explosive in blastholes based on measuring the frequency of an electrical resonant circuit and as it is consumed by the heat of detonation.

SLIFER method, a method to measure the detonation velocity of an explosive (c_d). It is based on measurement of the frequency of a shorted coaxial cable that is continuously consumed by the detonation front in an explosive. The shorted coaxial cable embedded in the explosive forms part of the resonant circuit as it is consumed. The name is an acronym for "shorted **l**ocation **i**ndication of **f**requency of **e**lectrical **r**esonance".

slim hole drilling, a number of types of drilling rely on a small clearance between the drill pip/rods and the hole. *Auger drilling*, both flight and bucket fits this definition, but the main drilling types are *diamond drilling*, and *top-hole hammer drilling*.

slinger belt truck, a fast conveyor belt in combination with a rotating roller throws waste rock up to a distance of 14 m and a height of 8 m. The truck capacity is either 6 or 10 m^3 and the filling capacity is about 55 m^3/shift. This technique has been used in the Meggen Mine. Almgren, 1999.

slip surface or **slip**, surface along which movement is occurring or is supposed to occur at failure in a material. TNC, 1979.

slope, the ratio of the vertical rise or height to horizontal distance in describing the angle a bank or bench face makes with the horizontal. For example, a slope of 1,5:1 means there is a 1,5 m rise to each 1 m of horizontal distance.

slope failure, the breakage and movement of the rock mass due to gravity forces where the relative position of rock mass considerably is changed. TNC, 1979.

slope mass rating SMR, (R_{SMR}), ($0 \leq R_{SMR} \leq 100$), (**dimensionless**), a quantification of the risk for slope mass failure. It is calculated from RMR by subtracting a factorial adjustment factor dependent on the joint-slope relationship and adding a factor depending on the method of excavation. Five classes are given with numerical values from 0–100, see Table 43.

Table 43 Tentative description of SMR classes proposed by Manuel R. Romana, Spain. Hudson, 1993.

Class	SMR	Description	Stability	Failures	Support
I	81–100	Very good	Complete stable	None	None
II	61–80	Good	Stable	Some blocks	Occasional
III	41–60	Normal	Partially stable	Some joints or many wedges	Systematic
IV	21–40	Bad	Unstable	Planar or big wedges	Important/ corrective
V	0–21	Very bad	Completely unstable	Big planar or soil-like	Re-excavation

SLOS (abbreviation), see '**sublevel open stoping**'.

slot-and-wedge bolt, rock reinforcement by a reinforcing bar and a wedge at the splitted end to fasten the bolt in the drillhole. Synonymous to '**wedge bolt**'.

slot blocking, a form of *block cave mining* in which the drawpoints are developed not as normally on a horizontal level but on an inclined plane that often is parallel to the footwall of an inclined orebody. The undercutting is made in a sublevel caving configuration and the crosscuts mined are stopped at the foot wall where the drawpoints are permanented. This minimizes the ore losses on the footwall. The method was developed by Mats Haglund at LKAB in Malmberget and used for blocking of a shaft pillar in the Captain orebody with great success.

slot drilling, drilling parallel holes with a spacing a little smaller than the hole diameter. The result is an open space used to shield ground vibrations.

slotted borehole, axial slots can be made mechanically or by water jet with the purpose of better control of the direction of radial cracks.

SLS (abbreviation), see '**sublevel stoping**'.

SLSC (abbreviation), see '**sublevel shrinkage caving**'.

sludge, rock fragments caused by drilling action. Sandvik, 2007.

sludge barrel, see '**calyx**'.

sludge box, is used to collect the coarse drill cuttings from water by filling a box with several thresholds at the bottom and letting the water leave the box at one short end, where the height is made about half of the other sides. Stein, 1997.

sludge box sampling, samples taken from a sludge box.

sludge bucket, see '**calyx**'.

sludge cutter, take fractional samples and deliver them to a sample bag. Stein, 1997.

slurry emulsified, see '**emulsion explosive**'.

slurry explosive, explosive based on a sensitizer in a high viscosity salt/water-solution. Any water-based explosive, either water gel or emulsion, which is poured or pumped into the blasthole. Usually, slurry explosives contain ammonium nitrate, sensitized with TNT, or for example gas, thickened and cross-linked to a gelatinous consistency.

slush, to fill mine workings with sand, culm (residues from coal), etc. by hydraulic methods. Fay, 1920.

slusher, a mechanically operated, chain pulled scraper which moves caved or broken ore from a row of drawpoints to a grizzly or ore pass.

SmartRig (tradename), a computer based control system intended for all kinds of automation in simple and advanced drill rigs used underground and on surface. The hardware is designed to operate in every possible weather condition, and the software can be upgraded at site. *SmartRig* has built in logging and monitoring functions, together with support for diagnostics and faultfinding. The *SmartRig* control system generates electrical signals to control the hydraulic valves. This introduces the concept of a 'dry cab', with no hydraulic pipe work and gauges, considerably reducing noise for the operator. Atlas Copco, 2006b.

SMB (abbreviation), see '**submarine blasting**'.

SME (acronym), see '**site mixed emulsion**'.

smoke, the airborne suspension of solid particles from the products of detonation or deflagration. IME, 1981.

smokeless propellant or **smokeless powder**, solid propellant, commonly called smokeless powder in the trade, used in small arms ammunition, cannons, rockets, propellant-actuated power devices, etc. Atlas Powder, 1987.

smooth blasting, is a controlled contour blasting method to reduce overbreak. Closely spaced holes are drilled at the contour of the excavation, and charged by explosives with low linear charge concentration, often centred in the middle of the blasthole (decoupled charges). The contour holes are fired at the same time delay and in the last sequence of the round.

smooth wall blasting, see '**smooth blasting**'.
SMR (acronym), see '**slope mass rating**'.
SMS (acronym), see '**site mixing system**'.
SMS (acronym), see '**site mixed slurry**'.
SN (acronym), see '**sodium nitrate**'.
snakehole, see '**snakeholing**'.
snail dynamite, see '**expanding agent**'.
snakeholing, a method of blasting boulders, drilling holes under a face or drilling holes slightly downwards in front of a vertical bench face, normally to remove toes formed by previous blasting. When used for boulder blasting a hole is drilled under the boulder and a charge is fired in the hole. This method is more efficient but slower than using plaster shots. See also '**plaster shooting**'.
snow charging method, a charging method for pulverized explosives in up holes where the charging hose includes an air bleeder tube and is pushed almost to the bottom of the hole. The collar of the hole must be plugged during charging. The borehole is filled from the ANFO falling down by gravity forces from the bottom to the top.
socket (material handling), a device fastened to the end of a rope by means of which the rope may be attached to its load. The socket may be opened or closed. Zern, 1928.
socket, butt, boot-leg, or *gun*, that portion or remainder of a shot hole in a face after the explosive has been fired incompletely after the blast and that still may contain explosives and is thus considered hazardous. A situation in which the blast fails to cause total failure of the rock because of insufficient explosives for the amount of burden, or owing to incomplete detonation of the explosives.
sodium nitrate (SN), a chemical composition consisting of $NaNO_3$. It is an oxidizer used in dynamites and sometimes in blasting agents. USBM, 1983.
soil (in geology), any loose natural surface material overlying solid rock. Soil is classified by different particle sizes, see '**particle size**'.
soil penetration test, by the penetration rate of a drill bit provides useful information, but the value of the data to an engineer is reduced because of the many variables affecting drill bit penetration. Specially designed penetration testing probes allow a fixed set of condition to be applied to the factors causing penetration. Penetration tests are performed most commonly by using: *electrometers*, which measure the energy required to force the tool into the ground and *probes*, which measure both the energy required for penetration and the frictional resistance on the sides of the probe. There are two basic forms of penetration testing, dynamic tests and static tests. Dynamic tests are those performed using the energy of a falling weight to drive the electrometer into the soil. Two types of dynamic tests are commonly used, 1) **Standard penetration test** STP, see '**soil strength test**' and 2) **Solid cone** STP or **Continuous penetration test** CPT. The solid cone is used in place of a standard STP tool in gravels or soil containing numerous rock fragments. The COPT does not provide a sample. Static tests sometimes called static probing, the usual test of this type is the **Dutch cone test**. The test probe is pushed steadily into the ground by the hydraulic head of a drill rib or by a special hydraulic thrust unit. Stein, 1997.
soil strength test, soil strength in situ is determined by the *standard penetration tests* and the values (**SPT**) are related to expressions of relative density for cohesionless soils. Standard penetration tests are also performed on weak rocks, the results being correlated with foundation-bearing capacity. SPT values have also been used in the design

of bearing piles socketed in weak rock. Classification of strength is done according to Table 44. Hudson, 1993.

Table 44 Soil strength test determined by standard penetration tests. SPT values.

Relative density of cohesionless soil	SPT values (blows/300 mm penetration)
Very loose	0–4
Loose	4–10
Medium dense	10–30
Dense	30–50
Very dense	>50

sole plate, the pressure-creosoted wood, steel, or precast concrete beams laid crosswise under the rails of a rail track and holding them at the correct rail gage. Also called 'sleeper'. Nelson, 1965.

solid drilling, the term is used in diamond drilling to define a *bit that grinds the whole face*, without preserving a core for sampling. Nichols, 1956.

solid rock volume (V_{sr}), (m³), in blasting it generally denotes the rock volume before blasting.

solution mining, the use of liquids to dissolve the minerals and transport them hydraulically to the concentrator. This can be done in situ after well fragmentation of the ore or after bringing the ore to surface where the minerals are dissolved after being sprayed by the liquid.

sonic velocities (c_P, c_S, c_L and c_R), (m/s), the velocity of sound waves in gases, liquids or solids. In solids the longitudinal waves also called primary waves or P-waves are the fastest, followed by S-waves with about half the speed and Love- and Rayleigh waves. Only P-waves can travel in gases. The sound velocity (P-wave velocity) in air is 340 m/s at normal temperature and air pressure. In water the sound velocity is 1 450 m/s and in solid bodies like aluminium and iron 5 100 m/s. In rock the sound velocity varies between 500–5 000 m/s.

sonic velocity measurement, the sonic source can consist of an electronically excited piezoelectric ceramic transducer and the excited pulses are measured when they arrive to geophones or accelerometers. The equipment can be used both on drill cores in the lab or to measure the velocity through mine pillars for checking of its support characteristic. The technique permits a non-destructive, repetitive, stable, shaped pulse to be used in place of explosive caps or hammer blows. Cannaday, 1966.

soufflé blasting, a mining method where the rock mass has only one free face admitting the blasted material to swell. The ore might swell only upwards like a baked soufflé and not more than the material is left in its geological stratification. The method is used normally when mining rich, narrow, irregular and stratified ore zones are occurring and where a dilution of ore with waste is not acceptable. After this, selective extraction by backhole loaders is carried out, resulting in maximum recovery of the rich, narrow ore-bearing veins. Fernberg, 2002.

sound intensity (I), (W/cm²), in a specified direction at any point, the average rate of sound energy transmitted in the specified direction through a unit area normal to this direction at the point considered. Hy, 1965. See '**sound pressure level**'.

sound level, see '**sound pressure level**'.

sound pressure (level), (p_s), (dB), the level, in decibels, of a sound is 20 times the logarithm to the base of the ratio of the pressure of this sound to the reference pressure. The reference pressure must be explicitly stated. Hy, 1965.

sound velocity (c_{son}), (m/s), see "**sonic velocity**".

sound waves (c_p) and (c_s), (m/s), P-waves (also called longitudinal-, compressional-and dilatational wave) and S-waves (also called transverse, distorsion- or shear wave). In liquids (e.g. water) and gases (e.g. air) only P-waves can be transmitted not S-waves, because water has no shear strength.

South Africa Institute of Mining and Metallurgy S.Af.I.M.M, see '**Institute of Mining and Metallurgy**' in London that is similar in objectives.

spacer, piece of non-explosive material, e.g. wood or ceramic interposed between charges to extend the column of explosives.

spacing (S), (m), distance between boreholes in a row. It is necessary to distinguish between *spacing in drilling* (S_d) and *spacing in blasting* (S_b).

spacing in blasting (S_b), (m), the distance between holes initiated on the same delay number. In some blasts when all holes are initiated with different delays the spacing in blasting is defined as the distance between holes detonated consecutively.

spacing in drilling (S_d), (m), the distance between adjacent holes in a row of holes located parallel to the blast front or free surface.

spacing/burden ratio at drilling (S_d/B_d), ($R_{Sd/Bd}$), (dimensionless), the ratio of drilled spacing to drilled burden determines the effectiveness of distribution of explosives in the rock mass. Due to the timing of the blastholes the spacing/burden ratio at blasting may be changed considerably.

spacing/burden ratio at blasting (S_b/B_b), ($R_{Sb/Bb}$), (dimensionless), the ratio of blasted spacing to blasted burden determines the fragmentation of the blast. Improved fragmentation has been observed up to $R_{Sb/Bb} = 8$ in model and full-scale tests in Sweden (the wide spacing technique), see Langefors and Sjölin, 1963. For good fragmentation $R_{Sb/Bb}$ should be larger than 2 according to Dojcár, 1991. In the USA blasting standard it is stated that $R_{Sb/Bb}$ should never be greater than 2. The optimal value depends on the drilling pattern (rectangular or staggered), the geology, and the method of timing. For staggered drilling patterns, $R_{Sb/Bb} = 1,15$ yields the best distribution of the explosive through the ground. Cunningham, 1993.

spall (noun), fragment of rock broken from a free surface by the tensile stress wave which is usually created by the reflection of a compressive wave at the free surface.

spallation, the process where rock pieces are detached off from a rock surface by ground vibration, heating, mechanical cutting etc.

spalling, a) longitudinal splitting in uniaxial compression b) breaking-off of plate-like pieces (spalls) from a free rock surface due to high tensile stresses caused by the reflection of a compressive wave at a free surface. See also '**rock burst**'.

spark detonator, the detonator was employed in the past to produce electric initiation of explosive charges. The priming charge itself, containing current conducting additives, served as current conductor through the priming pill itself. Relatively high voltage was required to produce the ignition, so that such devices were safe from stray currents. Meyer, 1977.

spark drill, an *exotic drilling method* where sparks between electrodes are created in water near the drill hole bottom. A pressure wave is created, up to 2 000 MPa, and this pressure is enough to fragment the rock. Russian tests indicate a drilling speed of 0,7 m/h for marble and 3 m/h for clay schist. KTH, 1969.

spearhead, latch located either on the core barrel or wire line tool used for recovering core. Atlas Copco, 2006.

spear point, a small jetted well. Stein, 1997.

specific boring (b_{sb}), **(1/m² or m/t)**, total bored meters in full-face boring per volume- or mass of rock broken.

specific charge (q), **(kg/m³)** or **(kg/t)**, consumption (planned or actual) of explosive per cubic meter or metric ton of rock. The specific charge varies for different kinds of rocks from 0,3–0,9 kg/m³. The specific charge can be correlated to the RQD according to Roy and Dahr (1993), to the rock quality index (drilling index) (RQI), see Leighton et al. (1982) or to the drilling index (I_p), see Jimeno and Muniz (1987). The specific charge can also be correlated to seismic wave velocity in the field, Broadbent (1975), Heinen and Dimock (1976), and Muftuoglu et al. (1991).

specific drilling (b_{sd}), **(1/m² or m/t)**, drilled meters (planned or actual) per cubic meter or metric ton of rock broken.

specific energy (w_{sp}), **(kJ/m³ or kJ/t)**, explosive or mechanical energy required to fragment a unit of volume or mass of rock (planned or actual).

specific energy of rupture (w_f), **(J/m³)**, see '**specific fracture energy**'.

specific energy pressure, (p_f), (MPa), theoretical calculated pressure of fumes from 1 kg of detonated explosive in a volume of 1/1 000 m³. DDR, 1978.

specific entropy (*s*), **(J/kgK)**, the change in heat energy of a mass divided by the temperature change in the mass.

specific extraction of rock broken (s_{ext}), **(m³/m, t/m, m³/kg, and t/kg)**, the ratio of broken rock expressed in volume or mass to the number of blasthole meters drilled. Also the ratio of broken rock expressed in volume or mass to the amount of explosive used in kilogram.

specific fracture energy (w_f), **(J/m³)**, the work done per unit volume in producing fracture. It is not practicable to establish a definite energy of rupture value for a given material, because the result obtained depends upon the form and proportions of the test specimen and the manner of loading. As determined by similar tests on similar specimens, the energy of rupture affords a criterion for comparing the toughness of different materials. Also called *specific energy of rupture* or *energy of rupture*. Roark, 1954. Minimum energy required to crate a certain (requested) level of fragmentation in a rock mass. The energy required depends on what loading mode is used, tensile, shear etc.

specific gas pressure (p_{spg}), **(MPa)**, is defined according to the following formula:

$$p_{spg} = 0,101 \cdot 10^{-3} v_e \left(\frac{T_{exp}}{273} + 1 \right) \quad (S.6)$$

where p_{spg} specific gas pressure in MPa, v_e is the specific gas volume of the explosive in m³/kg and T_{exp} the explosion temperature in °C. Trauzl, 1952.

specifick gas volume (v_e), **(m³/kg)**, volume of gas produced per mass of explosive.

specific gravity (γ), **(dimensionless)**, the ratio of the mass of a given volume of any substance to the mass of an equal volume of water.

specific heat (*C*), **(J/kgK)**, the amount of heat in Joules required to increase the temperature of 1 kg of a substance by 1 degree K in a constant volume or constant pressure process, respectively. The symbol for specific heat at constant volume is C_v as opposed to the specific heat at constant pressure C_p.

***specific splitting charge**, splitting factor (South Africa), charge factor, charge load*, **(qa), (kg/m²)**, mass of explosive in (kg) per square meter (m²) of splitting area. The quantity of explosive is of interest in controlled contour blasting. This is one criterion for estimating presplit charge mass per hole. Usually the specific splitting charge is ~0,5 kg/m². AECI, 1993.

specific surface (S_s), (1/m), the surface area per unit of volume of soil particles. ASCE P1826, 1958.

specific surface area per mass (S_{sm}), (m²/kg), the ratio of the total surface area of particles to their total weight. Taylor, 1965.

specific surface (broken rock), (S_{srb}), (1/m), the surface area of the rock fragments and particles per unit volume of the material.

specific surface energy (γ_s), (J/m²), the product of the surface tension and surface area or the work required to generate a unit of surface area. It can also be defined as the energy per unit surface area required to separate crystals into two parts along a cleavage plane.

specific surface of all open surfaces (O_{no}), (m⁻¹), the surface area of joints, cracks and the surface of fragments etc. per unit or bulk volume of the material.

specific surface (rock mass), (S_{sr}), (m⁻¹), the surface area of joints, cracks per unit volume of the material.

specific surface (soil), (S_{ss}), (m⁻¹), the surface area per unit of volume of soil particles. ASCE P1, 826.

specific volume (v), (m³/kg), the volume per unit mass of a material. Compact volume per unit mass is called *true specific volume* (when the material is fully compacted, no pores or space within the material) and bulk volume (rock after breakage) per unit mass is called *bulk specific volume*.

spectrum analysis, analysis of the energy or amplitude of the frequencies present in a ground vibration, e.g. by Fourier analysis.

Speedrod (trade mark), drill rod that has integrated couplings containing the female thread the other end of the rod has the male thread. Atlas Copco, 2006.

spherical button, cylindrical cemented carbide drill bit insert having a more *dome shape profile* when compared to ballistic buttons. Spherical buttons are primarily designed for great wear resistance in hard and abrasive rock formations. Atlas Copco, 2006.

sphericity, the degree in which the shape of a fragment approaches the form of a sphere. A.G.I., 1957 and 1960. One quantification of sphericity is the surface area of sphere to that of particles of same volume. Pryor, 1963. For quantification of sphericity, see '**shape factor**'.

spiling or **poling boards**, see '**poling boards**'. Bickel, 1982.

spiral blast, see '**spiral drill-and-blast concept**'.

spiral cut, a parallel hole cut where the blastholes are located on a spiral around the opening hole or holes. Hagan, 1979.

spiral drill-and-blast concept (in drifting), a drifting method where blastholes are drilled in rows radial to the centre of the drift, with each successive row at an increasing depth. When one row is blasted a pie-shaped segment is removed, see Fig. 60. If the total area is divided into 16 segments and the pull is 1,6 m per circle, each row must increase its depth by 0,1 m. A special machine has to be used to shield the blast area from the drilling and charging area. The advantage of the concept is that drilling, charging, initiation, blasting and mucking can be performed continuously. The concept was investigated by USBM. Peterson et al., 1977.

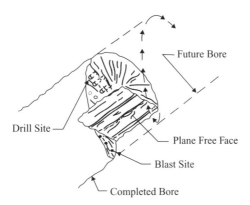

Figure 60 The spiral drill- and blast concept. After Peterson et al., 1977.

spiral reinforced hose, usually a high pressure hydraulic hose that has been reinforced with spirally wrapped wire. Atlas Copco, 2006.
spiral ribbed, cylindrical axle manufactured with a series of evenly spaced grooves and ridges in a spiral pattern. Used in stabilizers. Atlas Copco, 2006.
spitter cord, see '**igniter cord**'.
split, separation lengthwise, as along the direction of grain or layers. Synonym with parting.
split set bolt, a rock reinforcement bolt consisting of a steel tube (2,3 mm wall thickness) with a longitudinal 13 mm wide slot. The diameter of the tube is 3 mm larger than the diameter of the borehole where it is inserted by force and hold by friction. Used only for fast and light enforcement because of limited cross section area and friction. Brady, 1985. In general galvanized split-sets are used for wall bolting, while resin grouted rebar or mechanical bolts are used in the roof, and Swellex in sand fill. Atlas Copco, 2007.
splitting (blasting), the quasi-instantaneous firing of perimeter holes to form a clean split between them and minimize crack growth at right angles to the split. Presplitting means splitting prior to firing the main blast, such that cracks from the blast will terminate at the split. Post splitting or smooth blasting means firing the last line of holes as a split. These two techniques are fundamentally different in terms of applications and effects. AECI, 1993.
splitting (geology), the property or tendency of a stratified (sedimentary) rock of separating along a plane or surface of parting.
splitting factor (South Africa), see '**specific splitting charge**'.
spontaneous combustion, the heating and slow combustion of coal and coaly material, pyrotechnical material, turpentined cotton waste or sulphide dust produced in mining. The process is initiated by the absorption of oxygen to the material. The two factors involved are: 1) a coal of suitable chemical and physical nature; and 2) sufficient broken coal and air leaking through it to supply the oxygen needed. The heat generated is retained with a consequent rise in temperature. Nelson, 1965.

spot bolting, the technique to bolt only where it is absolutely necessary. The opposite is systematic bolting, see '**systematic rock bolting**'.

spot drilling, making an initial indentation in a work surface, with a drill, to serve as a centering guide in a subsequent machining operation. ASM Gloss, 1961.

spot facing, machining a flat seat for a bolt head, nut, or other similar element at the end of and at right angles to the axis of a previously made hole. ASM Gloss, 1961.

spring roll crusher, a crushing machine similar to the double roll crusher only with the difference that springs are fixed to the bearings of one roll. Nelson, 1965.

springing, the process of enlarging a portion of a blasthole (usually the bottom) by firing one or more small explosive charges; typically used in order that a larger charge of explosive material can be loaded in a subsequent blast in the same borehole.

spring rolls, crushing rolls used in ore breaking. Two parallel cylinders, mounted horizontally are held apart by shims, and pressed together by powerful springs. Crushable rocks falling between them are drawn down as the cylinders revolve, but unbreakable material causes the spring to yield and let it pass without damage. Pryor, 1963.

sprung hole, see '**springing**'.

SPT (acronym), see '**standard penetration test**'.

spud or spud-in, to start drilling a borehole. Also called '**collaring**'. Long, 1920.

spud drill, synonym for '**churn drill**'.

spudder, '**churn drill**'.

spudding, the operation in rope drilling of boring through the subsoil when starting the hole. B.S. 3618, 1963.

spudding bit, a broad dull drilling tool for working in earth down to the rock. Standard, 1964.

spudding drill or churn drill, a drill that makes a hole by lifting and dropping a chisel bit. Nichols, 1965.

spudding real, the churn drill winch that lifts and lowers the drill string. Also called *bull reel*. Nichols, 1965.

square drilling, making square holes by means of a specially constructed drill made to rotate and oscillate so as to follow accurately the periphery of a square guide bushing or template. ASM Gloss, 1961.

square-nose bit, see '**flat-face bit**'.

square set and fill (SSF), a 'cut and fill mining method', see '**square set stoping**'. Fay, 1920.

square set block caving, a method of block caving, in which the caved ore is extracted through drifts supported by square sets. A retreating system is adopted. Nelson, 1965.

square set slicing, see '**top slicing and cover caving**'. Fay, 1920.

square set stopes (old method), a method where square set timbering is used to support the ground as the ore is extracted. It is often used for mining wide ore bodies particularly where the conditions are irregular and where walls and ore masses are weak. The first tier of timber sets is termed the *sill floor* and the uppermost the *mining floor*. Nelson, 1965.

square hole pattern, pattern of parallel drilled holes positioned in a square grid.

square set stoping (SSS), (old method), a method of stoping in which the walls and back of the excavation are supported by regular framed timbers forming a skeleton enclosing a series of connected, hollow, rectangular prism in the space formerly occupied by the excavated ore and providing continuous lines of support in three direction at right angle to each other. The ore is excavated in small, rectangular blocks just large enough to provide room for standing a set of timber. The essential timbers comprising a standard square set are respectively termed *posts*, *caps*, and *girts*. The posts are the *upright members*, and the caps and girts are the horizontal members. The ends of the members are framed to give each a bearing against the other two at the corners of the sets where they join together. The stopes are usually mined out in floors or horizontal panels, and the sets of each successive floor are framed into the sets of the preceding floor. Sometimes, however, the sets are mined out in a series of vertical or inclined panels. USBM Bull. 390., 1936, pp. 10–12.

square set system, a method of mine timbering in which heavy timbers are framed together in rectangular sets, 1,8–2,1 m high, and 1,2 to 1,8 m square, so as to fill in as the orebody is removed by overhand stoping. Webster, 1961.

square set underhand, see '**square set stoping**'. Fay, 1920.

square work and caving, see '**sublevel stoping**'. Fay, 1920.

square work, pillar robbing, and hand filling, see '**sublevel stoping**'.

squeezing rock or ground, time dependent deformations around a rock excavation that are getting so large so the drift or tunnel cannot be used any longer. Examples are the settling or roof over a considerable area of working. Also called *creep*, *crush*, *pinch* or *nip*. The gradual upheaval of the floor of a mine, due to the weight of the overlying strata. Fay, 1920. Very common in mines with soft rock like coalmines.

squib, a firing device that burns with a flash and is used for igniting black powder or pellet powder. Atlas Powder, 1987.

squib shot, a blast with a small quantity of high explosives fired at some point in the borehole for the purpose of dislodging some foreign material which has fallen into it. Fay, 1920.

SR (acronym), for '**short rod**' test, see '**fracture toughness test**'.

SRAP (acronym), see '**step room and pillar**' mining.

SRM (acronym), see ' **synthetic rock mass modelling**'.

SRP, (p_{os}), (Pa or dB), (acronym), for '**stemming release pulse**'.

SS (acronym), for '**shrinkage stoping**'.

SSE (acronym), see '**site sensitised emulsion**' explosive.

SSF (acronym), for '**square set and fill**' a 'cut and fill mining method', see '**square set stoping**'. Fay, 1920.

SSS (acronym), for '**square set stoping**'.

stab detonator, blasting cap used in a stab hole.

stab hole, at the top of the bench short blastholes are drilled in between long blastholes, to improve fragmentation or to deal with hard strata located close to the collar region. Occasionally used in surface blast design.

stability (of explosive), the ability of an explosive material to retain the chemical and physical properties specified by the manufacturer when exposed to specific environmental conditions over a particular period of time. Atlas Powder, 1987.

stability (of rock), the condition of a structure or mass of material when it is able to support the applied stress for a long time without suffering any significant deformation or movement that is not reversed by the release of the stress. ISRM, 1975.

stacking, the collection or stacking of caved or broken ore or rock on the undercut level or on the major apices.

stage testing of discontinuities, the strength properties of discontinuities can be tested in a triaxial cell. A specimen is tested at a low confining pressure. When it appears that slip on the discontinuity has just been initiated (represented by flattening of the axial load-axial displacement curve that must be continuously recorded throughout each test), loading is stopped, the cell pressure is increased to a new value, and loading is recommenced. By repeating this process several times, a number of point on the peak strength envelope of the discontinuity can be obtained from one specimen. However, this approach exacerbates the major difficulty involved in using the triaxial test to determine discontinuity shear strengths, namely the progressive change in the geometry of the cell-specimen system that accompanies shear displacement on the discontinuity. Brady, 1985. The problem is illustrated by Fig. 61.

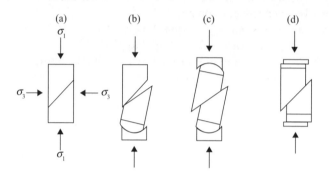

Figure 61 Discontinuity shear testing in a triaxial cell. After Jaeger, 1969.

It is clear from Figure 61 that, if relative shear displacement of the two parts of the specimen is to occur, there must be lateral as well as axial relative translation. If, as is often the case, one spherical seat is used in the system, axial displacement causes the configuration to change to that of Figure 61a, which is clearly unsatisfactory as shown in Figure 61c, the use of two spherical seats allows full contact to be maintained over the sliding surfaces, but the area of contact changes and frictional lateral forces are introduced at the seats. Figure 61d illustrates the most satisfactory method of ensuring that the lateral component of transition can occur freely and that contact to the discontinuity surface is maintained. Pairs of hardened steel discs are inserted between the platens and either end of the specimen. No spherical seats are used. The surfaces forming the interfaces between the discs are polished and lubricated with a molybdenum disulphide grease. In this way, the coefficient of friction between the plates can be reduced to the order of 0,005 which allows large amounts of lateral displacement to be accommodated at the interface with little resistance. Jaeger, 1969.

standard safety fuse, a safety fuse which burns at a rate of 100 ± 10 mm/s. Sen, 1992.

staggered hole pattern, a borehole pattern where the boreholes in each following blasted row are shifted by a half spacing and therefore are located facing the middle between the holes in the preceding row. Often, this borehole pattern yields improved fragmentation.

staking, the positioning of wooden stakes in the ground to mark the planned location of blastholes to be subsequently drilled.

stamp (indentation) test index (σ_{st}), **(MPa)**, a semi empirical test method for rock drillability. The tested rock sample is grouted into a steel cylinder which is open at both ends. The top surface of the rock at one end is made smooth. A metal stamp is pressed into the rock. The diameter of the stamp is 4 mm which corresponds to the contact areas between the used button on roller and percussive bits and the rock at the bottom of the borehole at drilling. The stamp test strength index is defined as;

$$\sigma_{st} = \frac{F_{max}}{\pi r^2 10^6} \qquad (S.7)$$

where σ_{st} is the stamp test index in (MPa), F_{max} is the maximum force before breakage in (N), r is radius of the stamp (indenter) in (m). Wijk, 1982.

standard ignition test, a method developed for testing coal dust to obtain the limits of explosion. Rice, 1960.

standard rheological body, (also called Poynting-Thomson model or the Zener model) has properties which are common to the majority of rocks
1. It possesses ability for "immediate" elastic deformation.
2. It possesses ability for limited creep, specific to rocks at low stress levels. The creep curve of the standard model represents stage I of the generalized creep curve, the stage of creep with a decreasing strain rate.
3. It exhibits a characteristic of partial retarded elasticity.
4. The creep strain of the standard model is completely reversible; this is the only property of this model, which is different from the actual behaviour of rocks. The standard model is characterized by an ability for partial stress relaxation. Hudson, 1993.

Standard Temperature and Pressure (STP), atmospheric pressure at 0,101 MPa (1 atm or 1 bar) and at 0°C. Pryor, 1963.

stand-off distance (l_{so}), (m), the distance between an explosive and an object, more specifically, between a shaped charge and the target for demolition.

stand-up time, originally proposed by Lauffer in the 1950s and modified by Linder in the 1960s this system has received a considerably popularity in the past. It is a one-parameter system, *stand-up time*, which is defined as the time during which an underground excavation can remain unsupported without serious deterioration. Stand-up time is influenced by the orientation of the geological structure, the shape of the tunnel cross-section, the type of excavation and the type of support procedure. Hudson, 1993.

start drill rod, short drill rod used when starting a drill hole. Atlas Copco, 2006.

starter, see '**non-electric detonating device**' and '**mushroom starter**'.

static electricity, electrical energy stored on a person or object in a manner similar to that of a capacitor. Static electricity may be discharged into electrical initiators, thereby detonating them. USBM, 1983.

stay, prop, strut, or **tie**, a diagonal brace or tie bar to stiffen or prevent movement of a structural component. Ham, 1965.

steady state velocity of detonation (c_{ds}), (m/s), the steady state velocity of detonation is attained once run-up or run-down is completed at a given distance from the booster. The chemically compounded rate of detonation of an explosive. It is governed by the diameter, degree of confinement, temperature, etc.

steel arch, a reinforcement method for drifts and tunnels.

steel cable, a flexible rope, the strands of which are steel wires instead of plant fibres. Long, 1960.

steel collapsible forms, see '**steel reducable forms**'. Bickel, 1982.

steel fibres, with the length of 50 mm and diameters of 0,4–0,8 mm are used as reinforcement in shotcrete. The steel fibres can have different shapes straight, wavy and stepped.

steel grit, steel grit is a product manufactured by crushing and grading *steel shot*. It is angular, presenting many sharp corners and is very hard and strong, these features being important towards its use as a sandblasting abrasive. Steel grit is produce in sizes from about 0,25 to 1,78 mm.

steel mesh, a square mesh made of reinforcement iron and with a diameter of ~5 mm. The steel mesh is fastened to the rock surface by rock bolts. It is used together with shotcrete for surface reinforcement of rock. Hudson, 1993.

steel reducable forms, these are designed to be sufficiently reducable to permit stripping of the forms. Bickel, 1982.

steel ribs, with a cross section like an I were used for rock support of the crown in the Gotthard road tunnel. Hudson, 1993.

steel shaft rings, are used as rock support in shafts. They can be yieldable. Hudson, 1993.

steel shot drilling, the use of small steel balls impacting the surface, which is going to be drilled. Fraenkel, 1952.

steel straps, 6 mm thick reinforcement bar in 100×100 mm mesh can be used together with rock bolts to reinforce rock surfaces. Straps are used where rock conditions are poor and a ravelling failure around the collars of the rock bolts has been experienced. Stillborg, 1994.

steel support, a straight or curved length or steel, usually of H- or channel section, used for support purposes in mine roadways, faces, or shafts.

steel tube confinement, see '**steel tube test**'.

steel tube test, the velocity of detonation (c_d) is often determined in steel tubes which offers a higher confinement than detonating the explosive in air but still the confinement is smaller than in ordinary rock blasting.

steel or **anchor prop**, a steel or timber prop fixed firmly between roof and floor at the end of a longwall face and from which a coal cutter is hauled by rope when cutting. A steal prop may also be used as part of a belt-tensioning arrangement, or a return sheave. Nelson, 1965.

stem-induced fracturing, fracture initiation from the reflected shock pressure when the air pressure from the detonation of a bottom charge in a blasthole, without any column charge, reaches the stemming in a blasthole. The technique was suggested for use in oil well stimulation. Fourney et al., 1981. See also '**airdeck blasting**'.

stem lock gas bag, a plastic bag that when being inflated can be used as a plug in the blasthole.

stemming, the inert material of dense consistency, such as drill cuttings, gravel, sand, clay or water in plastic bags, which is *inserted in the collar of the drill hole* after charging and used to seal the hole temporarily in order to prevent venting of gas, increase blasting efficiency, to reduce air shock waves or dampen any open flames. In coal mining water stemming cartridges work very well as they also contribute to minimise dust and fires. Stemming is also used as a material to *separate explosive charges in a borehole* (decks). Stemming can also be used to *seal off open cracks intersecting the blasthole*. Principally the size of the stemming material should be as large as possible but less than 1/3 of the diameter of the hole, to avoid pieces getting stuck in the hole when filling the blasthole, and to obtain the maximum bridging effects at blasting. Too large pieces of stemming material may damage the initiation material in the blasthole and should therefore be avoided. In coal mines it is preferable to have fine material because of the risk of fire. For large hole diameters the shape of the rock pieces used for stemming should be angular. The length of the stemming should be about the same size as the burden.

stemming length (l_s) **(m)**, the length of stemming varies from 12 times the diameter of the blasthole in case of hard competent rock (uniaxial compressive strength >210 MPa) to 30 times the diameter of the blasthole for soft competent rock (uniaxial compressive strength 30 MPa). Pradhan, 1996.

stemming plug, see '**borehole plug**'.

stemming material, see '**stemming**'.

stemming release pulse (SRP), (p_{sr}), **(Pa or dB)**, air blast overpressure generated by gas escaping from the blown-out stemming, see also '**air blast**'. Atlas Powder, 1987.

stemming retention, the use of stemming in the blastholes increases the time of the blasthole pressure in the blasthole and improves fragmentation and reduces air blast. Efficient stemming demands a certain size distribution of the stemming material and that the length of stemming is not less than 12 times the diameter of the blasthole.

stemming rod, a non-metallic rod used to push explosive cartridges into position in a blasthole and to tamp (ram) tight the stemming. B.S. 3618, 1964.

step length, see '**fractal dimension**'.

step room and pillar (SRAP), a room and pillar mining method that adapts the inclined orebody footwall for efficient use of trackless equipment in tabular deposits with thickness from 2 m to 5 m and dip ranging from 15 to 30 degrees. Stopes and haulage ways cross the dip of the orebody in a polar coordinate system, orienting the stopes at angles across the dip that can comfortably be travelled by trackless vehicles. Parallel transport roads cross the orebody to establish roadway access to stopes and for trucking blasted ore to shaft. Stopes are attacked from the transport drifts, branching out at the predetermined step-room angle. The stope is advanced forward, in a mode similar to drifting, until breakthrough into the next parallel transport drive. Next step is excavation of a similar drift, or side slash, one step down dip, adjacent to the first drive. This procedure is repeated until the full roof span is achieved, and an elongated pillar is left parallel with the stopes. The next stope is attacked in the same way, and mining continues downwards, step by step. Atlas Copco, 2007.

step velocity of front fragments, when high-speed films from bench blasting are analysed fragments already having a certain velocity will suddenly increase its velocity in steps. The initial velocity of the burden is generally only a fraction or the ultimate velocity; velocity apparently increasing discontinuously or in steps due apparently to collision from behind by faster moving fragments as a result possibly of rebounding pulses, seen clearly in the case of underwater blasts. Cook, 1965.

stereology, mathematical methods relating three-dimensional parameters defining the structure to two-dimensional measurements obtainable for sections of the structure. Weibel, 1989. This relatively new field of science is of importance when fragment size distributions are estimated from photographs or video camera recordings or blasted contours are going to be mapped.

stereo photographing technique, see '**stereology**'. Stereographic technique can be used to find the contours after blasting on surface. Moser, 2006. An alternative method is to do a laser scanning survey.

stick, see '**cartridge**'.

stick, is a timber post for enforcement of a working face are widely used to assist in supporting the mined out area behind the stope face in the mining of narrow, reef-like metalliferous ore bodies. The sticks are usually 100–200 mm in diameter and are designed to support, or help support, the dead weight of the first 1–2 m of rock in the hanging wall. A typical 200 mm diameter stick can have a short-term load carrying capacity of 60 t, but for long-term use and low rates of load application, lower design capacities should be used. Pipe sticks are also used, see '**pipe stick**'. Brady, 1985.

stick count, see '**cartridge count**'.

stiffness (*k*), (N/m), the ratio between force and displacement. ISRM, 1975.

stone, in geology a piece of rock having a size between 20 mm to 200 mm.

stone dresser and **surfacing hammer**, percussive hammer designed to use an assortment of tools to chip a rock surface to a desired form and texture. Atlas Copco, 2006.

stone dressing rig, rig mounting of a stone dressing hammer. Atlas Copco, 2006.

stone drift, hard heading or **rock drivage**, a drift excavated in rock, as from the surface down to a coal seam. Nelson, 1965.

stone dust barrier, a device erected at strategic points in mine roadways for the purpose of arresting explosions. Sinclair, 1958.

stone quarry, a surface excavation where crushed stone is produced.

stone size of 30/50 spc., measurement indicating the size of diamonds set in a diamond bit. Atlas Copco, 2006.

stope, the part of an orebody from which ore is currently being mined, or broken, by stoping (drilling and blasting). Gregory, 1980.

stoop and room or *stop-and-room*, see '**bord and room**'.

stope (verb), the act of drilling parallel blastholes along the free surface and blasting the holes. Sandvik, 2007.

stope mining with fill, see '**rill mining**'.

stoping, enlargement of an area by blasting holes parallel and along a free surface and having a free opening to move against. Synonymous to 'slashing'. TNC, 1997.

stoping, a term that includes all operations of breaking rock or mineral subsequent to development. Fraenkel, 1954.

stoping and filling, see '**overhand stoping**'. Fay, 1920.

stoping methods, the classification of stoping methods adopted by the USBM devised largely on the basis of rock stability is as follows 1) *stope naturally supported*, this includes open stoping with open stopes in small ore bodies, and sublevel stoping, and open stopes with pillar supports which includes casual pillars and room (or stope) and pillar (regular arrangement), 2) *stopes artificially supported*, this includes shrinkage stoping, with pillars, without pillar, and with subsequent waste filling, cut and fill stoping, stulled (timber supported stopes) in narrow veins, and square set stoping, 3) *caved stopes*, this includes caving (ore broken by induced caving), block caving including caving to main levels and caving to chutes or branched raises, sublevel caving, and top slicing (mining under a mat which, together with caved capping, follows the mining downward in successive stages) and 4) combinations of supported and caved stopes (as shrinkage stoping with pillar caving, cut and fill stoping width top slicing of pillar, etc.) USBM Bull. 390., 1936.

stoping underhand, mining a stope downward in such a series that presents the appearance of a flight of steps. Fay, 1920.

storage of explosives, the safe keeping of explosive materials, usually in specially designed structures called magazines. IME, 1981. There is no international standard for storage of explosives. Normally the explosives should be stored in a well-locked steel magazine. The European Federation of Explosives Engineers (EFEE) is working on a standard for the European Union.

stopper, device used to plug a hole. Atlas Copco, 2006.

STP (acronym), see '**Standard Temperature and Pressure**'.

straight bit, a flat or ordinary chisel for boring. Fay, 1920.

straight chopping bit, synonym for *chisel-edge bit, chisel-point bit, log-shank, chopping bit and Swedish bit*. Long, 1920. See '**chisel bit**'.

straight edge crack round bars in bending (SECRBB), see '**fracture toughness**'.

straight hole pattern, a rectangular hole pattern.

straight shaft (in pier drilling), a straight cylindrical shaft drilled through soft or weak material and bedding on stiff clays or bedrock. Stein, 1977.

straight side (sided) core bit, synonym for '*straight wall bit*'. Long, 1920.

straight wall bit, an annular-shaped (core) bit the inner walls of which are parallel with the outer walls and not tapered to receive a core lifter. Long, 1920.

strain (normal ε and shear γ), is a derived quantity. There are two types of strain, i.e. normal strain, ε and shear strain γ. For a continuum, normal strain is defined as the ratio between the length of an element before and after application of load. Shear strain is defined as the change due to deformation of the angle between two line segments.

strain cell, is used for measurement of rock stresses and based on strain gauges technology.

strain energy (W_s), **(MJ)**, the work done in deforming a body within the elastic regime up to the elastic limit of a material. It is more properly called elastic strain energy and can be recovered as work rather than heat. A.S.M. Gloss., 1961. Strain energy can be as high as 5% of the explosive's total energy in dense rocks and 20% in soft rocks. Hagan and Harries, 1979.

strain energy around a blasthole (W_{se}), **(J)**, due to the quasi static pressure in the borehole the surrounding rock mass will be under strain and the energy stored and which is recoverable is called the strain energy. Lownds, 1986.

strain energy density (w_s), **(J/m³)**, strain energy per unit of volume. A crack will propagate in the direction along which the *strain energy density* is a minimum. Krantz, 1979.

strain energy factor (k_s), **(mkg$^{-1/3}$)**, is the proportional factor in the Livingston crater formula for calculation of the optimal depth of the charge or optimal breakage burden B_{ob}

$$B_{ob} = k_s \sqrt[3]{Q} \tag{S.8}$$

where B_{ob} is the optimum depth or optimal breakage burden, k_s is the *strain energy factor* (a constant for a certain rock type and joint set) and Q is the mass of explosive in kg. Note: Livingston did not publish his formula. Reference is given only to lecture notes from a course on "A blasting Theory and its Application" given in Georgetown, Colorado, USA. Persson, 1993.

strain energy release rate or **crack extension force** (G) **(J/m²)**; the elastic surface energy per unit area required for incremental crack extension

$$G = 2\gamma_s = \frac{\pi\sigma^2 l}{E_e} \tag{S.9}$$

where G is the strain energy release rate in (J/ m²), γ_s is the specific surface energy in (J/m²), σ is the stress in (Pa), l is the crack length in (m), and E_e is the effective modulus of elasticity in (Pa). There are two different effective modulus. One for plane stress and one for plain strain as follows;
$E_e = E$ for plane stress and $E_e = \frac{E}{(1-\upsilon^2)}$ for plane strain, where E is Young's modulus of the material (Pa) and υ is Poisson's ratio. Some extreme values of G are 0,8 for lithium fluoride and 50–50 000 for steel.

strainer, screen used to separate solid material from a liquid. Such as in the strainer mounted in a hydraulic oil reservoir. Atlas Copco, 2006.

strainer drum, drum like structure made from wire mesh or perforated metal, used to separate two materials. Atlas Copco, 2006.

strain hardening, an increase in hardness and strength caused by plastic deformation at temperatures lower than the re-crystallization range ASM Gloss, 1961. Also the process when strength continuously is increased after the yield point has been reached.

strain intercept, a constant used for calculation of peak strain around a blasthole. See '**peak strain at rock blasting**'.

strainmeter, is based on the interferometry principal using a laser source with monochromatic radiation. It is a common tool in *precision displacement measurements* and also for calibration of instruments. A linear resolution of 0,01 μm, or even better is achievable. Hudson, 1993.

strain rate ($\dot{\varepsilon}$), (s⁻¹), the time derivate of the strain ε. Increasing the strain rate increases the peak strength of rock and the rock will be more brittle but a higher loss of energy after the peak is reached and rupture starts. Hudson, 1993. The fracture stress increases with strain rate, see '**dynamic tensile strength**'. Model blast tests show a strain rate of 3×10^4. Siskind, 1974. Extrapolation from full-scale blast calculations gives a strain rate of 1×10^5. From full-scale blast it does not exist so far any measurements.

strain softening, when a rock mass is strained beyond the peak strength, it exhibits a strain softening down to the residual strength. This behaviour has been modelled by a tri linear stress-strain law. Hudson, 1993.

strand, a part of a cable, which is built by several strands (wires). A commonly used cable used for rock reinforcement in mining has a diameter of 15,2 mm and 7 strands (wires). Normally they are installed in two units.

strap, woven strap with a buckle at one end used to tie up things. Such as the straps put around a bundle of hydraulic hoses on a drill rig. Atlas Copco, 2006.

strap mining or **partial mining**, a room and pillar mining method used in coalmines. It is successful in reducing subsidence. The strap method involves extraction of alternate straps in a circular extraction area. The recovery ratio is in the range of 50–60%. From the existing data, where strap mining combined with hydraulic filling the maximum possible subsidence is only 2% of the extracted thickness of the coal seam, and if combined with the roof-caving method the maximum possible subsidence will be in the region of 3–10%. Strap mining is mainly used when there are some important buildings on the ground surface. Hudson, 1993.

strata scope, see '**borehole binocular**'.

stratification, a structural system of planes within sedimentary rock deposits formed by interruptions during the course of deposition of sediments.

stray current (I_{st}), (A), an electric current that is introduced in a circuit by leakage of industrial currents in the ground. Schieferdecker, 1959. This undesired electric current may enter or leave a blasting circuit due to inadequate insulation. If it leaves the blasting circuit, detonators may misfire due to current starvation, and if it enters the circuit from earth, premature detonation may occur.

stray current protection, the increasingly large consumption of electric current has resulted in intensified stray currents. The stray current safety of an electric detonator is the maximum current intensity at which the glowing wire just fails to attain the ignition temperature of the charge in the detonator. In order to improve protection against stray currents, the 'A' bridgewire detonators, which were formerly used in Germany have now been replaced by the less sensitive 'U' detonator. Meyer, 1977.

strength, the power to resist strain, stress etc., and also to resist attack, impregnation, etc.

strength of explosive, the strength of an explosive is a comparative measure of its rock breaking capacity. It is often referred to as the percentage of blasting gelatin. It can be described in various terms, such as cartridge or mass strength, seismic strength, shock or bubble energy, crater strength, ballistic mortar strength, etc. None of the given methods is sufficient for a precise calculation of the rock breaking capacity but new methods are under development. See '**single hole blasting tests**'.

strength size relation (drill cores), it is generally known that the smaller size of rock sample the larger will be the strength be. For drill cores of granite with diameter 19 to 58 mm the uniaxial compressive strength was reduced from 219 to 175 MPa and the Brazilian tensile strength from 259 to 161 MPa. The reduction follows the Weibull law. Weibull, 1939.

stress (σ), (MPa), a physical derived tensor quantity. In a uniaxial case it is defined as the force per unit area obtained within the limits of the force acting on an infinitesimal area element at a point. In general the magnitude and direction of the components of the stress tensor vary from point to point in a solid.

stress caving, a caving mechanism in which the *in situ* rock fractures and caves under the influence of the stresses induced in the cave back.

stress intensity factor (SIF) for mode I cracks (K_I), mode II cracks (K_{II}) and mode III cracks (K_{III}), (N/m$^{3/2}$), analytical quantity characterising the state and level of deformation of crack faces due to external loading. Popularly, the SIF serve as a measure for the 'danger' of a crack. Mathematically the SIF is the magnitude of the crack singularity.

stress or **strain tensor**, the second order tensor whose diagonal elements consist of the normal stress/strain components with respect to a given set of coordinate axes, and whose off-diagonal elements consist of the corresponding shear stress/strain components. ISRM, 1975.

stress reduction factor (SRF), (f_{SRF}), (dimensionless), see '**Q-factor**'.

stress release waves, created when the quasi-static load on a material suddenly ends in a breakage of the material. These waves are the same as those caused by rock bursts and in the case of blasting they contain more energy than the Primary P- and S-waves initially caused by the explosive upon detonation. Normally these waves occur in time when the primary P- and S-waves have been reflected in the free surface and passed the blasthole on its way into the rock behind the blastholes. It is believed that the fragmentation caused by these waves is responsible for a considerable part of the total fragmentation.

stress release wave fragmentation, after the compression wave has been reflected in the free surface at bench blasting and changed its sign to a tensile wave and this tensile wave have passed the blasthole the quasi static strain in the rock will be released causing secondary compression (P-waves) and shear waves (S-waves) and these waves will cause a considerable amount of fragmentation. The waves caused are also called *release waves* and they will be similar to the waves at a *rock burst*. This process is also called *release of load fragmentation* and is believed to be an important part of the total fragmentation. Clay, 1965.

stress relief blast, see '**destress blasting**'.

s*tress wave emission*, see '**acoustic emission**'.

stress wave joint interaction, see '**joint initiation fracturing**'.

strike, the course or bearing of the outcrop of an inclined bed or geologic structure on a level surface. Konya and Walter, 1990.

string charging, charging equipment using a hose to charge the more ore less horizontal holes in tunnelling and by which the linear charge concentration can be varied. The technique has been developed for emulsion explosives but can also be used for ANFO explosive but in the latter case the charge concentration cannot be so well controlled.

string emulsion explosive, see '**string charging**'.

string loading, the procedure of loading cartridges end to end in a borehole without deforming them. Used mainly in controlled blasting and permissible blasting. USBM, 1983.

strip mine, a surface operation with only one bench in which the overburden is removed from a near horizontal coal bed or other usable sedimentary rock before the coal or the sedimentary material is taken out.

stripping, the process of removing the overlying earth or barren surface rock from the resource being mined in an open pit or quarry.

structure, one of the larger features of a rock mass, like bedding, foliation, jointing, cleavage or brecciation. Also the sum total of such features as contrasted with texture. Also, in a broader sense, it refers to the structural features of an area such as anticlines or synclines. ISRM, 1975.

structure demolition blasting, demolition blasting of any manmade construction.

strut, a reinforcement element made of steel or timber with the purpose to take compressive forces. TNC, 1979.

stud, a bolt having one end firmly anchored. Nichols, 1962. A threaded rod or a bolt without a head. Ham, 1965.

stump blasting, removal of tree stumps using explosives.

stump or **underbreak**, unbroken rock within the final contour. Nitro Nobel, 1993.

sub, a *coupling* with different types and sizes of box or pin threads at either end. Used to connect unlike threaded members of drill string, casing, or drive pipe equipment together. Also called *adapter* or *substitute*. Long, 1920. A special sub called *shock-absorbing sub* is located at the top in the rotary drilling and with the purpose to absorb vibrations from the drill bit at drilling. A '*cross over sub*' is a short piece of drill rod with similar threads at both ends and to prevent wear on the threads of the rotation head spindle, a '*saver sub*'. Atlas Copco, 2006.

subdrill, length of blasthole drilled below the planned level of breakage at the floor in bench blasting. Because of the larger confinement in a bench bottom, it is necessary to drill under the bench floor to place more explosive as a bottom load. The length drilled is called the subdrilling length and it is dependent on the burden and the inclination of the blastholes. For vertical blastholes in a bench, the subdrilling length should be about 30% of the burden.

subdrilling length (l_{sub}), **(m)**, the length of subdrilling, see '**subdrill**'.

subgrade, length of blasthole drilled below the grade (bench floor) to avoid toes. See '**subdrill**'.

subgrading the bottom or **subgrade drilling**, drilling below the bottom of the bench in surface mining. Sen, 1987.

sublevel, drifts opened at different levels to exploit orebodies.

sublevel block caving (SLBC), a *block caving method* where the undercut is made by a sublevel caving layout and the ore is thereafter blocked by gravity see '**caving methods**'.

sublevel caving (SLC), a mining method where longitudinal crosscuts drifts or transverse crosscuts are driven through the orebody on successive levels. From the crosscuts the ore above is fan drilled. The ore is blasted burden for burden and the ore is moved by gravity to the drift, where it is loaded, transported to the foot wall and dumped into shafts.

sublevel open stoping (SLOS), a stoping method where the stope is emptied more or less after each blast. A little amount of ore is normally used at the bottom of the stope to protect the drawpoints from damage during the blasting of the next round. See also '**sublevel stoping**'.

sublevel shrinkage caving (SLSC), a mining method that is a combination of sublevel caving and shrinkage stoping. The orebody is developed, if it is wide enough, like in sublevel caving with cross cuts but only the swell volume of the blasted ore is loaded. The rest is left in the cave as a temporary ore rest. When several levels have been mined a loading level is arranged below all these preloaded levels and the ore is now loaded as if it where ore fragmented in block caving. Hansagi, 1977.

sublevel slicing, see '**top slicing combined with ore caving**'. Fay, 1920.

sublevel stoping (SLS), an underground mining method where the ore is blasted in near-vertical or horizontal slices or benches into an open or rock-filled stope. The ore is mucked from openings (craters) made at the bottom of the stope. The stope can be open or partly filled, depending on the need for rock support.

submarine blasting (SMB), blasting of rock under water with water proofed explosives and initiation materials. The specific charge is normally increased due to the fact that it is needed more energy to blast against water than air. The larger the water depth the higher the specific charge due to the increase of the water pressure.

subsidence, the gravitational induced downward displacement of the overburden (rock and/or soil) lying above an underground excavation or adjoining a surface excavation. Also the sinking of a part of the earth's crust. ISRM, 1975. There are two main groups of subsidence, *continuous subsidence* (through subsidence) and *discontinuous subsidence*. Continuous subsidence involves the formation of a smooth surface subsidence profile that is free of step changes. This type of subsidence is usually associated with the extraction of thin, horizontal or flat-dipping ore bodies overlain by weak, non-brittle sedimentary strata. Discontinuous subsidence is characterised by large surface displacements over limited surface areas and the formation of steps or discontinuities in the surface profile. Sublevel caving mining is one example of discontinuous subsidence. Subsidence may be associated with a number of mining methods, may involve a range of mechanisms, may develop suddenly or progressively, and may occur on a range of scales. *Active subsidence* is directly caused by mine workings underground. *Residual subsidence* is defined as the subsidence after the mining influence has ceased. This subsidence is time-dependent and due to phenomena such as consolidation or viscoelastic or viscoplastic behaviour of the strata, which continues to develop after the point is no longer in the zone of influence of the face. If chimney formation is sudden rather than progressive, the phenomenon is sometimes known as *plug subsidence*. Generally plug subsidence is controlled by some structural feature such as a dyke or a fault which provides a plan of weakness whose shear strength is overcome at some critical stage of mining. This form of subsidence generally produces underground air blasts overpressures. Brady, 1985.

subsidence angle (α_{sub}), (°), the lowest angle measured from the vertical and the mining front to the nearest point on the surface where the subsidence is zero. This angle should be measured to the vertical and is for European coal fields generally quoted as 35° or between 38–33°. Hudson, 1993.The overburden normally have small vertical movements due to frost north of the Arctic Circel and to be sure that the subsidence is dependent only on the mining activity, 2 cm vertical change was taken as the minimum movement to be sure it depends on mining activity at the Kirunavaara mine. MassMin, 2008. pp. 867–976.

subsidence caving, a caving mechanism in which a large mass of rock subsides rapidly as a result of shear failure on the vertical or near-vertical boundaries of a block.

subsonic speed (c_{sub}), **(m/s)**, speed lower than the speed of sound. IME, 1981.

sulphide dust explosion, see '**dust explosion**'.

sump, an excavation for the purpose of collecting or storing seepage water before pumping it to the surface. The bottom of a shaft is sometimes used for this purpose. Gregory, 1983.

sump fuse, see '**safety fuse**'. AS 2187.1, 1996.

sump pump, pump used to remove fluid from a sump. Atlas Copco, 2006.

sunflower cross sectioner, see '**telescope tunnel profiler**'.

sunk shaft, a shaft which is driven from the top downwards (vertical or inclined). Fraenkel, 1954.

superposition, a principle stating that if a body is subjected to several stresses acting simultaneously, then each stress produces its own strain or strains, and these strains may be superimposed to give the complete state of strain of the solid. Isaacson, 1962.

sump pump, pump used to remove fluid from a sump. Atlas Copco, 2006.

sunflower cross sectioner, see '**telescope tunnel profiler**'.

superposition, a principle stating that if a body is subjected to several stresses acting simultaneously, then each stress produces its own strain or strains, and these strains may be superimposed to give the complete state of strain of the solid. Isaacson, 1962.

supersonic speed (c_{sup}), **(m/s)**, speed greater than the speed of sound. IME, 1981.

support, structure designed to bear or hold up a load by different procedures and materials and used to improve the stability and maintain the load-carrying capability of rock near the boundaries of underground excavations. The primary objective of support is to mobilise and conserve the inherent strength of the rock mass so that it becomes self-supporting. The procedures and materials used are described as reinforcement. *Primary support* is applied during or immediately after excavation, to ensure safe working conditions during subsequent excavation, and to initiate the process of mobilising and conserving rock mass strength by controlling boundary displacements. The primary support or reinforcement required. Any additional support or reinforcement applied at a later stage is termed *secondary support*. It was once the custom to describe support as being temporary or permanent. *Temporary support* was that support or reinforcement installed to ensure safe working conditions during mining. For centuries, such support consisted of some form of timbering. If the excavation was required to remain open for an extended period of time, *permanent support* was installed subsequently. Quite often, the temporary support was partly or wholly removed to enable the permanent support to be installed. Brady, 1985.

support of weak rocks, weak rock is here defined as extremely poor to fair, corresponding to a variation in RMR from 0 to 65 and a variation in Barton's Q-factor from 0,001 to 10. Recommended support dependent on span varies from systematic bolting to cast concrete lining, see Fig. 62.

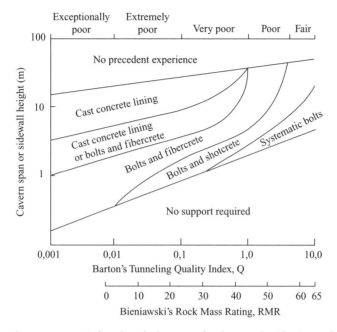

Figure 62 Experience on support of weak rocks in terms of rock mass classification. After Barton, 1989.

surface density (S_v), (1/m), the surface area of a phase (e.g. joints) contained in a unit volume of the structure. Weibel, 1989.

surface drilling, drilling of holes on the surface of the earth. The holes may be used for exploration, blasting, oil-, gas- or water extraction etc.

surface energy (γ_s), (J/m²), see '**specific surface energy**'.

surface fractal dimension (D_{fs}) (dimensionless), the *surface fractal dimension* is defined by $D_{fs} = 1 + D_f$ where D_f is the fractal dimension, see '**fractal dimension**'. Mandelbrot, 1984.

surface marks, are defined by the typical surface topography of a fresh crack. There are four kind of surface marks 1) *Rib-marks*, 2) *Hackle marks*, see Carrasco, 1977 and 3) *Striaes* and 4) *Steps*, Lutton, 1969. *Rib marks*, commonly found on a fracture surface caused by impact on a brittle material, are concentric with the point of origin of a tensile fracture and represent the transition region of increasing roughness. The spacing between each rib has been related to the velocity of the crack, the rate of loading and the brittleness of the material. *Hackle marks* are associated with tensile fractures concentric with the point of origin of the fracture. They are perpendicular to rib marks, and are typically arranged in a plumose pattern diverging from the point of origin of fracturing in the direction of propagation. The presence of hackle marks has commonly been interpreted as characteristic of shear fracture, however, the type of hackle structure appearing on a dynamic fracture surface is discussed in relation to the formation of adjacent cracks and is associated with fractures formed in tension. *Striae*, which are diagnostic features of *shear fractures*, occur in groups and consist of linear parallel grooves. In rock, the fractures' surfaces associated with shear generally show a strong abrasion and powdering, owing to the *crushing* or *shearing* of crystal or grain asperities on the failure surface. However, striae marks are well developed in soft materials such as silt and clay, or in hard material where the normal stress across the fracture was large. *Steps* are found on *shear surfaces* and are oriented approximately normal to the striae. Steps and hackle-marks are also typically arranged in a plumose pattern on *tensile fracture surfaces*. The step marks are considered to be one of the most prevalent of all surface markings.

surface mine, see '**open pit**'.

surface mining, excavation of minerals or organic matter (coal) from the surface of the earth.

surface morphology, see '**fractography**'.

surface-set bit, a bit containing single layer of diamonds set so that the diamonds protrude on the surface of the crown. Also called single layer bit. Long, 1920.

surface subsidence, the lowering of the ground surface following underground mining without backfill. See' **subsidence**'.

surface waves, Rayleigh and Love waves. See '**seismic waves**'.

suspension grouting, injection of suspension through a borehole to cracks and joints surrounding a drift, tunnel, cavern, room or rock slope. TNC, 1979.

S-wave velocity, (c_s), (m/s), shear-, transverse wave velocity. See '**seismic waves**'.

fragment size distribution function, this distribution function to represent rock fragments after blasting was searched for because the existing size distribution functions (Rosin-Rammler, Schumann, Gaudin-Meloy) did not simulate the fine part of the fragment size distribution very well. This distribution function has been tested on more than 100 real measured fragment size distributions from full scale, half scale and model blasts. The distribution function is,

$$P(x) = \frac{1}{1 + \left[\ln(k_{100}/x) / \ln(k_{100}/k_{50})\right]^b} \quad (S.10)$$

where $P(x)$ is the amount of material less than the sieve size x in m, k_{100} is the maximum fragment size in m, k_{50} is the mesh size where 50 weight% of the material is passing and b is a site specific constant. Ouchterlony, 2005.

Swedish bit, synonymous for 'chisel bit', also *chisel edge bit, chisel-point bit, chopping bit*.

swell (S), (%), the volumetric increase, normally expressed as the volume of open void space (V_{void}) in the material as a percentage of the volume of solid material (V_{solid}). Another definition is the inverse of swell multiplied by 100, see '**swell factor**'.

$$S = 100 \left[\frac{V_{void}}{V_{solid}}\right] \quad (S.11)$$

Swellex bolt (tradename), a friction rock bolt consisting of a folded steel tube with a longitudinal tuck. The bolt can be pressurized by water to 30 MPa and thereby the diameter of the bolt is increased, see Fig. 63. The Swellex bolt will be fixed to the rock by friction. Swellex bolt can be extended to any length by threaded coupling between elements and in roof heights larger than 2,4 m. The pressure can be reduced by 20–30% and thereby let the rock mass slide along the bolt where large rock deformations are expected. When appropriate it can be inflated to full pressure.

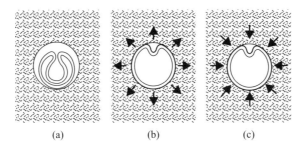

Figure 63 Principle of Swellex rock bolt. (a) Swellex bolt is placed in the hole, (b) high pressure water inflates the tube, causing a small elastic expansion of the borehole diameter, (c) water pressure is released and the surrounding rock contracts thereby providing the Swellex locking effect. After Stillborg, 1994.

There are three different Swellex rock bolts in the market today. Coated EXL Swellex (Corrosion protective coating), Super Swellex (200 kN rock bolt for hole diameters 43–52 mm), Yielding Super Swellex (160 kN) etc.

swell factor (u), (dimensionless), the ratio between the total volume of broken rock and the volume of solid rock before blast. The swell factor varies between 1,0 and 1,7 in rock blasting. The normal swell factor when blasting a few rows with sufficient swelling is normally ~1,4. When blasting more than 10 rows one after another on the surface, the swelling will be considerably reduced only 1,1 to 1,2.

swelling, a thermo-hygro-mechanical process. The constitutive mineralogy of rock is such that water is absorbed, causing a measurable increase in the volume of rock. Swelling can exert very large time-dependent forces on rock support systems, or can reduce the size of the openings. ISRM, 1975.

swelling *or bulking* due to fragmentation of rock (S), (%), the process of increase in volume of a mass of rock when it has caved or been blasted or otherwise broken and removed from its *in situ* state. See '*swell*'.

swelling clay, when a clay absorbs water it increases in volume. This volume increase may cause rock mass stability problems. Zones in the rock mass may consist of clay and when such zone is passed by a drift or tunnel the clay will start to be pressed into the drift or tunnel due to the rock pressure. This swelling is different from the earlier mentioned because it is due to the rock pressure on the clay. The zone might be reinforced by rock bolts, reinforced iron and shotcrete in combination.

swelling rock, a rock mass with joint filling material, which swells in the presence of water.

swing-hammer crusher, a rock breaker in which crushing force is generated by hammers loosely mounted on a rapidly revolving shaft. Rock entering the crushing chamber is hit, and rebounds against liner plates of walls or against other rock until small enough to escape through a grid. Pryor, 1963.

swivel (shortened form), for '*water swivel*' or '*swivel head*'. Long, 1960.

swivel head, **boring head** or **drill head**, the assembly that applies the drilling pressure and rotation to the drill rods. B.S. 3618, 1963. The assembly of a spindle, chuck, feed nut, and feed gears on a diamond-drill machine that surrounds, rotates, and advances the drill rods and drilling stem. On a hydraulic-feed drill the feed gears are replaced by a hydraulically actuated piston assembly. Also called '*drilling head*' and '*gate*'. Long, 1960.

swivel rod, is used to connect boring rods to the cable in rod percussion drilling. Stein, 1977.

sympathetic detonation, the detonation of an explosive material by means of an impulse from detonation of another charge in air, ground or water. When the charged holes are too close to each other, there is the possibility of the explosive shock wave from one blasthole being sufficiently intense as to initiate the explosive in an adjacent hole. This usually occurs with hole spacings of 57 mm and less and particularly in rock types which are effective transmitters of shock energy. Ground water present in open channels also encourages sympathetic detonation. By proper delay placement these effects are eliminated. Hole spacings of not less than 300 mm should be adopted as a general design basis. Sympathetic detonation is purposely used in some ditch blasting.

synthetic rock (SR), a detailed simulation of the rock mass properties by considering the rock mass joint fabric on the scale 10–100 m. The approach can be used to derive rock mass properties such as modulus, strength and brittleness for later use in larger-scale continuum or discrete-element models. Pierce et al., 2007.

synthetic rock mass modelling (SRM), use of synthetic rock mass to model the rock mass. ITASCA's *PFC^{2d}/PFC^{3d}* codes are used to create an assembly of bonded particles representing a large intact rock sample. A discrete fracture network is then generated which honours the joint measures derived from drilling an mapping on site (e.g. spacing, trace length, orientation of joints). The entire network is then embedded within the bonded assembly, which is subjected to stress changes expected in the field. Joints are inserted into the PFC particle assembly using a newly developed smooth joint model which allows slip and opening on internal planar surfaces.

systematic rock bolting, bolting performed after a special arrangement determined by the operator.

systematic rock reinforcement, rock reinforcement undertaken according to a certain plan. TNC, 1979.

systematic support, rock enforcement undertaken according to a certain plan. TNC, 1979.

Système International d'Unités (SI), the international system for units. It was established in 1960, today it is globally accepted.

tabular, the form of crystals which are flattened in one plane. Tabular crystals may occur in tables, plates, disks, foliae, and scales Schieferdecker, 1959.

tabular orebody, an orebody shaped like a tablet, relatively long in two dimensions and short in the third. Stokes, 1955.

Taby cut (Täby cut), a *cylinder cut* consisting of parallel holes with the same diameter and all charged with explosives and one or two larger uncharged swelling holes used to advance the drift. The first row of holes including six holes has a rhombic shape instead of quadratic like in normal burn cuts. The second row of holes has a rectangular shape instead of quadratic, like in burn cuts. The Täby cut advance is inferior to the *double hexagonal cut*. The benefits of the Täby cut are, however, that the holes are here located vertically below one another in each row. This facilities drilling and mechanisation of the drilling. Langefors, 1963.

tail, see '**capped fuse**'.

tail cracks, see '**horse tails**'.

tailored pulse loading, a blasting method where the pressure-time history of the gases produced by the explosive fulfils certain requirements.

tail rope, the rope which passes around the return sheave in main-and-return haulage or a scraper loader layout. Nelson, 1965. A counterbalance rope attached beneath the cage when the cages are hoisted in balance. Zern, 1928.

tamper, a person who is tamping. See '**tamping**'.

tamping, the process of compressing the stemming or explosive cartridges in a blasthole.

tamping bag, a cylindrical bag containing stemming material, used to confine explosive charges in boreholes. USBM, 1983.

tamping bar or **pole**, a wooden or plastic pole used to push and compact explosive cartridges and/or stemming in the blasthole. Plastic poles must be safe regarding static electricity. Iron or steel tools are not allowed to be used as tamping poles.

tamping rod, see '**tamping bar** or **pole**'.

tamping stick, see '**tamping bar or pole**'.

tandem thread, two threads after each other on a drill rod and with a little waist in between used on full sections rods. Sandvik, 2007.

tangential hoop strain at the blasthole wall (ε_t), (dimensionless), the strain caused from the pressure from blast gases on the blasthole wall ε_t is

$$\varepsilon_t = \frac{r_b}{r_d} - 1 \qquad (T.1)$$

where r_b is the radius of the blasthole at the equilibrium borehole pressure (p_{be}), or the maximum expansion of the blasthole diameter and r_d is the radius of the blasthole wall at the detonation pressure, before the expansion of the blasthole.

tangential force (F_t), (N), a force acting parallel to a surface. ISRM. 1990.

tape, see '**capped fuse**'. AS 2187.1, 1996.

tape extensometer, see '**extensometer**'.

taper, a gradual and uniform decrease in size, as a tapered socket, a tapered shaft, a tapered shank. Crispin, 1964.

taper angle, angle of a circular geometrical shape that tapers to a point. Some drill bits and drill rods are connected by mated tapered sections instead of threads. Different taper angles are used in different rock formations. Atlas Copco, 2006.

tapered rod, drill rod having one end shaped as a taper or partial cone on which the drill bit is pressed and the other formed to fit into the rock drill a rod with a conical end. Atlas Copco, 2006.

TBM (acronym), see '**tunnel boring machine**' and '**full-face boring**'.

TC (tungsten carbide) **insert type casing shoe bits**, drill bit like a casing shoe set with chips of tungsten carbide. *Casing shoes* are affixed to the bottom of a casing string and used to cut away any protrusions in the hole while installing casing. Atlas Copco, 2006.

TC (tungsten carbide) **-set bit**, coring bit set with chips of *tungsten carbide* rather than diamonds. Atlas Copco, 2006.

TDR (acronym), see '**time domain reflectometry**'

TE (acronym), see '**total energy**'.

tectonic stress, the stress state due to the relative displacement of lithosphere plates. Hyett, 1986.

tele remote mucking, wireless or normally radio controlled mucking of ore for example in open stopes that might collapse. Atlas Copco, 2007.

telescopic assembly, arrangement of tubes and rods fitting into one another that can be increased or decreased in length when a force is applied. Atlas Copco, 2006.

telescope tunnel profiler equipment to measure the contour of smaller drifts consisting of a telescopic rod with its base located in the middle of the drift on the floor. The telescopic rod is directed towards the periphery of the drift and perpendicular to the drift axes and elongated until the top end reaches the roof or wall of the drift. The distance to the periphery may be measured at regular angles e.g. 15° between the readings. From the values measured an approximate overbreak area can be calculated. Used e.g. in Indian coalmines. Also called '*sunflower cross sectioner*'.

temperature of explosion, see '**explosion temperature**'.

temperature sensitivity of explosives, the properties of explosives and initiation material changes with temperature. In cold areas, below 0°C, different additions of ingredients are made to make for example NG-based explosives less stiff with the aim to avoid accidental explosions when tamping the explosive by a pole. In India NG-based explosives are prohibited to be used in hot holes. Under these conditions slurries and emulsions are recommended to be used up to a temperature of +80°C. Also specialised initiation system components are needed.

template, a pattern device used as a guide to mark points at which boreholes are to be collared in ring drilling or drifting.

temporary rock reinforcement, reinforcement of the rock mass only for a shorter time (~3–5 years).

temporary rock support, support only needed for shorter time. TNC, 1979. See also '**support**'.

tendon, see '**cable reinforcement**'.

tenor, in mining, the percentage of metal in or average metallic content of an ore or impure metal.

tensile strength (σ_t), (MPa), a materials ability to withstand tensile loads. When testing the *mechanical strength of initiation materials*, e.g. safety fuse, detonating cord, shock tubes, etc. such strength is expressed in kg of load. The tensile strength of rocks ranges between 0 and 20 MPa. The static tensile strength is roughly 1/10 of the compressive strength but the scatter in the ratio is large for different rock types.

tensile stress (σ_t), (MPa), normal stress tending to lengthen the body in the direction in which it acts. ISRM, 1975.

tension drilling, drilling with part of the weight of the drill string supported by the drill swivel head or suspended on a drilling line, as opposed to drilling with the entire weight of the string imposed on the bit. Long, 1920.

terrestrial laser scanning (TLS) (acronym), a method to quantify the geometry of the surface from a land based position e.g. of a blasted bench before and after blasting by a laser scanner. Wetherelt, 2006.

tertiary explosives, tertiary explosives are extremely difficult to explode by fire alone, although a very large fire involving as much as several thousands of tons of these explosives can indeed lead to explosion and detonation, with devastating effects because of the large quantities involved. Persson, 1993. Tertiary explosives are insensitive to initiation by a standard No. 6 blasting cap. Most explosives in this category are the dry blasting agents (DBA's) or slurry explosives and blasting agents. Brady, 1985.

Terzaghi effective stress (σ_e), (MPA), see '**effective stress**'.

Terzaghi rock load classification, see '**history of rock mass classification**'.

test blasting cap No. 8., a blasting cap containing 0,40 to 0,45 g of PETN base charge at a specific gravity of 1 400 kg/m^3, and primed with standard masses of primer, depending on the manufacturer. USBM, 1983.

test drilling, 1) drilling through soil and rock in search for minerals, coal, oil water etc. 2) drilling to examine the strength of the rock mass for building underground caverns. TNC, 1979. See also '**prospect drilling** or **prospect exploration**'.

test mining, the last step in the exploration of a property by breaking a small amount of the ore and examining its suitability for mineral processing.

test of delay interval in detonators, measurement of the time lag between the initiation and the rupture of the detonator. Normally only one detonator is tested each time.

test of detonating cord, consists for example of visual inspection for smoothness, pliability and surface blemishes, measurement of diameter, test for flexibility, sensitivity after flexibility test, water proofness, detonation velocity (c_d), transmission of detonation and finally for breaking load. Pradhan, 2001.

test of permitted explosive, the tests are undertaken in a large **steel tank** (steel gallery) with the aim to initiate the explosive and examine if it will initiate an inflammable atmosphere containing methane and natural gas-air mixture. Pradhan, 2001.

test of slurry explosives and **water gel**, the following tests are normally used for testing water gels and slurry explosives; *rifle bullet test, thermal stability test, initiation sensitivity test, test for explosives being used in hot holes, drop and impact test* and *thermal behaviour test*. Pradhan, 2001.

texture, geometrical aspects of the component particles of a rock, including size, shape, and arrangement. A.G.I., 1962.

THD (acronym), see '**top hammer drilling**'.

thermal boring, use of high-temperature flame to fuse rock in drilling. Heat comes from ignition of kerosene with oxygen or other fuel system at bottom of drill hole, and water with compressed air may be used to flush out the products. Pryor, 1963.

thermal piercing, see '**jet piercing**' and '*fusion piercing*'.

thermal stability test, the explosive is pushed into a steel pipe and stemmed and this unit is put vertically into a furnace and surrounded by sand. The temperature of the explosive and the furnace are recorded. Cartridges are stored under the conditions specified below.

    I   At 27° ± 2°C under normal humidity condition.
    II   At 50°C in a thermostatically controlled chamber.
    III At 50°C and 25° alternately for a period of 12 hours in a thermostatically controlled chamber. Pradhan, 2001.

The most commonly referred thermal stability test is the UN test 3c), in which the substance is subjected to 75°C during 48 h. Sanchidrian, 2007.

thermic boring, see '**thermal boring**'. Streefkerk, 1952.

thermic drilling, see '**jet piercing**'. Nelson, 1965.

thickness (*t*), **(m)**, the mean width of a layer.

thimble, an oval iron ring around which a rope end is bent and fastened to from an eye. Zern, 1928.

thin walled bit, diamond drilling bit used with a thin wall drilling system that allows larger cores to be extracted without increasing the hole size. Atlas Copco, 2006.

thin walled tube, tube made from thin material. Such as thin walled diamond drilling tubes used to reduce the weight of the drill string so that deeper holes can be drilled with the same drill rig. Atlas Copco, 2006.

third degree of freedom, see '**presplitting**'.

third theory of comminution, states that the specific work input required for size reduction is inversely proportional to the square root of the product size, less the work required to form the feed. Pit and Quarry 53rd ed., Sec. B, p. 43.

thread cutter, a tool used to cut screw threads on a pipe or bolt. Long, 1920.

threaded detachable bit, drill bit having an internal thread that can be attached or removed from a drill rod. Atlas Copco, 2006.

thread gauge, tool to check the thread angles. Sandvik, 2007.

three cone bit, see '**roller rock bit**'. B.S. 3618, 1963.

three-dimensional rock strength ($\sigma_x, \sigma_y, \sigma_z$), **(MPa)**, the rock strength is dependent on small defects in the rock mass called microcracks, flaws, pores etc and their statistical distribution. Different models of failure of a material under three dimensional stress include von Mises, Tresca, Mohr-Coulomb and Lundborg criteria and can be calculated by the following formula

$$\left(\frac{\sigma_2}{\sigma_1}\right)^M = \sqrt{1-\mu^2} + \mu \quad \text{Lundborg failure envelope} \tag{T.2}$$

where σ_1 and σ_2 are the uniaxial and biaxial compressive strengths in MPa, M and μ are material constants. The four different theories are compared in Fig. 64.

three point disk test – tie, stay, prop or strut   381

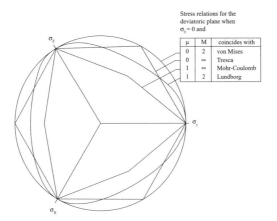

Figure 64 Comparison of a cut of the 3-D strength envelope perpendicular to its symmetry axis that starts at origo. Four different failure theories are compared; von Mises, Tresca, Mohr-Coulomb and Lundborg criteria. Von Mises represents a cone with its top at origo, Tresca a symmetric hexagonal cross section with sharp edges, Mohr-Coulomb represents a deformed hexagonal cross-section with also has sharp edges. The Lundborg model for $\mu = 1$, $M = 2$ yields a projectile shape with rounded edges which has features of all these criteria. After Persson, 1993.

three point disk test, a bending test using short cylindrical disks, the height of the disk about ¼ of the diameter, and using steel balls as span supports and a steel ball as the load point. This test is convenient for bedded rocks, with "cut-offs" from the preparation of compressive strength specimens providing suitable test pieces. Five or more test should be made. Hudson, 1993.

three section cut, a parallel hole cut in drifting where the first row has an hexagonal shape and the second row a pentagon shape and using a 110 mm hole (first section) for swelling. The advance is 15% larger compared to the Taby (Täby cut). Langefors, 1963.

throat of crusher, point at which the rock is discharged. It short dimension varies depending upon on the setting of the closed position.

through cut, rock excavation through a hill leaving a slope on both sides. The slope of the bank depends on the rock type and the minimum slope is usually 20° form the vertical. In very deep cuts the rock is excavated in benches of between 6 and 12 m in height. Sen, 1995.

throw blasting, see '**blast casting**'.

throw or **heave**, the displacement of rock during blasting due to the detonation and the resulting expansion of gases. Konya and Walter, 1990. Typically burden velocities are about 20 m/s or 72 km/hour and maximum 27 m/s or 97 km/hour. Chiapetta, 1983.

throw distance (R), (m), see '**distance of throw**'.

throw length (R), (m), the maximum length of throw in rock blasting, see '**distance of throw**'.

thrust (F), (N), force applied to a drill in the direction of penetration. ISRM, 1975.

thrust borer, mechanism for forcing a hole through an embankment for the insertion of pipes or cables. Ham, 1965.

tie, stay, prop or strut, a diagonal brace or tie bar to stiffen or prevent movement of a structural component. Ham, 1965.

tie back anchoring, long steel bar anchored into an earth cut or bank and attached to some form of retaining plate to prevent movement. Atlas Copco, 2006.

tie tamper, petrol powered breaker, which can alternately be used for tamping work. A railway tie is also known as a sleeper in some countries. Atlas Copco, 2006.

tie tamping, process of using a percussive tool to compact earth, gavel or other material around a railway tie. Atlas Copco, 2006.

TIGER, *a computer code* for modelling *detonation performance*. Persson, 1993.

till, that part of glacial drift consisting of material deposited by and underneath the ice, with little or no transportation and sorting by water. It is generally an unstratifyed, unconsolidated, heterogeneous mixture of clay, sand, gravel, and boulders. Fay, 1920.

tiltmeter, in earthquake seismology, a device for observing surface disturbances on a bowl of mercury, employed in an attempt to predict earthquakes. A.G.I., 1957.

timbering, the action of supporting drifts by timber, two timers at the wall and one at the top.

timber post, timber used for support at the face. Brady, 1985.

time (t), **(s)**, a physical quantity defined by the full rotation of the earth around it axes as one day. One day is divided into 24 hours (h), one hour into 60 minutes and one minute into 60 seconds.

time domain reflectometry (TDR) (acronym), a method to measure physical deformation of explosive substances at detonation or larger movements in the rock mass. A length of coaxial cable, grouted into a borehole, serves as the system sensor. Electronic pulses are sent down the cable and they are reflected where there is a break in the cable or at pre-established reference points (crimps). The system is commonly used to monitor deformation associated with landslide, mining, and construction activities. Also called 'electric time domain reflectometry' (ETDR).

time intervals of firing (t), **(s)**, delay time between the different delays.

tippler, a power-operated appliance of discharging the coal or mineral from a mine car. It consists of a steel structure with a rail track to receive a mine car. On rotation, it discharges the material (usually) sideways on to a primary screen or bunker. Also called '*dumper*', '*unloader*'. Nelson, 1965.

TLD (acronym), see '**trunk line delay**'.

TLS (acronym), see '**terrestrial laser scanning**'.

TNT (acronym), see '**trinitrotoluene**'.

toe, the shortest distance between the blasthole and the free face measured at the floor elevation of the pit or quarry. Often this burden is larger than the drilled burden due to failure of vertical holes to break cleanly at the bottom.

toe charge (m_b), **(kg)**, see '**bottom charge**'.

toe ditch, a ditch at the toe of a slope with the purpose to keep fallen rocks out of the road when failures are wedges, planes and/or minor topples. Hudson, 1993.

toe drains, drainage of water from the toe in a slope of soil or rock with the purpose to avoid landslides. Bored horizontally (with a very small inclination) from the slope toe. Many soil landslides can be fully corrected (or at least slowed) by internal drainage only. In rock slopes internal drainage must bee used in conjunction with other support measures (anchoring and/or walls). Hudson, 1993.

toe hole, a blasthole, usually drilled horizontally or at a slight inclination into the base of a bank, bench, or slope of a quarry or open pit mine. USBM, 1968.

toe of the bench, the bottom of the bench.

toe of the hole, the bottom of the hole.
toe space explosive loading, an air space is left in the bottom of the blasthole when charging the hole. This improves the fragmentation. The method has been tested both in small scale crater and bench blasting and in full scale. Zhang, 1996.
tomography, in mining, a technique using geophysical methods to investigate the rock properties in two or three dimensions within large rock volumes, with the help of mechanical or electromagnetic waves generated from boreholes and measured in surrounding boreholes.
tonne, in the metric system a unit of mass equal to 1 000 kilograms (kg) or approximately to 2 204,62 pound mass. McGraw-Hill, 1984.
tool shank, part of a chisel, moil point or drill rod that is inserted into a breaker or drilling machine. Atlas Copco, 2006.
toothed roller bit, synonym for '**roller bit**'. Long, 1920.
top hammer, rock drill mounted on a drill feed that transfers percussive energy to the drill bit through one or more drill rods. As opposed to a down-the-hole drill that is attached to the bottom of a drill sting and transfers percussive energy directly to the drill bit and follows it into the drill hole. Atlas Copco, 2006.
top hammer drifter, rig mounted top hammer rock drill designed specifically for drilling horizontal, parallel holes in mining applications. Atlas Copco, 2006.
top hammer drilling, percussive drilling (impact and rotation) with the drilling machine outside of the hole.
top heading, the upper area when larger drifts or tunnels are driven. The top heading is drilled and blasted like regular drifts and the bottom is excavated through benching or slushed upwards by horizontal holes.
top heading-underhand stoping, method of excavation where the upper part or top heading is driven to its full length, before the enlargement of the rest of the section is carried out. (Belgian method). This blasting method is used for adits, tunnels and drifts. Fraenkel, 1954.
top initiation, in a borehole the initiation of the charge close to the collar of the hole.
top load, see '**column charge**'.
top slice, a horizontal block of ore extracted by top slicing. The dimensions vary in different mines. Nelson, 1965.
top slicing (TS), a method of stoping in which the ore is extracted by excavating, beginning at the top of the orebody and working progressively downward. Timbered supported drifts are driven in ore horizontally (sometimes inclined) alongside each other and caved afterwards. The next slice is made below the first slice and having mats to protect the cave to penetrate downwards. The slices are caved by blasting out the timbers, bringing the capping or overburden down upon the bottom of the slices which have been previously covered with a floor or mat of timber to separate the caved material from the solid ore beneath. Succeeding lower slices are mined in a similar manner up to the overlying mat or gob, which consists of an accumulation of broken timbers and lagging from the upper slices and of caved capping. As the slice are mined out and caved, this mat follows the mining downward, filling the space occupied previously by the ore. USBM Bull. 390, 1936.
top slicing and cover, see '**top slicing and cover caving**'. Fay, 1920.
top sub, short threaded tube forming the upper part of a DTH hammer. Top sub thread is the thread of the upper part of the hammer. Atlas Copco, 2006.

toppling, the sinking and breakage of the hanging wall into large blocks which successively bends towards the mined out area.

toppling of hanging wall, bending of a rock mass consisting of a major set of persistent and parallel discontinuities parallel to the dip of the orebody and placed in the hanging wall. Brady, 1985.

top slicing, a mining method where timbered drifts are driven in ore and caved afterwards. The next slice is made below the first slice and having mats to protect the cave to penetrate downwards.

torque (M), (Nm), tangential force F applied for rotation of a body on a radial distance (r) from the rotation axis. Torque is the product of force times the perpendicular radial distance to the rotation axis $M = Fr$.

torsion modulus of elasticity (G), (MPa), see '**modulus of rigidity**'.

torsion strength (σ_{tor}), (MPa), a body is under torsion strength when subjected to force couples acting in parallel planes about the same axis or rotation but in opposite senses. A.G.I., 1957.

total bit load (m_{bitT}), (kg), the mass applied on the drill bit.

total borehole deviation (D_T), (°, m or %), see '**drilling accuracy**'.

total cap lag, the total time between application of a initiation current and the time of detonation for a blasting cap.

total energy (TE) release of charge (W_T), (MPa), is composed of shock (strain) energy (SE), and heave (bubble) energy (HE). The function of shock energy is to create cracks in the rock, which is then fragmented further and dislodged by the gas expansion of the heave energy of the explosive.

total mass of charge (m_T), (kg), the total charge mass of charge in a blasthole or for a whole round.

total porosity or **true porosity (n_T), (dimensionless)**, the ratio of the total volume of the open and sealed (closed) pores to the bulk volume expressed as a percentage. Dodd, 1964.

tough hardening, a process to make steel harder and tougher. Sandvik, 2007.

toughness, a property of a material that denotes, nominally, an intermediate value between softness and brittleness. Tensile tests show a tough material to have a fairly high tensile strength accompanied by moderate values of elongation and reduction of area. Generally a tough material also shows high values in the notched bar impact test. Henderson, See '**fracture toughness**'.

trace, when used for vibration records, the curve printed on the vibration record. Konya and Walter, 1990.

tracer blasting, reduction of the linear charge concentration of a blasthole with the help of a low strength detonating cord placed along the explosive column. The detonation cord is used for the initiation of the primer at the bottom of the hole and is detonated first and the compressive wave desensitizes the explosive by increase in pressure (increase in explosive density) such that when the primer is initiated, the explosive will detonate with a lower detonation velocity (c_d). The method can be used in controlled contour blasting. The effect will also be found to a greater or smaller extent in production blastholes if detonating cord is used as an initiation line through the explosive column. Adapted from Singh, 1994.

track diamonds, diamonds set in the face or lead portion of the drill-bit crown. Long, 1920.

track mounted jumbo, carrier for hydraulic booms used to drill blastholes in headings running on tracks.
tractor loader, a tractor equipped with a bucket which can be used to dig, and to elevate and dump at truck height. Also called '*tractor shovel*'. Nichols, 1962.
tractor shovel, see '**tractor loader**'.
train of cartridges, cartridges charged into a blasthole.
tramming, term used to define the movement of a mobile, powered drill rig around a drill site or in a mine, or the movement of ore carrying trucks and loaders underground in a mine. Atlas Copco, 2006.
transducer, a measurement device which operates on the basis of transformation of energy from one form to another, e.g. mechanical energy into electrical energy, to record for example a ground vibration or air blast.
transgranular crack, a microcrack lying on the cleavage plain in a single grain.
transient velocity (c_t), **(m/s)**, a wave movement with rapid varying particle velocity.
transition zone, the transition zone characterizes the zone between the crushing around a blasthole and the zone where cracks emerge from the borehole vicinity and new developed joint initiated cracks and opening of weakness planes occur. Plastic and elastoplastic deformation occurring in this zone (but no crushing) will lead to cracking. The zone outside the transition zone is called the elastic or seismic zone in which only elastic deformation will occur. Atchison, 1964.
transmission coefficient (T), **(dimensionless)**, the ratio of the amplitudes between the transmitted wave and the incidence wave with respect to a discontinuity in the rock mass.
transmitted pressure (p_t), **(Pa)**, the pressure which is transmitted over a border characterized by a change in acoustic impedance.
transportation of dangerous goods, see '**transport regulations**' and/or '**dangerous goods classification (United Nations)**'.
transportation of explosives, there is no universal standard for transportation of explosives. Normally the explosives should be transported in a safe manner within the presence of people and structures and equipment of interest. The European Federation of Explosives Engineers (EFEE) is working on a transportation standard for the European Union.
transport regulations, see '**ADR (road)**', '**RID (rail)**', and '**IMCO (shipping)**'.
transverse wave, S-wave, *shear wave, or distortion wave* (c_s), **(m/s)**, see '**seismic waves**'.
trapdoor, a door in a mine passage to regulate or direct the ventilating current. Also called weather door. See also '**trap**'. Fay, 1920.f
Trauzl lead block test, see '**lead block test**'.
travelling block, arrangement of pulleys that hang freely in the loops of a rope and travel up or down when the rope is lengthened or shortened. Atlas Copco, 2006.
trench, usually a long, narrow, near-vertical side cut in rock or soil, such as is made for utility lines. ISRM, 1975.
trend of a line (α), **(°)**, is the azimuth, measured by clockwise rotation from north, of the vertical plane containing the line. The direction is measured in the direction of the plunge. $0° \leq \alpha \leq 360°$. Basic term used in hemispherical projection. Brady, 1985.
trestle, a framework of timbers, carrying tram tracks. Weed, 1922.
triangle cut, the characteristic feature of this drifting cut lies in the fact that the drillholes are planned in zigzag. In this way a larger opening is obtained as the drill holes can brake out between the preceding row of holes. Each vertical row of holes breaks out a layer. If the front holes do not break out to the full depth, the burnout holes indicate the direction

of break for the following row of holes since the holes are in zigzag. The name triangle cut, is due to the distribution of the holes at the working face and the form of the initial opening. Fraenkel, 1952.

triangular burn cut, a parallel hole cut used for drifting consisting of 35 mm diameter holes, where three center holes are placed in a triangular patter and left uncharged as swelling volume. The first row of blastholes consists of three holes and has a triangular shape rotated 45° compared to the empty holes. The second row has a also a triangular shape rotated 45° compared to the first row, see Fig. 65. Langefors, 1963.

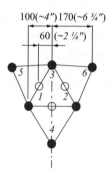

Figure 65 Triangular burn cut. After Langefors, 1963.

triaxial cell, an apparatus used to test the triaxial compressive strength of rocks. The core is placed inside a pressure vessel and a fluid pressure, $\sigma_3 = \sigma_2$, is applied to its surface. A jacket, usually made of a rubber compound, is used to isolate the specimen from the confining fluid which is usually oil. The axial stress, σ_1, is applied to the specimen via a ram passing through a bush in the top of the cell and hardened steel end caps. Pore pressure, p_p, may be applied or measured through a duct which generally connects with the specimen through the base of the cell. Axial deformation of the specimen may be most conveniently monitored by linear variable differential transformers (LVDTs) mounted inside or outside the cell, but preferably inside. Local axial and circumferential strain may be measured by electric resistance strain gauges attached to the surface of the specimen. Brady, 1985.

triaxial compression, compression caused by the application of normal stresses in three perpendicular directions. ISRM, 1975.

triaxial compression strength (σ_{ctri}), (MPa), the strength at which a material will fail when the normal stresses are applied in three perpendicular directions on the material.

triaxial compression tests, see '**triaxial cell**'.

triaxial extension test, the axial stress is reduced under the confining pressure. Brady, 1985.

tricone (rolling cone) bit, rotary drill bit with crushing and ripping action, made of three cones with steel teeth or hard metal inserts.

trim blasting, a method for controlled contour blasting technique in surface mining where the large size production holes are used also for contour blasting. The (linear) charge concentration is reduced in steps when reaching the contour row. The idea is to eliminate costly small diameter blasthole work, along with the associated hole loading difficulties. The recommended spacing between the holes generally ranges from 12 to 16 times the hole diameter. Kennedy, 1990.

trim row, the final row in trim blasting (contour row). The recommended decoupling of charge in the trim row is 40% to 50% void. Kennedy, 1990.

trinitrotoluene (TNT), an explosive with the chemical composition $C_7H_5N_3O_6$. This explosive consists of pale yellow crystals granulated or in flakes. The heat of explosion is 5,07 MJ/kg, the volume of detonation gases 620 l/kg, its density when molten 1 470 kg/m³ and solidification point +80,8°C. It is a military explosive compound used industrially as a sensitizer for slurries and as an ingredient in Pentolite and composition B. Once used as a free-running pelletized powder. USBM, 1983.

Tripod (old), a three-legged support for a rock drill used at quarries and opencast pits. See also 'air-leg support'. Nelson, 1965.

trough, the maximum amplitude of a wave in the downward (negative) direction. Konya and Walter, 1990.

trough (waves), tool used at hand loading consisting of a small steel box with three sidewalls and one end open to facilitate the filling the trough with the help of a pick-axe and thereafter emptying the trough. On two opposite standing sides are two handles fastened to make it possible to lift and empty the trough. LKAB, 1964.

truck, a vehicle used for transportation of goods, ore and waste.

truck haulage in ramp, this method is used normally for shorter distances but has also been used as the main haulage method at a small underground mine. See for example the today abandoned Viscaria mine at Kiruna in Sweden.

true crater, the normally conical cavity caused by crater blasting and after that all material broken or loosened by the explosion has been removed.

true density, (ρ), (kg/m³), the ratio of the mass of the material to its true volume. A term used when considering the density of a porous solid, e.g. a silica refractory. Sometimes referred to as 'powder density'. Dodd, 1964. See '**bulk density**'.

true depth or **true vertical depth**, the actual depth of a specific point in a borehole measured vertically from the surface in which the borehole was collared. Long, 1920.

true porosity or **total porosity (n_T), (dimensionless)**, the ratio of the total volume of *open and sealed* (*closed*) *pores* to the bulk volume expressed as a percentage. See also '**open porosity**' and '**sealed porosity**'. Dodd, 1964.

true specific volume (v), (m³/kg), volume per unit mass of a material when the material is fully compacted, no pores or space within the material.

true vertical depth, see '**true depth**'.

trunkline (trunk), the line of detonating cord or non-electric shock tube on the ground surface that connects the downlines or branch lines (detonating cords or non-electric shock tubes) in the blastholes.

TS (acronym), see '**top slicing**'.

tube, hollow structure made from steel or other hard substance used to transport a gas or fluid to a desired point. Drill tubes are used to transfer rotation and force to a drill bit or DTH machine. Atlas Copco, 2006.

tube charge, see '**pipe charge**'.

tube drilling, any rock or soil drilling where rotation torque is transferred to the drill bit through relatively thin wall tubes rather than rods with a minimum sized flushing fluid canal. The terms "rod" and "tube", when used in the drilling industry, have no clear technical definition. The diamond drilling industry tends to refer to all in hole hollow metal connecting tubes as *rods* while the percussive drilling industries use the term for hollow bars less than 70 mm in diameter. Atlas Copco, 2006.

tube handler, device used for moving drill pipe such as found on drill rigs. Atlas Copco, 2006.

Tubex method, casing tubes used for drilling for water. Sandvik, 2007.

tunnel, a horizontal or nearly horizontal passage underground, open at both ends to the atmosphere.

tunnel boring machine, method of excavating a tunnel by cutting away or boring out the rock in a single pass without the use of explosives Atlas Copco, 2006. A machine which can drill the full area of a drift or tunnel. Normally the shape is circular but elliptical shape is also possible to bore. See also '**full face boring**'. This machine is mainly used in self-supporting rock where rotary disc cutters or a combination of rotary disc cutters and drag bits can be used. Atlas Copco, 1982.

tunnel confinement (f_t), (m^{-1}), is defined as the blast hole depth divided by the drift area.

tunnel crown, the highest point of a tunnel arch. Nitro Nobel, 1993.

tunnel face, the working face at the end of a tunnel. Nitro Nobel, 1993.

tunnel pit, see '**tunnel shaft**'.

tunnel profile or **section**, (A), (m^2), the area of the tunnel in a plane perpendicular to the length axis of the drift. TNC, 1979.

tunnel profiler, an instrument used to determine the contour of an excavation underground, see '**blast damage measurement**'.

tunnel shaft or **tunnel pit**, a shaft sunk, as in a hill to meet a horizontal tunnel. Standard, 1964.

tungsten carbide (WC), black or grey, hexagonal, chemical composition of wolfram and coal WC with molecular weight 195,86, specific weight of 15,63 t/m³ (at 18°Celsius), melting point 2 780°C, boiling point 6 000°C. Hardness approaches that of diamond, or 9 on Mohs'scale, insoluble in water, and soluble in nitric acid and aqua regia. Used in *cemented carbide tools* and in cermets (a material or consisting of ceramic particles bonded with a metal). CCD 6d, 1961,

tungsten carbide abrasion test (abrasive hardness), test of the abrasive wear on a piece of tungsten carbide (length 30 mm, width 10 mm and top radius 15 mm) which is abraded by the rock material of interest placed in form of powder on a rotating disc. The abrasion value AV is defined as the measured tungsten carbide sliding wear weight loss in milligram per 100 m sliding distance and under a load of 10 kg. The wear is heavily dependent on the quartz content in the rock material. Persson et al., 1994. See also '**abrasive hardness**'.

tungsten carbide bit, drill bits having inserts or cutting surfaces made from tungsten carbide. Atlas Copco, 2006.

tungsten carbide insert, wolfram carbide metal plates or buttons inserted in drill bits to make them last longer against wear at drilling.

turbo compressor, a machine to supply compressed air based on a turbo wheel for the compression. Fraenkel, 1952.

turbo drill, a drill developed in the U.S.S.R. for drilling deep oil wells. It was patented in 1873. One type is designed as a turbine and driven by the drilling fluid, which is circulated at high pressure. Nelson, 1965 and Stein, 1997.

turbo driven wear drill, a diamond and tungsten set disc is driven by a turbo with a rotation of 5 000–1 000 rpm and an effect of 300 kW. Drilling speed measured was half to conventional methods at that time. KTH, 1969.

turned vertical shaft, a shaft sunk vertically in the hanging wall block until it intersects the reef after which it continues down at an angle in the footwall parallel to the reef.

This unusual practice is sometimes adopted on the Rand (South Africa) as it enables the mine to become productive at an earlier stage. Nelson, 1965.

turntable, a revolving platform on which cars or locomotives are turned around. Zern, 1928.

turret coal cutter, a coal cutter in which the horizontal jib can be adjusted vertically to cut at different levels in the seam, for example an over cut. The center of gravity of such a machine makes it top heavy and less stable than the ordinary under cutter. Nelson, 1965.

twist drill, a drill made by twisting a length of steel of rectangular or oval section into a spiral form, hence, the term twist drill. Many hand operated coal drills are of this type and the rotation of the drill spiral removes the cuttings from the hole. See also '**auger** or **coal auger**'. Nelson, 1965.

two-component explosive, see '**binary explosive**'.

two-cone bit, see '**roller bit**'. B.S. 3618, 1963.

U

U-bolt, U shaped rod with threads on either end. Atlas Copco, 2006.

UCAF (acronym), for 'undercut and fill' mining see '**cut and fill mining**'.

UCL (shortened form), for '**undercut level**'.

UCS (acronym), for '**unconfined compressive strength**' see '**confined compressive strength**'.

UDEC, a 2-D distinct element code. Hudson, 1993.

U-detonator, a detonator with a bridgewire which needs a high electric impulse, 22 mWs/Ω, for initiation. Earlier detonators require 4 mWs/Ω. These detonators are manufactured and used for mining in Germany. Meyer, 1977.

ultrasonic drilling, a *vibration drilling technique*, which can be used in drilling, cutting, and shaping of hard materials. In this method, ultrasonic vibrations are generated by the compression and extension of a core of electrostrictive or magnetostrictive materials in a rapidly alternating electric or magnetic field. The most easily assembled is a magnetostrictive transducer and the most common magnetostrictive materials, which change in dimension when magnetized, are nickel and vanadium permandur. Mining Minerals Eng. Vol. 1, No. Jan 1965, p.178.

ultrasonic longitudinal and transversal pulse examination of rock, an ultrasonic wave with a frequency of 80.000 Hz has been used to determine the P- and S-wave velocities in *rock cores with joints*. The data are used to calculate the ultrasonic dynamic constants: *Young's modulus*, *Poisson's ratio*, *modulus of rigidity* and *bulk modulus*. Krzyszton, 1986. Ultrasonic longitudinal pulse examinations of the anisotropy of different rock types have also been undertaken in *polished spheres* of different rock types (diameter about 8–10 cm dependent on the size of the grains). The longitudinal wave velocity is registered and the result is presented in equal area projections. Bur, 1969.

ultrasonic wave, a mechanical wave in the frequency range 50 kHz to 100 MHz.

ultrasound drill, a magneto restrictive ultrasound drill causes fragmentation by cavitation and wear. The wear is done at microscopic scale and the cavitation can, however, not be found at fluid pressures larger than 70 MPa and therefore the heat losses are large and efficiency about 50–100 times less conventional drilling methods. KTH, 1969.

unbarricaded, the absence of a natural or artificial barricade around explosive storage areas or facilities. IME, 1981.

UN-classification (dangerous goods), a standard for Storage Classes of Explosives. Dangerous goods are divided into 9 classes. Class 1–3 are described as follows. Class 1-Explosives and objects with explosives, Class 2-Gases and finally Class 3-Flamable liquid substances.

unconfined charge, a charge placed on the surface and surrounded by air.

unconfined compression, see '**uniaxial compression**'.
unconfined velocity of detonation (c_{du}), (m/s), velocity of detonation of an explosive on the surface and surrounded by air.
underbreak, the rock which remains unbroken inside the theoretical contour line in a tunnel, drift, stope, bench, etc. after firing a round.
underbreaking, see, '**underhand stoping**'.
undercut, to remove a horizontal section or kerf in the bottom of a block or coal to facilitate its fall. B.C.I., 1947. See, also 'undercutting'. Also, the approximately horizontal slot mined to initiate caving of a block or panel. Inclined undercuts can also be used in block caving.
undercut-and-fill mining, a stoping method variant of cut and fill mining, working downwards slice by slice, under an artificial roof of stabilized backfill. Oberndorfer, 1993.
undercut level, the level at the bottom of a block from which the mining of the undercut takes place.
undercutter, or '*eccentric bit*' is a drill bit operated in holes where the casing must follow closely behind the bit. The drill bit has two functions, one to make a pilot hole and the other to ream the hole to a larger diameter so the casing can be inserted. Stein, 1977.
undercutting rate (A_u) (m²/month), the rate, expressed as an area per unit time at which the undercut front is advanced.
under drilling, drilling of contour holes, by mistake, inside the theoretical contour.
underground core drilling rig, core drilling done in an existing mine, sometimes to search for new ore bodies, but more often to delineate known ore bodies which often is referred to as 'definition drilling'. Atlas Copco, 2006.
underground mining, practice of excavating openings in rock as a means of extracting and transporting minerals and coal. Atlas Copco, 2006.
underground storage, underground rooms used for storage of various materials. Atlas Copco, 2006.
underground overshot, tool used on an underground diamond drill to recovering lost drill rods. Atlas Copco, 2006.
underhand cut and fill mining, see '**cut and fill mining**'.
underhand stope, a stope made by working downward from a level. USBM, 1990.
underhand stoping, mining ore form an upper level to a lower, underhand. Ballard, 1955.
underhand work, picking or drilling downward. Fay, 1920.
underhole, to cut away or mine out the lower portion of a coal seam or a part of the underclay so as to win or get the overlying coal. Craigie, 1938.
underlier, see '**underlay shaft**'.
underlay shaft, a shaft sunk in the footwall and following the dip or a vein. Also called underlier. Fay, 1920.
underlier, see '**underlay shaft**'.
under ream, to enlarge or ream a borehole. Long, 1920.
under reamer, a tool or device having cutters that can be expanded or contracted by mechanical or hydraulic means and used to enlarge or ream a borehole below the casing or drivepipe. Also called 'expansion bit', 'expansion reamer'. Long, 1920.
underreamer cutter, see '**underreamer lug**'.
underreamer lug, a diamond set or other type of expansible or contractible jaw on an underreamer bit. Also called undeream cutter. Long, 1920.
underreaming bit, an expanding bit used to enlarge the diameter of the hole below the casing to allow the casing to be lowered further down the borehole. B.S. 3618, 1963.

undersea mining, the working of economic deposits (usually coal) situated in strata or rocks below the seabed. Nelson, 1965.

under shot, a condition resulting from employing any insufficient amount or quality of explosive. Usually characterized by poor fragmentation and lack of movement.

underwater blasting, the use of explosives for under water blast work.

underwater detonation test, see '**underwater (explosion) test**'.

underwater explosive test, the assessment of an explosive's performance by firing charges underwater and monitoring the pressure pulse in the water. A series of pulses is produced by the charge after the initiation and the pressure may be measured by electromechanical devices, e.g. piezeo quartz crystals, and the signals recorded on an oscillograph. The amplitude of the first pulse indicates the component of '*shock energy*', while the time period between this and the next pulse gives the '*bubble energy*'. Shock energy is estimated by extrapolation of the amplitude of the pulse to the explosive position. This test is not representative for the behaviour of explosives working in rock, but it is useful for comparing formulations and examining their sensitivity under different conditions. Bubble energy can be measured reliably and accurately, however, the measurement of the shock energy may include many sources of error. One advantage, however, of the underwater test is that more realistic size of charges can be detonated. Also ground vibrations may be measured but the amplitudes of the ground vibrations shows less difference for different explosives. Condon, 1970. The principle for underwater explosive testing is shown in Fig. 66.

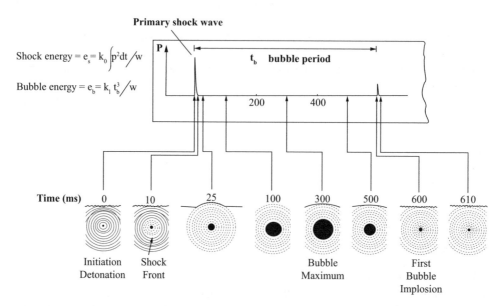

Figure 66 Underwater detonation test. Pressure time history from the detonation of a 100 kg charge at a depth of 20 m. After Persson, 1993.

underwater shock wave, a shock wave in water which may have been produced by explosive energy. Given the low compressibility of water the explosive energy is transmitted with great efficiency, which mean that the shock wave has high destructive power even over large distances. Only P-waves can be transmitted because water has no shear resistance that can build up a shear wave. The propagation velocity decreases as the distance increases from the detonation point, until it reaches the sonic speed 1 435 m/s. Underwater shock waves may damage nearby structures or vessels to divers close to the blasts, as well as to the existing aquatic fauna.

uniaxial compression, compression caused by the application of normal force in one direction.

uniformity coefficient (C_U), (dimensionless), measure of the shape of the grading curve within the range from k_{10} to k_{60}. $C_U = k_{60}/k_{10}$ where k_{10} and k_{60} are the particle sizes corresponding to the ordinates 10% and 60% by mass of the percentage passing. See also '**uniformity exponent**'. SS-EN ISO 14688-2:2004, 2003.

uniformity exponent or *-index* (*n*), **(dimensionless)**, a numerical value expressing the variety in sizes of grains or fragments that constitute a granular or fragmented material. The uniformity coefficient for a Rosin-Ramler-Sperrling fragment size distribution is the slope coefficient for the distribution. See also '**Rosin-Rammler-Sperrling distribution**'. Empirical formulas for the calculation of *n* dependent on geometrical parameters have been developed, see Cunningham, 1983 and 1987, and for the calculation of *n* dependent on the acoustic impedance of rock, see Rustan and Vutukuri, 1983. The uniformity coefficient for the Rosin-Rammler distribution varies between 0,7–1,98 for rock fragments after blasting. The higher the value of *n* the more uniform distribution. Harries, 1979.

uniformity index (*n*), **(dimensionless)**, see '**uniformity exponent**'.

unloading the stemming, remove the stemming from the blasthole.

untensioned bolt, a bolt that is not prestressed. TNC, 1979.

upcast shaft, a shaft used for the discharge of exhaust air from the mine to the surface.

up hole, a borehole collared in an underground working place and drilled in a direction pointed above the horizontal plane of the drill machine swivel head. Long, 1920.

uphole drilling, holes drilled upwards from a horizontal plane.

upsetting test, a brisance test, see '**Hess brisance test**' and '**Kast brisance test**'.

USBM (acronym), see '**U.S. Bureau of Mines**'.

U.S. Bureau of Mines, a bureau of the Department of the Interior active in promoting safety in coal mines and in carrying out broad programmes in mining and related fields.

useful energy of an explosive, the energy delivered to the rock mass when the pressure of the gaseous products is above 100 MPa.

V

VCR (*acronym*), see **'vertical crater retreat'**.
V-cut, a cut with holes in a V-layout, also called **'wedge cut'**, see Fig. 67.

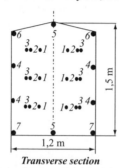 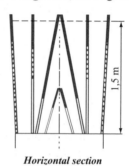

Transverse section *Horizontal section*

Figure 67 V-cut. After Langefors and Kihlström, 1978.

VDE (*acronym*), for **'variable density explosive'**.
vein, a thin orebody. Veins can be classified according to their width: *micro vein* = 0,1–1,0 m, *mini vein* 1,0–2,5 m, and *narrow vein* 2,5–4,0 m. Mairena, 1991.
vein mining, use of different mining methods like *cut and fill stoping, sublevel stoping* or *shrinkage stoping* to mine thin, flat and steep ore bodies. Mass mining methods like *sublevel caving* and *block caving cannot be used*.
velocity (*c*), **(m/s)**, a vector quantity which indicates the time rate of motion. The symbol c is used in blasting for wave propagation velocities. For displacement velocities use v.
velocity of burden (v_b), **(m)**, see **'burden velocity'**.
velocity of detonation (VOD), (c_d), **(m/s)**, the velocity at which the detonation wave travels through a column or mass of explosive. The detonation velocity of an explosive depends on the type of explosive, particle size, density, diameter, packing, confinement and initiation. VOD is an acronym for velocity of detonation, but it should not be used as a symbol in formulas. It may be measured under confined (c_d) or unconfined (c_{du}) conditions. For low explosives the c_d has a range of 1 500–2 500 m/s and for high explosives 2 500–7 000 m/s.
velocity of detonation test, there are several methods for measuring the velocity of detonation, c_d or (VOD), e.g. 1) High speed camera in the laboratory 2) c_d probes (electric wire or optical fibres) and 3) D'Autriche method (historical method), etc.
velocity of sound (c_p), **(m/s)**, see **'seismic waves (P-waves)'**.
venturi loader, see **'jet loader'**.

vertical crater retreat (VCR), a mining method where vertical or inclined parallel holes are drilled and successively detonated with point charges. The blasting direction is against the free face, normally downwards. Craters are formed and new charges are then placed in the holes near the blasted face. The technique is used in underground mines for making raises and blasting pillars and stopes. The fragmentation is normally finer than in bench blasting and under favourable circumstances can be used for automatic loading and transportation of ore or rock. One disadvantage is the large consumption of explosive.

vertical pressure (p_v), (Pa or MPa), pressure parallel to the direction of the field of gravity. It is better to use vertical stress.

vertical retreat mining (VRM), mining methods where the vertical crater retreat blasting method is used. The method is usually used in smaller ore bodies in combination with sublevel shrinkage stoping.

vertical retreat stoping, see '**vertical crater retreat**'. Atlas Copco, 1982.

vibrating wire sensor (stress gauge, stress cell, stress meter or wire gauge), a sensor for measurement of changes in tensile stress applied to a vibrating wire. The frequency of the vibrating wire depends on the tensile stress applied to the wire. Brady, 1985.

vibration, a continuing periodic change in a displacement with respect to a fixed reference point in solid, liquid or air. The mobility of a single mass point on any structure contains six degrees-of-freedom: three translational and three rotational. Fig. 68 defines the system of coordinate axes, the directions and terms used, and it may be noted that 75% of the terms in the mobility matrix involve either rotational velocities or moment-induced frequency response functions. Sanderson and Fredö, 1992.

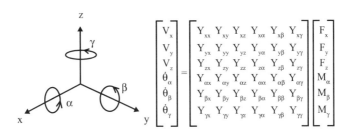

Figure 68 Six degree-of-freedom mobility matrix of ground vibrations. After Sandersson et al., 1992.

vibration control (*ground vibration*), in the near field the main controlling parameters are the linear charge concentration, strength of explosive and the decoupling ratio. In the far field the main controlling parameters are the maximum charge per delay, explosive strength and decoupling factor, the delay time between charges (must be larger than 8 ms to not to be regarded as cooperating charges), type of settlement of buildings, type of building and building construction, number of repeated blast, distance from the blast. The vertical particle velocity must be measured on the basement of the sensible buildings or other constructions to check that they don't overrun set values. In *Germany* the particle velocity must be measured in *three orthogonal directions* and the resultant peak particle velocity calculated. Examples of threshold values for maximum allowed vibration velocities set by authorities varies down to 2 mm/s in Western Europe and 50 mm/s in the US.

vibration damping by presplitting, in principal a presplit crack can reduce the amplitude of ground vibrations. The damping is, however, very much depending on the width of the presplit crack. Devine, 1965.

vibration drilling, drilling in which a frequency of vibration in the range of 100 to 20 kHz is used to fracture rock. Ultrasonic drilling is one of the better-known methods of vibration drilling. Mining and Mineral Engineering, 1965.

vibration measurement, recording of the displacement (u), velocity (v) and acceleration (a) of a particle and/or the frequency (f) and attenuation (Q) of a vibration. Vibration measurements serve to assess the likelihood of damage to structures from blasting.

vibration parameters, those physical quantities that are used to describe the ground vibration; e.g. particle displacement (u), particle velocity (v), particle acceleration (a), frequency (f) and attenuation (Q).

vibration velocity (v), (mm/s), wave induced velocity of a particle in the media.

vibrograph, a vibration measurement instrument used for graphical recording of mechanical vibrations (blast vibrations) in the form of particle displacement (u), particle velocity (v) or particle acceleration (a).

Vickers diamond hardness test, a small impression machine, capable of testing very hard metals, finished components and very thin sheets. The diamond is similar to that used in the *diamond pyramid hardness test*. The duration of application of the load is controlled automatically, being always applied and removed in exactly the same manner. This machine may also be used with a ball indenter for the Brinell hardness test. Ham, 1965.

A comparison between Moh's Hardness scale and Vickers micro hardness scale is made in Table 45.

Table 45 Comparison of two systems for hardness.

Moh's no.	Mineral	Vickers' micro hardness mean (kgmm^{-2})
1	Talc	14
2	Gypsum	40
3	Calcite	125
4	Fluorite	188
5	Apatite	530
6	Feldspar (orthoclase)	788
7	Quartz	1413
8	Topaz	1745
9	Corundum	2428
10	Diamond	–

Other examples of measured values are pyrite 1 300, hematite and rutile 1 000, magnetite and ilmenite 550, and finally silver and galena 500. Indumess, 1988.

Vickers hardness test, see 'Vicker diamond hardness test'. Vickers hardness scale (N/mm^2), The value range is from 0 to 1 800 N/mm^2.

video scope, a video camera, so small in size that it can be inserted into boreholes or cavities. The borehole can be viewed both perpendicularly and axially to the borehole axis. The information recorded is transferred via a fibre optic cable to the collar of the borehole.

virgin stress, natural stress *or* **paleostress (σ_{vir}), (MPa)**, the stress in ground before the start of an excavation. The vertical stress is increasing linearly with depth. The two horizontal stresses often show a difference in magnitude.

viscoelasticity, property of materials which strain under stress partly elastically and partly viscously, i.e. whose strain is partly dependent on time and magnitude of stress. ISRM, 1975. Rock behaves viscoelastic in the seismic zone. Persen, 1975.

viscoelastic model (*Kelvin model*), a rheological model (a spring and a dashpoint element in parallel) used to simulate time-dependent response of rock around supported drifts or tunnels. There is an analytical solution developed for this case. Hudson, 1993.

viscoelastic theory, the theory which attempts to specify the relationship between stress and strain in a material displaying viscoelasticity. Lapedes, 1978.

viscoplastic model (*Bingham model*), a rheological model (a spring in series with a dashpoint and frictional-cohesive element in parallel) used to simulate time-dependent response of rock around supported drifts or tunnels including permanent deformations. There is an analytical solution developed for this case. Hudson, 1993.

Vixen detonation model, a computer detonation model to simulate the energy release of an explosive. Cunningham, 2006.

VOD, (c_d), **(m/s)**, (*acronym*), for '**velocity of detonation**'. Acronyms should not be used in formulas and therefore a symbol for velocity of detonation has been introduced 'c_d'.

void, see '**pore**'. TNC, 1979.

void diffusion, a mechanism of the irregular flow and intermixing of ore or rock from adjacent draw zones which involves the upwards migration, filling and/or collapse of voids formed by large, angular particles.

void ratio (*e*), **(dimensionless)**, the ratio of volume of intergranular voids to volume of solid material in a sediment or sedimentary rock. A.G.I. Supp., 1960. The ratio of the volume of void space to the volume solid particles in a given soil mass. ASCE, 1958.

volcanic rock, effusive rock or **extrusive rock**, any igneous rock derived from a magma or from magmatic materials that was poured out or *ejected at the earths surface*, as distinct from an intrusive or plutonic igneous rock which has solidified from a magma that has been ejected into older rocks at depth without reaching the surface. A.G.I., 1957.

Volt (*U*), **(V)**, the difference in potential required to make an electric current flow of 1 Amp through a resistance of 1 Ω. It is the unit of electromagnetic force.

volume (*V*), **(m³)**, the space enclosed by a closed surface.

volume coupling ratio (R_{cv}), **(dimensionless)**, ratio of the explosive volume in the blasthole to the total volume of the blasthole possible to charge. This term may be used when there are both axial and radial coupling in a blasthole.

volume decoupling ratio (R_{dcv}), **(dimensionless)**, ratio of the blasthole volume to the explosive volume charged into the blasthole. This term may be used when there are both axial and radial decoupling in a blasthole. Bergman et al., 1973.

volume density (V_d), **(dimensionless)**, the fractional volume of a phase in a unit volume of a structure. Weibel, 1989.

volume density of joints (f_{j3D}), **(No./m³)**, the number of joints per unit of volume.

volume of apparent crater (V_{ac}), **(m³)**, see '**crater blasting**'.

$$V_{ac} = \pi d_{ac} \frac{L_{ac}}{8} \qquad (V.1)$$

where V_{ac} is the volume of apparent crater in (m³) assuming a parabolic shape of crater, d_{ac} is the diameter of apparent crater in (m) and L_{ac} is the apparent crater depth in (m).

volume of explosion, the volume occupied by the gases of one kilo of explosives under normal conditions (temperature and pressure).

volume property (*rock mass*), rock properties depending on discontinuities such as, modulus of deformation, secondary porosity, permeability of the rock mass, discontinuity frequency, RQD and rock mass classification indexes. Hudson, 1993.

volume rate of flow (q_v), **(m³/s)**, volume flow of fluid or gas per unit of time.

volume strength or *cartridge strength*, see '**bulk strength**'.

volumetric joint count (J_v), **(No. of joints/m³)**, the sum of the numbers of joints per meter for each joint set present. ISRM, 1978. When a drill core is unavailable, the RQD can be estimated from the numbers of joints per unit volume (volumetric joint count), in which the numbers of joints per meter for each joint set is added. A simple relation can be used to convert this number into RQD for the case of clay-free rock masses:

$$RQD \approx 115 - 3,3\, J_v \qquad (V.2)$$

where J_v is the volumetric joint count in (No./m³). After Palmström, 1985.

volumetric RQD (R_{RQDV}), **(dimensionless)**, is defined as

$$R_{RQDV} = 100 \sum_{i=1}^{n} \frac{V_i}{V} \qquad (V.3)$$

where R_{RQDV} is the volumetric RQD (dimensionless), V_i is the volume of the intact matrix block equal to or exceeding 0,001 m³ in size in m³, and V is the volume of the rock mass in m³. Kazi, 1985.

volumetric strain, (*e*), **(dimensionless)**, see '**dilatation**'.

VRM (*acronym*), see '**vertical retreat mining**'.

wagon drill, see '**wagon drill for bench drilling**'.
wagon drill (bench drilling), a drilling machine mounted on a light, wheeled carriage. B.S. 3618, 1964. Simple type of drill rig used for drilling down holes in mining and small benches. Atlas Copco, 2006.
wall control blasting, controlled contour blasting of the walls in a drift, tunnel or open pit. See '**controlled contour blasting**'.
wall plate, footing placed on the rock shelf to temporarily support the arch ribs of tunnel sets. Bickel, 1982.
wall rock, the rock forming the walls of a vein or lode. Gregory, 1980.
warning signal, a visual or audible signal that is used for warning personnel in the vicinity of the blast area of the impending explosion. Atlas Powder, 1987.
warwick, see '**derrick**'. Mason, 1951.
wash boring, drilling by the use of water jet applied inside a casing pipe, in unconsolidated ground. Pryor, 1963. A test hole from which samples are brought up mixed with water. Nichols, 1962.
washer, plate used to transfer forces from the rock bolt to the surface of rock.
waste, the barren rock in a mine. It is also applied to the part of the ore deposit that is too low in grade to be of economic value at the time, but this material may be stored separately in the hope that it can be profitably treated later. Lewis, 1964.
waste to ore ratio ($R_{w/o}$), **(dimensionless)**, a term used in surface mining used to describe how much waste rock has to be removed to be able to mine the ore. It is an important ratio in surface mining economy. Normally the ratio is between 1 to 4 but it could be as high as 5.
waste rock, barren or sub-marginal rock which has been mined but is not of sufficient value to warrant treatment, and is therefore removed ahead of the milling process. Pryor, 1963.
water-cement ratio ($R_{w/c}$), the mass ratio between water and cement in a concrete- or cement slurry.
water content (w_V), **(dimensionless)**, the ratio of the water volume in a rock to the total volume of the rock.
water content (w_m), **(%)**, mass of water which can be removed from soil, usually by drying, expressed as a percentage of the dry mass. SS-EN ISO 14688-2:2004, 2003.
water gel blasting agent, see '**slurry explosive**'.
water gel explosive, see '**slurry explosive**'.
water hammer drill (machine), a drilling machine using water as a medium for transport of energy to the drilling machine and the drilling piston. The machine is manufactured in

the same diameter as the hole being drilled. The advantages are many; fast penetration, no dust delivered to the air, less noise at the drilling platform because the machine is located in the borehole and straighter holes.

water head-, hydraulic- or **permeability measurement**, a method utilized to determine the hydraulic conductivity of rock. Single packers may be used for the pressurization of the whole blasthole and double packers for pressurization of only parts of the blasthole. The method can be used to determine the physical connections between two or several blastholes. The connections are dependent on cracks and joints, and can be determined by activating a water pressure in one borehole or part of the borehole and determining the leakage to the surrounding boreholes. The method is useful for quantification of blast induced rock damage. Rustan, 1993.

water infusion gun, a special tube which acts as a borehole seal in the water infusion process. The tube has two separate passages; one for the infusion water and the other admits hydraulic fluid to actuate the piston expanding the seal in the borehole. The hydraulic fluid is supplied by a hand pump and the infusion water by a power pump. Nelson, 1965.

water in oil emulsion explosive, see '**emulsion explosive**'.

water jet cutting, pressurized water is applied to rock through a fine nozzle at high velocity. The penetration effect may induce rock fragmentation. The degree of penetration is improved a lot by adding some abrasive material to the water. Water jets can be used to make slots or to cut rock boulders to smaller pieces.

water-jet drilling or **erosion drill**, the drilling of boreholes in unconsolidated or earthy formations using the erosive power of a small-diameter stream of water forcefully ejected as the cutting tool.

The pressure used in the jet is about 15 to 40 MPa. A laboratory test gave a drilling speed of 10 m/h for a 80 mm hole in granite. KTH, 1969.

water-jet notching, water jet can be used to create two diametrically opposite longitudinal notches along the borehole. In controlled contour blasting this will facilitate the cracking between the blastholes. It is also possible to create a disc crack at the bottom of the borehole to improve the breakage at the bottom.

water jet slitting, see '**water jet notching**'. Honda, 1984.

water Leyner, a type of rock drill in which water is fed into the drill hole through the hollow drill steel, to remove the drill cuttings, and at the same time allay the dust. Also known as Leyner-Ingersoll drill. Fay, 1920.

water loss measurement, see '**water head measurement**'.

water mist system, a system consisting of a water pressure tank that is pressurized by the drill rig air system through a hose and with an adjustable nozzle equipment and used to *suppress dust*. It is also useful in *stabilizing the drill hole* walls when drilling through the first meters of soft rock or overburden. Atlas Copco, 2006b.

water powered drilling, mechanical drill machine powered by high pressure water instead of compressed air or oil. The machines are normally used for in-the-hole drilling. The drilling costs are reduced and the environment improved because of less oil in the air. Jeansson, 2002.

water pressure test, see '**water head test**'.

water reel, drum like structure, on a drill rig, around which the water hose is wrapped when tramming the rig. Atlas Copco, 2006.

water resistance, the ability of an explosive to withstand the desensitizing effect of water penetration. The water resistance of an explosive depends on the density of the explosive, the exposure time to water, and the water pressure. The water resistance can be quantified by a water resistance number characterizing the period of time during which the explosive will still detonate.

water resistance number (USA), the ability of an explosive to withstand the exposure to water without being deteriorated or becoming desensitized. In the USA, the classification scheme shown in Table 46 is used (after the explosive has been set under a 25,4 mm high water column). Meyer, 1977.

Table 46 Classification of water resistance of explosives. After Meyer, 1977.

Water resistance	Time (hours)
Satisfactory	24
Acceptable	8
Poor	<2

A more detailed classification, Atlas Powder, 1987, is compiled in Table 47.

Table 47 Classification of water resistance of explosives. After Atlas Powder, 1987.

Class	Hours
1	Indefinitely
2	32–71
3	16–31
4	8–15
5	4–7
6	1–3
7	<1

water-resistant detonator, a detonator supplied with seals to prevent water from penetrating into the detonator even under elevated water pressure.

water separator, device that is used to separate water from air. Atlas Copco, 2006.

water shock wave, pressure wave in water caused by one or several explosive charges being suspended in water or placed in boreholes under water. As the shear resistance of water is zero, only longitudinal waves (P-waves) can be transmitted through water. The pressure pulse from an explosive suspended in water is considerably higher than if the explosive is confined in a borehole. The maximum water pressure is about 10 times higher for a charge placed in water compared to placed inside a borehole in rock. The influence of water pressure waves on the displacement in underwater constructions such as dam gates is more dependent on the impulse (pressure-time history) than the maximum water pressure. Bergman, 1989.

water swivel, device used to connect a water supply to a rotating rod or shaft. A water swivel is used in diamond drilling to connect the pressure hose from the flush pump to the rotation drill string. Atlas Copco, 2006.

waterways, grooves or faces manufactured into drill bits to direct the flow of flushing fluid in such a way that cuttings are effectively removed. Atlas Copco, 2006.

water well casing tubes, casing tubes are used for the '*Tubex method*' when drilling for water. Sandvik, 2007.
water well drill bit, bit often used for water well drilling. Sandvik, 1983.
water well drilling rig, drill rig normally mounted on a commercial truck, designed specifically for drilling water wells. Atlas Copco, 2006.
water well rig, see '**water well drilling rig**'.
water stemming, plastic tubes or bags filled with water and inserted into the blasthole after the explosive has been charged into the blasthole. Water is used as a stemming material instead of drill cuttings, clay or sand. The use of water stemming increases the breaking efficiency of the explosive. This technique is used in permissible, boulder and crater blasting.
water stemming bag, a plastic bag containing a self-sealing device which is filled with water. Classified as a permissible stemming device by the Mine Safety and Health Administration (MSHA) in the USA. USBM, 1983.
watt (*P*), (**W**), energy per unit of time.
wave length, (λ), (**m**), the distance between two points having the same phase in two consecutive cycles of a periodic wave, along a line in the direction of propagation.
wave parameters, those physical quantities that are used to describe wave motion. These parameters are amplitude, period, frequency, wave length, etc.
weakness plane or **weakness zone**, see '**plane of weakness**'.
wear, a process by which material is removed from one or both of two surfaces moving in contact with one another, for example, abrasion. Most wear phenomena have the common characteristic, namely, the mechanical overstressing of the surface material. wear may be classified according the manner in which the overstressing occurs. In single-sided wear or erosion, the contacting medium is a fluid, while double-sided wear is characterized by the presence of two mating solid surfaces. Osborne, 1956.
wear gauge, cemented carbide inserts are worn differently by different kind of rock formations. *Gauge wear* is used to explain that inserts are worn on parts parallel to the direction of drilling. Opposite is *frontal wear*. Sandvik, 2007.
wear plate, replaceable plate intended to sustain wear while protecting a structure. Such as the wear plates on scooptram buckets. Atlas Copco, 2006.
wear resistance, see '**abrasion hardness**'.
weathering, the process of disintegration and decomposition of rock as a consequence of exposure to the atmosphere, to chemical action and to the action of frost, water and heat.
wedge anchor, see 'wedge bolt'.
weather-resistant, construction designed to offer reasonable protection against weather.
web, rock mass between presplitting holes.
wedge bit, a tapered-nose non-coring bit used to ream out the borehole alongside the steel-deflecting wedge in hole-deflection operations. Also called *bull-nose bit*, *wedge reaming bit*, *wedging bit*. Long, 1920.
wedge bolt, a bolt designed for 'the use in roof bolting. It consists of a 19 mm diameter rod, at one end of which is a cold rolled threaded portion, the other end being shaped to form a solid wedge forged integrally with the bolt. Over this wedge is fitted a loose split sleeve of 38 mm external diameter. The anchorage is provided by the bolt being placed in the hole and the bolt is pulled downwards while the sleeve is held by the thrust tube. Split by the wedge head of the bolt, the sleeve expands until it grips the side of the hole.

Nelson, 1965. Another construction of a wedge bolt is to make a slot at the inner end of the rock bolt. A wedge is placed into the slot and when the wedge hits the bottom the slot will expand and fasten the rock bolt. Also called 'sleeve bolt'. See also '**grouted wedge bolt**'.

wedge cut, see '**V-cut**'.

wedge guides, an arrangement to arrest the cage or skip in the event of an overwind with a multirope friction winder. The cage is forced by its momentum into the wedge guides. The frictional force, gradually increased by virtue of the wedging action of the guides, is relied on to bring the cage to rest and hold it in a stationary position with the aid of jack catches. See also '*detaching hook*'. Nelson, 1965.

wedge plug, wedge that is driven between the feathers to mechanically split rock. Atlas Copco, 2006.

wedge reaming bit, see '**wedge bit**'.

wedge scaling, loosening of blocks in the wall and roof by wedges instead of using impact hammers.

wedge set, set consisting of two feathers and a wedge used for mechanically splitting of rock. Atlas Copco, 2006.

wedging, a method used in quarrying whenever the object is to obtain large, regular blocks of building stone such as syenite, granite, marble, sandstone, etc. In this method a row of holes is drilled either by hand or by pneumatic drills close to each other so that a longitudinal crevice is formed into which a gently sloping steel wedge is driven. Usually several wedges are driven in, and the block of stone can be detached without shattering. Stoces, 1954.

wedging bit, see '**wedge bit**'.

Weibull distribution, a mathematical distribution function used to describe rock fragment distributions (3 parameters are needed). The Rosin-Rammler-Sperrling distribution, the RRS-distribution, often used for rock fragments distributions, is a special kind of the Weibull distribution (2 parameters are needed). The Weibull distribution function is recommended to be used together with the Kuznetsov k_{50} fragmentation formula to calculate the fragment size distribution function instead of the Rosin Ramler distribution being advised by Cunningham for the KuzRam formula. The advantage is that the amount of fines will increase with this function and that is good because existing distribution functions underestimate the amount of fines.

$$x_c = \frac{k_{50}}{\gamma(1+1/n)} \tag{W.1}$$

where x_c is the characteristic size in m, k_{50} is the mean fragment size in m, γ is the gamma function, see Zhang and Jin, 1996 and n is the uniformity coefficient. Spathis, 2004.

Weibull distribution function (2 parameter), the cumulative distribution function reads as follows

$$F(x; k, \lambda) = 1 - e^{-\left(\frac{x}{\lambda}\right)^k} \quad \text{for } x \geq 0 \tag{W.2}$$

and $F(x; k, \lambda) = 0$ for $x < 0$. k is the *shape parameter* and λ is the *scale parameter*. Weibull, 1939.

Weibull distribution function (3 parameter), the cumulative distribution function reads as follows

$$F(x;\ k,\ \lambda,\ \theta) = 1 - e^{-\left(\frac{x-\theta}{\lambda}\right)^k} \quad for\ x \geq 0 \tag{W.3}$$

and $F(x;\ k,\ \lambda,\ \theta) = 0$ for $x < 0$. k is the *shape parameter* and λ is the *scale parameter* θ. Weibull, 1939.

weight of explosive charge **(Q), (N)**, *erroneously measured in* (kg), see and use '**mass of explosive charge**'.

weight rod, thick walled tubes mounted directly above the drill bit or core barrel to provide the force required to make the bit drill. *Weight rods* or *drill collars* are used on conventional surface, deep hole diamond drilling rigs employing a draw works and running sheave arrangement. Also called a drill collar. Atlas Copco, 2006.

weight strength or *absolute weight strength AWS*, (s_w), *erroneously measured in* (MJ/kg), should be called '**mass strength**' respectively '**absolute mass strength**' (AMS), see '**absolute mass strength**'.

weight strength (defined by Langefors), (s_{wL}), **(dimensionless)**, the sum of weighted explosive energy and explosive gas volume according to the following formula;

$$s_{wL} = \frac{5}{6}\frac{Q}{Q_r} + \frac{1}{6}\frac{v_e}{v_{er}} \tag{W.4}$$

where s_{wL} is the weight strength defined by Langefors (dimensionless), Q is the heat of explosion for the studied explosive in MJ/kg and $Q_r = 5$ MJ/kg is the heat of explosive for the reference explosive (Swedish LFB Dynamite) and v_e is the specific gas volume for the studied explosive in m³/kg and $v_{er} = 0,85$ m³/kg the specific gas volume for the reference explosive (Swedish LFB Dynamite) both at standard temperature and pressure.

weir board, is a box with an opening at on end in form of a V where the water can leave the box. The higher flow of drillwater into the box the higher will the level in the box be. After calibration the flow rate can be determined. Stein, 1977.

weld mesh, a more or less rigid steel net consisting of reinforced bars with a diameter from about 4,2 mm laid in a squared or a rectangular net and welded together forming meshes from about 15 × 15 cm and used for rock and concrete reinforcement. Is used in combination with rock bolts and shotcrete.

well, a shaft or hole sunk into the earth to obtain oil, gas, water etc. Webster, 1961.

well drill, see '**churn drill**'.

wet blasting, shot firing in wet hole. Special explosives are available for wet conditions and the detonator wires must be well insulated to prevent short-circuiting and misfires.

wet muck flow, the sudden collapse and rapid run-out of wet granular material, usually from a drawpoint.

wheel loader, a loader with a bucket large enough to be used for loading, haulage and dumping **(LHD)**, see '**load, haul and dump unit**'.

wheel mounted jumbo, a carrier on rubber tires with one or several hydraulic booms used to drill blastholes in headings. Fraenkel, 1952.

whipstock, a wedge used to deflect the drill bit in the drilled hole. Stein, 1997.

whizzer mill or *Jeffrey crusher*, a '**swing hammer crusher**', used to break soft rocks. Pryor, 1963.

wide spacing blasting (R_{S_b}/B_b), **(dimensionless)**, a blasting method where the ratio between blasted spacing S_b and blasted burden B_b is large, *about* 4–8.

width of crack (*b*), **(mm)**, the width of a crack is defined by the perpendicular distance between the crack walls for parallel cracks. ISRM, 1975. For diverging a crack it is the shortest distance between the two walls.

width (*W*), **(m)**, width of fragment, see '**fragment size**'.

winch, a small hand- or power-operated drum haulage for light duties on the surface or underground in mines. A heavy-duty power winch fitted to the rear of a tractor. Nelson, 1965.

winder, see '**winding engine**'.

winding engine, the steam or electric engine at the top of a shaft which rotates the winding drum and thus hoists and lowers the cage or skip by means of a winding rope. In metal mining, the winding engine is usually called a hoist. Also called '*winder*'. Nelson, 1965.

winding guides or **shaft guides**, the purpose of winding guides is to permit winding to proceed safely at relatively high speeds by preventing collisions between the cage and between cage and side of fittings in the shaft. They must be rigid enough to prevent material deviation of the cages or skips from the vertical, strong, since a broken guide causes danger from damage, smooth, so as to offer as little resistance to the movement of the cages as possible, and firmly supported and maintained vertical. They may be of two types, rigid or flexible. The former may be of timber, and in new shafts have generally been replaced by steel channels, steel rails or angles, the latter are steel ropes of round or semi locked section steel rods.

windlass, a device used for hoisting limited to small scale development work and prospecting because of its small capacity. Lewis, 1964.

wing coupling, hose coupling having two protrusions that are struck with a hammer to tighten or loosen it. Atlas Copco, 2006.

wing crack, a crack together with short angled extensions at the ends. Reyes and Einstein, 1991.

wing nut, nut having two protrusions allowing it to be screwed on and off by hand without the need for a tool. Atlas Copco, 2006.

winze, a vertical or inclined opening sunk from a point inside a mine for the purpose of connecting with a lower level and exploring the ground for a limited depth below a level. Atlas Powder, 1987.

winzing, the process of making a downwards shaft within an underground mine.

wireless initiation, see '**electromagnetic initiation**'.

wireline core barrel, type of core barrel used mainly in diamond drilling in which the rock cores can be recovered by running a wire line down the inside of the drill tubes. The wire line is equipped with a special tool that engages the inner barrel which is extracted to surface. Wireline core barrels save time as cores can be recovered several times before it is necessary to pull the rods and bit. Atlas Copco, 2006.

wire mesh, a method used to avoid the fall of larger rock pieces from a rock surface both underground and on road cut slopes on the surface. The size of the smallest piece that can fall out is dependent on the mesh size. The diameter of the wire mesh steel is ~2–3 mm and the mesh is delivered on rolls. The wire mesh has no rock reinforcement capacity if it is not combined with shotcrete.

wire mesh reinforcement, expanded metal, wire or welded fabric used as reinforcement for concrete or mortar. In general a reinforcement mat with diameter ~5 mm will be about 3,3 m long and 1,5 m wide and is installed in both roof and walls down to floor level. It is hang up on rock bolts or special bolts. The spacing between rock bolts is about 75 cm.

wire meshing or **screening**, the process of reinforcing the rock surface by a steel net which can hold smaller stones in place or by using reinforcement steel mats 5 mm in diameter which are fastened to the rock by rock bolts and shotcreted. It is a time consuming operation. Common practise in Canada and Australia. Atlas Copco, 2007.

wire rope, thin steel strands wound around a central core to form a rope. Atlas Copco, 2006. A steel wire rope used for winding in shafts and underground haulages. Wire ropes are made from medium carbon steels. Nelson, 1965.

wire rope lacing with Swellex, Swellex bolts can be combined with cables in the Swellex hole and connected to another Swellex bolt. A steel net can be attached inside the cables on the outside to prevent smaller stones to drop. Stillborg, 1994.

wood cribs, see '**cribs**'.

Woods test, a heat sensitivity test of an explosive where a small amount of the explosive is placed in a 75°C melt of Woods metal. The temperature is increased until the reaction occurs. The purpose of the test is to determine the minimum temperature needed for a reaction to start in the explosive. This test is compulsory for the United Nations classification. Persson et al., 1994.

workbar rig, a small diamond or other rock drill designed to be mounted and used on a bar. Also called '**bar drill**'. Long, 1920.

working face, the face of a drift or tunnel. TNC, 1979.

working level, the level where active work is done.

work or **energy** (W), (J or (Nm)), a derived quantity defined by force times distance.

work of fracture, **specific fracture energy,** *specific energy of rupture* or *energy of rupture* (w_f), (J/m^3), the work done per unit volume in producing fracture.

workbar rig, a small diamond or other rock drill designed to be mounted and used on a bar. Also called '**bar drill**'. Long, 1920.

working face, the face of a drift or tunnel. TNC, 1979.

working level, the level where active work is done.

world stress map (WSM), a map showing the direction of the maximum horizontal stress in the ground at different parts of the world. Usually more measurements have been made in the tectonic active areas. Stephansson, 1993.

WR (acronym), for '**water resistant**'.

wrapper, the cartridging material, e.g. paper, plastic or cardboard.

WSM (acronym), see '**World stress map**'.

X-bits, percussion drill bit with four hard metal inserts and 75° respectively 105° angle between the inserts. Commonly used for diameters ≥64 mm diameter to ensure circular holes, since in some cases cross bits tend to produce grooves and give pentagonal holes. Atlas Copco, 1982.

yield, the condition when a material cease to be elastic and when further deformation of the material becomes irrecoverable. Brady, 1985.

yield angle, see '**angle of yield**'.

yield loss, the difference between the actual yield of a product and the yield theoretically possible (based on the reconstituted feed) of a product with the same properties (usually percentage of ash). B.S. 3552, 1962. Also called washing error (undesirable usage).

yield of broken material m **(mass)** or V **(volume), (kg** or **m³)**, the mass or volume of broken material.

yield of debris, of mass or **of rock** m **(mass)** or V **(volume), (kg** or **m³)**, see '**yield of broken material**'.

yield pillar system, a method of roof control in coal mining whereby the natural strength of the roof strata is maintained by the relief of pressure in working areas and the controlled transference of load to abutments, which are clear of the workings and roadways. The method consists in making certain coal pillar yield in small amount. See also double packing Nelson, 1965. This method can also be used in other mining.

yield point or **stress**, the point or stress where plastic deformation of the material starts.

yielding arches, steel arches installed in underground openings as the ground is removed. They are employed to support load caused by changing ground movement or faulted and fractured rock. They are designed so that when the ground load exceeds the design load of the arch as installed, yielding takes places in the joint of the arch, permitting the overburden to settle into a natural arch of its own and thus tending to bring all force into equilibrium. The arc is stronger after this yielding than before because of increased joint overlap. Lewis, 1964.

yielding floor, a soft floor in coal mining which heaves and flows into open spaces when subjected to heavy pressure from packs or pillars. See also '**creep**'. Nelson, 1965.

yielding prop, a steel prop in coal mining which is adjustable in length and incorporates a sliding or flexible joint which comes into operation when the roof pressure excess a set load or value. See also '**hydraulic chock**', '**hydraulic prop**'. Nelson, 1965.

yielding rock bolt, see '**Swellex bolt** and **Roofex bolt**'.

yielding support, support that can be deformed without breakage.

yield strength (σ_y), (MPa), the stress at which a material exhibits a specified deviation from proportionality of stress and strain. An offset of 0,2% is used for many metals. ASM Gloss, 1961.

yield stress (σ_y), (MPa), the stress beyond which the induced deformation is not fully annulled after complete destressing. ISRM, 1975.

Young's modulus (E), (GPa), the ratio of stress to strain induced for a solid material under given loading conditions, numerically equal to the slope, (tangent) of the linear part of the stress-strain curve. The modulus of elasticity is defined for materials that deform in accordance with Hook's law, and the modulus of deformation for materials that deform otherwise. The modulus of elasticity for rock ranges between 5 GPa (in coal seams) and up to 150 GPa (in dense hematite iron ore). The modulus of elasticity can be calculated from;

$$E = 2\mu(1+v) = c_p^2 \rho = 2\rho c_s^2(1+v) \tag{Y.1}$$

where E is the elastic modulus in (Pa), μ is the shear modulus, ρ is the density of the material (kg/m³), c_p and c_s are the P- and S-wave velocities., and v is the Poisson's ratio. Extreme values for Young's modulus are 20 GPa for cement paste and 1 000 GPa for diamond. For hard rock, Young's modulus ranges in the order of 50–100 GPa. Four different Young's modulus can be measured 1) The tangent to the stress strain curve starting at the origin of coordinates also called '*tangent Young's modulus*' (E_t) 2) *Tangent modulus at a certain stress*, normally half the failure stress (E) 3) '*Secant modulus*' (E_s) is defined as the straight line between origin of coordinates and any chosen point on the stress-strain curve and 4) Effective Young's modulus, see '**effective Young's modulus**'.

ytterbium stress gauge, a stress sensor based on the change of the resistance of the rare earth metal *ytterbium* at different pressures. The sensors are used to measure the pressure in the rock at different distances from a blasthole. Grady, 1980b. See also '**piezoresistance dynamic pressure transducers**'.

Zener rheological model, this *linear viscoelastic model* is a combination of the Hook and the Kelvin rheological models. See also '**rheological models**'.
zero degree of freedom, see '**presplitting**'.
zone of disturbance (R_{ezd}), (m), see '**excavation zone of disturbance**'.
zone of influence, generally, the zone surrounding an excavation in which the pre-existing stresses are perturbed by a given amount (often taken as 5%) by the presence of the excavation. In block caving, the term is also used to describe the zone surrounding the cave in which the rock mass response is influenced by the presence of the cave.

List of symbols

Symbols

Note: The most important symbols are marked with an asterix (*).

A

	A	abrasiveness, abrasive hardness or hardness (mm or m)
	A	advance (m/day or m/month)
*	A	amplitude (mm or m)
*	A	area (m²)
*	A	cross sectional-, tunnel profile-, tunnel section-, drift profile-, drift section-area (m²)
	A	rock mass factor (dimensionless) in the Kutznetsov formula for k_{50} calculations (dimensionless)
	A	second moment of area (Nm)
	A_a	amplitude of particle acceleration (m/s² or mm/s²)
	A_b	drill boom coverage area (m²)
	A_d	amplitude of particle displacement (m or mm)
	A_{dc}	ground vibration amplitude from a decoupled charge (mm)
	A_{fc}	ground vibration amplitude from a fully coupled charge (mm)
	A_{max}	boring stroke in full face boring or maximum amplitude of a wave (m)
	A_o	overbreak area (m²)
	A_{oa}	average overbreak area (m²)
	A_r	advance per round or pull (% of drilled depth or m/round)
	A_u	undercutting rate (horizontal undercut area) (m²/month)
	A_v	amplitude of particle velocity (mm/s or m/s)
*	a	acceleration (mm/s² or m/s²)
	a_{max}	maximum acceleration (mm/s² or m/s²)

B

*	B	burden (m)
*	B_b	blasted burden *true or effective burden*
*	B_c	critical burden in bench blasting or burden (depth) in crater blasting (m)
*	B_d	drilled burden (m)
*	B_{max}	maximum burden (m)
*	B_{opt}	optimum burden (m)
*	B_{optb}	optimum breakage burden (m)

Symbols

*	B_{optf}	optimum fragmentation burden (m)
*	B_p	practical burden (m)
*	B_r	reduced burden (m)
	b	width of a crack or crack width (mm or m)
	b_m	crack mouth opening displacement (mm)
	b_{max}	maximum width of a crack (mm or m)
	b_{mt}	crack opening displacement at time t (mm)
	b_{sb}	specific boring (bored $m/m^3 = 1/m^2$ or m/t)
	b_{sd}	specific drilling (drilled $m/m^3 = 1/m^2$ or m/t)

C

	C	coefficient of consolidation (soil), (m^2/s)
*	C	cohesion, cohesive stress or intrinsic shear strength (MPa)
	C	specific heat (MJ/kgK) where K is the temperature in degree Kelvin.
	C_c	coefficient of curvature (dimensionless)
	C_m	mean specific heat (MJ//kgK) where K is temperature in degree Kelvin.
	C_p	specific heat at constant pressure (MJ/kgK) where K is temperature in degree Kelvin
	C_T	specific heat at constant temperature T (MJ/kgK) where K is temperature in degree Kelvin
	C_U	uniformity coefficient (dimensionless)
	C_V	specific heat at constant volume (MJ/kgK) where K is temperature in degree Kelvin
*	c	electromagnetic or mechanical wave propagation velocity (m/s)
*	c	rock constant (kg/m^3). A blastability factor used in formulas by Langefors and Kihlström, 1978
	c_c	crack-, fracture- or fissure propagation velocity and fracture growth velocity (m/s)
*	c_d	velocity of detonation (m/s)
	c_{dc}	confined velocity of detonation (m/s)
	c_{di}	ideal velocity of detonation (m/s)
	c_{dni}	non-ideal detonation velocity (m/s)
	c_{ds}	steady state velocity of detonation (m/s)
	c_{du}	unconfined velocity of detonation (m/s)
*	c_h	hydraulic conductivity or hydraulic permeability (m/s)
*	c_L	Love wave velocity (m/s)
*	c_P	P-wave-, sonic- or longitudinal wave velocity (m/s)
	c_{Pi}	P-wave velocity measured in situ (field) (m/s)
	c_{Pl}	P-wave velocity measured in the laboratory often in drill cores (m/s)
	c_{Po}	P-wave velocity in homogeneous material free of prestresses and without natural physical discontinuities or blast induced cracks (m/s)
	c_R	Rayleigh wave propagation velocity (m/s)
	c_{ra}	radar wave propagation velocity (m/s)
	c_S	S-wave – or transverse wave propagation velocity (m/s)
	c_{sh}	propagation velocity of shock wave (m/s)
	c_{son}	sonic wave velocities (m/s)

c_{sub} subsonic speed (m/s)
c_{sup} supersonic speed (m/s)
c_t transient velocity (m/s)

D

* D drillhole- or borehole deviation (degree, % or m/m)
D modulus of deformation (MPa)
D_a alignment deviation error in drilling (angular or in-the-hole), (degrees, mm/m or %)
D_b bending deviation error in drilling (mm/m or %)
D_c collaring deviation error in drilling (m or %)
D_e equivalent support dimension (m)
D_f fractal dimension (dimensionless)
D_{fs} surface fractal dimension (dimensionless)
D_i blast damage index (BDI) defined Yu and Vongpaisal, (dimensionless)
D_{ib} blast damage index or damage index at blasting defined by Paventi (dimensionless)
D_{scaled} scaled distance (units depend on the definition of scaled distance)
D_{set} setting-out deviation (error) in drilling (m)
D_{str} damage parameter (based on strength) $0 < D_{str} < 1$ (dimensionless)
D_T total borehole deviation or drilling accuracy (degrees, m or %)
D_{vol} damage parameter (based on volume) $0 < D_{vol} < 1$ (dimensionless)
* d diameter (mm)
d diameter of borehole or blasthole (mm or m)
d_{ac} diameter of apparent crater in crater blasting (m)
d_c diameter of cartridge (mm or m)
d_{cc} critical diameter of cartridge (mm or m)
d_e diameter of explosive charge (mm or m)
d_{ex} external diameter (mm or m)
d_i inside diameter of a bit or set inside diameter also called minor- or box diameter (mm)
d_{lip} diameter of lip of crater (m)
d_o outside diameter or set outside diameter of a bit (mm)
d_{opt} optimal blasthole diameter (mm or m)
d_p permissible diameter (mm or m)
d_{pz} diameter of plastic zone after crater blasting (m)
d_r diameter of relief hole or relief hole diameter in a cut (mm or m)
d_{rup} diameter of rupture zone after crater blasting (m)
d_s diameter of shaft, winze or raise (m)
d_{tc} diameter of true crater after crater blasting (m)

E

E east (cardinal point)
* E modulus of elasticity or Young's modulus (MPa)

	E_d	dynamic modulus of elasticity or dynamic Young's modulus (MPa)
	E_{dec}	decrease Young's modulus (MPa)
	E_e	effective Young's modulus or effective modulus of elasticity (MPA)
	E_F	flexural modulus of elasticity (MPa)
	E_{FC}	flexural modulus of elasticity for core specimen (MPa)
	E_{FR}	flexural modulus of elasticity for rectangular specimen (MPa)
	E_{qd}	quasi dynamic modulus of elasticity or quasi dynamic Young's modulus (MPa)
	E_s	secant modulus of elasticity (MPa)
	E_t	modulus of rupture in torsion (MPa)
	e	dilatation or volumetric strain (dimensionless)
*	e	natural logarithm (dimensionless)
*	e	void ratio (dimensionless)
	e_c	critical void ratio (dimensionless)

F

*	F	force, load or thrust (N or MN)
	F	shape factor (dimensionless)
	F_b	force of blow from drill hammer on the drill steel or rod (N or MN)
	F_{bi}	force on the drill bit or bit load (N)
	F_c	crack extension force (N)
	F_f	feed force on drill machine at drilling (N)
	F_{max}	maximum force (N or MN)
	F_n	normal force (N or MN)
	F_t	tangential force (N or MN)
*	f	confinement, degree of confinement, fixation factor (dimensionless)
	f	factor or coefficient (dimensionless)
*	f	frequency (Hz)
	f_b	degree of confinement dependent on the bench slope in blasting (dimensionless)
	f_{bi}	inclination factor for the blastholes (dimensionless)
	f_{bo}	boulder frequency (No./1000 t or No./1000 m^3)
	f_c	frequency of cracks (No./m)
	f_d	frequency of discontinuities (No./m)
	f_{dm}	mean frequency of discontinuities (No./m)
	f_H	hardness (dimensionless)
	f_i	frequency of incoming blast wave (Hz)
	f_j	frequency of joints (No./m)
	f_{JCS}	joint wall compressive strength (MPa)
	f_{j3D}	3-D volume density of joints (No./m^3)
	f_{JPO}	joint plane orientation (dimensionless)
	f_{JPS}	joint plane spacing (dimensionless)
	f_{RMD}	rock mass description (dimensionless)
	f_{RQD}	RQD-value (Rock quality designation) (dimensionless)
*	f_n	natural frequency or natural vibration frequency of a structure (Hz or cycles/s)

f_{nb}	natural frequency of a building (Hz)	
f_{nrc}	the natural frequency for reinforced concrete (Hz)	
f_o	overbreak factor (%)	
f_{sei}	seismicity (No./year and 100 km^2)	
f_{SGI}	specific gravity influence (dimensionless)	
f_{SRF}	stress reduction factor (dimensionless)	
f_t	tunnel-, drift- or drivage confinement (m^{-1})	
f_x	ground vibration frequencies induced by a blast (Hz)	

G

*	G	shear modulus of elasticity, modulus of elasticity in shear, rigidity modulus or modulus of rigidity (MPa)
	G	strain energy release rate (crack resistance, crack resistance energy, energy of rupture, fracture energy or fracture surface energy) (J/m^2)
	G_c	critical strain energy release rate (J/m^2 or MJ/m^2)
	G_m	approximate potential strain energy release rate (J/m^2)
	G_I	strain energy release rate for a mode I loaded crack (J/m^2 or MJ/m^2)
	$G_{Ic}, G_{IIc}, G_{IIIc}$	critical strain energy release rate for a mode-I, mode-II and mode-III loaded crack, respectively (J/m^2 or MJ/m^2)
*	g	acceleration due to gravity (m/s^2)

H

*	H	height (m)
	H	height of specimen or beam (mm or m)
	H_a	height of arch (m)
	H_{ab}	height of abutment (m)
	H_b	height of bench (m)
	H_d	height of tunnel, drive, crosscut, adit, incline or decline (m)
	H_{draw}	height of draw (m)
	H_l	distance between underground levels (m)
	H_{lip}	height of lip of crater or crater lip height (m)
	H_{lm}	distance between main levels underground (m)
	h	draw height (m)
	h_{loose}	height of loosening drawbody in caving methods (m)

I

*	I	electrical current (A)
*	I	impulse or momentum (Ns)
*	I	movement of inertia of mass (kgm^2)
	I	sound pressure or acoustic intensity (W/cm^2)
	I_a	point load anisotropy (dimensionless)
	I_b	block size index (m)

	I_{BI}	blasting index (dimensionless)
	I_c	consistency index (%)
	I_{cd}	coal dust index (%)
	I_{cru}	crushing zone index (dimensionless)
	I_D	density index (dimensionless)
	I_{jA}	joint intensity, number of joints per unit of area (No./m²)
	I_{jAV}	joint intensity, total joint area per unit of volume (1/m)
	I_{jtlA}	joint intensity, total joint trace length per unit of area (1/m)
	I_{jV}	joint intensity, number of joint per unit of volume (No./m³)
	I_L	liquidity index (dimensionless)
	I_{max}	maximum (recommended) firing current (A)
	I_{min}	minimum (recommended) firing current (A)
	I_p	drilling index
	I_P	plasticity index (mass-%)
	I_{rc}	rock quality index regarding drillability (kPa·min/m)
*	I_s	point load strength or point load index (Pa or MPa)
	I_{st}	stray current (A)

J

J_a	joint alternation number (dimensionless)
J_d	joint length density (1/m²)
J_n	joint set number (dimensionless)
J_r	joint roughness number (dimensionless)
J_v	number of joints per m³ or volumetric joint count (No. of joints/m³)
J_w	joint water reduction factor (dimensionless)

K

*	K	bulk modulus or incompressibility (the inverse of compressibility) (Pa)
	K	Fischer constant (dimensionless)
	K	number of cycles at breakage of structure (No.)
	K_a	adiabatic bulk modulus (MPa)
	K_c	scaled strain intercept (dimensionless)
	K_d	dynamic bulk modulus (MPa)
	K_o	number of cycles when damage of a structure starts (No.)
	K_s	shear stiffness (MPa)
	K_I, K_{II}, K_{III}	stress intensity factor for mode I, II and III loaded cracks (MN/m$^{3/2}$)
*	K_{Ic}	fracture toughness or critical stress intensity factor for a mode I loaded crack (MN/m$^{3/2}$)
	K	*coefficient* of permeability or permeability *coefficient* (m/s). Not the same quantity as hydraulic conductivity
	K	joint persistence (m)
	K	stiffness (N/m)

	K	strain intercept (dimensionless)
	k_{JCS}	joint wall compressive strength JCS (MPa)
	k_{JRC}	joint roughness coefficient JRC (dimensionless)
	k_s	strain energy factor, used in the crater formula (kgm$^{-1/3}$)
	k_x	square mesh size through which x mass-% of the fragmented material will pass (m)
	k_{50}	mean fragment size (m)
	k_{100}	maximum fragment size (m)

L

	L	depth drilled per bit, drilled depth per bit or meters drilled (m)
	L	length of a fragment of a structure (m)
	L	shape factor (dimension varies with type of shape factor selected)
	L	shape factor for charge in Livingstone crater formula (dimensionless)
	L	width of support span for specimen (m)
	L_{ac}	apparent crater depth in (m)
	L_{dis}	dispersiveness (dimensionless)
	L_i	total length of visible half casts after blasting (m)
	L_j	total drilled length of a wall and roof holes in the blast (m)
	L_T	total drilled length of the wall and roof holes in a blast (m)
	L_{true}	depth of true crater (m)
	L_x	shape factor where x is a number denoting the type of shape factor. Units see 'shape factor'
*	l	length (m)
*	l_b	length of bottom charge (m)
*	l_{bh}	length of a borehole or blasthole (m)
*	l_c	length of column charge (m)
*	l_{ch}	length of charge (m)
*	l_{co}	length from collar to column charge (m)
	l_{cr}	length of crack (m)
	l_o	overbreak or backbreak depth (m)
	l_{oa}	average overbreak depth or backbreak depth (m)
	l_p	pore (microcrack or flaw) length (m)
	l_s	length of stemming (m)
	l_{sd}	length of stemming between deck charges (m)
	l_{so}	stand-off distance (m)
*	l_{sub}	subdrilling length (m)

M

*	M	moment of force (torque) (Nm)
	M	oedometric modulus (MPa)
	M_i	material index, defined by the quota between the optimum crater volume V_o divided by the optimum crater depth (optimum burden) B_o, $M_i = V_o/B_o$

*	m	mass (kg). (Eng. unit, 1 pound (lb) = 0,454 kg)
	m_{bh}	total mass of explosive in a blasthole (kg)
	m_{bit}	mass on bit (kg)
	m_{bitT}	total bit load or mass on drill bit (kg)
	m_d	draw rate (t/day or t/shift)
	m_r	total mass of the explosion reaction products (kg)

N

	N	distance exponent for peak strain (dimensionless)
*	N	number (No.)
*	n	porosity (dimensionless)
*	n	revolutions per second (1/s or rps)
*	n	uniformity index, uniformity exponent or fragmentation gradient appearing in the Rosin-Rammler-Sperrling mathematical distribution function (dimensionless)
	n_e	effective porosity (dimensionless)
	n_o	open porosity (dimensionless)
	n_s	sealed porosity (dimensionless)
	n_T	total porosity (dimensionless)

O

	O	open

P

*	P	effect, power or duty (J/s or W)
	P	degree of packing of explosives in a blasthole. Volume of blasthole for an intended fully charged hole occupied by explosives (dimensionless)
	P	Fischer distribution function (dimensionless)
	P	perimeter distance (m)
*	P	probability (dimensionless)
	P_b	brisance value (MJ/s or MW)
	P_r	mass reaction rate (kg/s)
*	p	pressure (Pa)
	p_a	adiabatic pressure or explosion pressure (MPa)
	p_{am}	ambient noise levels (dB, Pa or MPa)
	p_{atm}	atmospheric pressure in (Pa)
*	p_b	borehole pressure (explosion pressure) (MPa)
*	p_{be}	equilibrium borehole pressure (MPa)
	p_{bit}	pressure on the drill bit (MPa)
*	p_d	detonation pressure (MPa)
	p_f	specific energy pressure (Pa or MPa)
	p_{feed}	feed pressure (kPa)

	p_h	horizontal pressure (MPa)
	p_i	incidence pressure (MPa)
	p_o	overpressure (Pa or dB)
	p_{oa}	air pressure pulse (Pa or dB)
	p_{og}	gas release pulse (Pa or dB)
	p_{or}	rock pressure pulse (Pa or dB)
	p_{os}	stemming release pulse (Pa or dB)
	p_p	pore pressure (Pa or MPa)
*	p_{qs}	quasi-static pressure (MPa)
	p_r	reflected pressure (Pa or MPa)
	p_{res}	residual pressure in the borehole (Pa or MPa)
	p_s	sound pressure (level) (dB or Pa)
	p_{spg}	specific gas pressure (Pa or MPa)
	p_{sr}	stemming release pulse (Pa or dB)
	p_{sup}	NGI classification formula for roof support pressure (MPa)
	p_t	transmitted pressure (MPa)
	p_v	vertical pressure (MPa)

Q

Note: Q is used instead of m for mass of explosive because the energy content in the explosive (detonation heat Q_e) is more interesting in blasting than the mass of the explosive itself.

*	Q	heat or quantity of heat (J)
*	Q	mass of explosive charge (kg). In North America W is commonly being used
	Q	rock quality factor (for attenuation of blast waves, seismic) (dimensionless)
	Q	rock quality factor (for rock mass support, after Barton) (dimensionless)
	Q_b	mass of bottom charge (kg)
	Q_{bh}	mass of the total explosive charge in a blasthole (kg)
	Q_c	mass of column charge (kg)
	Q_{co}	heat of combustion (MJ/kg)
	Q_{cv}	total heat released at constant volume (MJ)
	Q_d	mass of deck charge (kg)
	Q_{del}	mass of charge per delay (kg)
	Q_e	heat of detonation (detonation heat, detonation energy). See 'heat of explosion' (MJ/kg)
*	Q_e	heat of explosion (explosion energy, explosion heat) (MJ/kg)
	Q_f	heat of formation (heat of reaction) (MJ)
	Q_{max}	maximum charge per delay (kg)
	Q_{pri}	primer (*bursting charge*) (kg)
	Q_T	explosive charge in a blast (kg)
*	q	specific charge (powder factor, explosive factor, explosive ratio or explosive specific) (kg/m³ or kg/t of rock)
*	q_a	specific splitting energy, *charge distribution per area* (theoretical) or *charge load*, (kg/m²)
*	q_{cell}	cell specific charge (cell powder factor), (CSC), (kg/m³)
*	q_l	linear charge concentration or explosive load (kg/m)

	q_{lb}	linear charge concentration in the bottom of the blasthole (kg/m)
	q_{lc}	linear charge concentration in the column of the blasthole (kg/m)
	q_m	mass rate of flow or mass flow per unit of time (kg/s)
*	q_p	'perimeter specific charge' or 'perimeter charge factor' (kg/m³)
	q_v	volume rate of flow (m³/s)

R

*	R	correlation coefficient (dimensionless)
*	R	distance from blast hole or blast site to the vibration measuring point normally located in the far field (m)
*	R	distance- or length of throw, throw distance or throw length (m)
*	R	gas constant (Pa·m³/mole K)
	R	reflection coefficient (dimensionless)
*	R	resistance (electrical current) (Ω)
	R_A	amplification factor (due to ground vibrations) for a structure (dimensionless)
	R_a	aspect ratio (dimensionless)
	R_{af}	advance factor (dimensionless)
	R_{ani}	coefficient of anisotropy (dimensionless)
	R_{burst}	burst-proneness index (dimensionless)
	R_c	coupling ratio (dimensionless)
	R_{ca}	axial coupling ratio (dimensionless)
	R_{cr}	radial coupling ratio (dimensionless)
	R_{cra}	crack aspect ratio (dimensionless)
	R_{cv}	volume coupling ratio (dimensionless)
	R_{dc}	decoupling ratio (dimensionless)
	R_{dca}	axial decoupling ratio (dimensionless)
	R_{dcc}	critical decoupling ratio (dimensionless)
	R_{dcr}	radial decoupling ratio (dimensionless)
	R_{dcv}	volume decoupling ratio (dimensionless)
	R_{dec}	decrease Young's modulus index (dimensionless)
	R_{df}	decoupling factor (dimensionless)
	R_{elast}	elastic ratio (dimensionless)
	R_{elo}	elongation ratio (dimensionless)
	R_{end}	endurance ratio (dimensionless)
	R_{ene}	energy ratio defined by Crandall (m²/s²)
	R_{ener}	energy ratio defined by Konya (m/s)
	R_{ESR}	excavation support ratio or equivalent support ratio (dimensionless)
	R_{eun}	volume at optimum breakage V_o or $R_{eun} = V/V_o$, (dimensionless)
	R_{ezd}	excavation zone of disturbance (m)
	R_f	felicity ratio (dimensionless)
	R_{fla}	flatness ratio (dimensionless)
	R_g	gas ratio (dimensionless)
	R_{hcf}	half cast factor (dimensionless)

	R_i	rock mass classification system where 'i' is the symbol for the specific system
	$R_{opt/c}$	optimum (breakage) burden (depth) ratio (dimensionless)
	R_{rib}	rib ratio (dimensionless)
	R_{RMR}	rock mass rating RMR (dimensionless)
	R_{RQDV}	volumetric RQD (dimensionless)
	R_{MRMR}	mining rock mass rating MRMR (dimensionless)
	R_{RSR}	rock support ratio RSR (dimensionless)
	R_s	safe distance, regarding throw of fragments and ground vibrations (m)
	R_{S_b/B_b}	ratio between blasted spacing and blasted burden (dimensionless)
	$R_{Sd/Bd}$	ratio between drilled spacing and drilled burden (dimensionless)
	R_{sf}	safety factor (dimensionless)
	R_{sfrb}	safety factor for rock bolts (dimensionless)
	R_{SMR}	slope mass rating (dimensionless)
	R_{sr}	ratio of size reduction (dimensionless)
	R_v	relative vibration velocity (dimensionless)
	$R_{w/c}$	water-cement ratio (dimensionless) ς
	$R_{w/o}$	waste to ore ratio (dimensionless)
*	r	distance from blast in the near field zone (m)
*	r	radius of borehole or blasthole (m)
	r_a	apparent crater radius (m)
	r_{bdz}	blast damage zone radius (m)
	r_{be}	radius of the blasthole wall when expanded to the equilibrium borehole pressure (m)
	r_c	radius of the equivalent sphere from the cylindrical charge used in the blasthole (m)
	r_{cr}	radius of crushing zone (m)
	r_h	hydraulic radius (m)
	r_{mac}	radius of macrocracks zone (m)
	r_{mic}	radius of microcracks zone (m)
	r_p	radius of plastic deformation zone (m)
	r_r	radius of radial cracks zone (m)

S

	S	spacing (distance) between blastholes in a row (m)
	S	maximum vertical displacement in subsidence of the earth surface (m)
	S	swell (dimensionless)
*	S_b	blasted spacing (m)
	S_c	spacing (mean) between natural cracks in the rock to be blasted (m)
*	S_d	spacing in drilling (m)
	S_{dis}	discontinuity spacing (m)
	S_j	joint spacing (m)
	S_p	practical spacing (m)
	S_s	specific surface area (per volume), (m²/m³ = m⁻¹)
	S_{sajcf}	specific surface area of joints, cracks and surface of fragments (m⁻¹)
	S_{sm}	specific surface area (per mass), (m²/kg)

424 Symbols

	S_{so}	specific surface area of all open surfaces (m^{-1})
	S_{sr}	specific surface (rock mass) (m^{-1})
	S_{srb}	specific surface (rock broken) (m^{-1})
	S_{ss}	specific surface (soil) (m^{-1})
	S_v	surface density (for in situ rock), (m^{-1})
	S_{20}	brittleness value (friability value) (dimensionless)
*	s	specific entropy (J/kg K)
	s	vertical displacement in subsidence of the earth surface (m)
	s_b	bulk strength (MJ/m^3)
	s_{ba}	absolute bulk strength (ABS), (MJ/m^3)
	s_{br}	relative bulk strength (dimensionless)
	s_{ext}	specific extraction of rock broken (m^3/m, t/m, m^3/kg, and t/kg)
	s_{ma}	absolute mass strength or mass strength (AMS), (J/kg or MJ/kg)
	s_{mr}	relative mass strength (dimensionless)
	s_{va}	absolute volume strength or volume strength (AVS), (MJ/m^3)
	s_{vr}	relative volume strength (dimensionless)
	s_{wa}	absolute weight strength or weight strength (AWS), (J/m^3 or MJ/m^3)
	s_{wL}	weight strength (defined by Langefors = L), (dimensionless)
	s_{wr}	relative weight strength (dimensionless)

T

*	T	absolute temperature, degree Kelvin (K)
*	T	time for one period of vibration (s)
	T	transmission coefficient for wave energy propagating across a closed crack or joint (dimensionless)
	T_e	explosion temperature (K)
	T_r	rock temperature (C°) or (K)
	t	detonator factor (t/detonator)
	t	thickness (m)
*	t	time (s)
	t	degrees Celsius or degrees centigrade (C°)
	t_b	bursting time (s)
	t_{inter}	interhole delay time (ms)
	t_{opt}	optimal delay time between blastholes (ms)
	t_r	(pulse) rise time (μs)
	t_{relax}	relax time for a wave (s)

U

	U	electrical voltage (V)
*	u	displacement (mm or m)
	u	swell factor (dimensionless)
	u_{max}	max (peak) particle displacement (dimensionless)

V

	V	gas ratio (dimensionless)
*	V	volume (m³)
	V_{ac}	volume of apparent crater in crater blasting (m³)
	V_b	bulk volume (m³)
	V_b	volume of blasthole before detonation of the explosive in (m³)
	V_c	coefficient of volume compressibility (1/Pa)
	V_d	volume density (dimensionless)
	V_D	damage volume (m³)
	V_e	volume of the blasthole measured at the time instant of maximum expansion of the blasthole (m³)
	V_{eg}	volume of explosive gas (m³)
	V_{er}	volume of explosive gas for reference explosive (m³)
	V_g	volume of a grain (inclusive its open- and closed pores) (mm³ or m³)
	V_h	volume of rock broken per blasthole (m³)
	V_o	overbreak volume (m³)
	V_{ob}	volume at optimum breakage (m³)
	V_p	volume of pores (m³)
	V_{scaled}	scaled volume (m³/kg)
	V_{solid}	volume of solid fragments in a rock pile (m³)
	V_{sr}	solid rock volume (m³)
	V_{void}	volume of void between fragments in a rock pile (m³)
	V_T	total volume (m³)
	v	particle velocity, vibration velocity (mm/s or m/s)
	v_a	rate of advance (m/h, m/day)
	v_b	burden velocity (m/s)
	v_d	drilling rate or penetration rate (boring rate in full-face boring), (m/min or m/h)
	v_f	feed rate in drilling (m/min)
	v_{fs}	free surface velocity (m/s)
	v_{hmax}	maximum (peak) horizontal component of particle velocity (m/s or mm/s)
	v_m	mining rate (m/day)
	v_{max}	maximum (peak) particle velocity often measured in vertical direction (m/s or mm/s)
	v_{maxP}	maximum particle velocity of the P-wave (m/s)
	v_R	resultant ground vibration particle velocity (m/s or mm/s)
	v_{Rmax}	maximum resultant of mutually perpendicular components of ground vibration particle velocity (m/s or mm/s)
	v_{Smax}	maximum S-wave component of the particle velocity (m/s or mm/s)
	v_{vmax}	maximum (peak) vertical component of particle velocity (m/s or mm/s)

W

*	W	energy or work (J)
	W	width of beam, specimen or fragment (m)

W_a	activation energy (J)
W_b	bubble energy (MJ)
W_{bi}	burst energy release index (J)
W_k	kinetic energy (J)
W_L	fractional loss of energy in a seismic wave (J)
W_p	potential energy (J)
W_s	strain energy under uniaxial compression (J)
W_{sd}	strain energy dissipated (J)
W_{smax}	max strain energy under uniaxial compression (J)
W_{sh}	shock energy (J)
W_{sr}	strain energy retained (J)
W_t	throw energy of rock fragments (J)
W_T	total energy release of charge (MJ)
w_{Bond}	Bonds working index (J/kg, J/m^3 and J/t)
w_f	specific fracture energy (specific energy of rupture, energy of rupture or work of fracture) (J/m^3)
w_i	specific internal energy (J/kg)
w_L	liquid limit (mass-%)
w_m	water content (mass-%)
w_P	plastic limit (mass-%)
w_s	strain energy density (J/m^3)
w_{se}	elastic strain energy density (J/m^3)
w_{sp}	specific energy (kJ/m^3 or kJ/t)
w_{sur}	surface energy (J/m^2)
w_V	water content (dimensionless)

X

x	digging depth for LHD loader in sublevel caving (m)

Y

open

Z

Z	acoustic impedance (kg/m^{2}s)
Z_e	acoustic impedance of an explosive (kg/m^{2}s)
Z_i	acoustic impedance for the material of the incidence wave (kg/m^{2}s)
Z_m	mechanical impedance (Ns/m)
Z_R	acoustic 'impedance ratio' or 'acoustic impedance mismatch' (dimensionless)
Z_t	acoustic impedance for the material of the transmitted wave (kg/m^{2}s)

Greek symbols

	α	angle of dip (dip of orebody) or angle of blasthole (°)
	α	charge-related attenuation factor for blast-generated waves
	α	constant
	α	covolume (m³/kg)
	α	trend of a line (°)
	α_b	angle of breakage or *breakout angle* (°)
	α_{bi}	blasthole inclination (°)
*	α_{cav}	angle of caving, -break -failure or limit angle (°)
*	α_{lb}	angle of limit border (limit border angle) (°)
	α_{lim}	limit angle, angle of caving, - break or - failure (°)
	α_{orer}	angle of ore rest (°)
	α_{sub}	angle of subsidence (°)
	β	(beta), angle of shear (visual) in uniaxial compression (°)
	β	(beta), constant
	β	(beta), distance-related attenuation factor for blast-generated waves
	β	(beta), plunge of a line (°)
	γ	adiabatic exponent or specific heat ratio (dimensionless))
	γ	shear strain (dimensionless)
*	γ	specific gravity (dimensionless)
	γ	polytropic exponent (dimensionless)
	γ_s	specific surface energy (J/m²)
	δ	small increase
	δ_{dump}	angle of dumping, - deposition, - pouring, - rest, natural slope angle or angle of repose (°)
	δ_{fekin}	angle of kinematic external friction (wall friction kinematic) (°)
	δ_{load}	angle of loading, angle of draining, angle of slide or (angle of repose in loading or draining) (°)
	δ_{ferest}	angle of external friction at rest (wall friction at rest) (°)
	ε	strain or elastic strain (dimensionless)
	ε'	strain rate (s⁻¹)
	ε'	creep (s⁻¹)
	ε_a	axial strain (dimensionless)
	ε_d	dynamic strain (dimensionless)
	ε_d	(epsilon), deviator strain (dimensionless)

Greek symbols

	ε_f	deviatoric strain (dimensionless)
	ε_{max}	peak strain at rock blasting (dimensionless)
	ε_{res}	residual strain (dimensionless)
	ε_{rmax}	maximum radial strain (dimensionless)
	ε_t	tangential hoop strain at the blasthole wall (dimensionless)
	$\Delta\varepsilon$	differential strain analysis (dimensionless)
	ζ	open
	η	energy dissipation (s)
	η	blasting efficiency or any other efficiency
	η_b	rock burst efficiency ratio (dimensionless)
	η_d	efficiency of the detonation reaction (dimensionless)
	θ	(theta), constant (°)
	ι	(jota), open
	κ	compressibility inverted value of the bulk modulus K, (1/Pa)
	κ	permeability or mobility coefficient (m^2/Pa · s)
	λ	Lame's constant
	λ	wave length (m)
	μ	coefficient of friction
	μ	fluid dynamic viscosity (Pa · s)
	μ	Lame's constant
	μ	magnetic permeability in (m^{-2}s^2)
	ν	(ny), Poisson's ratio
	ν	(ny), specific volume or true specific value (m^3/kg)
	ν_b	(ny), bulk specific volume (m^3/kg)
	ν_e	specific gas volume of explosive gases (m^3/kg)
	ν_{er}	specific gas volume of reference explosive gases (m^3/kg)
	ν_d	(ny), dynamic Poisson's ratio
	ν_{qd}	(ny), quasi dynamic Poisson's ratio
	ξ	(Xi), fraction of some original quantity $0 \leq \xi \leq 1$ (dimensionless)
	o	open
	π	(pi), 3,14159
	ρ	density or true density (kg/m^3)
	ρ_a	apparent density or *compact density* (kg/m^3)
	ρ_c	critical density (kg/m^3)
	ρ_{dry}	density of a dry sample (kg/m^3)
*	ρ_e	(bulk) density of explosive before charging (kg/m^3)
	ρ_{ec}	bulk density of explosive after charging or charged explosive density (kg/m^3)
	ρ_f	density of fluid (kg/m^3)
*	ρ_r	density of rock (kg/m^3)
	ρ_{RC}	density of reinforced concrete (kg/m^3)
	ρ_s	density of stemming (kg/m^3)
*	ρ_w	density of water (kg/m^3)
	ρ_{wet}	density of a wet sample (kg/m^3)
	σ	strength or stress (Pa)
	σ_a	applied stress (MPa)
	σ_c	uniaxial compressive strength (MPa)
	σ_{cd}	dynamic compressive strength (MPa)

Greek symbols

σ_{cj}	joint compressive strength (MPa)
σ_{ctri}	triaxial compression strength (MPa)
σ_d	deviator stress or flow stress (MPa)
σ_e	(sigma), effective stress (MPa)
σ_f	(sigma), flexural strength (MPa)
σ_{fail}	failure or rupture stress in (MPa)
σ_{fb}	flexural strength in bending (modulus of rupture in bending) (MPa)
σ_{ft}	flexural strength in torsion (MPa)
σ_h	horizontal stress (MPa)
σ_{cw}	joint compressive strength (MPa)
σ_{cwj}	joint wall compressive strength (MPa)
σ_{max}	the maximum earlier stress level applied in a material (MPa)
σ_n	normal stress (MPa)
σ_p	stress caused by the fluid pressure in pores (MPa)
σ_{nf}	near-field stress (MPa)
σ_{onset}	the stress level when acoustic emission starts (MPa)
σ_{res}	residual stress (MPa)
σ_{st}	stamp test index (MPa)
σ_t	tensile stress (MPa)
σ_T	total or applied stress (MPA)
σ_{td}	dynamic tensile strength (MPa)
σ_{thyd}	disruptive strength (tensile strength at hydrostatic tension) (MPa)
σ_{tor}	torsion strength (MPa)
σ_v	vertical stress (MPa)
σ_{vir}	virgin-, natural- or paleostress (MPa)
σ_x	longitudinal horizontal stress (MPa)
σ_y	latitudial horizontal stress (MPa)
σ_y	yield strength or stress (MPa)
σ_z	vertical stress (MPa)
σ_1	major principal stress (MPa)
σ_2	intermediate principal stress (MPa)
σ_3	minor principal stress (MPa)
σ_{1e}	effective major principal stress (MPa)
σ_{2e}	effective intermediate principal stress (MPa)
σ_{3e}	effective minor principal stress (MPa)
σ_{1ef}	effective major principal stress at failure (MPa)
σ_{2ef}	effective intermediate principal stress at failure (MPa)
σ_{3ef}	effective minor principal stress at failure (MPa)
ς	crack density (mm^{-1})
ς_{mic}	microcrack density (mm^{-1})
τ	shear stress (MPa)
τ_l	limit value for shear strength at very high confinements (MPa)
τ_s	shear strength (MPa)
τ_{sd}	dynamic shear strength (MPa)
τ_{sj}	joint shear strength (MPa)
τ_{so}	shear strength at zero normal pressure (MPa)
υ	(ypsilon), Poisons ratio (dimensionless)

υ	specific volume (m³/kg)
υ_b	bulk specific volume (m³/kg)
υ_{eg}	specific volume of the gaseous explosive reaction products at temperature 273° K (m³/kg)
υ_d	dynamic Poisons ratio (dimensionless)
υ_t	true specific volume (m³/kg)
$\Phi^{(i)}(l)$	Abels equation, a density function to quantify the statistical distribution of the trace length l of joints in m on the x_i plane (m)
Φ	resistivity (Ωm)
ϕ	angle of internal friction or shear angle (°)
φ	angle of friction
χ	flexural rigidity (N²/m)
Ψ	rock bolting density (No./m³)
ω	angular frequency, $\omega = 2\pi f$ (o/s or rad/s)

References

A

ACSG, 1961. *Tentative ceramic glossary, Part I*. American Ceramic Society, Alfred, N.Y.
ACSG, 1963. Van Schoick, Emily C. 1963. *Ceramic glossary*. American Ceramic Society, Columbus, Ohio, 1st ed., 1963, 31 pp.
AECI, 1984. Explosives today. *AECI Series 2, No. 35, 1st quarter*.
AECI, 1993. Perimeter blasting. *In-house course given at AECI Explosive Limited, 12 pp.*
A.G.I., 1957 and 1960. *Glossary of Geology and Related Sciences*. American Geological Institute. Washington, D.C., 1957, 325 pp. Supplement, 1960, 72 pp.
AIME, 1960. *Industrial Minerals and Rocks (Non-metallic Other Than Fuels)*. American Institute of Mining, Metallurgical, and Petroleum Engineers, New York, 3rd ed., 934 pp.
Akaev, M.S., Tregubov, B.G. and Krutilin, A.A. 1971. Optimal parameters of gapped borehole charges. *Fiziko-Tekhnicheskie Problemy Razrabotke Polenznykh Iskopaemykhl, No. 4, pp. 42–46, July Aug. Sovjet Mining Science*.
Almgren, Gunnar, 1999. Mining methods excluding caving methods. *Division of Mining Engineering Luleå University of Technology, Luleå, Sweden in cooperation with CENTEK, Luleå, Sweden. 324 pp.*
Anderson, Basil W., 1964. *Gem testing*. Temple press books, Ltd. London, 7th ed., 377 pp. Includes a glossary, pp. 349–354.
Anderson, Douglas, 1993. *Blast monitoring: regulations, methods and control techniques*. Comprehensive Rock Engineering, Vol. 4, pp. 95–110. Editor-in-Chief John Hudson. Pergamon Press Ltd.
Anderson, Douglas, 2008. Signature hole blast vibration control—twenty years hence and beyond. The Journal of Explosives Engineering, Vol. 25, No. 5, Sept/Oct 2008.
Anon., 1941. *The seismic method for dam site and tunnel investigations*. AB Hasse W. Tullberg, Esselte AB, Stockholm.
Anon., 1985. Ingeniørsgeologi Berg, Håndbok. *Norsk Bergmekanikkgruppe, Tapir Förlag. English-Norwegian Glossary, pp. 35–39*.
Anon., 1986. Erschütterungen im Bauwesen—Einwirkung auf Bauliche Anlagen. DIN 4150, Teil 3.
Anon., 1990. Nordtest scheme for testing commercial high explosives. *Nordtest Doc Gen 024. Approved Feb, 1990. P.O. Box 11, SF-02101 Esbo, Finland. Fax: 358 0 455 4272, 4 pp.*
Anon., 1992. *Dictionary of Science and Technology*. Academic Press, Inc.
Anon., 1996. Blast initiation via ultra low frequency communication system. *Engineering and Mining Journal, July, p. 38.*
Anon., 2006. Pre-conditioning with W100. *Information in Wassara Newsletter No 2, October 2006. Stockholm. pp 4, 8 pp.*
Anon., 2007. *A Guideline for Surface Rock Blasting*. Ministry of Land Transport and Maritime. (In Korean).

API 1953. *Glossary of Terms used in Petroleum Refining.* American Petroleum Institute, New York. 188 pp.

Arkell, W.J. and Tomkeieff, S.I., 1953. *English Rock Terms Chiefly as Used by Miners and Quarrymen.* Oxford University Press, London. 139 pp.

ASA C42.85, 1956. Definitions of Electrical Terms; Group 85: Mining (C42.85: 1956). *American Standards Association, American Institute of Electrical Engineers, New York, 12 pp.*

ASA MH4.1, 1958. *Conveyor Terms and Definitions.* The Conveyor Equipment Manufacturers Association, Washington D.C. 78 pp.

ASCE, 1958. Glossary of Terms and Definitions in Soil Mechanics. American Society of Civil Engineers. *Proceedings, Vol. 84, No., Paper 1826, Oct., 43 pp.*

Ash, Richard L., 1985. *Flexural rupture as a rock breakage mechanism in blasting.* In Fragmentation by Blasting (Edited by W.L.. Fourney, R. Boade and L. Costin), pp. 24–29. Society of experimental Mechanics, Bethel, CT.

A.S.M. Gloss., 1961. Metals Handbook. Volume 1. Properties and Selection of Metals. Includes a glossary of definitions relating to metals and metalworking, 41 pp. *American Society for Metals, Metals Park, Ohio, 8th ed., 1300 pp.*

ASTM, 1982. ASTM Publication Catalogue 1982. *European Edition, American Technical Publishers Ltd., Herts, England, 65 pp.*

Atchison, T. et al., 1964a. Comparative studies of explosives in granite. *Second series of tests. USBM RI 6434, 26 pp.*

Atchison, Thomas C. and Pugliese, Joseph M., 1964b. Comparative studies of explosives in limestone. *USBM RI 6395. pp 1–25.*

Atlas Copco, 1982. Atlas Copco Manual. *4th. Ed., Ljungföretagen AB, Örebro, Sweden., 652 pp.*

Atlas Copco, 1997. Coprod's quarry crusade. *Mining and Construction, Mechanized Rock Excavation with Atlas Copco, No. 1,* pp. 10–13.

Atlas Copco, 2001. Ballast South is drilling so you cant almost hear it. *Borrmästarn No. 2. A Customer Journal from Atlas Copco CMT Sweden AB. 2 pp. (In Swedish).*

Atlas Copco, 2006. Definitions delivered to Agne Rustan by Atlas Copco Sept 2007.

Atlas Copco, 2006a. Finns find silenced Smartrig smart. *Mining and Construction, Mechanized Rock Excavation with Atlas Copco, No 2, p. 9.*

Atlas Copco, 2006b. Surface drilling in open pit mining. *First edition 2006. pp. 148.*

Atlas Copco, 2007. Loading and haulage in underground mining. *First edition 2007. pp. 108.*

Atlas Copco, 2007b. Money in the bank. *Mining and Construction, Mechanized Rock Excavation with Atlas Copco, No. 1,* pp. 28–29.

Atlas Copco, 2009a. Face drilling. *Reference Booklets from Atlas Copco Rock Drills AB. Fourth edition 2009.* www.atlascopco.com.

Atlas Copco, 2009b. New solution for rapid micropiling. *Mining and Construction. Mechanized Rock excavation with Atlas Copco. No. 3, pp. 20–21.*

Atlas Copco Craelius, 2006. Chips or cores. *Nordic Mining Review, Bergsmannen med Jernkontorets Annaler,* No. 3, pp. 101–103.

Atlas Powder, 1987. *Explosives and Rock Blasting.* Atlas Powder Company, Dallas, Texas, USA, 662 pp.

Austin, J., 1964. Dragline excavators for strip mining. *Mining Magazine, Vol. 3, No. 5, Nov.*

Australian Standard, 2006. AS 2187.2 Explosives—Storage, -, Transport and Use—Use of Explosives. *Standards Australia, Sydney.*

B

Ballard, Thomas J. and Conklin, Quentin, E., 1955. *The Uranium Prospector's Guide.* Harper & Brothers, New York, 251 pp. Includes a glossary, pp. 178–205.

Barker, L.M., 1977. A simplified method for measuring plane strain fracture toughness. *Eng. Fract. Mech. Vol. 9, pp. 361–369.*

Barton, N., Lien, R. and Lunde, J., 1974. Engineering classification of rock masses for the design of tunnel support. *Norwegian Geotechnical Institute. Publication No. 106, Oslo, 115 pp.*

Barton, Nick, R. and Choubey, V., 1977. The shear strength of rock joints in theory and practice. *Rock Mech. Vol. 8*, pp. 1–54.

Barton, Nick, 1989. Cavern design for Hong Kong rocks. *Proc. Rock Cavern Seminar—Hong Kong. Edited by A.W. Malone and P.F.D. Whiteside. Institution of Mining and Metallurgy, London.* pp.179–202.

Bates, R.L. and Jackson, J.A., 1987. *Glossary of Geology.* Third Edition. American Geological Institute, Alexandria, Virginia, USA., 788 pp.

Bauer, A. and Calder, P.N., 1978. Open pit and blast seminar. *Course No. 63321, Mining Engineering Department, Queens University, Kingston, Ontario, Canada.*

B.C.I., 1947. *Bituminous Coal Institute Glossary of Current and Common Bituminous Coal Mining Terms.* Washington D.C., January. 26 pp.

Beer, F.P., Johnson, E.R., Eisenberg, E.R. and Clausen, W.E., 2004. *Vector Mechanics for Engineers—Statics and Dynamics.* 7th Edition. New York McGraw Hill.

Bellairs, Peter, 1987. The application of geological and downhole geophysical data to blast pattern design. *Second Int. Symp. on Rock Fragmentation by Blasting. Keystone, Colorado, 23–28 Aug.* pp. 398–411.

Bennett, H. (ed.), 1962. Concise Chemical and Technical Dictionary. *Chemical Publishing Co., Inc., New York, 2nd ed., 1039 pp.Addenda, 119 pp.*

Berger, P. and Froedge, D.T., 1983. A new technique for recording, predicting and controlling blast vibrations. *Journal of Explosives Engineering. Vol. 6, No. 3. pp. 38–43.*

Bergman, G.A., Lindström, G., Rundquist, H., Sjöberg, C. and Widing, S., 1989. Influence on gates of water of shock waves caused by rock blasting under water. *Swedish State Power Board. Report No. U (B) 1990/1.*

Bergmann, O.R. and Wu, F.C., 1973. Model rock blasting effect of explosives properties and other variables on blasting results. *Int. J. Rock Mech. Min. Sci. and Geomechanics Abstr. Vol. 10.,* pp. 585–612.

Besançon, R.M., 1985. *The Encyclopaedia of Physics.* 3rd edition Reinhold Co., New York.

Bhandari, S. and Badal, R., 1990. Relationship of joint orientation with hole spacing parameter in multihole blasting. *3rd Int. Symp. on Rock Fragmentation by Blasting. Brisbane, Australia, 26–31 Aug.,* pp. 225–231.

Bickel, John O. and Kuesel, T.R., 1982. *Tunnel Engineering Handbook.* Van Nostrand Reinhold Company Inc. 670 pp.

Bieniawski, Z.T., 1973. Engineering classification of jointed rock masses. *Trans S. African. Inst. of Civil Engrs, Dec. pp. 335–344.*

Bieniawski, Z.T., 1976. Rock mass classifications in rock engineering. *In Exploration for Rock Engineering. Z.T. Bienawski (ed.), 1, pp. 97–106. Cape Town: A.A. Balkema.*

Bilgin, H.A., 1991. Single hole test blasting at an open pit mine in full scale: A case study. *Int. Journ. of Surface mining and Reclamation No. 5, pp. 191–194.*

Bilgin, H.A. and Paþamehmetoðlu, A.G. & Ozkahraman, H.T., 1993. Optimum burden determination and fragmentation evaluation by full scale slab blasting. *Fourth Int. Symp. on Rock Fragmentation by Blasting, Vienna, Austria, A.A. Balkema, pp. 337–344.*

Billings, Marland P., 1954. *Structural Geology.* Prentice-Hall, Inc, New York 2nd ed.

Birkhoff, G., D.P. and Mac Doughall, E. M and Pugh, G. Taylor., 1948. Explosives with lined cavities. *Journal of Applied Physics, Vol. 19, No. 6, June, pp. 563–582.*

Bjarnholt, G., 1981. A System for Contour Blasting with Guided Crack Initiation. (In Swedish). *Bergsprängningskommitténs Diskussionsmöte i Stockholm, Jan. 29, 1981, pp. 313–386.*

Bjarnholt, G., 1987. Initiation and propagation of a disc fracture in a plane perpendicular to the borehole axis. *Second Int. Symp. on Rock Fragmentation by Blasting, Keystone, Colorado, USA, Society of Experimental Mechanics, pp. 147–158.*
Brady, B.H.G. and Brown, E.T., 1985. *Rock Mechanics for Underground Mining.* George Allen & Unwin (Publishers) Ltd, Park Lane, Hemel Hempstead, Herts, UK. 527 pp.
Brady, B.H.G. and Brown, E.T. 1993. *Rock Mechanics for Underground Mining.* 2nd edition, Prentice and Hall, London. 571 pp.
Brantly, J.E., 1961. *Rotary Drilling Handbook.* Palmer Publications, New York 6th ed. Includes a glossary of mud drilling terms, pp. 312–317.
Bridgman, P.W., 1911. Physics above 20.000 kg/cm^2. *Bakerian Lecture. Proceedings of American Academy of Arts and Sciences, Vol. 47.*
Briggs, Henry, 1929. *Mining subsidence.* Edward Arnold & Co., London, 1929, 215 pp.
Broadbent, C.D. 1974. Predictable blasting with in-situ seismic surveys. *Min. Eng., Vol. 26, No. 4, pp. 37–41.*
Broch, E. and Franklin, J.A., 1972. The point-load strength test. *Int. J. of Rock Mech. and Min. Sci. Vol. 9., pp. 669–667.*
Broch, E., 1974. The Influence of Water on Some Rock Properties. Advances in Rock Mechanics. *National Academy of Sciences, Washington, D.C., pp. 33–38*
Brown, E.T., 1981. Rock Characterization Testing and Monitoring. *ISRM Suggested Methods.* Pergamon Press Ltd.
Brännfors, Sten, 1973. Bergsprängningsteknik. (Rock Blasting Technology). Kungl. Boktryckeriet P.A. Norstedt & Söner, 197 pp. (In Swedish).
B.S. 3552, 1962. *Glossary of Terms used in Coal Preparation.* B.S. 3552, Institution, London. 44 pp.
B.S. 3618, 1964. *Glossary of Mining Terms. Section 6, Drilling and Blasting.* British Standards Institution (London), 20 pp.
BuMines Bull., 1936. *Bureau of Mines Bulletin No. 390.*
BuMines Bull., 1965. *Bureau of Mines Bulletin No. 630.*
Bur, Thomas R., Lyle, Colburn W, Nicholls, Harry R. and Slykhouse, Thomas E., 1967. Comparison of two methods for studying relative performance of explosives in rock. *USBM RI 6888, 40 pp.*
Bur, Thomas R., Hjelstad, Kenneth E. and Thill, Richard, 1969. An ultrasonic method for determining the elastic symmetry of materials. *USBM RI 7333. pp. 1–23.*
By, Tore Lasse, 1980. Sjokkbølger i fjell—eksempel fra forsøk utført i Franzefoss bruks steinbrudd på Steinskogen. *National Conference on "Fjellsprengningsteknikk, Bergmekanikk/Geoteknikk". (In Norwegian). pp. 22.1–22.17.*

C

Cameron, Alan, Forsyth, Bill and Steed, Chuck, 1996. Blasting. How to access performance and minimize damage in open-pit and underground mines. *Engineering and Mining Journal, Jan. pp. 26–32.*
Camm, Frederick James ed., 1940. *A Dictionary of Metals and Their Alloys.* George Newes, Ltd., London. 244 pp.
Cannaday, Francis and Leo, Gary M., 1966. Piezoelectric pulsing equipment for sonic velocity measurements in rock samples from laboratory size to mine pillars. *USBM RI 6810.* 23 pp.
Carrasco, L.G. and Saperstein, L.W. 1977. Surface morphology of presplit fractures in Plexiglas models. *Int. J. Rock Mech. Min. Sci. and Geomech. Abstr. Vol. 14, pp. 261–275.*
Carson, A. Brinton., 1961. General excavation methods. F.W. Dodge Corp., New York. *Coal Age V. 71, No. 8, Aug 1966.*CCD, 1961. Condensed Chemical Dictionary. *New York, 6th ed. 1961, 1256 pp.*
CCD 6d, 1961. *Reinhold Publishing Corp. Condensed Chemical Dictionary.* New York, 6th ed., 1256 pp.
Celio, Tino and Matthias, Herbert, 1983. PMS2 A fast automatic profile recorder for underground surveys. *Int. Symp. on Field Measurements in Geomechanics. Zurich, Sept 5–8. pp. 999–1005.*

Chernigovskii, A.A., 1976. *Application of Directional Blasting in Mining and Civil Engineering.* Oxonian Press Pvt. Ltd., New Delhi, 318 pp.

Chiapetta, Frank and Borg, David, 1983. First International Symposium on Rock Fragmentation by Blasting. Luleå Sweden, 22–26 August, pp. 301–331.

Chiapetta, F.R. and Mammele, M.E., 1987. Analytic high-speed photography to evaluate air decks, stemming retention and gas confinement in presplitting, reclamation and gross motion applications. *Sec. Int. Symp. on Rock Fragmentation by Blasting, Keystone, USA, Society of Experimental Mechanics, pp. 257–301.*

Chiapetta, F.R., 1991. Generating site specific blast designs with state-of-the-art blast monitoring instrumentation and PC based analytical techniques. *Proc. 17th Society of Explosives Engineers Conf., Las Vegas.*

Choquet, P. and Charette, F., 1988. Applicability of rock mass classifications in the design of rock support in mines. *Proc. 15th Can. Symp. Rock Mech., Toronto, pp. 39–48.*

Clarke, Lewis., Elliot, Ron.J., Gobl, Blair, Fulop, Ed, Singh, Neil K. and Frank Huber, 2010. Explosive compaction of foundation soils for the seismic upgrade of the Weymour Falls Dam. *The Journal of Explosives Engineers. Vol. 28, No. 2, March/April. pp. 6–17.*

Clay, Robert B., Cook, Melvin A., Cook, Vernon O., Keyes, Robert T. and Udy, Lex L., 1965. Behaviour of rock during blasting. *VII Symp. on Rock Mech., The Pennsylvania State University, University Park, Pennsylvania, June 14–16. pp. 438–461.*

Cook, Melvin A., Cook, V.O., Clay, R.B., Keyes, R.T. and Udy, L.L., 1965. Behaviour of rock during blasting. *Transactions, Society of Mining Engineers, Dec, pp. 183–392.*

Cook, M., 1974. The Science of Industrial Explosives. *Graphic Service & Supply, Inc., 1974, pp. 449.*

Cooper, T., 1963. *An Introduction to Mining Chemistry. Chapter I, Chemical Definitions and Terminology, 27 pp.* Leonard Hill Books, Ltd., London, 439 pp.

Condon, Joseph L., Murphy, N. John and Fogelson, David E., 1970. Seismic effects associated with an underwater explosive research facility. *USBM RI 7387. 20 pp.*

Constable, John, 2009. LKAB shifts to advanced underground crushing technology. *Nordic Mining and Steel Review. The International Issue of Bergsmannen JKA No: 3, Vol. 193 and p. 52.*

Craigie, William A., and Hulbert, James R. eds. 1938, 1940, 1942, 1944. *A Dictionary of American English on Historical Principles.* University of Chicago Press, Chicago, Ill. 4 Volumes.

CRC, 1976. Handbook of Chemistry and Physics. *Chemical Rubber Co., Cleveland, Ohio, 76th ed., 2512 pp.*

Crispin, Frederic, 1964. *Dictionary of Technical Terms.* The Bruce Publishing Co., Milwaukee, Wis., 10th ed.

C.T.D., C.T.D. Supp., 1958. *Chamber's Technical Dictionary.* Tweney, C.F. and L.E.C. Hughes (eds.). MacMillan Co., New York, 3d ed., 1028 pp. Supplement, pp. 952–1028.

Cundall, Peter A. and Hart, Roger, D., 1993. *Numerical modelling of discontinua.* Comprehensive rock engineering. Vol. 2, pp. 231–243. Editor John Hudson. Pergamon Press Ltd.

Cunningham, C.V.B., 1983. The Kuz-Ram model for prediction of fragmentation by blasting. *First Int. Symp. on Rock Fragmentation by Blasting, Luleå, Sweden, pp. 439–453.*

Cunningham, C.V.B., 1987. Fragmentation estimations and the Kuz-Ram model—four years on. *2nd Int. Symp. on Rock Fragmentation by Blasting. Keystone, USA, Society of Experimental Mechanics, pp. 475–487.*

Cunningham, C.V.B., 1988. Control over blasting parameters and its effect on quarry productivity. *Institute of Quarrying Conference, 11 March, Durban, South Africa, pp. 1–12.*

Cunningham, C.V.B. and Goetzsche, A.F. 1990. The specification of blast damage limitations in tunnelling contracts. *Tunnelling and Underground Space Technology. Vol. 5, No. 3, pp. 193–198.*

Cunningham, C.V.B., 1991. The assessment of detonation codes for blast engineering. *Third High-Tech Seminar on Blasting Technology: Instrumentation and Explosives Application, San Diego.*

Cunningham, C.V.B., Braithwaite, Martin and Parker, Ian., 2006. Vixen detonation Codes: Energy input of the HSBM. *Fragblast 8. Eight International Symposium on Rock Fragmentation by Blasting. May 7–11 2006, Santiago, Chile. pp. 169–174.*

D

Daehnke, A., Rossmanith, H.P. and Kouzniak, N., 1996. *Dynamic fracture propagation due to blast-induced high pressure gas loading.* Rock Mechanics. Balkema, Rotterdam. pp. 619–626.

D`Andrea, D.V., Fisher, R.L. and Fogelson, D.E., 1965. Prediction of compressive strength from other rock properties. *USBM RI 6702. 23 pp.*

Day, P.,R., Webster, W.K., 1982. Controlled blasting to minimize overbreak with big boreholes underground. *CIM Bulletin, Vol. 75, No. 839, pp. 112–121.*

Daw, A.W. and Daw Z.W., 1898. *The blasting of rock: in Mines, Quarries, etc.,* p. 8. Spon, London.

DDR, 1978. DDR-Standard, Sprengwesen, Begriffe. TGL 23779. Gruppe 921100. *VEB Autobahnbaukombinat Magdeburg. Bestätigt: Aug 31th, 1978, Amt für Standardisierung, Messwesen und Warenprüfung, Berlin, 28 pp. (In German).*

Deere, D.U. and Miller, R.P., 1966. Engineering classification and index properties for intact rock. *Technical report, Air Force Weapons Laboratory, No. AFNL-TR-65-116, New Mexico, USA, 300 pp.*

Deere, D.U., Hendron, A.J., Jr., Patton, F.D. and Cording, E.J., 1967. Design of surface and near-surface construction in rock. *In Proc. 8th U.S. Symp. Rock Mech., Minneapolis, MN, editor C. Fairhust, pp. 237–302. Port City Press, Baltimore, MD.*

Deere, D.U., 1968. *Geological considerations. Rock Mechanics in Engineering Practice.* Eds K.G. Stagg and O.C. Zienkiewicz. John Wiley & Sons, London, pp. 1–20.

Dershowitz, W.S. and Einstein, H.H., 1988. Characterizing rock joint geometry with joint system models. *Rock Mech. and Rock Eng., Vol. 21.*

Devine, James F., Beck, Richard H., Meyer, Alfred V.C., and Duvall, Wilbur I., 1965. Vibration levels transmitted across a presplit facture plane. *USBM RI 6695. 1–29 pp.*

Dick, R.A., 1970. Effects of type of cut, delay, and explosive on underground blasting in frozen gravel. *USBM Report of Investigations No. 7356, pp. 1–17.*

Dick, R.D., Weaver, T.A. and Fourney, W.L., 1990. An alternative to cube-root scaling in crater analysis. *Third Int. Symp. on Rock Fragmentation by Blasting, Brisbane, Australia, The Australasian Institute of Mining and Metallurgy, pp. 167–170.*

Dobrin, Milton B., 1964. *Introduction to Geophysical Prospecting.* McGraw-Hill Book Co, New York, 2d ed., 446 pp.

DoD, 1993. Ammunition and Explosives Safety Standards. *US Department of Defence Explosives Safety Board (DoD). Report DoD 6055.9-STD. Washington, USA.*

Dood, A.E., 1964. *Dictionary of Ceramics: Pottery, Glass, Enamels, Refractories, Clay Building Materials, Cement and Concrete, Electroceramics, Special Ceramics.* Philosophical Library Inc., New York, 327 pp.

Dojcár, O., 1991. Investigation of blasting parameters to optimize fragmentation. *Trans. Inst. Min. Metall. Sect. A; Min. Industry, 100, Jan.-April, pp. A31- A41.*

Duffy, John, 1991. Field tests and evaluation of rock fall restraining nets. *Int. Cong. on Rock Mechanics. Aachen 15–19 Sept 1991, Germany. Balkema, Rotterdam.*

Du Pont, 1966. Du Pont de Nemours & Company (E.I.). Blasters' Handbook; a Manual Describing Explosives and Practical Methods of Use. *Wilmington, Del., 15th ed., 1966, 524 pp.*

Duvall,Wilbur I. and Pugliese, Joseph M., 1965. Comparison between end and axial methods of detonating an explosive in granite. *USBM RI 6700. pp. 1–11.*

E

El-Saie, Ahmed, 1980. Rock slotting by high pressure water jet for use in tunnelling. *Proceedings 21st Symp. on Rock Mechanics. University of Missouri, Rolla. May. 123–131 pp.*

English, George Letchworth, 1892–1938. 1939. *Descriptive List of the New Minerals.* Containing All New Mineral Names not Mentioned in Dana's System of Mineralogy, 6th ed., 1892. McGraw-Hill Book Co., New York.

Esen, S., Onederra, I., Bilgin, H.A. 2003. Modelling the size of the crushed zone around a blasthole. *Int. J. of Rock Mech. & Mining Sci. 40, pp. 485–495.*

F

Favreau, R.F., 1969. Generation of strain waves in rock by an explosion in a spherical cavity. *Journal of Geophysical Research. Vol. 74, August 15, pp. 4267–4280.*

Fay, A.H.A., 1920. Glossary of the Mining and Mineral Industry. *BuMines Bull. 95, 754 pp.*

Fernberg, Hans, 2002. New trends in open pits. *Nordic Steel and Mining Review. The Annual International Issue of Bergsmannen with Jernkontorets Annaler. No. 3, p. 106.*

Fischer, R., 1953. Dispersion on a sphere. *Proc. R. Soc. London, Ser. A Vol. 217, pp. 295–305.*

Fogelson, D.E., D'Andrea, D.V. and Fisher, R.L., 1965. Effects of decoupling and type of stemming on explosion generated pulses in mortar. A laboratory study. *USBM RI 6679, pp. 1–8.*

Ford, M.J.R. and Bonneau, M.D.J., 1991. Developments in cast blasting using high bulk strength explosives at Rietspruit Opencast Services. *Proc. of the Conference on Explosive and Blasting Techniques, Feb. 3–7, Las Vegas, Society of Explosives Engineers, Vol. I.*

Forrester, James Donald, 1946. *Principles of Field and Mining Geology.* John Wiley & Sons, Inc., New York, 647 pp.

Forsyth, W.W. and Moss, A.E., 1990. Observations of blasting and damage around development openings. *Proceedings of 92nd Canadian Institute Mining and Metallurgy Annual General Meeting, Ottawa, pp. 245–251.*

Fourney, W.L., Dally, J.W. and Holloway, D.C., 1976. Attenuation of stress waves in core samples of three types of rock. *Experimental Mechanics, April, pp. 121–128.*

Fourney, W.L. and Baker, D.B., 1979. Effect of time delay on fragmentation in a jointed model. *Mechanical Engineering Department, Univ. of Maryland, College Park Campus, USA, Aug. National Science Foundation, USA., pp. 1–12.*

Fourney, W.L., Barker, D.B. and Holloway, D.C., 1981. Model studies of explosive well stimulation techniques. *Int. Journal of Rock Mechanics and Mining Sciences, Vol. 18, pp. 113–127.*

Fourney, W.L., Barker, D.B. and Holloway, D.C., 1984. Fracture control blasting. *Proc. of the Tenth Conf. on Explosive and Blasting Technique. Jan. 29–Feb. 2, Buena Vista, Florida, Society of Explosives Engineers, pp. 182–196.*

Fourney, W.L., 1992. Fragmentation by blasting. *Int. Society for Rock Mechanics News Journal, Vol. 1, No. 1, Sept. 1992, pp. 35–39.*

Fourney, William W.L., 1993. *Mechanisms of rock fragmentation by blasting.* Comprehensive Rock Engineering. Vol. 4, pp. 39–69. Pergamon Press Ltd.

Fowell R.J. and Xu C., 1991. The CCNBD test for cutting performance prediction. *International Conference on Rock Mechanics. Aachen 15–19 Sept 1991, Germany. Balkema Rotterdam. pp. 467–490.*

Fowler, C.M.R., 2005. *The solid earth.* Cambridge Univ. Press, 685 pp.

Fraenkel, K.H., 1954. Manual on Rock Blasting. Section on Terminology. *Chapter 4:00, Esselte AB, Stockholm, pp. 1–35.*

Franklin, J.A., Broch, E. and Walton, G., 1971. Logging the mechanical character of rock. *Trans. Inst. Min. and Metallurgy, V. 80, pp. A1-A9.*

Franklin, J.A., Ibarra, J. and Maerz, H., 1989. Blast overbreak measurement by light sectioning. *Int. J. of Min. and Geological Eng., No. 7, pp. 323–331.*

Franklin, J.A. and Ibarra, J., 1991. Overbreak in relation to rock quality and blasting methods. *Proc. 9th Canadian Tunnelling Conf., Oct. 30–Nov. 2, Montréal, Québec.*

Franklin, J.A. and Katsabanis, T., 1996. Measurement of Blast Fragmentation. *Proceedings of the Fragblast 5 Workshop on Measurement of Blast Fragmentation, Aug. 23–24, Montreal, Canada,* A.A. Balkema, 315 pp.

Fröström, Jan, 1970. *Examination of equivalent model materials for development and design of sublevel caving.* Royal Institute of Technology in Stockholm, Department of Mining, Bachelor of Science Thesis E 80, 66 pp.

G

Gaudin, A.M. and Meloy, T.P., 1962. *Model and comminution distribution equation for single fracture.* AIME, 223, pp. 40–43.

Giltner, S.G., 1993. The penetration and fracturing mechanisms generated in brittle rock by the impingement of a high velocity jet. *PhD. Thesis submitted to the Faculty of Engineering, University of the Witwatersrand, Johannesburg, Republic of South Africa, 250 pp.*

Gony, J., Guillemin, C., Ragot J.P. and Sima, A., 1970. Methods d'étude du champ microfissural des mineraux et des roches au cours de leur alteration. *Rev. Indus. Minér., No. Special, pp. 40–51.*

Goodman, R., 1976. *Methods of Geological Engineering in Discontinuous Rock.* St. PaulWest. (Ref given in Brady 1985).

Grady, D.E., Kipp, M.E. and Smith, C.S., 1980. Explosive fracture studies on oil shale. *Soc. Petroleum Engineers J., pp. 249–260.*

Grady, Dennis E., Kipp, Marlin E., 1980a. Continuum modelling of explosive facture in oil shale. *Int. J. Rock Mech. Min. Sci. & Geomech. Abstr. Vol. 17, pp. 147–157. Pergamon Press Ltd.*

Grady, Dennis E., Kipp, Marlin E. and Smith, Carl S., 1980b. Explosive fracture studies on oil shale. *Society of Petroleum Engineers Journal. Okt. pp. 349–356.*

Gregory, C.E., 1980. *A Concise History of Mining.* Pergamon Press. 259 pp.

Gregory, C.E., 1983. *Rudiments of Mining Practice.* Trans. Tech. Publications. First edition.

Grove, George W., 1968. *A Dictionary of Mining, Mineral and Related Terms. Ed. Trush, Paul W. and the staff of Bureau of Mines.* Definitions furnished by George W. Grove, U.S. Bureau of Mines, Pittsburgh, Pa. US Department of the Interior, 1269 pp.

G.S.A., 1949. *Geological Society of America Memoirs 39.* (No page number available).

H

Hackh, I.W.D., 1944. *Hackh's Chemical Dictionary,* rev. and ed. by Julius Grant. The Blakiston Co., Philadelphia, Pa., 3rd ed., 1944 (reprinted with changes and additions in 1946), 925 pp.

Hagan, T.N. and Harries, G., 1979. The effects of rock properties on the design results of blasting. *Workshop Course Manual, Chapter 2, Adelaide, Australia, pp. 16–60.*

Hagan, T.N., 1979. Understanding the burn cut—a key to greater advance rates. *The Second Int. Symp. on Tunnelling, 12–16 March, London, U.K, pp. 1–8.*

Ham, R., 1965. *Dictionary of Civil Engineering.* George Newnes, Ltd., London. 253 pp.

Hammelmann, F., 1995. Gerichtetes hydraulisches impulssprengen. *Ein methodischer Weg zu Innovativen Technologien, Tagungsband zur zweitägigen Vortragsveranstaltung am 29 und 30 Juni 1995. Verlag der Augustinus Buchhandlung, pp. 163–171.*

Hanko, 1967. Kaivossanasto, Gruvterminologi, Glossary of Mining Terms. *Hangon Kirjapaino Oy,* 155 pp.

Hansagi, Imre, 1965. Practical Rock Mechanics and Rock Support. Almquist & Wiksell, Stockholm. 169 pp.

Hansagi, Imre and Hermansson, Lars, 1977. The next step after sublevel caving—sublevel shrinkage caving. Internal report LKAB. 5 pp.

Harries, Gwynn, 1973. A mathematical model of cratering and blasting. *Proc. Nat. Symp. on Rock Fragmentation, Adelaide, Australia. pp. 41.*

Harries, G. and Hengst, B., 1977. The use of computer to describe blasting. *15th APCOM Symp. Brisbane, Australia, pp. 317–324.*

Harries, H.D. and Mellor M., 1974. Cutting rock with water jets. *Int. J. Rock Mech. Min. Sci. & Geomech. Abstr. Vol.II, pp. 343–358.*

Harries, G. and Gribble, D.P., 1993. *Development of a low energy shock energy explosive.* Proc. 4th Int. Symp. on Rock Fragmentation by Blasting, Vienna, Austria, 5–8 July, pp. 379–386.

Hartman, H.L., 1961. *Mine Ventilation and Air Conditioning.* The Ronald Press Co., New York, 398 pp.

Hearst, J.R., Butkovich, T., Laine, E., Lake, R., Leach, D., Lytle, J., Sherman, J., Snoeberger, D. and Quong, R., 1976. Fractures induced by a contained explosion in Kemmerer coal. *Int. J. Rock Mech. Min. Sci. & Geomech. Abstr. Vol. 13, pp. 37–44.*

Heinen, R.H. and Dimock, R.R., 1976. The use of seismic measurements to determine the blastability of rock. *Proc. 2nd Conf. on Explosives and Blasting Technique, Louisville, Kentucky, Society of Explosives Engineers, pp. 234–248.*

Heltzen, A.M. and Kure, K. 1980. *Blasting with ANFO/Polystyrene mixtures.* Proc. of the 6th Conf. on Explosives and Blasting Techniques, Tampa Florida, 5–8 February, Cleveland, OH. International Society of Explosives Engineers, pp. 105–116.

Hendersson, G. and Bates, J.M., 1953. *Metallurgical Dictionary.* Reinhold Publishing Corp., New York. 396 pp.

Hertelendy, Andor, 1983. Automatic measurement of tunnel profiles. *Tunnels & Tunnelling, Sept.*

Hess, Frank L., 1968. *A Dictionary of Mining, Mineral and Related Terms.* Ed. Trush, Paul W. and the staff of Bureau of Mines. Definitions furnished by Fran L. Hess. Mining Engineer U.S. Bureau of Mines, College Park, Maryland.. US Department of the Interior, 1269 pp.

Heusinkveld, M., Bryan, J., Burton, D. and Snell, C., 1975. Controlled blasting calculations with the TENSOR74 Code. *California Univ., Livermore, Lawrence Livermore Lag. Contract No. W-7405-eng-48.* 68 pp.

Heuze, F.E., Walton, O.R., Maddix, D.M., Shaffer, R.J. and Butkovich, T.R., 1990. Analysis of explosions in hard rocks: The power of discrete element modelling. *Int. Conf. on Mechanics of Jointed and Faulted Rocks Vienna, Austria April 18–20. Lawrence Livermore National Laboratory Report No. UCRL-JC-103498.* 68 pp.

Higham, S., 1951. *An Introduction to Metalliferous Mining.* Charles Griffin & Co., Ltd., London. 337 pp.

Hinzen, K.G., 1988. Modelling of blast vibrations. *Int. J. Rock Mech. Min. Sci. & Geomech. Abstr. Vol. 25, No. 6, pp. 439–445.*

Hellberg, T. and Jansson, L.M., 1983. *Alfred Nobel.* Lagerblads Tryckeri AB, Sweden, 339 pp. (In Swedish).

Hoek, E., 1974. Progressive caving induced by mining an inclined orebody. *Trans. Inst. Min. Metall. 83. A 133–139.*

Holloway, D.C., Bjarnholt, Gjert and Wilson, W.H., 1986. A field study of fracture control techniques for smooth wall blasting. *Proc. 27th Symp. on Rock Mech. pp. 456–463.*

Holmberg, R. 1984. Improved stability through optimized rock blasting. *Proc. of the Tenth Conf. on Explosive and Blasting Tech. Orlando Florida, Jan. 28 to Feb. 2., pp. 166–181.*

Holmes, Arthur, 1928. *The Nomenclature of Petrology; With reference to selected literature.* Thomas Murby & Co., London. 284 pp.

Hommert, P.J., Kuszmaul, R. and Parrish, L. 1987. Computational and experimental studies of the role of stemming in cratering. *Second Int. Symp. on Rock Fragmentation by Blasting, Keystone, Colorado, USA, Society of Experimental Mechanics, pp. 550–562.*

Honda, H., 1984. Application of water jet cutting to economical tunnel excavation. *7th Int. Symp. on Jet Cutting to Economical Tunnel Excavation. Ottawa, Ontario, June 26–28. pp. 551–558.*

Hoshino, Kenzo, 1980. Application of water jet cutting on the smooth blasting. *Proc. of the 5th Int. Symp. on Jet Cutting Technology. Hannover Germany, June 2–4, pp. 165–180.*

Houpert, R. and Hommand-Etienne, F., 1979. Influence de la temperature sur le comportement mecanique des roches: *Proc. 4th Cong. Int. Soc. of Rock Mech., Montreux. pp. 177–180. (In French).*

Hudson Coal, 1932. The Story of Anthracite. (Includes a glossary of Mining Terms, pp. 401–409). *New York, 425 pp.*

Hudson, John, A., 1989. *Rock mechanics principles in engineering practice.* Butterworths, London, p. 72.

Hudson, John A., 1991. Rock engineering mechanisms information technology (REMIT). *Int. Cong. on Rock Mechanics. Aachen 15–19 Sept 1991, Germany. Balkema Rotterdam. pp. 1113–1119.*

Hudson, John A., 1992. Rock Engineering Systems. *Theory and practice. Ellis Horwood Ltd., 185 pp.*

Hudson, John A. (Editor in Chief) and Brown, E.T. (Volume editor), 1993. *Comprehensive Rock Engineering. Principles, Practice and Projects. Volume 1. Fundamentals. Pergamon Press Ltd, pp. 752.*

Hudson, John A., Editor-in-Chief, 1993. *Comprehensive Rock Engineering. Principles, Practice & Projects.* Imperial College of Science, Technologic & Medicine, London, UK. Pergamon Press Ltd. Vol. 1–5, 4.407 pp.

Hunt, M.H. and Groves, D.G. (editors), 1965. *A Glossary of Ocean Science and Undersea Technology Terms.* Compass Publications, Inc., Arlington, Virginia, 172 pp.

Hy, 1965. *Glossary of Hydrospace Projects, Terms, Abbreviations, and Weaponry.* Hydrospace Buyers' Guide. Data Publications, Washington, D.C., Vol. 3, pp. 85–116.

Hökmark, Harald, 1990. Distinct element method modelling of fracture behaviour in near field rock. *SKB Technical report Stripa Project 91–01. 90 pp.*

I

I.C. 8137, 1963. *Bureau of Mines Information Circular 8137.*

IME, 1981. Glossary of Industry Terms. *Publication No. 12. Institute of Makers of Explosives, 28 pp.*

Ingelstam, Erik and Sjöberg, Stig, *ELFYMA-tabellen.* Third edition. Sjöbergs Förlag Stockholm. Tables, Formulas, Nomograms within Mathematics, Physics and Electronics. (In Swedish).

Ingraffea, Anthony R., Boone, Thomas J. and Swenson, Daniel V., 1993. Computer simulation of fracture process. *Comprehensive Rock Engineering. Principles, Practice & Projects. Pergamon Press Ltd. Editor John Hudson. Vol. 1, pp. 545–575.*

Innaurato, N., Mancini, R. and Cardu, M., 1998. On the influence of rock mass quality on the quality of blasting work in tunnel driving. *Tunnelling and Underground Space Technology. Vol. 13. No. 1, pp. 81–89.*

Institute of Petroleum, 1961. *A Glossary of Petroleum Terms. London, 3rd ed.*

Isaac, I.D., Bubb, C., 1981. A study of blast vibrations Part 2. *Tunnels and Tunnelling, Sept., pp. 61–65.*

Isakov, A.L., 1984. Directed fracture of rocks by blasting. *Soviet Mining Science, Sept., pp. 479–488.*

Issacson, E. de St. Q., 1962. *Rock Pressure in Mines.* Mining Publications, Ltd., London, 2nd ed., pp. 39.

ISO 31-0, 1993. Quantities and Units. Part 0: General principles, Section 3.3.2 Decimal sign. *Swedish Standard SS-ISO 31-0, p. 12. (In Swedish).*

ISRM, 1975. Terminology (English, French, and German). *Commission on Terminology, Symbols and Graphic Representation. Final draft: July. 250 terms, 83 pp.*

ISRM, 1978. Commission on Standardization of Laboratory and Field Tests. *Suggested methods for the quantitative description of discontinuities in rock masses, pp. 319–368.*

ISRM, 1978. Suggested Methods for Determining Hardness and Abrasiveness of Rocks. *Int. J. Rock Mech. Min. Sci. and Geomechanic Abstr. Vol. 15. No. 3, pp. 97–103.*
ISRM, 1981. Rock Characterization, Testing and Monitoring, ISRM Suggested Methods. *Pergamon Press, Ed. by E.T. Brown, 221 pp.*
ISRM, 1985. Suggested Methods for Determining Point Load Strength. *April, pp. 51–60.*
ISRM, 1988. Suggested Methods for Determining the Fracture Toughness. *April, pp. 71–96.*
ISRM, 1992. Suggested Methods for Blast Vibration Monitoring. *Int. J. Rock Mech. Min. Sci. and Geomechanic Abstr. Vol. 29, No. 2., pp. 143–156.*
ISSE, 2007. Suggested methods for the measurement of the velocity of detonation of an explosive in a blast. Chairman and editor, Alex Spathis, Australia. In preparation for publication.

J

Jana, S., 1991. Non-explosive expanding agent—an aid for reducing environmental pollution in mines. *Indian Mining and Engineering Journal, Oct., pp. 31–35.*
Janelid, Ingvar and Kvapil, Rudolf, 1966. Sublevel caving. *Int. J. Rock . Mech. Min. Sci. 3, pp. 129–153.*
Jeansson, Johan, 2002. Water powered drilling—a new drilling technology. *Nordic Steel and Mining Review. The Annual International Issue of Bergsmannen with Jernkontorets Annaler. No. 3., pp. 122–123.*
Jimeno, C.L., Jimeno, E.L., Carcedo, F.J.A. and Y.V. De Ramiro., 1995. *Drilling and Blasting of Rocks.* A.A. Balkema, Rotterdam, 391 pp.
Jimeno, E.L. and Muniz, H.E., 1987. A new method for the design of bench blasting. *Proc. 2nd Int. Symp. on Rock Fragmentation by Blasting, Aug. 23–26, pp. 302–309, Keystone, Colorado, USA. Society of Experimental Mechanics, pp. 302–307.*
Johansson, C.H. and Persson, P.A., 1970. *Detonics of High Explosives.* Academic Press London and New York, 330 pp.
Jones, Donald C. and Hunt, Joseph W., 1949. *Coal Mining.* Pennsylvania State College, 3rd ed. Vol. 1 and 2. Vol. 1. Includes a glossary of mining terms, pp. 322–354.
Jonsson, B.A., 1992. Personal communication with the European Federation of Explosive Engineers, (EFEE). *Björn Jonsson, Nitro Consult AB, Kilabergsvägen 8, Box 32058, 126 11 STOCKHOLM.*
Jönsson, Jan. 2006. Chips or cores. *Nordic Steel and Mining Review, Bergsmannen med Jernkontorets Annaler. No. 3, pp. 101–103.*

K

Kahriman, A. and Ceylanodlu, A., 1996. Blast design and optimisation studies for a celestite open-pit mine in Turkey. *Mineral Resources Engineering. Vol. 5, No. 2, pp. 93–106.*
Kavetsky, A., Chitombo, G.P.F. and McKenzie, C.K., Yang, R.L., 1990. A model of acoustic pulse propagation and its application to determining Q for a rock mass. *Int. J. Rock Mech. Min. Sci. and Geomech. Abstr., pp. 32–41.*
Kaye, S.M. and Herman, H.L., 1983. *Encyclopaedia of Explosives and Related Items.* US Army Armament Research and Development Command. Large Caliber Weapon Systems Laboratory. Dover, New Jersey. 10 Volumes. PATR 2700.
Kazi, A. and Sen, Z., 1985. Volumetric RQD: An index of rock quality. *Proceedings of Int. Symp. on Fundamental of Rock Joint. Björkliden, 15–20 Sept. Luleå University of Technology. pp. 95–102.*
Keller, P.A., 2001. Six Sigma development. *A Guide for Implementation of Sig Sigma in Your Organization.* Tucson, Arizona. Quality Publishing.

Kennedy, B.A., 1990. *Surface Mining. 2nd edition.* Society for Mining and Metallurgy, and Exploration, Inc., Littleton, Colorado. 1194 pp.

Kennedy, J.E. and Jones, D.A., 1993. Modelling shock initiation and detonation in the non-ideal explosive PBXW-115. *Tenth International Symposium on Detonation. July 16, Boston, Massachutes, USA.*

Kentucky, Mayo, 1952. *Elements of Practical Coal Mining.* Mayo State Vocation School (Paintsville), and the Kentucky Mining Institute. 436 pp.

Keough, Douglas D., De Carli, Paul S., Hall and Lee B., 1979. Ytterbium piezoresistant gage for measurement of air shocks to 2 GPa. *Airspace Industry Vol. 25., pp. 659–666.*

Kikuchi, Kohkichi, 1991. Investigation and estimation on discontinuity (including geothermography). *Rock Mechanics in Japan Volume VI. Japanese Committee for ISRM.*

Kirby, I.J. and Leiper, G.A., 1985. A small divergent detonation theory of intermolecular explosives. *Proceedings of the Eight International Symposium on Detonation. Albuquerque, New Mexico, USA, July 15–19, pp. 176–186.*

Kiser, A.B., 1929. *Coal Cutters, Loaders, and Conveyors; Trackwork.* International Textbook Co., Scranton, Pa. 97 pp. In two parts. Part. 1 Coal cutters, Loaders, and Conveyors, 39 pp. Part 2, Trackwork, 58 pp.

Kjartansson, E., 1979. Constant Q wave propagation and attenuation. *J. Geophys. Research, Vol. 84, pp. 4737–4748.*

Koenen, H. and Ide, K.H., 1956. Ermittlung der Empfindlichkeit Explosiver Stoffe gegen Thermische Beanspruchung (Stahlhulsen verfahren). *Explosivstoffe No. 4, pp. 119–143. (In German).*

Konya, C.J. and Walter, E.J., 1990. *Surface Blast Design.* Prentice Hall Int. USA.

Konya, C.J., 1996. Problems with deck-loaded blastholes. *Eng. and Min. Journal, July, p. 73.*

Kornfält, Karl-Axel, Vikman, Hugo, Nordlund, Erling and Chunlin, Li, 1991. Blasting damage investigations in access ramp section 0/526–0/565 m. No.4. Optical examination of microcracks in thin sections of core samples and acoustic emission of core samples. *Progress report 25-91-15. SKB, Äspö hard rock laboratory., pp. 1–15.*

Korson, George, 1938. *Minstrels of the Mine Patch, Songs and Stories of the Anthracite Industry.* University of Pennsylvania Press, Philadelphia, Pa. 332 pp. Includes a glossary of anthracite technical and colloquial word and phrases, pp. 311–320.

Korzak, G., 1989. *Dictionary of Fracture Mechanics. English/German and German/English.* VCH Publishers (UK) Ltd., ISBN 0-89573-896-1, 277 pp.

Kou, S.Q. and Rustan, A., 1992. Burden related to blasthole diameter in rock blasting. *Int. Journ. Rock Mech. Min. Sci. and Geomech. Abstr. Vol. 29, No. 6, pp. 543–553.*

Krantz, Robert, 1979. Crack-crack and crack-pore interactions in stressed granite. *Int. J. Rock Mech. Min. Sci. & Geomechanics Abstr. Vol. 16, pp. 37–47.*

Krzyszton, Danuta, 1986. Ultrasonic investigations of mechanical properties of jointed rock block. *Archiwum Gornictwa. Tom 31, Zeszyt 2. pp. 311–358.*

KTH, 1969. Bergborrning Kompendium.(Rock drilling Compendium). Department of Mining, Royal Institute of Technology, Stockholm, Atlas Copco Printing Office, pp. 173. (In Swedish).

Kuchta, Mark E., 2002. A revised form of the Bergmark-Roos equation for describing the gravity flow of broken rock. *Mineral Resources Engineering, Vol. 11, No. 4 (2002), pp. 349–360.*

Kutter, H.K., 1969. The electrohydraulic effect: Potential application in rock fragmentation. *USBM, RI 7317.* 35 pp.

Kuznetsov, V.M., 1973. The mean diameter of the fragments formed by blasting rock. *Soviet Mining Science, Vol. 9, pp. 144–148.*

Kuzyk, G.W., Onagi, D.P., Keith, S.G. and Karklin, G.R., 1995. The development of long blast rounds at AECL's underground research Laboratory. *Mineral Resources Engineering, Vol. 4, No. 3, pp. 225–235.*

Kvapil, Rudolf, 2004. Gravity flow in sublevel and panel caving—a common sense approach. Separate print given to the participants at MassMin 2008. Luleå University of Technology Press, Luleå, Sweden. 151 pp.

L

Lama, R.D. and Vutukuri, V.S., 1978. *Handbook on Mechanical Properties of Rocks. Vol. II*. Clausthal, Trans. Tech Publications, 481 pp.

Langefors, U. and Sjölin, T., 1963. Relation between fragmentation, spacing to burden ratio and initiation in model tests. *Bergsprängningskommitténs diskussionsmöte i Stockholm, Feb. 28, pp. 48–51. (In Swedish).*

Langefors, U. and Kihlström, B., 1978. *The Modern Technique of Rock Blasting.* John Wiley & Sons, Inc., New York. 405 pp.

Lapedes, D.N., Editor in Chief, 1978. *McGraw-Hill Dictionary of Physics and Mathematics.* McGraw-Hill Company, 1074 pp.

Latham, J.P. and Lu, Ping, 1999. Development of an assessment system for the blastability of rock masses. *Int. J. of Rock Mech. and Mining Sci. Vol. 36, pp. 41–55.*

Laubscher, D.H., 1976. Geomechanics classification of jointed rock masses—mining applications. *Trans. Inst. Min. Metall. 86, Sect A1-A8.*

Lauritzen, E.K. and Schneider, J. editors, 1992. First International Concrete Blasting Conference. *Copenhagen, Denmark, 18–19 June 1992, 189 pp.*

Leet, L. Don, 1960. *Vibrations from Blasting Rock.* Harvard University Press, Cambridge, Mass.

Leighton, J.C., Brawner, C.O. and Stewart, D., 1982. Development of a correlation between rotary drill performance and controlled blasting powder factors. *CIM Bulletin, Vol. 75, No. 844, pp. 67–73.*

Leiper, G.A. and Cooper, J., 1990. Reaction rates and the diameter effect in heterogeneous explosives. *Proceedings of the Ninth International Symposium on Detonation. Portland, Oregon, pp. 197–208.*

LeJuge, G.E., Jubber, L., Sandy, D.A. and McKenzie, C.K., 1993. Plenary lecture not in documentation. *4th Int. Symp. on Rock Fragmentation by Blasting in Vienna, Austria, 5–8 July.*

Lewandowski, Thomas, Luan Mai, V.K. and Danell, Richard E., 1996. Influence on discontinuities on presplitting effectiveness. *5th Int. Symp. on Rock Fragmentation by Blasting. Montreal, Quebec, Canada, 25–29 Aug. pp. 217–225.*

Lewis, R.S. and Clark, G.B., 1964. *Elements of Mining.* John Wiley & Sons, Inc., New York, 3rd ed., 768 pp.

Lien, R., 1961. En indirekte undersökelsesmetode for bestemmelse av bergartenes borbarhet. *Avhandling for den tekniske licentiatgrad i ingeniörgeologi ved N.T.H. (In Norwegian).*

Liinamaa, M.A. and Reed, L.D., 1988. Light and charge emissions from rock during fracture: Applications to mining. *Mining Science and Technology, 7, pp. 277–283. Elsevier Science Publishers B.V., Amsterdam, The Netherlands.*

Li Gonbo and Xu, Xiaohe, 1993. Fractal dimension as morphology and size parameters of fractured particles of rock. *Transactions of Nonferrous Metals Society of China. Vol. 3 No. 1, Feb. pp. 6–9.*

Lislerud, A., 1990. Rock blasting technique. *University of Trondheim, PhD Thesis No. 1990:85, 229 pp. (In Norwegian).*

Liu, Qian, 1994. A study of blasthole stope—blast damage control. *CANMET. Vol. 1; Chapters 1–6, pp. 146.*

Livingstone, C.W., 1956. Fundamentals of rock failure. *Quarterly of the Colorado School of Mines. 51(3), July 1956.*

Lizotte, Yves C., 1990. Empirical procedures for prediction of rock fragmentation by blasting. *CANMET Mining Research Laboratories Report MRL 90.06 (TR), Jan. 59 pp.*

Lizotte, Yves C., 1994. Geological control over blast fragmentation. *CIM Bulletin, Sept.*

Long, A.E.A., 1960. Glossary of the Diamond Drilling Industry. *BuMines Bull. 583, 98 pp.*

Lovely, B.G., 1973. A study of the sizing analysis of rock particles fragmented by a small explosives blast. *1st National Symposium on Rock Fragmentation. Adelaide, Australia, Feb 26–28. pp. 24–34.*

Lownds, C.M., 1975. Prediction of the performance of explosives in bench mining. *J. of the South African Inst. of Min. and Met., Feb., pp. 165–180.*

Lownds, C.M. and du Plessis, M.P., 1984. The double pipe test for commercial explosives. Description and results. *Propellants, Explosives and Pyrotechnics 9.*

Lownds, C.M., 1986. The strength of explosives. *The Planning and Operation of Open-Pit and Strip Mines. Ed. Deetlefs, J.P. Johannesburg, SAIMM, pp. 151–159.*

Lundborg, Nils, 1967. The strength-size relation of granite. *Int. J. Rock Mech. Min. Sci. Vol. 4, pp. 269–272.*

Lundborg, Nils, 1972. A statistical theory of the polyaxial compressive strength of materials. *Int. J. Rock Mech. Min. Sci. 9, pp. 617–624.*

Lutton, R.J., 1969. Systematic mapping of fracture morphology. *Geol. Soc. Am. Bull. Vol. 80, pp. 2001–2066.*

M

Mader, C.L., 1979. *Numerical Modelling of Detonations.* University of California Press, Berkeley, CA.

Magnusson, Nils H., Lundqvist G. and Regnéll, Gerhard. 1963. *Sverige Geologi.* Svenska Bokförlaget. (In Swedish).

Mainiero, Richard J. and Rowland, James H., 2009. A review of recent accidents involving explosives transport. NIOSH, Pittsburgh Research Laboratory, Pittsburgh, PA, USA. *The Journal of Explosives Engineering, Vol. 26, No. 2, March/April.*

Mairena, Hector, 1991. New techniques for planning and costs estimation concerning mechanized mining in narrow orebodies. *Licentiate Thesis 1991:20 L, Luleå University of Technology, Division of Mining and Rock Excavation. 92 pp.*

Mandelbrot, B.B., Passoja, Dann. E. and Paullay, Alvin J. 1984. Fractal character of fracture surfaces in metal. *Nature Vol. 308 (19/4 1984), pp. 721–722.*

Mason, E. ed., 1951. *Practical Coal Mining for Miners.* Virtue & Co. Ltd, London, 2nd edition.

MassMin 2008. Proceedings of the 5th International Conference and Exhibition on Mass Mining, Luleå, Sweden, 9–11 June 2008. Editors Håkan Schunnesson and Erling Nordlund. Luleå University of Technology Press, Luleå, Sweden.

Maynard, Brian, 1990. A blast design model using the inherent fragmentation of a rock mass. *CIM Bulletin, Aug.*

McAdam, R. and Westwater, R., 1958. *Mining Explosives.* Oliver and Boyd, London, 187 pp.

McGraw-Hill, 1984. *Dictionary of Scientific and Technical Terms.* McGraw-Hill, 5 ed., 2194 pp.

McHugh, S.L., Curran, D.R. and Seaman, L., 1980. The NAG-FRAG computational fracture model and its use for simulating fragmentation and fracture. *Society of Experimental Stress Analysis. Fall meeting, Ft. Lauderdale, Florida.*

McKenzie, C.K., Stacey, G.P. and Gladwin, M.T., 1982. Ultrasonic characteristics of a rock mass. *Int. J. of Rock Mech. and Min. Sci. and Geomech. Abstr. Vol. 19, pp. 25–30.*

McKinstry, Hugh Exton., 1948. *Mining Geology.* Prentice-Hall, Inc, New York.

Mellor, Malcom, 1975. Controlled perimeter blasting in cold regions. Corps of Engineers, U.S. Army. *Cold Regions Research and Engineering Laboratory, Hanover, New Hampshire. CRREL Techn. Report 267. pp. 24.*

Mero, John L., 1964. *The mineral resources of the sea.* Elsevier Publishing Co., Amsterdam, Netherlands. 312 pp.

Meyer, R., 1977. *Explosives.* Verlag Chemie, Weinheim, New York. 358 pp.

Miguel, Luis and del Rio, Suarez, 1983. Effects of pores and cracks on mechanical and physical properties of rocks: Instrumental procedures to locate the onset of micro fissures in rocks under compressive loads. *Course literature in the "Workshop on microcracks", April 25–26 1983, Div. of Rock Mechanics, Luleå University.*

Militzer, Heinz and Petzold, Hellfried, 1978. Möglichkeiten, Stand und einige Ergebnisse zur Kluftigkeitsbestimmung im untertägigen Bergbau. *Neu Bergbautechnique. 8 Jg.Heft 1, Jan., pp. 28–32.*

Miron, Y., 1992. Blasting hazards of gold mining in sulphide-bearing ore bodies. *USBM IC 9335.*

Mohanty, B., 1991. Precision delay detonators and their effect on blasting performance in quarry blasts. *Proc. of the Conference on Explosives and Blasting Techniques, Feb. 3–7, 1 Society of Explosives Engineer, Vol. I.*

Mojtabai, N. and Beatti, S.G., 1996. Empirical approach to prediction of damage in bench blasting. *Trans. Inst. Min. and Metall. Sect A, Vol. 105, Jan.-April, pp. A 75–A 80.*

Montolo, Luis, 1982. Digital multi-image analysis: application to the quantification of rock microfractography. *IBM Journal of Research and Development, Vol. 26, No. 6, Nov.*

Montoto, Modesto, 1983. Petrophysics: The petrographic interpretation of the physical properties of rocks. *Proc 3rd Int. Soc, of Rock Mech. Conf. Melbourne, Australia. April.*

Morrey, W.B., 1982. The Magnadet electric initiation system. *CIM Bulletin, Nov. 1982, pp. 61–66.*

Moser, P., Gaich, A., Zechmann, E. and Grasedieck, A., 2006. The SMX Blast Metrix—A new tool to determine the geometrical parameters of a blast based on 3D imaging. *8th Symposium on Rock Fragmentation by Blasting, Fragblast 8, Santiago, Chile, May 7–11, pp. 80–84.*

Muftuoglu, Y.V., Paþamehmetoðlumehmetoglu, A.G. and Karpuz, C., 1991. Correlation of powder factor with physical properties and rotary drill performance in Turkish surface coal mines. *7th Int. Cong. on Rock Mechanics, Aachen, Germany, A.A. Balkema, pp. 1049–1051.*

Müller, Bernd, 1990. Erarbeitung der Sprengtechnologie für verschiedene Festgebirge und die beurteilung des Sprengergebnisses. *Nobel Hefte, Juli-Dez, pp. 92–102.*

Müller, Bernd, 1991. Praktische Ergebnisse von Großbohrlochsprengungen mit doppelter Zündung. *Spreng-Info 1991/2. Vortrag in Siegen 1991.*

Müller, Bernd., 1996. Minimierung von Erschütterungen bei Großbohrloch-sprengungen. *Glückauf 132, No. 6, pp. 257–262.*

Müller, L. and Hoffman, H., 1970. Selection, compilation and assessment of geological data for the slope problem. Planning open pit mines, ed. P.W.J. van Rensburg. *A.A. Balkema, Cape Town (1970), pp. 153–170.*

N

Nantel, J.H. and Kitzinger, F., 1990. Plasma blasting techniques. *Third Int. Symp. on Rock Fragmentation by Blasting, Brisbane, Australia, pp. 79–82.*

NCB, Great Britain, 1964. *A Glossary of Automation and Remote Control as Applied to the Coal Mining Industry.* London. 34 pp.

Nelson, A., 1965. *Dictionary of Mining.* Philosophical Library Inc., New York, 523 pp.

NEMA MB1, 1961. *NEMA Standards Publication for mining belt conveyors.* National Electrical Manufacturers Association Pub. MB1-1961.

Neugebauer, Erich, 2008. Ready for Roofex. A new way to tackle safety in underground mining operations. *Mining and Construction. Mechanized Rock Excavation with Atlas Copco. No. 3, pp. 12–13.*

New South Wales, 1958. *Department of Mines. Prospector's Guide.* Sydney, Australia, 7th ed., 1958. Includes a glossary of mining terms, pp. 191–196.

Newton, Joseph, 1959. *Extractive Metallurgy.* John Wiley and Sons, Inc., New York. 532 pp.

Nichols, H., 1956. *Modern Techniques of Excavation.* North Castle Books, Greenwich, Conn., USA.

Nichols, Herbert L., Jr., 1962. *Moving the Earth; the Workbook of Excavation.* North Castle Books, Greenwich, Conn., 2nd ed., v.p. Includes a glossary, pp. G-1 to G-27.

Nichols, H.R. et al., 1965. Comparative study of explosives in granite. Third series of tests. *USBM RI 6693, 46 pp.*

Nie, Shu Lin, 1988. New hard rock fragmentation formulas based on model and full-scale tests. *Luleå University of Technology, Luleå, Sweden, Licentiate Thesis No. 1988:02 L, 138 pp.*
Nitro Nobel, 1991. Explosive News. *Information Journal from Nitro Nobel, No. 2, pp. 18–19.*
Nitro Nobel, 1993. Introduction. 2. Nomenclature. *Nitro Nobel Blasting Techniques Department. Publication No. C 731, 23 pp.*
Noguchi, K. and Goto, S., 1991. In-situ experimental study on resistivity tomography in tunnels. *International Conference on Rock Mechanics. Aachen 15–19 Sept 1991, Germany. Balkema Rotterdam. pp. 577–580.*
Noon, David A., Stickley, Glen F. and Longstaff, Dennis Editors, 2000. *Eighth International Conference on Ground Penetrating Radar. SPIE (The International Society for Optical Engineering), Bellingham, Washington, USA. Vol. 4084.*
Nord, Gunnar, Olsson, Per and By, Lasse Tore, 1991. Geophysical ground probing in TBM tunnelling. *International Conference on Rock Mechanics. Aachen, 19–23 Sept 1991, Germany, pp. 581–585. Balkema Rotterdam.*
Nordenström, G., 1880. Om nitroglycerinhaltiga sprängämnen. *Aftryck av Jernkontorets Annaler, 1880. K.L. Beckman, pp. 84. (In Swedish)*
Nordic Mining Review, 2006. Raise boring at LKAB in Kiruna. *Bergsmannen med Jernkontorets Annaler, No. 3, p. 111.*
Nordyke, Milo D., 1962. Nuclear and chemical explosive cratering experience applicable to plowshare. *Int. Symp. on Mining Research. University of Missouri, Feb 1961. Editor George B. Clark, Pergamon Press. Vol. 1 (2), pp. 211–226.*
Noren, C.H., 1974. A comparison of theoretical explosive energy and energy measured underwater with measured rock fragmentation. *Proceedings of the Third Congress of the Int. Society for Rock Mechanics. Denver 1974. Vol. II. Part B., pp. 1371–1376.*

O

Obara, Yuzo, Sugawara, Katsuhiko, and Sakaguchi, Kiyotoshi, 1991. Application of hemispherical-ended borehole technique to hot rock. *International Conference on Rock Mechanics. Aachen, 19–23 Sept 1991, Germany, Balkema Rotterdam, pp. 587–594.*
Oberndorfer, T., 1993. Computer-aided mining method decision with special emphasis on computer-oriented mining method description. *Scientific Publications, Department of Mining Engineering and Mineral Economics, Montan-University, Leoben, Austria. Volume 1, 328 pp.*
O'Keefe, S.G. and Thiel, D.V., 1991. Electromagnetic emission during rock blasting. *Geophysical Research Letters. Vol. 18. No. 5., pp. 889–892.*
Olofsson, S.O., 1990. *Applied explosives technology for construction and Mining.* Nora boktryckeri AB, Sweden, pp. 303.
Olsson, James J., Eillard, Robert J., Fogelson, David E. and Hjelmstad Kenneth E., 1973. Rock damage from small scale charge blasting in granite. *USBM RI 7751, 44 pp.*
Olympus Industrial, (1990). Video image scopes. *Pamphlet from Olympus Industrial Endoscopy System, Tokyo, Japan. Printed in U.K. GR5019028/289, 2 pp.*
Onederra, I., 2004. Estimation of fines generated by blasting—applications for the mining and quarrying industries. Mining Technology. Trans. Min. Metall. A. Dec 2004, Vol. 113, pp. A 237–247.
Onodera, T.F., 1963. Dynamic investigation of foundation rock in situ. *In Proc. 5th U.S. Rock Mech., Minneapolis, MN. Editor C. Fairhurst. Pergamon Press, Oxford, pp. 517–533.*
Osborne, A.K. An Encyclopaedia of the Iron & Steel Industry. *Philosophical Library Inc., New York, 1957, 188 pp.*
Ouchterlony, Finn, 1983. New methods of measuring fracture toughness on rock cores. *First Int. Symp. on Rock Fragmentation by Blasting, 22–26 Aug. 1983, Luleå. pp. 199–223.*

Ouchterlony, Finn., Niklasson, Bengt. and Abrahamsson, Sten, 1990. Fragmentation monitoring of production blasts at Mrica. *Third Int. Symp. on Rock Fragmentation by Blasting. Brisbane, Australia, pp. 283–289.*

Ouchterlony, Finn, 2005. The Swebrec function linking fragmentation by blasting and crushing. *Mining Technology. (Trans. Inst. Min. Metall. A. March. Vol. 114. pp. A29-A44.*

P

Paine, G.G., 1983. The use of computer blast modelling to optimise blast design. *ICI, Australia. pp. 196–242.*

Palmström, A., 1985. Application of the volumetric joint count as a measure of rock mass jointing. *Proceedings of the Int. Symp. on Fundamentals of Rock Joints, Björkliden, 15–20 Sept., pp. 103–110.*

Park, R.G., 1989. *Foundations of Structural Geology.* Blackie & Son Ltd., 148 pp.

Paventi, M., Lizotte, Y., Scoble, M. and Mohanty, R., 1996. Measuring rock mass damage in drifting. *Fifth Int. Symp. on Rock Fragmentation by Blasting, Aug. 26–29, Montreal, Canada, pp. 131–138.*

Peck, Jonathan and Hendricks, Carl, 1997. Application of GPS-based navigation systems on mobile mining equipment in open-pit mines. *CIM Bulletin. Vol. 90, No. 1011, June, pp. 114–119.*

Pegden, M., Birch, W.J. and White, T.J., 2006. The multilevel modelling of blast vibration data. *Eight International Symposium on Rock Fragmentation by Blasting, May 7–11 2006, Santiago, Chile. pp. 226–231.*

Persen, L.N., 1975. *Rock Dynamics and Geophysical Exploration.* Elsevier, Amsterdam, Oxford and New York.

Persson, P.-A., Holmberg, R. and Lee, J., 1994. *Rock Blasting and Explosives Engineering. A textbook for students and a handbook for scientists and engineers.* CRD Press, Inc. Florida, USA. 540 pp.

Peterson, C.R., Fisk, A.T., Brooks, R.E. and Olson, J.J., 1977. Spiral drill-and blast concept may speed up tunnelling advance rate. *Mining Engineering, June, pp. 29–32.*

Petrosyan, M.I., 1994. *Rock breakage by Blasting.* Russian Translations series 105. A.A. Balkema, 141 pp.

Pierce, M., Mas, D., Cundall, P. and Patyondy, D., 2007. A synthetic rock mass model for jointed rock. *1st Canada-US Rock Mechanics Symposium, May 27–31, 2007, Vancouver, Canada.*

Pit and Quarry, 1960. *Pit and Quarry Handbook and Purchasing Guide for the Non-metallic Minerals Industries.* Pit and Quarry Publications, Inc. Chicago, III, 53rd ed.

Poli, R., Langdon, W.B. and McPHee, N.F. 2008. *A Field Guide to Genetic Programming.* Springer Verlag.

Porter, Hollis, 1930. *Petroleum Dictionary for Office, Field and Factory.* Gulf Publishing Co., Houston, Tex., 234 pp.

Pradhan, G.K., 1996. *Explosives and Blasting Techniques.* Mintech Publications, Bhubaneswar, India, 388 pp.

Pradhan, Gyanindra Kumar, 2001. *Explosives and Blasting Techniques.* Shreema Printers, Bhubaneswar, India. 386 pp.

Preston, Christopher J. and Tienkamp, Norman J., 1984. New Techniques in blast monitoring and optimization. *CIM Bulletin, July, 43–48*, pp. 43–48.

Priest, S.D. and Hudson, J.A., 1981. Discontinuity spacing in rock. *Int. J. Rock Mech. and Min. Sci. 13, pp. 135–148.*

Pryor, E.J., 1963. *Dictionary of Mineral Technology.* Mining Publications Ltd., London, 1963, 437 pp.

Pryor, Edmund J., 1965. *Mineral Processing.* Elsevier Publishing Co, Ltd, London, 3d ed., 1965, 844 pp. Includes a glossary. Appendix A., pp. 809–818.

Pusch, Roland, 1974. *Geoteknik.* Allmqvist och Wiksell, Uppsala, 326 pp. (In Swedish).
Pusch, R. and Börgesson, L., 1992. Performance assessment of bentonite clay barrier and nearfield rock assuming intact canisters. A PASS-project on Alternative Systems. *Swedish Nuclear Fuel and Waste Management Co (SKB). Technical Report.*
Pusch, R. and Stanfors, R., 1992. The zone of disturbance around blasted tunnels at depth. *Int. J. Rock Mech. Min. Sci. and Geomech. Abstr. Vol. 29, No. 5, pp. 447–456.*

Q

Qian Liu and Katsabanis, P.D., 1993. A theoretical approach to the stress waves around a borehole and their effect on rock crushing. *Fourth Int. Symp. on Rock Fragmentation by Blasting, 5–8 July, Vienna, Austria.* pp. 9–23.
Qin, J.F., Qin, R.X. and Li B.H. Study and application of an elliptical bipolar linear shaped charge. *9th Int. Symp. on Rock Fragmentation by Blasting. Granada 15–17 Sept. pp. 165–170.*

R

Ramana, Y.V. and Venkatanarayany, B., 1973. Laboratory studies on Kolar rocks. *Int. J. Rock Mech. Min. Sci. & Geomech. Abstr. Vol. 10.*
Reyes, O. and Einstein, H.H., 1991. Failure mechanisms of fractured rock-a fracture coalescence model. *Seventh Int. Cong. on Rock Mechanics. Aachen, Germany, pp. 233–340.*
Rice, C.M., 1960. Dictionary of Geological Terms (Exclusive of Stratigraphic Formations and Paleontologic Genera and Species). *Edwards Brothers, Inc., Ann Arbor, Mich., 465 pp.*
Riley, W.F., and Dally, J.W., 1966. A photoelastic analysis of stress wave propagation in layered model. *Geophysics Vol. 35, pp. 881–899.*
Rinehart, J.S., 1970. Fractures and strain generated in joints and layered rock masses by explosions. *Proc. Symp. Mechanism of Rock Failure by Explosions, Oct 1970. Fontinebleau, France.*
RISI, 1992. Technical topics. Topic: *Detonator diagnostics. Issue No. 11-92, 2 pp.*
Roark, Raymond J., 1954. *Formulas for Stress and Strain.* McGraw-Hill Book Co., New York, 3rd ed. Includes a definition chapter pp. 3–15.
Roberts, A. ed., 1960. *Mine Ventilation.* Cleaver-Hume Press, Ltd., London. 363 pp.
Roegiers, Jean-Claude and Liang Shao, Xuan, 1991. *The measurement of fracture toughness of rocks under simulated downhole conditions.* Int. Conf. on Rock Mech., Aachen, 19–23 September 1991, Germany. Balkema Rotterdam. pp. 60–608.
Rollins, R.R. and Clark G.B., 1973. Penetration in granite by jets from shaped-charge liners of six materials. *Int. J. Mech. Min. Sci. & Geomech. Abstr. Vol. 10, pp. 183–207.*
Rossmanith, H.P., 1978. Dynamic fracture in glass. *Prepared for National Science Foundation by Photmechanics Laboratory, Mechanical Engineering Department, University of Maryland, College Park, Maryland 20742. 78 pp.*
Rossmanith, H.P., Knasmillner, R.E., and Fourney, William, 1987. Fracture initiation by stress wave diffraction at cracked interfaces. *Second Int. Symp. on Rock Fragmentation by Blasting. Keystone, Colorado, Aug 24–26 1987. pp. 172–191.*
Rossmanith, H.P., 2000. The influence of delay timing on optimal fragmentation in electronic blasting. *First Conf. on Explosives & Blasting Technique. Editor Roger Holmberg. pp. 141–147.*
Rotschlau, Horst, 1994. Sprengtechnologie in untertägigen Uranerzbergbau der AG/SDAG Wismut (II). *Erzmetall 47, Nr. 8., pp. 480–488.*
Roy, P.P. and Dhar, B.B., 1993. Rock fragmentation due to blasting-a scientific survey. *The Indian Mining and Engineering Journal. Vol. 32, No. 9, Sept., pp. 27–32.*

Rustan, A., Vutukuri, V.S. and Naarttijärvi, T., 1983. The influence from specific charge, geometric scale and physical properties of homogeneous rock on fragmentation. *First Int. Symp. on Rock Fragmentation by Blasting, Aug. 22–26, Luleå, Sweden, pp. 115–142.*

Rustan, Agne, 1983. Linear shaped charges for contour blasting or stone cutting. Full scale tests at Gårdlidenberget, Storsund. *Forskningsrapport TULEA 1983:12, Avd för Bergteknik, Tekniska Högskolan i Luleå. pp. 84.*

Rustan, P.A., Naarttijärvi, T. and Ludvig, B., 1985. Controlled blasting in hard intense jointed rock in tunnels. *CIM Bulletin, Dec., pp. 63–68.*

Rustan, A. and Nie, Shu-Lin., 1987. New method to test the rock breaking properties of explosives in full scale. *Second Int. Symp. on Rock Fragmentation by Blasting, Keystone, Colorado, USA, Society of Experimental Mechanics, pp. 36–47.*

Rustan, A., 1990. Blasting against fill in rill mining. Part 1 literature review. *Division of Mining, Technical report No. 1990:05 T, Luleå University of Technology. Mining 2000 Fragmentation report No 90:31.* 26 pp.

Rustan, A., 1993. Minimum distance between charged boreholes for safe detonation. *The Fourth Int. Symp. on Rock Fragmentation by Blasting. Vienna, Austria, 5–8 July., pp. 127–135.*

Rustan, Agne, 2000. Gravity flow of broken rock—What is known and unknown. MassMin 2000, Brisbane, 29 Oct- 2 Nov 2000, Queensland. *The Australasian Institute of Mining and Metallurgy, Victoria, Australia, pp. 557–567.*

Rustan, Agne, 2007. Weight strength, energy, gas volume, and VOD—passé? *SprängNytt, Vol. 21, No. 1, March,* pp. 22–27. (In Swedish).

Rustan, P.A., Hagan, T. and Ouchterlony, F., 1989. International list of symbols in drilling and blasting. *Luleå University of Technology, March, 9 pp.*

Rustan, P.A., 1990. Burden, spacing and borehole diameter at rock blasting. *The Australasian Institute of Mining and Metallurgy. Third Int. Symp. on Rock Fragmentation by Blasting, Brisbane, Australia, 26–31 Aug., pp. 303–309.*

Rustan, P.A., 1993. Quantification of physical connections between blastholes by water loss measurements. *Int. Cong. on Mine Design, Aug. 23–26, Kingston, Canada, pp. 863–869.*

Rustan, P.A., 1996. Micro-sequential contour blasting—Theoretical and empirical approaches. *Fifth Int. Symp. on Rock Fragmentation by blasting, Montreal, Canada, pp. 157–165.*

Rustan, P.A., 2009. A new prinicipal formula for the determination of explosive strength in combination with the rock mass strength. *9th Int. Symp. on Rock Fragmentation by Blasting. Granada, 13-17 Sept. CRC Press. pp. 155–164.*

Ryncarz, Tadeusz, 1989. Some remarks on the content of a modern course in mining geomechanics. *Mineral Resources Engineering, Vol. 2, No. 4, pp. 299–307.*

S

Sanderson, M. and Fredö, C., 1992. Moment Mobility. *SVIB Vibrationsnytt. Year 10, No. 2, June,* pp. 12–21. (In Swedish).

Sandhu, M.S. and Pradhan, G.K., 1991. *Blasting Safety Manual.* IME publications, Calcutta, India, 271 pp.

Sandström, Gösta, 1963. *Tunnels.* Holt, Rinehart and Winston, New York 427 pp. Includes a glossary, pp. 208–250.

Sanford, Samuel and Stone W. Ralph, 1914. Useful minerals of the United States. *U.S. Geological Survey, Bull. 585, 250 pp. Includes a Glossary, pp. 218–250.*

Sarracino, R., 1993. Personal communication. Technical Department. *AECI Explosives and Chemicals Limited, P/Bag X2, Modderfontein 1645, South Africa.*

Sauna, M. and Peters, T., 1982. The Cherchar abrasivity index and its relation to rock mineralogy and petrography. *Rock Mechanics, 15, pp. 1–8.*

Schach, R., 1971. Nye erfaringer med fjellbolter. *Konferense i Fjellsprengningstekkniikk, Oslo.*

Schatz, J., 1974. A one-dimensional wave propagation code for rock media. *UCRL-51689. Lawrence Livermore Laboratory, Livermore CA.*

Schieferdecker, A.A.G. (ed.), 1959. Geological Nomenclature. *J. Noorduijn en Zoon, Royal Geological and Mining Society of the Netherlands, 521 pp.*

Selmer-Olsen, R. and Blindheim, O.T., 1970. On the drillability of rock by percussive drilling. *Proc. 2nd Cong. Int. Soc. Rock Mech., Belgrade, 21–26 Sept., Yugoslavia, Publication sector of the 'Jaroslav Cerni' Institute for Development of wer Resources, Belgrade, Yugoslavia, pp. 65–70.*

Sen, G.C., 1995. *Blasting Technology for Mining and Civil Engineers.* UNSW Press, Sydney, Australia, 146 pp.

Sen, G.C., 1992. Personal communication. *University of New South Wales, School of Mines, Department of Mining Engineering, Kensington, New South Wales, Australia.*

Seto, M., Nag, D.K. and Vutukuri, V.S., 1996. Evaluation of rock mass damage using acoustic emission technique in the laboratory. *Proceedings of the Fifth International Symposium on Rock Fragmentation by Blasting. Fragblast 5. Editor Mohanty. Montreal, Quebec, Canada, Aug. 23–24, pp. 139–145.*

Shi, G-H., 1990. Forward and backward discontinuous deformation analyses of rock systems. *Proc of the Int. Conf. on Rock Joints, Loen, Norway. Balkema, pp. 731–743.*

Shipley, Robert Morrill., 1945. *Dictionary of Gems and Gemnology Including Ornamental, Decorative, and Curio Stones.* Gemnological Institute of America, Los Angeles, Calif. 2nd ed.

Silvia, G. and Scherpenisse, C. 2009. Development of low density reactive agents. *9th Int. Symp. On Rock Fragmentation by Blasting. Granada, 13–17 Sept. CRC Press. pp. 117–126.*

Simha, K.R.Y., 2001. *Fracture Mechanics for Modern Engineering Design.* University Press, Hyderabad, India. 182 pp.

Simmons, G. and Richter, D. 1976. *Microcracks in Rock.* The Physics and Chemistry of Minerals and Rocks, R.G.J. Strens (Editor). Wiley, New York, N.Y., pp. 105–137.

Sinclair, John, 1958. *Environmental Conditions in Coal Mines (Including fires, Explosions, Rescue, and Recovery Work).* Sir Isaac Pitman & Sons, Ltd., London. 341 pp.

Singh, P.K. and Roy, M.P., 2009. Blast vibration damage threshold for the safety of residential structures in mining areas. *Vibrations from Blasting. Spathis A.T. and Noy M.J. (eds). Taylor Francis Goup, London.*

Singh, R.N. and Appo Rao, Y.V., 1979. Influence of physico-mechanical properties and geological discontinuities on rock blasting. *Colliery Guardian, Nov., pp. 631–634.*

Singh, R.N., Hassani, F.P. and Elkington, P.A.S., 1983. The application of strength and deformation index testing to the stability assessment of coal measures excavations. *24 US symposium on Rock Mechanics, June, pp. 97–106.*

Singh, R.N., Elmherig, A.M. and Zunu, M.Z., 1986. Application of rock mass characterization to the stability assessment and blast design in hard rock surface mining excavations. *27th US Rock Mech. Symp., pp. 471–478.*

Singh, S.P., 1989. Classification of mine workings according to their rock burst proneness. *Mining Science and Technology. 8, pp. 253–262.*

Singh, S.P. and Wondrad, M.W., 1989. Drift blasting practices and problems in Canadian underground mines. *Mining Research Engineering, Vol. 2, No. 4, pp. 341–352.*

Singh, S.P., 1994. Blast damage control during underground mining. *Proceedings or the Twentieth Annual Conference on Explosives and Blasting Technique, Society of Explosive Engineers. Jan. 30–Feb. 3., pp. 329–341.*

SIS, 1991. Vibration and impact—Guidelines for blastinduced vibrations in buildings. Vibration och stöt—Riktvärden för sprängningsinducerade vibrationer i byggnader. *SIS Standardiseringskommissionen i Sverige. SS 460 48 66.* Dec 4th 1991. (In Swedish).

Siskind, D.E. and Fumanti, Rober R., 1974. Blast-produced fractures in Lithonia granite. *USBM RI 7901, pp. 1–38.*

Siskind, D.E., Stagg, M.S. and Kopp, I.W. and Dowding, C.H., 1980. Structure response and damgae produced by ground vibrations from surface mine Blasting. *USBN RI 8507.*

Siskind, D.E. and Stagg, M.S., 1994. Surface mine blasting near transmission pipelines. *USBM RI 9523.* 51 pp.

Skinner, B.J. and Porter, S.C., 1989. *The dynamic earth: an introduction to physical geology.* John Wiley & Sons, 541 pp.

Smith, Mike, 1987. Dimensional stone blasting in Finland. *Mining Magazine, Oct., pp. 312–317.*

South Australia, 1961. *South Australia, Department of Mines Handbook on Quarrying.* W.L. Hawkes, Government Printer, Adelaide, South Australia. 185 pp.

Spalding, Jack, 1949. *Deep Mining, and Advanced Textbook for Graduates in Mining and for Practicing Mining Engineers.* Mining Publications, Ltd., London. 405 pp. Includes a glossary, pp. 389–392.

Spathis, Alex, 1988. Evidence of rock micro structure from seismic wave propagation. *Eng. Fracture Mech.,* Vol. 35, No. 1/2/3. Selected papers presented at the Int. Conf. on Fracture and Damage of Concrete and Rock. A special seminar on large concrete dam structures. pp. 377–384.

Spathis, Alex, 2004. A correction relating to the analysis of the original Kuz-Ram model. Fragblast, *The Int. J. for Blasting and Fragmentation Vol. 8 No. 4, Dec., pp. 201–205.*

Spathis, Alex, 2007. Fragblast/ISEE Suggested methods for the measurement of the velocity of detonation of an explosive in a blast. (In preparation).

SS-EN ISO 14688-1:2002, 2003. Geotechnical investigation and testing—Identification and classification of soils. Publ. Swedish Standard Institute, 2003. CEN, 24 of June 2002.

SS-EN ISO 14688-2:2004, 2004. Geotechnical investigation and testing—Identification and classification of soils. Publ. Swedish Standard Institute, 2003. CEN, 24 of June 2002.

Standard, 1964. *Funk & Wagnalls New Standard Dictionary of the English Language.* Funk & Wagnalls Co., New York, 1964, 2816 pp.

Stauffer, David McNeely, 1906. *Modern Tunnel Practice.* Engineering News Publishing Co., New York. 314 pp. Includes a glossary of some of the more unusual terms used in tunnelling, pp. 301–307.

Stecher, F.P. and Fourney, W.L., 1981. Prediction of crack motion from detonation in brittle materials. *Int. J. Rock Mech. Min. Sci. & Geomechanic Abstr. Vol. 18, Pergamon Press Ltd. pp. 23–33.*

Stein, Joel (ed.), 1997. *Drilling. The manual of Methods, Applications, and Management.* Australian Drilling Committee Limited. Lewis Publishers, CRC Press. pp. 615.

Stephansson, Ove, 1993. Rock stress in the Fennoscandian shield. *Comprehensive Rock Engineering.* Vol. 3. Editor-in-chief John Hudson. *pp. 445–459. Map pp. 454–455.*

Stillborg, Bengt, 1994. *Professional users Handbook for Rock Bolting.* Trans Tech Publications. 164 pp.

Stoces, Bohuslav, 1954. *Introduction to Mining.* Lange, Maxwell & Springer, Ltd., London, 2 volumes. (Page no. lacking).

Stokes, W.L. and Varnes, D.J., 1955. *Glossary of Selected Geological Terms with Special Reference to Their Use in Engineering.* The Colorado Scientific Society, Denver, Colorado, 165 pp.

Stout, K.S., 1980. *Mining Methods and Equipment.* McGraw Hill, New York. 218 pp.

Strandh, Sigvard, 1983. Alfred Nobel. *The man, his work and the age in which he lived.* A book for the 150-year celibration of Alfred Nobel, 21 of October 1983. Natur och Kultur. 339 pp. (In Swedish).

Streefkerk, H., 1952. *Quarrying Stone for Construction Projects.* Uitgeverij Waltman, Delft, The Netherlands, 159 pp.

Student-Bilharz, B., 1988. *Sprengtechnologie (German). Dictionary of Blasting Technology (English). Dictionnaire de la Technologie du Tir (French). Deutsch-English-Französisch, English-German-French, and Fransçais-Allemand-Anglais.* Verlagsgesellschaft GmbH, D-6940 Weinheim (Federal Republic of Germany), 331 pp.

Swan, Graham, 1975. The observation of cracks propagating in rock plates. *Int. J. Rock Mech. Min. Sci. & Geomechanic Abstr. Vol. 12, pp. 329–334.*

T

Tamrock, 1983. *Handbook of Underground Drilling.* 307 pp.
Tamrock, 1989. *Surface Drilling and Blasting.* Published by Tamrock. Editor: Naapuuri, 479 pp.
Taylor, W.H., 1965. *Concrete Technology and Practice.* American Elsevier Publishing Co., New York, 639 pp. Includes a glossary of terms, pp. 611–623.
Thum, W and Leins, W., 1972. Der Energiebedarf bei der Gewinnung and Zerkleinerung von Gestein durch Sprengen. *Verlag Chemie, GMBH, Weinheim/Bergstrasse. Dritte Europäischen Symposions 'Zerkleinern' in Cannes, 5–8 Oct., 1971, pp. 313–345.*
Thum, Wolfgang, 1978. *Sprengtechnik im Steinbruch and Baubetrieb.* Bauverlag GmbH, Wiesbaden und Berlin. 400 pp.
TNC, 1979. *Glossary of Rock Engineering. Swedish, English, French, Spanish, German, Danish, Norwegian, and Finnish Glossary.* Swedish Centre of Technical Terminology Publication (TNC) No. 73, Stockholm, 306 pp.
Trauzl, Karl, 1959. *Explosives.* Handbok i bergsprängningsteknik, Manual on Rock Blasting Editor K.H. Fraenkel. AB Atlas Diesel, Stockholm and Sandviken Jernverks AB, Sandviken. Esselte AB. Stockholm. First published in 1952 and continuously expanded to 3 Volumes. Vol. 3, Chapter 16:03-1 to 16:03-28, dated 15 Jan 1959.
Tunstall, A.M., Djordjevic, N. and Villalobos, H.A., 1997. Assessment of rock mass damage from smooth wall blasting at El Soldado mine, Chile. *Trans. of the Institution of Mining and Metallurgy, Section A, Vol. 106*, pp. A42-A46.
Tweney, C.F. and Hughes, L.E.C. Eds, 1958. *Chambers Technical Dictionary.* MacMillan Co., New York, 3rd ed.

U

United Nations, 1990. *Recommendations on the Transport of Dangerous Goods. Tests and Criteria.* Second Edition, United Nations, New York, ST/SG/AC.10/11/Rev.1, June.
USBM, 1966. Instructions for Disaster, Fatal-accident, and Miscellaneous Health and Safety Reports, April. *Chapter 5.1, p. 45.*
USBM, 1968. *A Dictionary of Mining, Mineral and Related Terms.* Ed. Trush, Paul W. and the Staff of Bureau of Mines. U.S. Department of the Interior, 1269 pp.
USBM Staff, 1968. See 'USBM 1990'.
USBM, 1983. Explosives and Blasting Procedures Manual. *Bureau of Mines Inf. Circ. 8925, 105 pp.*
USBM, 1990. *Dictionary of Mining Terms.* Editor Paul W. Thrush and Staff of Bureau of Mines. Washington, U.S. Dept. of the Interior, Bureau of Mines. Reprinted January 1990. ISBN 0-92531-11-6. 1269 pp.

V

Villegas, T. and Nordlund, Erling, 2008. Numerical analysis of hanging wall failure at the Kiirunavaara mine. *MassMin 2008, 9-11 June 2008, Luleå, Sweden.*
Vijay, M.M., 1990. Evaluation of abrasive-entrained water jets for slotting for rocks. *Min. Res. Eng., Vol. 3, No. 2*, pp. 143–154.
von Bernewitz, M.W., 1931. *Handbook for Prospectors.* McGraw-Hill Book Co., New York, 2nd ed., 359 pp. Includes a glossary of terms used in mining, pp. 304–401.

W

Wahlström, E.E., 1955. *Petrographic Mineralogy.* New York.

Walker, P.M.B. (General editor), 1988. Chambers Science and Technology Dictionary. *W. and R. Chambers Ltd and Cambridge. University Press, pp. 983.*

Wang, H., Latham, J.-P. and Poole, A.B., 1990. In situ block size assessment from discontinuity spacing data. *Proc. Sixth Int. Cong. Int. Assoc. of Eng. Geol., 6–10 Aug., Amsterdam, Netherlands, pp. 117–127.*

Wang, H., Latham, J.P., and Poole, A.B., 1991. Producing Armour Stone Within Aggregate Quarries. *ASCE Conference Proceedings, 11 pp.*

Watson, Richard W., 1967. Gauge for determining shock pressures. *Review of Scientific Instrument, Vol. 38, nr 7,* pp. 978–980.

Webster, Noah, 1960. *Webster's New International Dictionary of the English Language, Second Edition, Unabridged.* G. & C. Merriam Co., Springfield, Mass. 3194 pp.

Webster, N., 1961. *Webster's Third New International Dictionary of the English Language,* Unabridged. G. & C. Merriam Co., Springfield, Mass, 2662 pp.

Webster, 1986. *Webster New World Dictionary of the American Language.* Editor D.B. Guralnik. Simon and Schuster, New York. 1275 pp.

Weed, Walter Harvey, 1922. *The Mine Handbook*; Succeeding the Copper handbook, Volume XV. The Mines Handbook Co., Tuckahoe, N.Y., 1922, 2248 pp. Includes a glossary of mining terms, pp. 1–22.

Weibel, E.R., 1989. *Stereological Methods. Volume 1- Practical Methods for Biological Morphometry. Volume 2- Theoretical Foundations.* Academic Press Ltd, 415 pp.

Weibull, Waloddi, 1939. A statistical theory of the strength of materials. *Ing. Vetenskaps Akad. Handl. 151, Stockholm, Sweden. 45 pp.*

Weiss, E.S., Morrison, J.S, Beattie, G. and Sapko, M.J., 1992. Low incentive explosives for sulphide ore blasting. *Explosives Engineering, Volume 9, No. 5, Jan./Feb., pp. 15–27.*

Wetherelt, A. and Williams, D.C., 2006. Using high definition surveying (HDS) to quantify tunnel hole burdens and fragmentation. *8th Int. Symp. on Rock Fragmentation by Blasting, Santiago, Chile. Editec S.A.* pp. 55–60.

Whimmer, Matthias, 2008. Experimental investigation of blastability. *MassMin 2008, 9–11 June 2008, Luleå, Sweden.*

Whittaker, B.N., Singh, R.N. and Sun. G., 1992. *Rock Fracture Mechanics, Principles, Design and Applications.* Elsevier Science Publishers B.V., Amsterdam, 570 pp.

Wijk, G., 1982. The stamp test for rock drillability classification. *Swedish Detonic Research Foundation, Report DS 1982:1.*

Wikipedia, 2006. The free encyclopaedia. 2 Nov. http//en.wikipedia.org/wiki/brittle.

Williams, Howard R. and Meyers, Charles J., 1964. *Oil and Gas Terms,* Matthew Bender & Co, New York, 2nd . 449 pp.

Winzer, S.R. and Ritter, A.P., 1980. The role of stress waves and discontinuities in rock fragmentation: a study of fragmentation in large limestone blocks. *Proceedings 21st Symp. on Rock Mechanics, University of Missouri, Rolla, May 28–30, pp. 362–367.*

Wood, W.W. and Kirkwood, J.G., 1954. Diameter effect in condensed explosives. The relation between velocity and radius of curvature of the detonation wave. *Journal Chemical Physics Vol. 22, p. 1920.*

Woodruff, Seth D., 1966. *Methods of Working Coal and Metal Mines.* Pergamon Press, Oxford, England, 3 Vol. (Page No. not given).

Worsey, P., 1996. Report by Working Group on Rock Assessment. Measurement of Blast Fragmentation. *Proceedings of the Fragblast-5 Workshop on Measurement of Blast Fragmentation, Aug. 23–24, Montreal, Canada, pp. 9–11.*

X

Xuguang, Wang; Guoli, Wang; Xiaozhi, Zhang; Tingzhang, Kang; Zhigiang, Li and Hifeng, Cao, 2002. Study on powdered emulsified explosive. 7th Int. Symp. on Rock Fragmentation by Blasting, Beijing, China, 11–15 of August. pp. 47–49.

Y

Yoshikawa, Sumio and Mogi, Kiyoo, 1981. A new method for estimation of the crustal stress from cored rock samples: laboratory study in the case of Uniaxial compression. *Tectonophysics, Vol. 74*, pp. 323–329.

Young, Chapman and Fourney, William L., 1983. Fracture control blasting techniques for oil shale mining. *Eastern Oil Shale Symp., Nov. 13–16, pp. 363–369.*

Yu, T.R. and Vongpaisal, S., 1996. New blast damage criteria for underground blasting. *CIM Bulletin, March, pp. 139–145.*

Z

Zern, E.N., 1928. *Coal Miners' Pocketbook.* McGraw-Hill Book Co., New York, 12th ed., 1273 pp. Includes a Glossary of Mining Terms, pp. 1201–12551, Glossary of Rope Terms, pp. 755–757.

Zhang, G.J. 1996. A study of free toe-space explosive loading and its application in open pit blasts. *Fifth Int. Symp. on Rock Fragmentation by Blasting, Aug. 26–29, Montreal, Canada, pp. 313–318.*

Zhang, S. and Jin, J., 1996. *Computation of Special Functions.* Wiley, New York.